T0178297

Universitext

Universitext

Universitext is a series of textbooks that presents material from a wide variety of mathematical disciplines at master's level and beyond. The books, often well class-tested by their author, may have an informal, personal even experimental approach to their subject matter. Some of the most successful and established books in the series have evolved through several editions, always following the evolution of teaching curricula, into very polished texts.

Thus as research topics trickle down into graduate-level teaching, first textbooks written for new, cutting-edge courses may make their way into *Universitext*.

More information about this series at http://www.springer.com/series/223

Christer Bennewitz • Malcolm Brown
Rudi Weikard

Spectral and Scattering Theory for Ordinary Differential Equations

Vol. I: Sturm–Liouville Equations

 Springer

Christer Bennewitz
Lund University
Lund, Sweden

Malcolm Brown
Cardiff University
Cardiff, UK

Rudi Weikard
University of Alabama at Birmingham
Birmingham, AL, USA

ISSN 0172-5939   ISSN 2191-6675   (electronic)
Universitext
ISBN 978-3-030-59087-1   ISBN 978-3-030-59088-8   (eBook)
https://doi.org/10.1007/978-3-030-59088-8

Mathematics Subject Classification (2010): 28, 33, 34, 46, 47, 81

This Springer imprint is published by the registered company Springer Nature Switzerland AG
The registered company address is: Gewerbestrasse 11, 6330 Cham, Switzerland

Preface

The aim of this book is to give a modern presentation of the spectral and scattering theory for Sturm–Liouville equations, including some results on inverse theory. We plan a second volume to deal with more general ordinary differential equations and first order systems. Our starting point were lecture notes that have been used for many years by one of us, and some of our publications from the last few decades.

There are not many recent books in this field, so we believe there is a need for one that deals with the relevant issues from a modern point of view. We use the fact that the resolvent of a self-adjoint operator is an 'operator-valued Nevanlinna function' to prove the spectral theorem, as well as many different more concrete eigenfunction expansion theorems. Although our proofs of these results are based on very classical analytic function theory we use modern functional analytic methods where appropriate, and Schwartz distributions to deal with equations with low coefficient regularity.

We have tried to write the book assuming minimal prerequisites, and much of the background material required beyond the most basic can be found in the appendices, which include a substantial treatment of one-dimensional measures and some elementary theory for Schwartz distributions. We do expect the reader to be familiar with basic one-dimensional analysis, including uniform convergence. We also assume familiarity with linear and matrix algebra and a little about determinants. Finally, we assume knowledge of complex analysis up to and including the residue theorem.

To set the stage and whet the appetite we present a very elementary approach to regular Sturm–Liouville equations based on variational methods already in Chapter 1. This requires no preparation beyond elementary one-variable analysis, and gives an eigenfunction expansion theorem with little effort. As we move on to the later chapters the level of mathematical sophistication is gradually increased, but we try to keep this within bounds. In Chapter 2 we give a brief presentation of the theory of bounded and unbounded operators, as well as relations, in Hilbert space, and in Chapter 3 the general spectral theorem for unbounded self-adjoint operators and the associated functional calculus is proved, before moving on to the classical Sturm–Liouville equations, regular or singular, in Chapter 4.

A spectral theory for left-definite Sturm–Liouville equations is given in Chapter 5, including a full theory of eigenfunction expansions. We give first a traditional theory. Even so, to our knowledge there are few publications and no books dealing with this. Starting in Section 5.5 we present a more general theory requiring minimal coefficient regularity which has not been expounded in the literature at all. It should be pointed out that in some respects the left-definite theory is more satisfactory than the classical right-definite theory. One can for example give explicit necessary and sufficient conditions for essential self-adjointness in this case. The very common case of a semi-bounded operator may be viewed as either right or left-definite.

Chapter 6 contains some of Sturm's classical results on zeros of solutions and moves on to discuss basic results on spectral asymptotics for right and left-definite Sturm–Liouville equations. The chapter ends with a brief discussion of special functions by calculating the spectrum for the classical quantum model of the hydrogen atom.

In Chapter 7 some uniqueness theorems for inverse spectral theory are proved, starting with the classical Borg–Marčenko theorem. Most of the other results, dealing both with right-definite and left-definite problems, have not appeared in book form before.

Chapter 8 is devoted to scattering. We discuss scattering for right and left-definite general Sturm–Liouville equations on $\mathbb{R}$. Some uniqueness theorems for inverse scattering, where we concentrate on the left-definite case, are then given. The scattering and inverse theory for the left-definite case is most important for the solution of the Cauchy problem for the Camassa–Holm equation.

We include a number of appendices for easy reference to results needed but not covered in the main text, starting with a few fundamental results from functional analysis in Appendix A. In Appendix B we give a fairly complete theory of one-dimensional integration geared to our needs, and in Appendix C a brief introduction to distribution theory. Appendix D deals with fundamental questions concerning existence and uniqueness of solutions to ordinary differential equations in various circumstances, and Appendix E deals with some topics in analytic function theory important to the rest of the book. The final Appendix F gives an outline of some facts about integrable systems, in particular, the Camassa–Holm equation, as a background and motivation for some of our results on inverse theory.

At the end of each chapter we have added notes and remarks expanding a bit on the historical comments which typically appear in the introductions to those chapters. It would, however, be presumptuous to pretend that we have given justice to the vast amount of literature which has accumulated on our subject over the last two centuries.

CB is grateful to Gudrun for putting up with him for the many months of preparing the book, as well as for the 45 years preceding them.

BMB thanks the Leverhulme Trust for the award of an Emeritus Professorship during the tenure of which this book was written.

RW thanks Minh Nguyen and Steven Redolfi for critical remarks which led to a number of improvements of the book. He also thanks his wife Claudia for her unwavering support during the last 40 years.

Contents

Chapter 1
Introduction

1.1 Background

Although the simplest version of spectral theory is found in matrix theory this is not its historic origin; that lies in the field of differential equations. Spectral theory for differential equations originates with the method of separation of variables, used to solve many of the (linear) equations of mathematical physics and leading directly to the problem of expanding an 'arbitrary' function in terms of eigenfunctions of the reduced equation. This is the central problem of spectral theory.

A simple example is that of heat flow in a thin rod, insulated from the environment. We introduce coordinates in the rod so that it corresponds to the interval $(0, \ell)$ where ℓ is the length of the rod, and denote the temperature at the point x at time t by $u(x, t)$. The temperature is controlled by two laws. The first is *Fourier's law*, which says that the heat flow through a point is proportional to the temperature gradient (in this case u_x). The negative of the constant of proportionality is the *heat conductivity* of the rod. The second law says that the temperature at a point is proportional to the energy density; the constant of proportionality is the *heat capacity* of the material of the rod.

If the heat capacity and the heat conductivity are independent of x and t and the temperature distribution at time 0 is known this leads to the following problem:

$$\begin{cases} u_t = u_{xx} & \text{(heat equation)} \\ u_x(0, t) = u_x(\ell, t) = 0 & \text{(boundary conditions)} \\ u(x, 0) \text{ given in } (0, \ell) & \text{(initial condition)}, \end{cases} \quad (1.1.1)$$

where for simplicity we suppose units chosen so that the heat conductivity equals the heat capacity. The boundary conditions express the fact that no heat flows through the endpoints of the rod.

The idea behind separating variables is first to disregard the initial condition and try to find solutions of the differential equation that satisfy the boundary conditions which are *separated*, i.e., of the special form $u(x, t) = f(x)g(t)$. The linearity and

© Springer Nature Switzerland AG 2020
C. Bennewitz et al., *Spectral and Scattering Theory for Ordinary Differential Equations*,
Universitext, https://doi.org/10.1007/978-3-030-59088-8_1

homogeneity of the equation and boundary conditions imply that sums of solutions are also solutions (the *superposition principle*), so if we can find enough separated solutions there is the possibility that *any* solution might be a superposition of separated solutions.

By substituting $f(x)g(t)$ for u in (1.1.1) it follows that $f''(x)/f(x) = g'(t)/g(t)$. The left-hand side does not depend on t and the right-hand side not on x, so both sides in fact equal a constant $-\lambda$. The general solution of $g' + \lambda g = 0$ is $g(t) = Ce^{-\lambda t}$, and for a non-trivial separated solution we must have

$$\begin{cases} -f'' = \lambda f & \text{in } (0, \ell) \\ f'(0) = f'(\ell) = 0 \,. \end{cases} \tag{1.1.2}$$

As is easily seen, (1.1.2) has non-trivial solutions only when λ is an element of the sequence $\lambda_0, \lambda_1, \lambda_2, \ldots$, where $\lambda_j = (j\pi/\ell)^2$. These numbers are the *eigenvalues* of (1.1.2), and the corresponding solutions (non-trivial multiples of $\cos((\lambda_j)^{1/2} x)$) are the *eigenfunctions* of (1.1.2). The set of eigenvalues is called the *spectrum* of (1.1.2).

A superposition of separated solutions is consequently of the form

$$u(x,t) = \sum C_j e^{-\lambda_j t} \cos(\lambda_j^{1/2} x) \,. \tag{1.1.3}$$

The initial condition of (1.1.1) therefore requires that

$$u(x,0) = \sum C_j \cos(\lambda_j^{1/2} x)$$

equals a given function. The question of whether (1.1.1) has a solution which is a superposition of separated solutions for arbitrary initial conditions is thus seen to amount to the question of whether an 'arbitrary' function may be written as a series $\sum u_j$ where each term is an eigenfunction of (1.1.2).

The technique described above was used systematically by Fourier in his *Théorie analytique de la chaleur* (1822) [74] to solve problems of heat conduction, which in the simplest cases (like our example) lead to what are now called *Fourier series expansions*.

Fourier was never able to give a satisfactory proof of the *completeness* of the eigenfunctions, i.e., the fact that essentially arbitrary functions can be expanded in *Fourier series*. This problem was solved by Dirichlet in 1829 [143], and somewhat later (mid 1830s) Sturm [208, 209] and Liouville [151, 152, 153, 154] independently but simultaneously showed various completeness results for more general eigenvalue problems of the form $-(pu')' + qu = \lambda wu$, with boundary conditions of the form $Au + Bpu' = 0$, to be satisfied at the endpoints of the given interval.

Here $p > 0$, q and $w > 0$ are given real-valued and sufficiently regular functions, and A, B given real constants, not both 0 and possibly different in the two interval endpoints. The Fourier cases correspond to $p \equiv w \equiv 1$, $q \equiv 0$ and A or B equal to 0.

For the Fourier equation (1.1.1), the distance between successive eigenvalues decreases as the length of the base interval increases, and as the base interval approaches the whole real line, the eigenvalues accumulate everywhere on the positive real line. The *Fourier series* is then replaced by a continuous superposition, i.e., an integral, and we get the classical *Fourier transform*. Thus a *continuous spectrum* appears, and this is typical of problems where the base domain is unbounded, or the coefficients of the equation behave sufficiently badly at the boundary.

In 1910 Hermann Weyl [222] gave the first rigorous treatment, in the case of an equation of Sturm–Liouville type, of problems where continuous spectra can occur. Weyl's treatment was based on Hilbert's then recently proved *spectral theorem* (Hilbert [101], [103]). Hilbert's theorem was a generalization of the usual diagonalization of a quadratic form to the case of infinitely many variables. This theorem is not directly applicable to differential operators, since these are 'unbounded' in a sense we shall discuss in Chapter 2.

With the creation of quantum mechanics in the 1920s, these matters became of basic importance to physics. Mathematicians, who had not advanced much beyond the results of Weyl, then took the matter up again. The outcome was the general spectral theorem, generally attributed to John von Neumann [171] (1928), [172] (1930), although essentially the same theorem had been proved by Torsten Carleman [44] in 1923, in a less abstract setting.

Von Neumann's theorem is an abstract result about operators in Hilbert space, and detailed applications to differential operators of reasonable generality had to wait until the early 1950s, although expansion theorems for the one-dimensional Schrödinger equation were given in the groundbreaking book [207] by M. H. Stone. In the meantime many independent results about expansions in eigenfunctions had been given, particularly for ordinary differential equations. Consequently many treatises on spectral theory for ordinary differential equations do not use von Neumann's spectral theorem.

In the following sections we shall review the simplest aspects of the theory of linear spaces with scalar product and then give a theory of eigenfunction expansions for the equations considered by Sturm and Liouville based on very elementary considerations. By contrast, the eigenfunction expansions given in later chapters are all based on von Neumann's theorem, so beginning in the next chapter we shall give a condensed version of the abstract theory before moving on to more concrete results concerning differential equations.

Exercise

Exercise 1.1.1 Show, by elementary means, that the boundary value problem $-u'' = \lambda u$, $u'(0) = 0$, $u'(\ell) = hu(\ell)$ has infinitely many eigenvalues and determine their asymptotic behavior. Here h is a fixed real non-zero constant (the case $h = 0$ was treated above).

Hint: You will encounter a transcendental equation. Consider the graphs of the functions involved.

1.2 Linear spaces

This section contains a quick review of the basic facts about linear spaces. In the definition below the set $\mathbb{K}$ can be any *field*, although to us only the fields $\mathbb{R}$ of real numbers and $\mathbb{C}$ of complex numbers are of interest.

Definition 1.2.1. A *linear space* or *vector space* over the *scalar field* $\mathbb{K}$ is a set $\mathcal{V}$ provided with an addition +, which to every pair of elements $u, v \in \mathcal{V}$ associates an element $u + v \in \mathcal{V}$, and a multiplication by a scalar, which to every $\lambda \in \mathbb{K}$ and $u \in \mathcal{V}$ associates an element $\lambda u \in \mathcal{V}$. The following rules for calculation hold:

(1) $(u + v) + w = u + (v + w)$ for u, v and w in $\mathcal{V}$.   (associativity)

(2) There is a unique element $0 \in \mathcal{V}$ such that $u + 0 = 0 + u = u$ for every $u \in \mathcal{V}$.
                                                                        (neutral element)

(3) For every $u \in \mathcal{V}$ there exists a unique $v \in \mathcal{V}$ such that $u + v = v + u = 0$. One denotes v by $-u$.   (additive inverse)

(4) $u + v = v + u$ for all $u, v \in \mathcal{V}$.   (commutativity)

(5) $\lambda(u + v) = \lambda u + \lambda v$ for all $\lambda \in \mathbb{K}$ and all $u, v \in \mathcal{V}$.

(6) $(\lambda + \mu)u = \lambda u + \mu u$ for all $\lambda, \mu \in \mathbb{K}$ and all $u \in \mathcal{V}$.

(7) $\lambda(\mu u) = (\lambda \mu)u$ for all $\lambda, \mu \in \mathbb{K}$ and all $u \in \mathcal{V}$.

(8) $1u = u$ for all $u \in \mathcal{V}$.

These axioms are not independent, but stated this way have a convenient structure. If $\mathbb{K} = \mathbb{R}$ we have a *real* linear space, if $\mathbb{K} = \mathbb{C}$ a *complex* linear space. Axioms 1–3 above say that $\mathcal{V}$ is a group under addition, Axiom 4 that the group is *Abelian* (or commutative). Axioms 5 and 6 are distributive laws and Axiom 7 an associative law related to the multiplication by scalars, whereas Axiom 8 gives a normalization for the multiplication by scalars.

Note that by restricting oneself to multiplying only by real numbers, any complex space may also be viewed as a real linear space. Conversely, every real linear space can be 'extended' to a complex linear space (Exercise 1.2.1). We shall therefore mostly consider complex linear spaces in the sequel.

Let M be an arbitrary set and $\mathbb{C}^M$ the set of complex-valued functions defined on M. Then $\mathbb{C}^M$, provided with the obvious definitions of the linear operations, is a complex linear space (Exercise 1.2.2). In the case when $M = \{1, 2, \ldots, n\}$ one writes $\mathbb{C}^n$ instead of $\mathbb{C}^{\{1,2,\ldots,n\}}$. An element $u \in \mathbb{C}^n$ is of course given by the values $u(1), u(2), \ldots, u(n)$ of u so one may also regard $\mathbb{C}^n$ as the set of ordered n-tuples of complex numbers. The corresponding real space is the usual $\mathbb{R}^n$.

If $\mathcal{V}$ is a linear space and V a subset of $\mathcal{V}$ which is itself a linear space, using the linear operations inherited from $\mathcal{V}$, one says that V is a *linear subspace* of $\mathcal{V}$.

Proposition 1.2.2. *A non-empty subset V of $\mathcal{V}$ is a linear subspace of $\mathcal{V}$ if and only if $u + v \in V$ and $\lambda u \in V$ for all $u, v \in V$ and $\lambda \in \mathbb{C}$.*

The proof is left as an exercise (Exercise 1.2.3). Most linear spaces met with in this book are subspaces of $\mathbb{C}^M$, where M is some real interval. If $u_1, u_2, \ldots, u_k$ are elements of a linear space $\mathcal{V}$ we denote by $\mathrm{Sp}(u_1, u_2, \ldots, u_k)$ the *span* or *linear hull* of $u_1, u_2, \ldots, u_k$, i.e., the set of all linear combinations $\lambda_1 u_1 + \cdots + \lambda_k u_k$, where $\lambda_j \in \mathbb{C}$, $j = 1, \ldots, k$. It is not hard to see that linear hulls are always subspaces (Exercise 1.2.5). One says that $u_1, \ldots, u_k$ *generate* $\mathcal{V}$ if $\mathcal{V} = \mathrm{Sp}(u_1, \ldots, u_k)$, and any linear space which is the span of a finite number of its elements is called *finitely generated* or *finite-dimensional*. A linear space which is *not* finitely generated is called *infinite-dimensional*.

It is clear that if, e.g., u_k is a linear combination of $u_1, \ldots, u_{k-1}$, then it follows that $\mathrm{Sp}(u_1, \ldots, u_k) = \mathrm{Sp}(u_1, \ldots, u_{k-1})$. If none of $u_1, \ldots, u_k$ is a linear combination of the others one says that $u_1, \ldots, u_k$ are *linearly independent*. A single vector is called linearly independent if it is not 0. Note that the space $\{0\}$ is generated by 0 but doesn't have any linearly independent elements. It is easy to see that $v_1, \ldots, v_j$ are linearly independent if and only if $\lambda_1 v_1 + \cdots + \lambda_j v_j = 0$ only for $\lambda_1 = \cdots = \lambda_j = 0$ (Exercise 1.2.6). The following lemma is fundamental.

Lemma 1.2.3. *Suppose $u_1, \ldots, u_k$ generate $\mathcal{V}$, and that $v_1, \ldots, v_j$ are linearly independent elements of $\mathcal{V}$. Then $j \le k$.*

Proof. Since $u_1, \ldots, u_k$ generate $\mathcal{V}$ we have $v_1 = \sum_{s=1}^{k} x_{1s} u_s$, for some coefficients $x_{11}, \ldots, x_{1k}$ which are not all 0 since $v_1 \ne 0$. By renumbering $u_1, \ldots, u_k$ we may assume $x_{11} \ne 0$. Then $u_1 = \frac{1}{x_{11}} v_1 - \sum_{s=2}^{k} \frac{x_{1s}}{x_{11}} u_s$, and therefore $v_1, u_2, \ldots, u_k$ generate $\mathcal{V}$. In particular, $v_2 = x_{21} v_1 + \sum_{s=2}^{k} x_{2s} u_s$ for some coefficients $x_{21}, \ldots, x_{2k}$. We cannot have $x_{22} = \cdots = x_{2k} = 0$ since v_1, v_2 are linearly independent. By renumbering $u_2, \ldots, u_k$, if necessary, we may assume $x_{22} \ne 0$. It follows as before that $v_1, v_2, u_3, \ldots, u_k$ generate $\mathcal{V}$. We can continue in this way until we run out of either v's (if $j \le k$) or u's (if $j > k$). But if $j > k$ we would get that $v_1, \ldots, v_k$ generate $\mathcal{V}$, in particular that v_j is a linear combination of $v_1, \ldots, v_k$ which contradicts the linear independence of the vectors $v_1, \ldots, v_j$. Hence $j \le k$.   □

The most important fact about finitely generated linear spaces $\mathcal{V}$ (except the space $\{0\}$) is that they have a set of linearly independent generators; such a set is called a *basis* for $\mathcal{V}$. A given finite-dimensional space $\mathcal{V} \ne \{0\}$ always has many different bases, but a fundamental fact is that all such bases have the same number of elements, called the *dimension* of $\mathcal{V}$ (by definition the space $\{0\}$ has dimension 0). This follows immediately from the following theorem.

Theorem 1.2.4. *Every finitely generated linear space not equal to $\{0\}$ has a basis, all bases of a given space have the same number of elements, and any set of linearly independent vectors can be extended to a basis.*

Proof. Lemma 1.2.3 shows that in a finitely generated space there is a bound on the number of elements in a linearly independent set of vectors.

Now suppose $e_1, \ldots, e_n$ are linearly independent but cannot be extended to a larger set of linearly independent vectors. Then given any $u \in \mathcal{V}$ one of $u, e_1, \ldots, e_n$ is a linear combination of the others, and in such an equation the coefficient of

u cannot be zero, since that would mean that $e_1, \ldots, e_n$ are linearly dependent. Dividing by the coefficient of u we see that every vector $u \in \mathcal{V}$ is a linear combination of $e_1, \ldots, e_n$ which is therefore a basis.

Given two different bases Lemma 1.2.3 shows that they must have the same number of elements, since a basis both generates $\mathcal{V}$ and is linearly independent. □

For a finite-dimensional space the existence and uniqueness of coordinates for any vector with respect to an arbitrary basis now follows easily (Exercise 1.2.9). As important for us is that it is also clear that $\mathcal{V}$ has infinite dimension if and only if every linearly independent subset of $\mathcal{V}$ can be extended to linearly independent subsets of $\mathcal{V}$ with arbitrarily many elements. This usually makes it quite easy to see when a given space has infinite dimension (Exercise 1.2.6).

If V and W are both linear subspaces of some larger linear space $\mathcal{V}$, then the *linear span* $\mathrm{Sp}(V, W)$ of V and W is the set

$$\mathrm{Sp}(V, W) = \{v + w : v \in V, \ w \in W\} \, .$$

This is obviously a linear subspace of $\mathcal{V}$. If in addition $V \cap W = \{0\}$, then for any $u \in \mathrm{Sp}(V, W)$ there are *unique* elements $v \in V$ and $w \in W$ such that $u = v + w$. In this case $\mathrm{Sp}(V, W)$ is called the *direct sum* of V and W and is denoted by $V \dotplus W$. The proof of these facts is left to Exercise 1.2.11.

Note that one sometimes also talks about the direct sum $V \dotplus W$ of completely unrelated linear spaces V and W. This is the set of pairs $\{(v, w) : v \in V, \ w \in W\}$ provided with the obvious definitions of addition and multiplication by scalars. One may identify V with the subset of pairs with second component zero and W with the subset with first component zero, and then this kind of direct sum is reduced to the kind of direct sum defined previously.

If W is a linear subspace of $\mathcal{V}$ we can create a new linear space $\mathcal{V}/W$, the *quotient space* of $\mathcal{V}$ by W, in the following way. We say that two elements u and v of $\mathcal{V}$ are *equivalent* if $u - v \in W$. It is immediately seen that any u is equivalent to itself, that u is equivalent to v if v is equivalent to u, and that if u is equivalent to v, and v to w, then u is equivalent to w. It then follows that we may split $\mathcal{V}$ into *equivalence classes* such that every vector is equivalent to all vectors in the same equivalence class, but not to any other vectors. The equivalence class containing u is denoted by $u + W$, and then $u + W = v + W$ precisely if $u - v \in W$.

We now define $\mathcal{V}/W$ as the set of equivalence classes, where addition is defined by $(u+W)+(v+W) = (u+v)+W$ and multiplication by a scalar as $\lambda(u+W) = \lambda u+W$. It is easily seen that these operations are well defined and that $\mathcal{V}/W$ becomes a linear space with neutral element $0 + W$ (Exercise 1.2.12). One defines $\mathrm{codim}\, W = \dim \mathcal{V}/W$. The fundamental fact about quotient spaces is the following.

Theorem 1.2.5. *If W is a linear subspace of the linear space $\mathcal{V}$, then*

$$\dim W + \mathrm{codim}\, W = \dim \mathcal{V} \, .$$

The proof is left to Exercise 1.2.13.

Exercises

Exercise 1.2.1 Let $\mathcal{V}$ be a real linear space, and let $\mathcal{V}_{\mathbb{C}}$ be the set of ordered pairs (u, v) of elements of $\mathcal{V}$ with addition defined componentwise.

Show that $\mathcal{V}_{\mathbb{C}}$ becomes a complex linear space if multiplication by a complex scalar $\lambda = x + iy$ with real x, y is defined by $\lambda(u, v) = (xu - yv, xv + yu)$.

Also show that $\mathcal{V}$ can be 'identified' with the subset of elements of $\mathcal{V}_{\mathbb{C}}$ of the form $(u, 0)$, in the sense that there is a one-to-one correspondence between the two sets preserving the linear operations (for real scalars).

Exercise 1.2.2 Let M be an arbitrary set and let $\mathbb{R}^M$ be the set of real-valued, and $\mathbb{C}^M$ the set of complex-valued, functions on M.

Show that, provided with the obvious definitions of the linear operations, $\mathbb{R}^M$ is a real and $\mathbb{C}^M$ a complex linear space.

Exercise 1.2.3 Prove Proposition 1.2.2.

Exercise 1.2.4 Let M be a non-empty subset of $\mathbb{R}^n$. Which of the following sets $\mathcal{V}$ are linear subspaces of $\mathbb{C}^M$?

(1) $\mathcal{V} = \{u \in \mathbb{C}^M : |u(x)| < 1 \text{ for all } x \in M\}$.

(2) $\mathcal{V} = C(M) = \{u \in \mathbb{C}^M : u \text{ is continuous on } M\}$.

(3) $\mathcal{V} = \{u \in C(M) : u \text{ is bounded on } M\}$.

(4) $\mathcal{V} = L(M) = \{u \in \mathbb{C}^M : u \text{ is Lebesgue integrable over } M\}$.

Exercise 1.2.5 Let $\mathcal{V}$ be a linear space and $u_j \in \mathcal{V}, j = 1, \ldots, k$. Show that the span $\text{Sp}(u_1, u_2, \ldots, u_k)$ is a linear subspace of $\mathcal{V}$.

Exercise 1.2.6 Show that $u_1, \ldots, u_k$ are linearly independent precisely if the linear combination $\lambda_1 u_1 + \cdots + \lambda_k u_k = 0$ only for $\lambda_1 = \cdots = \lambda_k = 0$.

Exercise 1.2.7 Let $\mathcal{V}$ be a finite-dimensional real linear space and $\mathcal{V}_{\mathbb{C}}$ be the complexification of $\mathcal{V}$ as defined in Exercise 1.2.1. Show that the (real) dimension of $\mathcal{V}$ equals the (complex) dimension of $\mathcal{V}_{\mathbb{C}}$, but the real dimension of a complex space is twice the complex dimension.

Exercise 1.2.8 Let M be an arbitrary set and $\mathbb{R}^M, \mathbb{C}^M$ be defined as in Exercise 1.2.2. Show that the dimensions of $\mathbb{R}^M$ as a real linear space and that of $\mathbb{C}^M$ as a complex linear space equal the number of elements of M if this is finite and that the dimension is otherwise infinite.

Exercise 1.2.9 Show that if $e_1, \ldots, e_n$ is a basis for $\mathcal{V}$ and $u \in \mathcal{V}$ there are uniquely determined scalars $x_1, \ldots, x_n$, called *coordinates* for u with respect to $e_1, \ldots, e_n$, such that $u = x_1 e_1 + \cdots + x_n e_n$.

A map L between linear spaces (with the same scalar field $\mathbb{K}$) is called *linear* if $L(u + v) = Lu + Lv$ and $L(\lambda u) = \lambda L u$ for scalars λ. Show that, introducing a basis in $\mathcal{V}$, the mapping $\mathcal{V} \ni u \mapsto (x_1, \ldots, x_n) \in \mathbb{K}^n$ mapping a vector onto its coordinates is an invertible, linear map (a linear isomorphism).

Exercise 1.2.10 Let M be a non-empty open subset of $\mathbb{R}^n$. Verify that $\mathcal{V}$ is infinite-dimensional for each of the choices of $\mathcal{V}$ in Exercise 1.2.4 for which $\mathcal{V}$ is a linear space.

Exercise 1.2.11 Prove all statements in the paragraph defining $Sp(V, W)$.

Exercise 1.2.12 Prove that if $\mathcal{V}$ is a linear space and W a subspace of $\mathcal{V}$, then $\mathcal{V}/W$ is a well defined linear space.

Exercise 1.2.13 Prove Theorem 1.2.5.

1.3 Spaces with scalar product

If one wants to do analysis in a linear space, some structure in addition to the linearity is needed. This is because one needs some way to define limits and continuity, and this requires an appropriate definition of what a neighborhood of a point is. Thus one must introduce a *topology* on the space. This means to define a class of subsets of the space satisfying some simple properties, the elements of which are called *open sets*.

A *topological vector space* is a linear space provided with a topology such that the linear operations are continuous. We shall not deal with the general notion of a topological vector space, but only the following particularly convenient way to introduce a topology in a linear space. This also covers most cases of importance to analysis (but see Appendix C for some exceptions).

A *metric space* is a set $\mathcal{M}$ provided with a *metric*, which is a function $d : \mathcal{M} \times \mathcal{M} \to \mathbb{R}$ such that for any $x, y, z \in \mathcal{M}$ the following holds.

(1) $d(x, y) \geq 0$ with equality if and only if $x = y$.   (positive definite)

(2) $d(x, y) = d(y, x)$.   (symmetric)

(3) $d(x, y) \leq d(x, z) + d(z, y)$.   (triangle inequality)

A *neighborhood* of $x \in \mathcal{M}$ is then a subset $\mathcal{O}$ of $\mathcal{M}$ such that for some $\varepsilon > 0$ the set $\mathcal{O}$ contains all $y \in \mathcal{M}$ for which $d(x, y) < \varepsilon$. An *open set* is a set which is a neighborhood of all its points, and a *closed set* is one with an open complement. A function $f : \mathcal{M}_1 \to \mathcal{M}_2$ between two metric (or more general topological) spaces is called *continuous* if the inverse image of every open subset of $\mathcal{M}_2$ is an open subset of $\mathcal{M}_1$. On $\mathbb{R}$ and $\mathbb{C}$ we will always use $d(x, y) = |x - y|$.

One says that a sequence $x_1, x_2, \ldots$ of elements in a metric space $\mathcal{M}$ *converges* to $x \in \mathcal{M}$ if $d(x_j, x) \to 0$ as $j \to \infty$. It is easy to see that a subset $A \subset \mathcal{M}$ is closed precisely if the limit of any convergent sequence in A is also in A (Exercise 1.3.1).

The most convenient, but not the only important, way of introducing a metric on a linear space $\mathcal{V}$ is via a *norm*. A norm on $\mathcal{V}$ is a function $\|\cdot\| : \mathcal{V} \to \mathbb{R}$ such that for any $u, v \in \mathcal{V}$ and $\lambda \in \mathbb{C}$

(1) $\|u\| \geq 0$ with equality if and only if $u = 0$.   (positive definite)

(2) $\|\lambda u\| = |\lambda|\,\|u\|$. (positive homogeneous)

(3) $\|u \pm v\| \leq \|u\| + \|v\|$. (triangle inequality)

If instead of (1) we only have the weaker statement

$(1')$ $\|u\| \geq 0$. (positive semi-definite)

one calls $\|\cdot\|$ a *semi-norm*.

As is well known (Exercise 1.3.3) the triangle inequality implies the *reverse triangle inequality*

$$\big|\|u\| - \|v\|\big| \leq \|u \pm v\| \ .$$

In a normed space one introduces a metric by setting $d(x, y) = \|x - y\|$. That this is a metric is immediately verified (Exercise 1.3.2). It is also clear that this metric is *translation invariant* and *homogeneous*, i.e., $d(u + w, v + w) = d(u, v)$ for any $w \in \mathcal{V}$ and $d(\lambda u, \lambda v) = |\lambda|\, d(u, v)$ for all $\lambda \in \mathbb{C}$.

Which sets are open and which functions are continuous on a normed linear $\mathcal{V}$ space naturally depends on the norm used. However, two norms $\|\cdot\|$ and $\|\cdot\|_*$ obviously define the same topology (the same open sets) on $\mathcal{V}$ if they are *equivalent*, i.e., if there are constants $0 < m \leq M$ such that $m\|u\| \leq \|u\|_* \leq M\|u\|$ for all $u \in \mathcal{V}$.

Theorem 1.3.1. *If* $\dim \mathcal{V} < \infty$ *all norms on* $\mathcal{V}$ *are equivalent.*

Proof. Let $e_1, \ldots, e_n$ be a basis in $\mathcal{V}$, so that given $u \in \mathcal{V}$ there are unique $x_1, \ldots, x_n$ for which $u = x_1 e_1 + \cdots + x_n e_n$. We may define a norm on $\mathcal{V}$ by setting $\|u\| = (\sum |x_j|^2)^{1/2}$ so that we may identify $\mathcal{V}$ with $\mathbb{C}^n$ provided with its standard norm. Note that $|x_j| \leq \|u\|$.

If $\|\cdot\|_*$ is another norm on $\mathcal{V}$ we have $\|u\|_* = \|x_1 e_1 + \cdots + x_n e_n\|_* \leq \sum |x_j|\,\|e_j\|_* \leq M\|u\|$ where $M = \sum \|e_j\|_*$. Since $\big|\|u\|_* - \|v\|_*\big| \leq \|u - v\|_* \leq M\|u - v\|$ the function $\mathcal{V} \ni u \mapsto \|u\|_*$ is continuous with respect to the metric defined by $\|\cdot\|$. Now, the unit sphere $S = \{u \in \mathcal{V} : \|u\| = 1\}$ is compact if $\mathcal{V} = \mathbb{C}^n$, and thus for general $\mathcal{V}$, so the function $u \mapsto \|u\|_*$ has a least value m on S, and $m > 0$ since $\|\cdot\|_*$ is a norm. If $u \neq 0$ we obtain

$$m\|u\| \leq \left\|\frac{u}{\|u\|}\right\|_* \|u\| = \|u\|_* \leq M\|u\| \ ,$$

so the norms are equivalent. It follows immediately that any two norms on $\mathcal{V}$ are equivalent. $\qquad\square$

An important corollary is the following.

Corollary 1.3.2 (Bolzano–Weierstrass). *If* $\mathcal{V}$ *is a normed, finite-dimensional linear space, then any bounded sequence in* $\mathcal{V}$ *has a convergent subsequence.*

This is well known in $\mathbb{C}^n$, which in this respect is the same as $\mathbb{R}^{2n}$, and therefore a consequence of Theorem 1.3.1.

Sometimes a topology is defined on a linear space by use of a family of semi-norms. If they are finitely many they may be added up to a single semi-norm, which will then define the topology. But if they are infinitely many they cannot always be replaced by a single semi-norm. However, as long as they are countably many, i.e., may be listed in a sequence, and there is no non-zero vector for which all semi-norms vanish, they can always be used to define a metric giving the topology, as in Exercise 1.3.20.

The usual norm on the real space $\mathbb{R}^3$ is of course obtained from the dot product $(x_1, x_2, x_3) \cdot (y_1, y_2, y_3) = x_1 y_1 + x_2 y_2 + x_3 y_3$ by setting $\|x\| = \sqrt{x \cdot x}$. For a complex linear space $\mathcal{V}$, one may similarly define a norm by setting $\|u\| = \sqrt{\langle u, u \rangle}$, where $\langle \cdot, \cdot \rangle$ is a *scalar product* on $\mathcal{V}$.

A scalar product on a complex linear space $\mathcal{V}$ is a function $\mathcal{V} \times \mathcal{V} \ni (u, v) \mapsto \langle u, v \rangle \in \mathbb{C}$ such that for all u, v and w in $\mathcal{V}$ and all $\lambda, \mu \in \mathbb{C}$ holds

(1) $\langle \lambda u + \mu v, w \rangle = \lambda \langle u, w \rangle + \mu \langle v, w \rangle.$ (linearity in first argument)

(2) $\langle u, v \rangle = \overline{\langle v, u \rangle}.$ (Hermitian symmetry)

(3) $\langle u, u \rangle \geq 0$ with equality only if $u = 0$.   (positive definite)

If $\mathcal{V}$ is real the definition is the same, where now $\lambda \in \mathbb{R}$, $\langle \cdot, \cdot \rangle$ is real-valued and the conjugation in (2) is redundant. In this case the property (2) is called *symmetry*. If instead of (3) only

(3′) $\langle u, u \rangle \geq 0,$ (positive semi-definite)

one speaks about a *semi-scalar product*. Note that (2) implies that $\langle u, u \rangle$ is real so that (3) and (3′) make sense. Also note that by combining (1) and (2) we have $\langle w, \lambda u + \mu v \rangle = \overline{\lambda} \langle w, u \rangle + \overline{\mu} \langle w, v \rangle$. One says that the scalar product is *anti-linear* in its second argument.[1] Together with (1) this makes the scalar product into a *sesquilinear* (one and a half-linear) form. In words: A scalar product is a Hermitian, sesquilinear and positive definite form.

In the case of a real space anti-linear is of course the same as linear, and the scalar product is *bilinear*. In this case a scalar product is a symmetric, bilinear and positive definite form.

We now assume that we have a (semi-)scalar product on $\mathcal{V}$ and define $\|u\| = \sqrt{\langle u, u \rangle}$ for any $u \in \mathcal{V}$. To show that this definition makes $\|\cdot\|$ into a (semi-)norm we need the following basic inequality.

Theorem 1.3.3 (Cauchy–Schwarz). *Suppose $\langle \cdot, \cdot \rangle$ is a semi-scalar product on $\mathcal{V}$. Then $|\langle u, v \rangle|^2 \leq \langle u, u \rangle \langle v, v \rangle$ for all $u, v \in \mathcal{V}$.*

Proof. For arbitrary $u, v \in \mathcal{V}$ and complex λ we have

$$0 \leq \langle \lambda u + v, \lambda u + v \rangle = |\lambda|^2 \langle u, u \rangle + \lambda \langle u, v \rangle + \overline{\lambda} \langle v, u \rangle + \langle v, v \rangle .$$

For $\lambda = -r \langle v, u \rangle$ with real r we obtain

$$0 \leq r^2 |\langle u, v \rangle|^2 \langle u, u \rangle - 2r |\langle u, v \rangle|^2 + \langle v, v \rangle .$$

---

[1] **Warning:** In the so-called *Dirac formalism* in quantum mechanics the scalar product is instead anti-linear in the first argument and linear in the second.

If $\langle u, u \rangle = 0$ but $\langle u, v \rangle \neq 0$ this expression becomes negative for large r, which is not allowed. Hence $\langle u, u \rangle = 0$ implies that $\langle u, v \rangle = 0$ so that the inequality is valid if $\langle u, u \rangle = 0$. If $\langle u, u \rangle > 0$ we multiply by $\langle u, u \rangle$ and set $r = 1/\langle u, u \rangle$ to obtain

$$0 \leq |\langle u, v \rangle|^2 - 2 |\langle u, v \rangle|^2 + \langle u, u \rangle \langle v, v \rangle$$

which proves the theorem. $\qquad\square$

We may also write the Cauchy–Schwarz inequality as $|\langle u, v \rangle| \leq \|u\| \|v\|$. If $\langle \cdot, \cdot \rangle$ is positive definite it is easy to see that there is equality in the Cauchy–Schwarz inequality precisely when u and v are linearly dependent. To see that $\|\cdot\|$ is a (semi-)norm on $\mathcal{V}$ the only non-trivial point is to verify that the triangle inequality holds; but this follows from the Cauchy–Schwarz inequality (Exercise 1.3.6).

Recall that in a finite-dimensional space with scalar product it is particularly convenient to use an *orthonormal basis*, i.e., a basis $e_1, \ldots, e_n$ for which $\|e_j\| = 1$ and $\langle e_j, e_k \rangle = 0$ if $j \neq k$. This makes it very easy to calculate the coordinates of any vector. In fact, if $x_1, \ldots, x_n$ are the coordinates of u in an orthonormal basis $e_1, \ldots, e_n$, so that $u = \sum x_k e_k$, then scalar multiplication by e_j shows that $x_j = \langle u, e_j \rangle$. Given an arbitrary basis it is easy to construct an orthonormal basis by use of the *Gram–Schmidt method* (see the proof of Lemma 1.3.4).

In an infinite-dimensional space one cannot find a (finite) basis.[2] However, if the space is provided with a norm one may hope for a sequence of vectors $e_1, e_2, \ldots$ such that each finite subset is linearly independent, and any vector is the limit in norm of a sequence of finite linear combinations of $e_1, e_2, \ldots$. To be really useful more is required of the sequence $e_1, e_2, \ldots$, and various notions of such bases exist in the literature. However, if the space is provided with a scalar product it will, as in the finite-dimensional case, turn out to be very convenient if $e_1, e_2, \ldots$ is an *orthonormal sequence*, i.e., $\|e_j\| = 1$ for all j and $\langle e_j, e_k \rangle = 0$ for $j \neq k$. The following lemma is easily proved by use of the Gram–Schmidt procedure.

Lemma 1.3.4. *Any infinite-dimensional linear space $\mathcal{V}$ with scalar product contains an orthonormal sequence.*

Proof. According to Section 1.2 we can find a linearly independent sequence in $\mathcal{V}$, i.e., a sequence $u_1, u_2, \ldots$ such that $u_1, \ldots, u_k$ are linearly independent for any k.

Now define $e_1 = u_1 / \|u_1\|$. If we have already found orthonormal $e_1, \ldots, e_k$ such that $e_j \in \mathrm{Sp}(u_1, \ldots, u_j)$, $j = 1, \ldots, k$, put

$$v_{k+1} = u_{k+1} - \sum_{j=1}^{k} \langle u_{k+1}, e_j \rangle e_j$$

---

[2] By use of Zorn's lemma one may prove the existence of a *Hamel basis* $\mathcal{E}$, which is a set of vectors such that each finite subset is linearly independent, and such that every vector is a linear combination of finitely many vectors in $\mathcal{E}$. For practical purposes Hamel bases in infinite-dimensional spaces are useless.

and $e_{k+1} = v_{k+1}/\|v_{k+1}\|$, so that $e_{k+1} \in \mathrm{Sp}(u_1, \ldots, u_{k+1})$. That this procedure will lead to a well-defined orthonormal sequence $e_1, e_2, \ldots$ is left for the reader to verify (Exercise 1.3.8).   □

Supposing we have an orthonormal sequence $e_1, e_2, \ldots$ in $\mathcal{V}$ a natural question is: How well can one approximate (in the norm of $\mathcal{V}$) an arbitrary vector $u \in \mathcal{V}$ by finite linear combinations of $e_1, e_2, \ldots$? Here is the answer:

Proposition 1.3.5. *Suppose $e_1, e_2, \ldots$ is an orthonormal sequence in $\mathcal{V}$ and put, for any $u \in \mathcal{V}$, $\hat{u}_j = \langle u, e_j \rangle$. Then we have*

$$\left\| u - \sum_{j=1}^{k} \lambda_j e_j \right\|^2 = \|u\|^2 - \sum_{j=1}^{k} |\hat{u}_j|^2 + \sum_{j=1}^{k} |\lambda_j - \hat{u}_j|^2 \qquad (1.3.1)$$

for arbitrary complex numbers $\lambda_1, \ldots, \lambda_k$.

The proof is a simple calculation (Exercise 1.3.9), but the interpretation of the proposition is very interesting. The identity (1.3.1) says that if we want to choose a linear combination $\sum_{j=1}^{k} \lambda_j e_j$ of $e_1, \ldots, e_k$ which approximates u well in norm, the best choice of coefficients is to take $\lambda_j = \hat{u}_j$, $j = 1, \ldots, k$. Furthermore, with this choice, the size of the error is given exactly by $\|u - \sum_{j=1}^{k} \hat{u}_j e_j\|^2 = \|u\|^2 - \sum_{j=1}^{k} |\hat{u}_j|^2$ (this may be interpreted as a form of Pythagoras' theorem).

One calls the coefficients $\hat{u}_1, \hat{u}_2, \ldots$ the (generalized) *Fourier coefficients* of u with respect to the orthonormal sequence $e_1, e_2, \ldots$. The following theorem is now a simple consequence of Proposition 1.3.5 (Exercise 1.3.10).

Theorem 1.3.6 (Bessel's inequality). *For any $u \in \mathcal{V}$ the series $\sum_{j=1}^{\infty} |\hat{u}_j|^2$ converges and one has*

$$\sum_{j=1}^{\infty} |\hat{u}_j|^2 \leq \|u\|^2 .$$

Another immediate consequence of Proposition 1.3.5 is the next theorem (cf. Exercise 1.3.11).

Theorem 1.3.7 (Parseval's formula). *The series $\sum_{j=1}^{\infty} \lambda_j e_j$ converges to u in norm if and only if $\lambda_j = \hat{u}_j$, $j = 1, 2, \ldots$ and*

$$\sum_{j=1}^{\infty} |\hat{u}_j|^2 = \|u\|^2 .$$

There is also a slightly more general form of Parseval's formula.

Corollary 1.3.8. *Suppose $\sum_{j=1}^{\infty} |\hat{u}_j|^2 = \|u\|^2$ for some $u \in \mathcal{V}$. Then $\sum_{j=1}^{\infty} \hat{u}_j \overline{\hat{v}_j} = \langle u, v \rangle$ for any $v \in \mathcal{V}$.*

Proof. Consider the following Hermitian form on $\mathcal{V}$:

$$[u, v] = \langle u, v \rangle - \sum_{j=1}^{\infty} \hat{u}_j \overline{\hat{v}_j} .$$

Since $\left| \hat{u}_j \overline{\hat{v}_j} \right| \leq \frac{1}{2}(|\hat{u}_j|^2 + |\hat{v}_j|^2)$ by the arithmetic-geometric inequality,[3] Bessel's inequality shows that the series is absolutely convergent. It follows that $[\cdot, \cdot]$ is a Hermitian, sesquilinear form on $\mathcal{V}$. Because of Bessel's inequality it is a semi-scalar product on $\mathcal{V}$. By the Cauchy–Schwarz inequality we obtain $|[u, v]|^2 \leq [u, u][v, v]$, but by assumption $[u, u] = \|u\|^2 - \sum_{j=1}^{\infty} |\hat{u}_j|^2 = 0$ so that the corollary follows. $\square$

It is now clear how to generalize the concept of an orthonormal basis to the case of an infinite-dimensional space.

Definition 1.3.9. An orthonormal sequence is called an *orthonormal basis* in $\mathcal{V}$ if the Parseval equality $\|u\|^2 = \sum_{j=1}^{\infty} |\hat{u}_j|^2$ is true for all $u \in \mathcal{V}$.

An orthonormal basis is often called a *complete orthonormal sequence* but we will avoid this in order not to confuse it with the concept of a complete space defined below. It is by no means clear that in a given space we can always find a sequence which is an orthonormal basis in the space. This requires the space to be *separable*.

Definition 1.3.10. A metric space $\mathcal{M}$ is called *separable* if it has a dense, countable subset. This means a sequence $u_1, u_2, \ldots$ of elements of $\mathcal{M}$, such that for any $u \in \mathcal{M}$, and any $\varepsilon > 0$, there is an element u_j of the sequence for which $d(u, u_j) < \varepsilon$.

The vast majority of spaces used in analysis are separable (Exercise 1.3.12), but there are exceptions (Exercise 1.3.14). The proof of the following proposition is left to the reader (Exercise 1.3.13).

Proposition 1.3.11. *An infinite-dimensional linear space provided with a scalar product is separable if and only if it contains a sequence which is an orthonormal basis.*

It should be remarked that even in a non-separable space with scalar product which is also *complete* (see below) there are always orthonormal bases, but these are not countable, i.e., they cannot be listed as a sequence. We shall never have occasion to use this.

Suppose $e_1, e_2, \ldots$ is an orthonormal basis in $\mathcal{V}$. We then know that any $u \in \mathcal{V}$ may be written as $u = \sum_{j=1}^{\infty} \hat{u}_j e_j$, where the series converges in norm, and that the numerical series $\sum_{j=1}^{\infty} |\hat{u}_j|^2$ converges to $\|u\|^2$.

The following question arises: Given a sequence $\lambda_1, \lambda_2, \ldots$ of complex numbers for which $\sum_{j=1}^{\infty} |\lambda_j|^2$ converges, does there exist an element $u \in \mathcal{V}$ for which $\lambda_1, \lambda_2, \ldots$ are the Fourier coefficients? Equivalently, does $\sum_{j=1}^{\infty} \lambda_j e_j$ converge in

---

[3] If $a, b \in \mathbb{R}$ then $ab \leq (a^2 + b^2)/2$, which is just a restatement of $(a - b)^2 \geq 0$.

norm to an element $u \in \mathcal{V}$? As it turns out, this is not always the case. The property required of $\mathcal{V}$ for a positive answer is that it is *complete* as a metric space.

To explain what a complete space is, we need a few definitions.

Definition 1.3.12. A *Cauchy sequence* in a metric space $\mathcal{M}$ is a sequence $u_1, u_2, \ldots$ of elements of $\mathcal{M}$ such that $d(u_j, u_k) \to 0$ as $j, k \to \infty$. More exactly: To every $\varepsilon > 0$ there exists a number ω such that $d(u_j, u_k) < \varepsilon$ if $j > \omega$ and $k > \omega$.

It is clear by use of the triangle inequality that any convergent sequence is a Cauchy sequence. Far more interesting is the fact that this implication may sometimes be reversed.

Definition 1.3.13.

(1) A metric space $\mathcal{M}$ is called *complete* if every Cauchy sequence in $\mathcal{M}$ converges.

(2) A normed linear space which is complete is called a *Banach space*.

That $\mathbb{R}^n$ and $\mathbb{C}^n$ provided with the standard norm are complete is a simple consequence of the Bolzano–Weierstrass theorem. This is also true for any finite-dimensional normed space by Corollary 1.3.2 (Exercise 1.3.15). It follows that any finite-dimensional subspace of a normed space is complete and therefore also closed.

Returning now to the question discussed before Definition 1.3.12, suppose the series $\sum_{j=1}^{\infty} |\lambda_j|^2$ converges and $e_1, e_2, \ldots$ is an orthonormal sequence in $\mathcal{V}$ and put $u_k = \sum_{j=1}^{k} \lambda_j e_j$. If $k < n$ we then have (the second equality is Pythagoras' theorem, the case $u = 0$ of Proposition 1.3.5)

$$\|u_n - u_k\|^2 = \left\| \sum_{j=k+1}^{n} \lambda_j e_j \right\|^2 = \sum_{j=k+1}^{n} |\lambda_j|^2 = \sum_{j=1}^{n} |\lambda_j|^2 - \sum_{j=1}^{k} |\lambda_j|^2 .$$

Since $\sum_{j=1}^{\infty} |\lambda_j|^2$ converges, the right-hand side $\to 0$ as $k, n \to \infty$. Hence $u_1, u_2, \ldots$ is a Cauchy sequence in $\mathcal{V}$. It therefore follows that if $\mathcal{V}$ is complete, then $\sum_{j=1}^{\infty} \lambda_j e_j$ converges in norm to an element of $\mathcal{V}$. On the other hand, if $\mathcal{V}$ is *not* complete and $e_1, e_2, \ldots$ is an orthonormal basis in $\mathcal{V}$, then $\lambda_1, \lambda_2, \ldots$ may be chosen so that $\sum_{j=1}^{\infty} |\lambda_j|^2$ is convergent but the series $\sum_{j=1}^{\infty} \lambda_j e_j$ does *not* converge in $\mathcal{V}$ (Exercise 1.3.17).

Exercises

Exercise 1.3.1 Show that a subset A of a metric space is closed if and only if the limit of every convergent sequence in A is also in A.

Exercise 1.3.2 Show that if $\|\cdot\|$ is a norm on $\mathcal{V}$, then $d(u, v) = \|u - v\|$ is a metric on $\mathcal{V}$. Also show that the maps $\mathcal{V} \times \mathcal{V} \ni (u, v) \mapsto u + v \in \mathcal{V}$ and $\mathbb{K} \times \mathcal{V} \ni (\lambda, u) \mapsto \lambda u \in \mathcal{V}$ are continuous.

Exercise 1.3.3 Prove the reverse triangle inequality for a normed linear space. What does it look like in a metric space?

Exercise 1.3.4 Show that $d(x, y) = |x - y| / (1 + |x - y|)$ is a metric on $\mathbb{R}$ which can be extended to a metric on the set of extended reals $\overline{\mathbb{R}} = \mathbb{R} \cup \{-\infty\} \cup \{+\infty\}$.

Exercise 1.3.5 Consider the linear space $C^1[0, 1]$, consisting of complex-valued, differentiable functions with continuous derivative, defined on $[0, 1]$. Show that the following are all norms on $C^1[0, 1]$.

(1) $\|u\|_\infty = \sup_{0 \le x \le 1} |u(x)|$.

(2) $\|u\|_1 = \int_0^1 |u|$.

(3) $\|u\|_{1,\infty} = \|u'\|_\infty + \|u\|_\infty$.

Invent some more norms in the same spirit.

Exercise 1.3.6 Find all cases of equality in the Cauchy–Schwarz inequality for a scalar product. Then show that $\|\cdot\|$, defined by $\|u\| = \sqrt{\langle u, u \rangle}$, where $\langle \cdot, \cdot \rangle$ is a scalar product, is a norm. Finally show that the scalar product is continuous in its arguments, i.e., $\langle u_j, v_j \rangle \to \langle u, v \rangle$ if $\|u_j - u\|$ and $\|v_j - v\| \to 0$ as $j \to \infty$.

Exercise 1.3.7 Show that $\langle u, v \rangle = \int_0^1 u(x)\overline{v(x)}\, dx$ is a scalar product on the space $C[0, 1]$ of continuous, complex-valued functions defined on $[0, 1]$.

Exercise 1.3.8 Finish the proof of Lemma 1.3.4.

Exercise 1.3.9 Prove Proposition 1.3.5.

Exercise 1.3.10 Prove Bessel's inequality.

Exercise 1.3.11 Prove Parseval's formula.

Exercise 1.3.12 It is well known that the set of step functions (see Section B.4) which are identically 0 outside a compact subinterval of an interval Ω are dense in $L^2(\Omega)$. Use this to show that $L^2(\Omega)$ is separable (for the definition of $L^2(\Omega)$ see Appendix B).

Exercise 1.3.13 Prove Proposition 1.3.11.

Hint: Use Gram–Schmidt as in the proof of Lemma 1.3.4.

Exercise 1.3.14 Let $\mathcal{V}$ be the set of functions u defined on $\mathbb{R}$ which are finite linear combinations of functions of the form $e^{i\alpha x}$ with $\alpha \in \mathbb{R}$. Clearly $\mathcal{V}$ is a linear space.

Show that $\langle u, v \rangle = \lim_{T \to \infty} \frac{1}{2T} \int_{-T}^{T} u\bar{v}$ defines a scalar product on $\mathcal{V}$, that the corresponding norm of $e^{i\alpha x}$ is 1 for any $\alpha \in \mathbb{R}$ and that $\langle e^{i\alpha x}, e^{i\beta x} \rangle = 0$ if $\alpha \ne \beta$. Conclude that $\mathcal{V}$ provided with this scalar product is *not* separable, while the set $\{e^{i\alpha x} : \alpha \in \mathbb{R}\}$ is an (uncountable) orthonormal basis in $\mathcal{V}$, as well as a Hamel basis.

Elements of the completion of $\mathcal{V}$ (see Section 2.1) are called *almost periodic functions*.

Exercise 1.3.15 Show that as a metric space the set $\mathbb{Q}$ of rational numbers is not complete but the set $\mathbb{R}$ of reals is. Also show that any finite-dimensional real or complex linear space is complete.

Exercise 1.3.16 Show that if a normed space is complete under two different norms, then the norms are equivalent.

Hint: The closed graph theorem of Appendix A.

Exercise 1.3.17 Suppose $\mathcal{V}$ is a space with scalar product which is *not* complete, and that $e_1, e_2, \ldots$ is an orthonormal basis in $\mathcal{V}$. Show that there exists a sequence $\lambda_1, \lambda_2, \ldots$ of complex numbers such that $\sum |\lambda_j|^2 < \infty$ but $\sum \lambda_j e_j$ does not converge to any element of $\mathcal{V}$.

Exercise 1.3.18 Suppose $\{u_j\}_{j=1}^{\infty}$ is a sequence in a normed space $\mathcal{V}$. The series $\sum_{j=1}^{\infty} u_j$ is called *norm convergent* if the numerical series $\sum_{j=1}^{\infty} \|u_j\|$ converges. Prove that $\mathcal{V}$ is complete if and only if every norm convergent series is convergent.

Hint: One direction is easy. For the other direction, see the proof of Theorem B.7.4.

Exercise 1.3.19 Let $C[a, b]$ be the continuous functions on the compact interval $[a, b]$, $-\infty < a < b < \infty$. Define $\|u\|_{[a,b]} = \sup_{[a,b]} |u|$ for $u \in C[a, b]$.

Show that $\|\cdot\|_{[a,b]}$ is a norm on $C[a, b]$ under which it is complete, i.e., normed this way $C[a, b]$ is a Banach space.

Exercise 1.3.20 Let $C(a, b)$ be the continuous functions on the open interval (a, b), $-\infty \le a < b \le \infty$.

Let $\{I_j\}_1^{\infty}$ be a sequence of compact intervals with union (a, b) and define $\|u\|_j = \sup_{I_j} |u|$. Show that this is a semi-norm on $C(a, b)$.

Next define $d(u, v) = \sum_{j=1}^{\infty} 2^{-j} \|u - v\|_j / (1 + \|u - v\|_j)$ and show that this is a metric on $C(a, b)$, that $d(u + w, v + w) = d(u, v)$ (translation invariance) and that $C(a, b)$ provided with this metric is a complete metric space.

Also show that $d(u_k, u) \to 0$ is equivalent to $\|u_k - u\|_j \to 0$ as $k \to \infty$ for every j and that convergence in this metric means *locally uniform convergence* or uniform convergence on every compact subset of (a, b).

A complete, metric and linear space with the metric defined as above by a sequence of semi-norms is called a *Fréchet space*.

1.4 The equations of Sturm and Liouville

Here we shall briefly consider equations of the type discussed by Sturm and Liouville. We shall see that quite elementary methods are capable of giving satisfying results about expansions in eigenfunctions. In particular it may be argued that this approach leads to some of the basic results for Fourier series with less effort than standard methods.

The approach taken is based on the so-called direct methods of the calculus of variations and is quite different from the methods of Sturm and Liouville. It was inspired by an analogous treatment of the Helmholtz equation in Courant–Hilbert [48, Chapter 7] and by the masters thesis Karlsson [120].

A minimization problem. Let $[a, b]$ be a compact interval and consider functions defined on $[a, b]$. To keep the discussion on an elementary level we assume these functions to be continuously differentiable. We allow them to be *complex-valued*, although everything said before page 25 will be true also if we consider only real-valued functions. This class of functions is a linear space of infinite dimension (Exercise 1.4.1) which we denote by $\mathcal{D}_0$. On $\mathcal{D}_0$ we define a scalar product $\langle u, v \rangle = \int_a^b u\bar{v}w$ and norm $\|u\| = \sqrt{\langle u, u \rangle}$, where w is a fixed, strictly positive function continuous on $[a, b]$. Now suppose we are also given an *energy integral*

$$H(u, v) = \int_a^b (u'\bar{v'} + qu\bar{v})$$

where q is a given real-valued and continuous function. Like the scalar product H is a *Hermitian form*, i.e., it has the following properties

(1) $H(\alpha u_1 + \beta u_2, v) = \alpha H(u_1, v) + \beta H(u_2, v)$ for u_1, u_2, v in $\mathcal{D}_0$ and α, β in $\mathbb{C}$.

(Linear in first argument)

(2) $H(u, v) = \overline{H(v, u)}$ for u, v in $\mathcal{D}_0$.

(Hermitian symmetric)

(3) $H(u, \alpha v_1 + \beta v_2) = \bar{\alpha} H(u, v_1) + \bar{\beta} H(u, v_2)$ for u, v_1, v_2 in $\mathcal{D}_0$ and α, β in $\mathbb{C}$.

(Anti-linear in second argument)

Here the third point is a consequence of the first two.

We consider the problem of minimizing $H(u, u)$ over $\mathcal{D}_0$ under the constraint $\langle u, u \rangle = 1$. One sees immediately that this is equivalent to minimizing the *Rayleigh quotient*, i.e., to determining

$$\lambda_1 = \inf_{0 \neq u \in \mathcal{D}_0} \frac{H(u, u)}{\langle u, u \rangle}. \tag{1.4.1}$$

This is a finite number since $H(u, u) \geq \int_a^b q|u|^2 \geq \min_{[a,b]}(q/w) \|u\|^2$ so that $\lambda_1 \geq \min_{[a,b]}(q/w)$. However, it is not at all clear that there exists a function $e_1 \neq 0$ in $\mathcal{D}_0$ for which $H(e_1, e_1) = \lambda_1 \langle e_1, e_1 \rangle$, a so-called *extremal* for (1.4.1). But if an extremal exists and φ is an arbitrary function in $\mathcal{D}_0$, then the (rational) function

$$\mathbb{R} \ni s \mapsto \frac{H(e_1 + s\varphi, e_1 + s\varphi)}{\langle e_1 + s\varphi, e_1 + s\varphi \rangle}$$

has a minimum for $s = 0$, so that its derivative vanishes there. Calculation shows that this condition means that $\mathrm{Re}(H(e_1, \varphi) - \lambda_1 \langle e_1, \varphi \rangle) = 0$. Replacing φ by $e^{i\theta}\varphi$ for a suitably chosen constant $\theta \in \mathbb{R}$ we obtain

$$H(e_1, \varphi) - \lambda_1 \langle e_1, \varphi \rangle = 0$$

for all $\varphi \in \mathcal{D}_0$. Thus $\int_a^b (e_1' \overline{\varphi'} + (q - \lambda_1 w) e_1 \overline{\varphi}) = 0$ for all functions $\varphi \in \mathcal{D}_0$. If e_1' is continuously differentiable integration by parts shows that

$$0 = e_1' \overline{\varphi}\big|_a^b + \int\limits_a^b (-e_1'' + (q - \lambda_1 w) e_1) \overline{\varphi}$$

for all $\varphi \in \mathcal{D}_0$. It is easily seen that this means that λ_1 is an eigenvalue and e_1 a corresponding eigenfunction for the eigenvalue problem

$$\begin{cases} -u'' + qu = \lambda wu \text{ in } [a, b] \\ u'(a) = u'(b) = 0 \,. \end{cases} \tag{1.4.2}$$

An *eigenfunction* to the eigenvalue λ is a non-trivial ($\neq 0$) solution of the differential equation satisfying the *boundary conditions* $u'(a) = u'(b) = 0$.

Conversely, integration by parts shows that if λ is an eigenvalue with eigenfunction u, then $\frac{H(u,u)}{\langle u,u \rangle} = \lambda$. The following proposition is an immediate consequence.

Proposition 1.4.1. *If λ_1 as defined by* (1.4.1) *is an eigenvalue for* (1.4.2), *then it is the* smallest *eigenvalue.*

If λ and μ are eigenvalues with eigenfunctions u and v use of the differential equation and integrating by parts twice shows that

$$\lambda \langle u, v \rangle = \int\limits_a^b (-u'' + qu) \overline{v} = \int\limits_a^b u(\overline{-v'' + qv}) = \overline{\mu} \langle u, v \rangle \,. \tag{1.4.3}$$

For $u = v$ this shows that eigenvalues are always real, so if $\lambda \neq \mu$ we must have $\langle u, v \rangle = 0$. We have proved the following proposition.

Proposition 1.4.2. *All eigenvalues are real, and eigenfunctions to different eigenvalues are orthogonal.*

If we define

$$\lambda_2 = \inf_{\substack{0 \neq u \in \mathcal{D}_0 \\ \langle u, e_1 \rangle = 0}} \frac{H(u, u)}{\langle u, u \rangle}$$

we therefore see in the same way as before that λ_2 is the next eigenvalue in order of size, and one may now continue in this way, finally obtaining an infinite, non-decreasing sequence of eigenvalues and corresponding mutually orthogonal eigenfunctions.

To rigorously carry out the above we must show that to each λ_j, defined as an infimum, there is a corresponding normalized function e_j realizing this infimum. Furthermore we must show that these functions solve the eigenvalue problem with eigenvalue λ_j. The main problem is then to show that they are twice differentiable.

We shall also show that the eigenvalues can have no finite point of accumulation and therefore tend to plus infinity. Finally we shall show that a more or less arbitrary function may be developed in a series where the terms are eigenfunctions. We shall then also have shown that the method of separation of variables solves the problem of heat conduction discussed in Section 1.1.

Existence of extremal. Let $c > -\inf(q/w)$, so that $\lambda_1 + c > 0$. Define $q_c = q + cw > 0$ and put $H_c(u, v) = \int_a^b (u'\overline{v'} + q_c u\overline{v})$ so that $\int_a^b |u'|^2 \le H_c(u, u)$.

By the fundamental theorem of calculus we have, if $u \in C^1[a, b]$, i.e., u has a continuous derivative in $[a, b]$, that $u(x) - u(y) = \int_y^x u'$ for $x, y \in [a, b]$ so that

$$|u(x) - u(y)| \le \left| \int_y^x dt \int_y^x |u'|^2 \right|^{1/2} \le \sqrt{|x - y|}\sqrt{H_c(u, u)} \tag{1.4.4}$$

if $x, y \in [a, b]$. We have used the Cauchy–Schwarz inequality for the semi-scalar product $B(f, g) = \int_x^y f\overline{g}$ with $f = u'$ and $g \equiv 1$.

Furthermore we obtain $|u(x)| \le |u(y)| + \int_a^b |u'|$ so multiplication by $q_c(y)$ and integration with respect to y gives

$$|u(x)| \int_a^b q_c \le \int_a^b q_c |u| + \int_a^b q_c \int_a^b |u'| \, .$$

Division by $\int_a^b q_c > 0$ and use of the Cauchy–Schwarz inequality (what is the semi-scalar product used here?) then shows that

$$|u(x)| \le C\sqrt{H_c(u, u)} \tag{1.4.5}$$

where $C = \sqrt{b - a + 1/\int_a^b q_c}$. We will show that the inequalities (1.4.4) and (1.4.5) imply the following theorem.

Theorem 1.4.3. *If $\{u_k\}_1^\infty$ is a sequence of functions in $C^1[a, b]$ for which the sequence $H_c(u_k, u_k)$, $k = 1, 2, \ldots$, is bounded, then the sequence $\{u_k\}_1^\infty$ has a uniformly convergent subsequence.*

To prove this we shall use the *Arzela–Ascoli theorem*. A sequence of continuous functions $\{u_k\}_1^\infty$ on an interval I is called *equi-continuous* if for each $\varepsilon > 0$ there is a $\delta > 0$ *independent of k, x* and y such that we have $|u_k(x) - u_k(y)| < \varepsilon$ if $|x - y| < \delta$. The sequence is *uniformly bounded* if there is a constant M, independent of x and k, such that $|u_k(x)| \le M$ if $x \in I$.

Theorem 1.4.4 (Arzela–Ascoli). *Suppose $\{u_k\}_1^\infty$ is a sequence of equi-continuous and uniformly bounded functions on a compact interval I. Then there is a subsequence that converges uniformly on I.*

In the proof we shall use the fact that the rational numbers in I may be *enumerated*, that is, they may be listed as a sequence. We shall also use the fact that the rational numbers in I are *dense* in I. This means that given any $\varepsilon > 0$ and $x \in I$ there is a rational number $r \in I$ such that $|x - r| < \varepsilon$. We shall finally use the *Bolzano–Weierstrass theorem* (Corollary 1.3.2), which asserts that a bounded sequence of numbers has a convergent subsequence.

Proof. Let $\{r_j\}_1^\infty$ be an enumeration of the rational numbers in I. By the Bolzano–Weierstrass theorem and since the sequence $u_1(r_1), u_2(r_1), u_3(r_1), \ldots$ is bounded (by the uniform bound) it has a convergent subsequence obtained by choosing a subsequence $\{u_{1,k}\}_{k=1}^\infty$ of $\{u_k\}_1^\infty$. Call the limit $u(r_1)$.

The sequence $u_{1,1}(r_2), u_{1,2}(r_2), u_{1,3}(r_2), \ldots$ is also bounded, so it has a convergent subsequence obtained by choosing a subsequence $\{u_{2,k}\}_{k=1}^\infty$ of $\{u_{1,k}\}_1^\infty$. Call the limit $u(r_2)$, and note that we also have $u_{2,k}(r_1) \to u(r_1)$ as $k \to \infty$, since $\{u_{2,k}(r_1)\}_1^\infty$ is a subsequence of $\{u_{1,k}(r_1)\}_1^\infty$.

We may continue in this way, obtaining a sequence of sequences of functions, each element of which is a subsequence of those coming before it, and with the property that $u_{j,k}(r_n) \to u(r_n)$ as $k \to \infty$ if $n \leq j$. It follows that the sequence $\{u_{k,k}(r_n)\}_{k=1}^\infty$ converges to $u(r_n)$ for every n, since it is a subsequence of $\{u_{n,k}(r_n)\}_{k=1}^\infty$ from $k = n$ on.

So far we have only used the uniform bound of our sequence, but now suppose $x \in I$ but not necessarily rational. Given $\varepsilon > 0$ let δ be chosen as in the definition of equi-continuity. We may then divide I into a finite number, say N, of non-overlapping intervals of length $< \delta$ and choose rational numbers $\rho_1, \ldots, \rho_N$, one in each subinterval. Given $x \in I$ one of these numbers, say ρ_j, is at a distance $< \delta$, and by the triangle inequality

$$|u_{n,n}(x) - u_{m,m}(x)|$$
$$\leq |u_{n,n}(x) - u_{n,n}(\rho_j)| + |u_{n,n}(\rho_j) - u_{m,m}(\rho_j)| + |u_{m,m}(\rho_j) - u_{m,m}(x)|.$$

In the second line the first and third terms are $< \varepsilon$ by equi-continuity, and so is the middle term if n and $m \geq \omega_j$ for some ω_j. Thus, setting $\omega = \max(\omega_1, \ldots, \omega_N)$, the middle term is $< \varepsilon$ independent of j if n and $m > \omega$. Thus $|u_{n,n}(x) - u_{m,m}(x)| < 3\varepsilon$ if $x \in I$ and n and $m > \omega$.

This means, first, that $\{u_{n,n}(x)\}_1^\infty$ is a Cauchy sequence for each $x \in I$ so that it has a limit $u(x)$. Secondly, letting $m \to \infty$, it follows that $|u_{n,n}(x) - u(x)| \leq 3\varepsilon$ for all $x \in I$ if $n > \omega$. Thus the convergence is uniform.   □

Proof (Theorem 1.4.3). The inequalities (1.4.4), (1.4.5) imply that the sequence $\{u_k\}_1^\infty$ is equi-continuous and uniformly bounded, so that the *Arzela–Ascoli theorem* shows that the sequence $\{u_k\}_1^\infty$ has a uniformly convergent subsequence.   □

Now set $\lambda_1 = \inf_{0 \neq u \in \mathcal{D}_0} \frac{H(u,u)}{\langle u, u \rangle}$, where $\mathcal{D}_0 = C^1[a, b]$. One sees immediately that it is sufficient to consider functions u which are unit vectors, i.e., for which $\|u\| = 1$, since the quotient is unchanged if $u \neq 0$ is replaced by $u/\|u\|$. It is then clear that there is a sequence $u_1, u_2, \ldots$ of unit vectors in $C^1[a, b]$ so that $H(u_k, u_k) = \frac{H(u_k, u_k)}{\langle u_k, u_k \rangle} \to \lambda_1$ as $k \to \infty$. But then the sequence $H_c(u_k, u_k)$, $k = 1, 2, \ldots$, is bounded, so that the sequence $\{u_k\}_1^\infty$ has a uniformly convergent subsequence according to Theorem 1.4.3. To simplify notation we may as well assume that the subsequence has already been selected, so that the sequence $\{u_k\}_1^\infty$ has a uniform limit on $[a, b]$, which we shall call e_1.

Because of the uniform convergence we obtain $1 = \langle u_k, u_k \rangle \to \langle e_1, e_1 \rangle$ so that the function e_1 is also a unit vector. If we knew that $e_1 \in C^1[a, b]$ and that $H(u_k, u_k) \to H(e_1, e_1)$ we would deduce that e_1 satisfies $\frac{H(e_1, e_1)}{\langle e_1, e_1 \rangle} = \lambda_1$, so that e_1 would be an extremal for our minimizing problem. To show this we shall need the *du Bois-Reymond lemma*.

The du Bois-Reymond lemma. This is a lemma (see du Bois-Reymond [31], Bliss [28]) which belongs to the calculus of variations, but there is no consensus as to exactly what it states. It is sometimes taken to state simply that if $\int_a^b uv = 0$ for all v in some sufficiently wide class of functions, then $u = 0$. Sometimes it states that if $\int_a^b uv' = 0$ for all v in some sufficiently wide class of functions, then u is constant. In du Bois-Reymond's paper [31] both statements are made. In this section we shall use the name in a slightly unorthodox way to mean Theorem 1.4.5 below. Both the latter variants are special cases of the more general Theorem C.11.

By $C^2[a, b]$ we denote those (complex-valued) functions defined on $[a, b]$ which have a second derivative continuous on $[a, b]$, and by $C_*^2[a, b]$ those functions $u \in C^2[a, b]$ for which $u(a) = u'(a) = u(b) = u'(b) = 0$. We shall prove the following theorem.

Theorem 1.4.5. *Suppose u is continuous on $[a, b]$ and that $\int_a^b u\overline{v''} = 0$ for all $v \in C_*^2[a, b]$. Then u is a polynomial of degree at most one.*

Proof. First note that if A and B are constants, then integration by parts shows that $\int_a^b (Ax + B)\overline{v''}(x)\, dx = 0$ for all $v \in C_*^2[a, b]$. Hence $\int_a^b (u(x) - Ax - B)\overline{v''}(x)\, dx = 0$ for all $v \in C_*^2[a, b]$ and all constants A, B. Now put

$$v(x) = \int_a^x \left(\int_a^t (u(s) - As - B)\, ds \right) dt \ .$$

Then $v(a) = 0$, v is differentiable and

$$v'(x) = \int_a^x (u(s) - As - B)\, ds \ ,$$

so that $v'(a) = 0$. Now $v''(x) = u(x) - Ax - B$, so that $v \in C^2[a, b]$. Suppose we can choose the constants A, B so that $v(b) = v'(b) = 0$. We would then have $v \in C_*^2[a, b]$ and would obtain

$$0 = \int_a^b (u(x) - Ax - B)\overline{v''}(x)\, dx = \int_a^b |u(x) - Ax - B|^2\, dx \ ,$$

implying that $u(x) = Ax + B$ (see Exercise 1.4.3) and the proof would be finished.

It remains to show that we may select the constants A, B so that $v(b) = v'(b) = 0$. But

$$v(b) = \int_a^b \int_a^t u \, dt - A \int_a^b \int_a^t s \, ds \, dt - B \int_a^b \int_a^t ds \, dt$$

and

$$v'(b) = \int_a^b u - A \int_a^b s \, ds - B(b-a),$$

so the conditions $v(b) = v'(b) = 0$ give a linear system of equations for A, B. It is easy to see that the determinant of this system equals $(b-a)^4/12 \neq 0$, so there is always a unique solution. The proof is now complete. $\quad\square$

Eigenfunctions. Now return to the sequence $\{u_k\}_1^\infty$ converging to e_1. Defining $B_1(u,v) = H(u,v) - \lambda_1 \langle u, v \rangle$ for $u, v \in \mathcal{D}_0$ we see that B_1 is a Hermitian form on $\mathcal{D}_0$ which is also positive semi-definite since $H(u,u) \geq \lambda_1 \langle u, u \rangle$ for all $u \in \mathcal{D}_0$. Thus the Cauchy–Schwarz inequality is valid for the form B_1 and we obtain

$$|B_1(u_k, v)|^2 \leq B_1(u_k, u_k) B_1(v, v)$$

for all $v \in \mathcal{D}_0$. But $H(u_k, u_k) \to \lambda_1$ and $\langle u_k, u_k \rangle = 1$ so $B_1(u_k, u_k) \to 0$. Thus the inequality shows that $B_1(u_k, v) \to 0$ for every $v \in \mathcal{D}_0$. If we have $v \in C^2[a,b] \subset \mathcal{D}_0$ we may integrate by parts and obtain, using the uniform convergence in the integral,

$$B_1(u_k, v) = u_k \overline{v'} \Big|_a^b + \int_a^b (-u_k \overline{v''} + (q - \lambda_1 w) u_k \overline{v})$$

$$\to e_1(b) \overline{v'}(b) - e_1(a) \overline{v'}(a) - \int_a^b (e_1 \overline{v''} - (q - \lambda_1 w) e_1 \overline{v}).$$

Since the limit is zero we obtain

$$\int_a^b (e_1 \overline{v''} - (q - \lambda_1 w) e_1 \overline{v}) = e_1(b) \overline{v'}(b) - e_1(a) \overline{v'}(a) \qquad (1.4.6)$$

for all $v \in C^2[a,b]$. This is in particular true if $v \in C_*^2[a,b]$, and in this case integrating by parts twice gives

$$\int_a^b \left(e_1(x) - \int_a^x \left(\int_a^t (q - \lambda_1 w) e_1 \right) dt \right) \overline{v''}(x) \, dx = 0$$

for all $v \in C_*^2[a,b]$. According to Theorem 1.4.5 this means that

$$e_1(x) = Ax + B + \int\limits_a^x \left(\int\limits_a^t (q - \lambda_1 w)e_1 \right)$$

for some polynomial $Ax + B$. Here the right member, and therefore also the left member e_1, is twice continuously differentiable and we obtain $e_1'' = (q - \lambda_1 w)e_1$.

We may now return to (1.4.6), where $v \in C^2[a, b]$, and integrate by parts, which gives

$$H(e_1, v) = \lambda_1 \langle e_1, v \rangle . \tag{1.4.7}$$

Since $e_1 \in C^2[a, b]$ we may choose $v = e_1$, which shows that e_1 is an extremal of our minimization problem. Another integration by parts in (1.4.7) now shows that

$$0 = -e_1' \bar{v} \Big|_a^b + \int\limits_a^b (e_1'' - (q - \lambda_1 w)e_1) \bar{v} = e_1'(a)\bar{v}(a) - e_1'(b)\bar{v}(b)$$

for all $v \in C^2[a, b]$. Since $v(a)$ and $v(b)$ are arbitrary it follows that $e_1'(a) = e_1'(b) = 0$. Thus we have shown that the extremal e_1 is an eigenfunction with eigenvalue λ_1 of the eigenvalue problem (1.4.2). According to Proposition 1.4.1 λ_1 is the *smallest* eigenvalue for (1.4.2).

It also follows that (1.4.7) is true even if we just have $v \in \mathcal{D}_0$, since then $H(e_1, v) = e_1' \bar{v} \Big|_a^b + \int_a^b (-e_1'' + q e_1)\bar{v} = \lambda_1 \langle e_1, v \rangle$. In particular, if $\langle e_1, v \rangle = 0$ then also $H(e_1, v) = 0$.

We now proceed by induction. Suppose we have found the first j eigenvalues $\lambda_1 \leq \lambda_2 \leq \cdots \leq \lambda_j$ and corresponding orthonormal eigenfunctions $e_1, e_2, \ldots, e_j$. Then set

$$\mathcal{D}_j = \{ u \in \mathcal{D}_0 : \langle u, e_k \rangle = 0, \ k = 1, \ldots, j \}$$

and also assume we know that $H(u, e_k) = 0$ if $u \in \mathcal{D}_j$ for $k = 1, \ldots, j$. We have just verified that this is true if $j = 1$. Note that since $\dim \mathcal{D}_0 = \infty$ also $\dim \mathcal{D}_j = \infty$.

Now define

$$\lambda_{j+1} = \inf_{0 \neq u \in \mathcal{D}_j} \frac{H(u, u)}{\langle u, u \rangle} . \tag{1.4.8}$$

Since $\mathcal{D}_j \subset \mathcal{D}_{j-1}$ we have $\lambda_{j+1} \geq \lambda_j$. Suppose $\{u_k\}_1^\infty$ is a minimizing sequence of unit vectors for (1.4.8). As before the inequalities (1.4.4) and (1.4.5) show that the sequence is equi-continuous and uniformly bounded, so replacing the sequence with an appropriate subsequence we may assume that it converges uniformly to a unit vector e_{j+1}.

Setting $B_{j+1}(u, v) = H(u, v) - \lambda_{j+1}\langle u, v \rangle$ the Hermitian form B_{j+1} is positive semi-definite on $\mathcal{D}_j$, and $B_{j+1}(u_k, u_k) \to 0$ as $k \to \infty$, so that, using the Cauchy–Schwarz inequality as before, we have $B_{j+1}(u_k, v_j) \to 0$ as $k \to \infty$ for every $v_j \in \mathcal{D}_j$.

For $v \in \mathcal{D}_0$ we define $\hat{v}_n = \langle v, e_n \rangle$ and put $s_j v = \sum_1^j \hat{v}_n e_n$. With $v_j = v - s_j v$ it follows that if $1 \leq i \leq j$, then $\langle v_j, e_i \rangle = \langle v, e_i \rangle - \sum_{n=1}^j \hat{v}_n \langle e_n, e_i \rangle = 0$ so that $v_j \in \mathcal{D}_j$. Since $H(u_k, e_n) = \langle u_k, e_n \rangle = 0$ for $n = 1, \ldots, j$, it follows that $B_{j+1}(u_k, v) = B_{j+1}(u_k, v_j) \to 0$ as $k \to \infty$ for every $v \in \mathcal{D}_0$.

Using now the du Bois-Reymond lemma as before, we see that e_{j+1} is in $C^2[a, b]$, is an extremal for (1.4.8) and is a normalized eigenfunction with eigenvalue λ_{j+1} to (1.4.1) such that $H(v, e_{j+1}) = 0$ if $\langle v, e_{j+1} \rangle = 0$.

This completes the induction step so that we obtain an infinite sequence of eigenvalues $\lambda_1 \leq \lambda_2 \leq \ldots$ with corresponding orthonormal eigenvectors $e_1, e_2, \ldots$. Note that we have not excluded that $\lambda_j = \lambda_{j+1}$ for[4] some values of j, but if so the eigenfunctions we have associated with λ_j and λ_{j+1} are still orthogonal to each other.

Finally, it follows that $\lambda_j \to \infty$ as $j \to \infty$, for the only alternative is that the eigenvalues have a finite limit λ. But then $H_c(e_j, e_j) \leq \lambda + c$ for all j, so appealing to Theorem 1.4.3 as before it follows that there is a subsequence $\{e_{j_k}\}_1^\infty$ of the eigenfunctions which converges uniformly to a unit vector e. We then obtain $0 = \langle e_{j_{k+1}}, e_{j_k} \rangle \to \langle e, e \rangle = 1$ as $k \to \infty$, which is a contradiction.

We have proved the following theorem.

Theorem 1.4.6. *The problem* (1.4.2) *has infinitely many eigenvalues. These may be listed as a non-decreasing sequence which tends to ∞, and the corresponding eigenfunctions may be chosen orthonormal.*

Expansion in eigenfunctions. The sequence of eigenfunctions $e_1, e_2, \ldots$ is an orthonormal sequence with respect to the scalar product $\langle \cdot, \cdot \rangle$. Assuming the function u to be continuous (what is needed is just that u is in the L^2-space with scalar product $\langle \cdot, \cdot \rangle$, see Appendix B), and setting $\hat{u}_j = \langle u, e_j \rangle$ the Bessel inequality $\sum_1^\infty |\hat{u}_j|^2 \leq \langle u, u \rangle$ is valid.

The form $H_c(u, v)$ is a scalar product on $\mathcal{D}_0$ for which $H_c(e_j, e_k) = 0$ if $j \neq k$. Since $H_c(u, e_j) = (\lambda_j + c)\langle u, e_j \rangle$, in particular for $u = e_j$, the Bessel inequality is replaced by the inequality

$$\sum_{j=1}^\infty (\lambda_j + c) |\hat{u}_j|^2 \leq H_c(u, u) . \tag{1.4.9}$$

For, if $f_j = (\lambda_j + c)^{-1/2} e_j$, then $\{f_j\}_1^\infty$ is an orthonormal sequence for the scalar product H_c. We have $H_c(u, f_j) = (\lambda_j + c)^{-1/2} H_c(u, e_j) = (\lambda_j + c)^{1/2} \hat{u}_j$, so the inequality (1.4.9) is just an instance of Bessel's inequality.

As in the proof of Theorem 1.4.6 put $s_j u(x) = \sum_1^j \hat{u}_k e_k(x)$ and $u_j(x) = u(x) - s_j u(x)$ so that $u_j \in \mathcal{D}_j$. We then have

$$H(u_j, u_j) \geq \lambda_{j+1} \langle u_j, u_j \rangle . \tag{1.4.10}$$

If $k \leq j$, then $H_c(u_j, e_k) = 0$ since $\langle u_j, e_k \rangle = 0$ so it follows that $H_c(u_j, s_j u) = 0$. Since $u = u_j + s_j u$ we obtain

$$H_c(u, u) = H_c(u_j, u_j) + H_c(s_j u, s_j u) \geq H_c(u_j, u_j) \geq (\lambda_{j+1} + c)\langle u_j, u_j \rangle$$

by (1.4.10), so

---

[4] With the present boundary conditions this cannot happen, but for the coupled boundary conditions discussed later it may.

$$0 \leq \langle u_j, u_j \rangle \leq H_c(u, u)/(\lambda_{j+1} + c) .$$

Since $\lambda_j \to \infty$ as $j \to \infty$ it follows that $\langle u_j, u_j \rangle \to 0$ as $j \to \infty$.

We have therefore shown that $s_j u \to u$ in the norm $\|\cdot\|$. But we easily obtain a stronger result, since it is clear that $H_c(u_j - u_n, u_j - u_n) = \sum_{k=n+1}^{j} (\lambda_k + c) |\hat{u}_k|^2$ if $n < j$. In view of (1.4.9) it follows that the sum tends to 0 as $n, j \to \infty$. If we now use our inequality (1.4.5) it follows, first that the numerical sequence $\{u_j(x)\}_1^\infty$ for each fixed $x \in [a, b]$ is a Cauchy sequence of numbers and hence has a limit $v(x)$, and then that $u_j \to v$ uniformly in $[a, b]$. It therefore follows that v is continuous and that $\langle u_j, u_j \rangle \to \langle v, v \rangle$. Since we know that the limit is zero it follows that $v = 0$.

In other words, we have proved the following theorem.

Theorem 1.4.7. *Suppose $u \in \mathcal{D}_0$. Then the generalized Fourier series $\sum_{j=1}^{\infty} \hat{u}_j e_j$ converges in norm and uniformly to the function u.*

We have thus also verified that the heat equation problem we discussed in Section 1.1 can be solved by separation of variables (Exercise 1.4.2).

Generalizations. By modifying either the space $\mathcal{D}_0$ or the form H, or both, one obtains in the same way as above results on eigenfunction expansions for eigenvalue problems with other boundary conditions than in (1.4.2).

A first example is obtained by replacing the function class $\mathcal{D}_0$ used previously by $\mathcal{D}_0 = C_*^1[a, b] = \{u \in C^1[a, b] : u(a) = u(b) = 0\}$. Repeating the previous investigation with the modified $\mathcal{D}_0$ we obtain Theorems 1.4.6 and 1.4.7 for this choice of $\mathcal{D}_0$ and the eigenvalue problem

$$\begin{cases} -u'' + qu = \lambda w u \text{ in } [a, b] , \\ u(a) = u(b) = 0 . \end{cases} \tag{1.4.11}$$

Instead of the *Neumann boundary conditions* $u'(a) = u'(b) = 0$ we now have *Dirichlet boundary conditions* $u(a) = u(b) = 0$.

Choosing for our basic function class $\mathcal{D}_0 = \{u \in C^1[a, b] : u(a) = 0\}$ we obtain an eigenvalue problem with a Dirichlet condition at a and a Neumann condition at b. A further variant is obtained by considering $\mathcal{D}_0 = \{u \in C^1[a, b] : u(a) = u(b)\}$. Considerations similar to the previous ones give the eigenvalue problem

$$\begin{cases} -u'' + qu = \lambda w u \text{ in } [a, b] , \\ u(a) = u(b) , \\ u'(a) = u'(b) . \end{cases}$$

The corresponding boundary conditions are called *periodic*. More generally one may use $\mathcal{D}_0 = \{u \in C^1[a, b] : u(a) = Ku(b)\}$ for some fixed constant $K \neq 0$. The boundary conditions obtained are then $u(a) = Ku(b)$, $u'(b) = \overline{K}u'(a)$. In the case $K = -1$ we obtain so-called *semi-periodic boundary conditions*. If K is not real it is, for the first time, essential to use complex-valued functions in $\mathcal{D}_0$. In all cases corresponding versions of Theorems 1.4.6 and 1.4.7 are obtained.

However, all this by no means exhausts all possibilities, since one may also replace the form H by the Hermitian form

$$H(u, v) = \int_a^b (u'\overline{v}' + qu\overline{v})$$

$$+ Au(a)\overline{v(a)} + Bu(b)\overline{v(b)} - Cu(b)\overline{v(a)} - u(a)\overline{Cv(b)}, \quad (1.4.12)$$

where A, B are real constants and C a constant which may be complex. One must prove that this form is bounded from below (Exercise 1.4.4), but may then use it with all the different choices of $\mathcal{D}_0$ mentioned above to obtain corresponding versions of Theorems 1.4.6 and 1.4.7. This gives a wide variety of eigenvalue problems.

We shall briefly look at the different possibilities, and then first note that the new terms in H vanish on $\mathcal{D}_0$ if we choose $\mathcal{D}_0 = C_*^1[a, b]$. Independent of A, B and C this choice of $\mathcal{D}_0$ therefore always gives rise to the eigenvalue problem (1.4.11).

If we choose $\mathcal{D}_0 = \{u \in C^1[a, b] : u(b) = 0\}$ all the boundary terms vanish on $\mathcal{D}_0$ except $Au(a)\overline{v(a)}$. It is easy to see that this choice gives the eigenvalue problem

$$\begin{cases} -u'' + qu = \lambda wu \text{ in } [a, b] , \\ u'(a) - Au(a) = 0 , \\ u(b) = 0 . \end{cases} \quad (1.4.13)$$

One may similarly consider problems with a Dirichlet condition at a, or more generally use $\mathcal{D}_0 = \{u \in C^1[a, b] : u(a) = Ku(b)\}$ where $K \in \mathbb{C}$. One may then replace all the boundary terms in H by $Du(b)\overline{v(b)}$ where $D = A|K|^2 - 2\operatorname{Re}(K\overline{C}) + B$ and one obtains the eigenvalue problem

$$\begin{cases} -u'' + qu = \lambda wu \text{ in } [a, b] , \\ u(a) = Ku(b) , \\ u'(b) + Du(b) = \overline{K}u'(a) . \end{cases} \quad (1.4.14)$$

Finally, if we use $\mathcal{D}_0 = C^1[a, b]$ we obtain the eigenvalue problem

$$\begin{cases} -u'' + qu = \lambda wu \text{ in } [a, b] , \\ u'(a) - Au(a) = -Cu(b) , \\ u'(b) + Bu(b) = \overline{C}u(a) . \end{cases} \quad (1.4.15)$$

We always obtain two different boundary conditions, i.e., two independent linear, homogeneous conditions on $u(a)$, $u'(a)$, $u(b)$ and $u'(b)$. These are of two kinds, namely either *separated*, where each boundary condition only concerns the values of u, u' in one of the two endpoints a or b, or else *coupled* boundary conditions, where both boundary conditions involve values at both endpoints.

A separated boundary condition at a is either $u(a) = 0$ or of the form $u'(a) - Au(a) = 0$ where A is a *real* constant. If we determine an angle $\alpha \in (0, \pi)$ so that

$\cot \alpha = -A$ one may express any separated boundary condition at a as a condition of the form $u(a) \cos \alpha + u'(a) \sin \alpha = 0$ where α is a fixed angle in $[0, \pi)$. In the same way a separated boundary condition at b is of the form $u(b) \cos \beta + u'(b) \sin \beta = 0$ for some $\beta \in [0, \pi)$. We obtain separated boundary conditions in the cases (1.4.11), (1.4.13), (1.4.14) if $K = 0$ and (1.4.15) if $C = 0$.

Coupled boundary conditions are obtained in the cases (1.4.14) if $K \neq 0$ and (1.4.15) if $C \neq 0$. These may both be expressed in the form

$$\begin{pmatrix} u(a) \\ u'(a) \end{pmatrix} = S \begin{pmatrix} u(b) \\ u'(b) \end{pmatrix}$$

where S is a 2×2-matrix. In case of (1.4.14) with $K \neq 0$ we obtain

$$S = \frac{K}{|K|} \begin{pmatrix} |K| & 0 \\ D/|K| & |K|^{-1} \end{pmatrix},$$

and in case of (1.4.15) with $C \neq 0$ we obtain

$$S = \frac{C}{|C|} \begin{pmatrix} B/|C| & 1/|C| \\ AB/|C| - |C| & A/|C| \end{pmatrix}.$$

We may summarize this by saying that S is an *arbitrary real* 2×2 *matrix with determinant 1 multiplied by a complex number of absolute value 1*. A 2×2 matrix of this kind is called *symplectic*. We shall show in Section 4.2 that if one looks for boundary conditions for the equation $-u'' + qu = \lambda wu$ on the interval $[a, b]$ such that there is a complete orthonormal sequence of eigenfunctions with real eigenvalues, then the boundary conditions must be either separated conditions of the form above, or else coupled boundary conditions given by a symplectic matrix. Theorem 1.4.7 remains true with essentially the same proof for all these choices of H and $\mathcal{D}_0$.

If we study a problem with separated boundary conditions, one of which is $u(a) \cos \alpha + u'(a) \sin \alpha = 0$, then all eigenfunctions for a certain eigenvalue λ are multiples of the solution of $-u'' + qu = \lambda wu$ which satisfies the initial conditions $u(a) = -\sin \alpha$, $u'(a) = \cos \alpha$. By Theorem D.2 there is precisely one such solution. It follows that the eigenspace to each eigenvalue is of dimension one. One says that all eigenvalues for separated boundary conditions are *simple*.

The same is true for coupled boundary conditions which are non-real, i.e., the matrix S is not real (Exercise 1.4.6), but there is no such restriction for real coupled boundary conditions, where it is quite possible that for a certain λ *all* solutions of the differential equation satisfy the boundary conditions; such an eigenvalue is called *double*.

We have shown the eigenfunction expansion of $u \in \mathcal{D}_0$ to converge uniformly and in norm to u. Convergence in norm is true even if u is only continuous (or even just in the appropriate L^2-space), which follows from the fact that any such u may be approximated in norm by functions[5] $v \in \mathcal{D}_0$. For if $\|u - v\| < \varepsilon$ we obtain

---

[5] $C_0^1(a, b)$ is dense in the L^2-space by Appendices B and C and contained in $\mathcal{D}_0$.

$$\|u_j\| \le \|u_j - v_j\| + \|v_j\| \le \|u - v\| + \|v_j\| \le \varepsilon + \|v_j\| \,,$$

where we used the triangle inequality and $\|u_j - v_j\| \le \|u - v\|$, as follows from $\langle u_j - v_j, s_j(u - v)\rangle = 0$ and $u - v = u_j - v_j + s_j(u - v)$. Thus if $\|v_j\| \to 0$ we have $\|u_j\| < 2\varepsilon$ for large j.

We have thus shown that the eigenfunctions yield an orthonormal basis in the linear space $\mathcal{D}_0$ provided with the norm $\|\cdot\|$. Anticipating some Hilbert space theory from Section 2.1 one easily obtains an orthonormal basis of eigenfunctions also with respect to the norm given by H_c. For if $u \in \mathcal{D}_0$ and $H_c(u, e_k) = 0$ for all k, then $\hat{u}_k = \langle u, e_k\rangle = 0$ for all k since $H_c(u, e_k) = (\lambda_k + c)\langle u, e_k\rangle$. Thus $u = \sum \hat{u}_k e_k = 0$.

The simplest example of a Sturm–Liouville equation is the equation $-u'' = \lambda u$ on the interval $[-\pi, \pi]$ with periodic boundary conditions. The smallest eigenvalue is 0, with non-zero constants as the only eigenfunctions. The other eigenvalues are $\lambda_k = k^2, k = 1, 2, \dots$. The eigenfunctions to λ_k are all linear combinations of $\sin(kx)$ and $\cos(kx)$, so all eigenvalues except the smallest one are double. This example, of course, gives rise to the ordinary (real) Fourier series expansions. It is also easily seen that e^{ikx} and e^{-ikx} are linearly independent orthogonal eigenfunctions with eigenvalue k^2, and this gives rise to the standard *complex* Fourier series.

Another example is given by the same equation considered on $[0, \pi]$ with Neumann or Dirichlet boundary conditions. In the first case the eigenvalues are $\lambda_k = k^2$, $k = 0, 1, 2 \dots$, with eigenfunctions $\cos(kx)$, $k = 0, 1, 2, \dots$, and in the second case the eigenvalues are $\lambda_k = k^2$ with eigenfunctions $\sin(kx)$, $k = 1, 2, 3 \dots$. Thus we also obtain the basic facts about cosine and sine series.

It should be remarked that with minimal extra effort the methods presented here can cope with the more general Sturm–Liouville equation $-(pu')' + qu = \lambda wu$ on the interval (a, b), where $1/p, q$ and w are given integrable and real-valued functions on (a, b) and $p > 0$. If $w \ge 0$ but $w \not\equiv 0$ as well there are only minor technical differences from what we have already dealt with, but one can also deal with the so-called left-definite equation when $q \ge 0, q \not\equiv 0$ but w may change sign arbitrarily. In this case one must use the form H as the scalar product, but very little else changes.

However, we shall leave this now and eventually turn to methods of greater generality based on the general spectral theorem in Hilbert space. These methods will handle all cases mentioned and much more.

Exercises

Exercise 1.4.1 Prove that all alternative choices for the space $\mathcal{D}_0$ give an infinite-dimensional space.

Hint: Show that for all choices of $\mathcal{D}_0$ there is a non-zero element of $\mathcal{D}_0$ which is zero outside an arbitrary subinterval of (a, b). Use this to show that we can find arbitrarily many linearly independent elements of $\mathcal{D}_0$.

Exercise 1.4.2 Consider the heat equation problem (1.1.2) for $u(x, 0) \in C^1(0, \ell)$. Show that the solution (1.1.3) is infinitely differentiable with respect to (x, t) for $t > 0$ and satisfies the heat equation there if $u(x, 0) \in \mathcal{D}_0$. Also show that $u(x, t) \to u(x, 0)$ uniformly as $t \downarrow 0$.

Hint: The series converges uniformly for $t \geq 0$ and the term by term differentiated series converges uniformly for $t > 0$.

Exercise 1.4.3 If f is continuous, non-negative and $\int_a^b f = 0$, then $f = 0$ on $[a, b]$. Prove this.

Exercise 1.4.4 Prove that the Hermitian form H of (1.4.12) is bounded from below. Then extend this to the case when $p > 0$ a.e., $1/p$, q and w are integrable in $[a, b]$ and $w \geq 0$ and not a.e. equal to zero while $H(u, v) = \int_a^b (pu'\overline{v'} + qu\overline{v})$.

Hint: The proofs of (1.4.4) and (1.4.5).

Exercise 1.4.5 Verify that with the choices of $\mathcal{D}_0$ and H on page 25 the eigenvalue problems (1.4.11) and (1.4.13)−(1.4.15) result.

Exercise 1.4.6 Prove that eigenvalues for coupled boundary conditions with a non-real symplectic matrix S are simple.

Hint: If λ_j is a double eigenvalue, then there are real-valued eigenfunctions.

1.5 Notes and remarks

The first problem where separation of variables was used in some form is probably that of the vibrating string, modelled by the equation $u_{tt} = c^2 u_{xx}$ on an interval $[0, \ell]$, with boundary conditions at 0 and ℓ. This was considered by d'Alembert who also gave a solution formula. Daniel Bernoulli gave another solution amounting to the solution obtained by separating variables. This caused a long-lasting controversy involving also Euler, who gave reasons for not believing that Bernoulli's solution could be generally applicable.

When Fourier submitted his first paper on the heat equation in 1807 it was criticized for similar reasons, but his methods were so successful that they were eventually accepted, and finally vindicated by the results of Dirichlet [143]. More details on the contributions of Sturm and Liouville may be found in Lützen [155].

A detailed exposition of the theory of linear spaces with and without scalar products may be found in Lax [141] or Halmos [98].

The elementary real analysis used in Section 1.4 may be found in any basic text on one-variable analysis, for example Rudin [195]. Sturm and Liouville seem to have considered only separated boundary conditions, and they never quite succeeded in proving that their series expansions converge to the given function.

The first rigorous proofs of such theorems, with varying assumptions, seem to be by Stekloff [205] in 1898 and Kneser [126] in 1904. They used methods inspired by

the work of Sturm and Liouville and the convergence proof for Fourier series given
by Dirichlet in 1829. This may seem a little surprising, since the characterization of
eigenvalues as minima of quadratic forms we used was well known at the time, as
were du Bois-Reymond's lemma and the Arzela–Ascoli theorem.

Chapter 2
Hilbert space

2.1 Complete spaces

A Hilbert space is a linear space $\mathcal{H}$ (we shall as always assume that the scalars are complex numbers) provided with a scalar product such that the space is also *complete* according to Definition 1.3.13. We denote the scalar product of u and $v \in \mathcal{H}$ by $\langle u, v \rangle$ and the norm of u by $\|u\| = \sqrt{\langle u, u \rangle}$. It is usually required, and we shall mostly follow this convention, that $\mathcal{H}$ is infinite-dimensional but separable. However, most results and proofs in this chapter remain correct for non-separable or finite-dimensional spaces. Recall that according to Proposition 1.3.11 $\mathcal{H}$ is separable precisely if it contains a sequence which is an orthonormal basis.

Example 2.1.1. The space ℓ^2 consists of all infinite sequences $u = \{u_j\}_1^\infty$ of complex numbers for which $\sum |u_j|^2 < \infty$, i.e., which are *square summable*. Multiplication by a scalar and addition of two sequences are defined in an obvious way, and since $2|u_j \overline{v_j}| \leq |u_j|^2 + |v_j|^2$ it is easily seen that ℓ^2 is a linear space.

The scalar product of u with $v = \{v_j\}_1^\infty$ is defined as $\langle u, v \rangle_{\ell^2} = \sum u_j \overline{v}_j$. By the inequality above this series is absolutely convergent. The space ℓ^2 is a Hilbert space, i.e., it is complete, separable and infinite-dimensional; see Exercise 2.1.1.

The space Hilbert worked with was ℓ^2, while the abstract concept of a Hilbert space was introduced by von Neumann in 1930 [172]. Actually, any separable Hilbert space of infinite dimension is unitarily equivalent to ℓ^2, i.e., there is a *bijective* (one-to-one and onto) linear map $\mathcal{H} \ni u \mapsto \hat{u} \in \ell^2$ such that $\langle u, v \rangle = \langle \hat{u}, \hat{v} \rangle_{\ell^2}$ for any u and v in $\mathcal{H}$; see Exercise 2.1.2.

This is the reason why any complete, separable and infinite-dimensional space with scalar product is called a Hilbert space. However, there are infinitely many isomorphisms that will serve, and none of them is 'natural', i.e., in general there is no reason to prefer one over any other, so the fact that all separable and infinite-dimensional Hilbert spaces are unitarily equivalent is not particularly useful in practice. For this reason the term Hilbert space is now often also used for complete

© Springer Nature Switzerland AG 2020
C. Bennewitz et al., *Spectral and Scattering Theory for Ordinary Differential Equations*,
Universitext, https://doi.org/10.1007/978-3-030-59088-8_2

spaces with a scalar product which are *not* separable and, more seldom, for finite-dimensional spaces.

Example 2.1.2. The most important example of a Hilbert space is $L^2(\Omega, \mu)$ where Ω is some domain in $\mathbb{R}^n$ and μ is a positive (Radon) measure defined there; often μ is simply Lebesgue measure, or Lebesgue measure multiplied by a positive, locally integrable weight function w. The space consists of equivalence classes of (usually) complex-valued functions on Ω, measurable with respect to μ and with integrable square over Ω with respect to μ. Such a space is separable and complete. See Appendix B for more information on such spaces.

Given a normed space one may of course ask whether there is a scalar product on the space which gives rise to the given norm in the usual way. Here is a simple criterion.

Lemma 2.1.3 (Parallelogram identity). *If u and v are elements of $\mathcal{H}$, then*

$$\|u + v\|^2 + \|u - v\|^2 = 2\|u\|^2 + 2\|v\|^2 .$$

Proof. A simple calculation gives

$$\|u \pm v\|^2 = \langle u \pm v, u \pm v \rangle = \|u\|^2 \pm (\langle u, v \rangle + \langle v, u \rangle) + \|v\|^2 .$$

Adding these yields the parallelogram identity.   □

The name parallelogram identity comes from the fact that the lemma has a geometric interpretation, namely that the sum of the squares of the lengths of the sides in a parallelogram equals the sum of the squares of the lengths of the diagonals. This is a theorem that may be found in Euclid's Elements.

Given a normed space, Lemma 2.1.3 shows that a necessary condition for the norm to be associated with a scalar product is that the parallelogram identity holds for all vectors u, v in the space. It was proved by Jordan and von Neumann [115] that this is also sufficient; see Exercise 2.1.4. We shall soon have another use for Lemma 2.1.3.

In practice it is quite common that one has a space with scalar product which is *not* complete, like the space $\mathcal{D}_0$ of Section 1.4. Such a space is often called a pre-Hilbert space. In order to use Hilbert space theory, one must then embed the space in a larger space which is complete. The process is called *completion* and is fully analogous to the extension of the rational numbers to the reals, which is also done to make the Cauchy convergence principle valid.

In very brief outline the process is as follows. Starting with a (not complete) normed linear space $\mathcal{V}$ let $\mathcal{V}_c$ be the set of all Cauchy sequences in $\mathcal{V}$. The set $\mathcal{V}_c$ is made into a linear space in the obvious way. We may embed $\mathcal{V}$ in $\mathcal{V}_c$ by identifying $u \in \mathcal{V}$ with the sequence $(u, u, u, \dots)$. In $\mathcal{V}_c$ we may introduce a semi-norm $\|\cdot\|$ by setting $\|(u_1, u_2, \dots)\| = \lim\|u_j\|$. Since $\big|\|u_j\| - \|u_k\|\big| \le \|u_j - u_k\|$ the numerical sequence $\{\|u_j\|\}_1^\infty$ is a Cauchy sequence so the limit exists.

Now let $\mathcal{N}$ be the subspace of $\mathcal{V}_c$ consisting of all elements with semi-norm 0, and put $\mathcal{B} = \mathcal{V}_c/\mathcal{N}$, i.e., elements in $\mathcal{V}_c$ are identified whenever the distance between them is 0. One may now prove that $\|\cdot\|$ induces a norm on $\mathcal{B}$ under which $\mathcal{B}$ is complete, and that through the identification above we may view the original space $\mathcal{V}$ as a dense subset of $\mathcal{B}$. If the original norm was derived from a scalar product, then so will the norm of $\mathcal{B}$. We leave it to the reader to verify the details, using the hints provided; see Exercise 2.1.6.

The process above is satisfactory in that it shows that *any* normed space may be 'completed' (in fact, the same process works in any metric space). Equivalence classes of Cauchy sequences are of course rather abstract objects, but in concrete cases one can often identify the elements of the completion of a given space with less abstract objects. So, for example, one may view $L^2(\Omega, \mu)$ as the completion, in the appropriate norm, of the linear space $C_0(\Omega)$ of functions which are continuous in Ω and 0 outside a compact subset of Ω; see Appendix B.

In the sequel $\mathcal{H}$ is always assumed to be a Hilbert space. There are two properties which make Hilbert spaces far more convenient to deal with than more general spaces. The first is that any closed, linear subspace has a topological complement which can be chosen in a canonical way (Theorem 2.1.7). The second, a Hilbert space can be identified with its topological dual (Theorem 2.1.9).

Our proofs are essentially those of F. Riesz [189] and show that these properties are true even if the space is not assumed separable (and of course if the space is finite-dimensional). However, it is essential that the space is complete. To prove these properties we start with the following definition.

Definition 2.1.4. A subset M of a linear space is called *convex* if it contains all line-segments connecting two elements of the set, i.e., if u and $v \in M$, then $tu + (1-t)v$ is in M for all $t \in [0, 1]$.

Recall that a subset of a metric space is closed if and only if all limits of convergent sequences contained in the subset are themselves in the subset (Exercise 1.3.1). This is equivalent to the complement of the subset being open, in the sense that it is a neighborhood of all its points.

Lemma 2.1.5. *Any closed, convex and non-empty subset K of a Hilbert space $\mathcal{H}$ has a unique element of smallest norm.*

Proof. Put $d = \inf\{\|u\| : u \in K\}$ and let $u_1, u_2, \dots$ be a minimizing sequence, i.e., $u_j \in K$ and $\|u_j\| \to d$. By the parallelogram identity we then have

$$\|u_j - u_k\|^2 = 2\|u_j\|^2 + 2\|u_k\|^2 - 4\|\tfrac{1}{2}(u_j + u_k)\|^2 \, .$$

On the right-hand side the two first terms both tend to $2d^2$ as $j, k \to \infty$. By convexity $\frac{1}{2}(u_j + u_k) \in K$ so the last term is $\geq 4d^2$. Therefore $u_1, u_2, \dots$ is a Cauchy sequence, and has a limit u which obviously has norm d and is in K, since K is closed. If u and v are both minimizing elements, replacing u_j by u and u_k by v in the calculation above immediately shows that $u = v$, so the minimizing element is unique. $\square$

Lemma 2.1.6. *Suppose M is a proper (i.e., $M \neq \mathcal{H}$) closed, linear subspace of a Hilbert space $\mathcal{H}$. Then there is a non-trivial normal to M, i.e., an element $u \neq 0$ in $\mathcal{H}$ such that $\langle u, v \rangle = 0$ for all $v \in M$.*

In general the normal direction is far from unique.

Proof. Let $w \notin M$ and put $K = w + M$. Then K is closed and convex (check!) so it has a smallest element u which is non-zero since $0 \notin K$. Let $v \neq 0$ be in M so that for any scalar a we have $u + av \in K$. Hence $\|u\|^2 \leq \|u + av\|^2 = \|u\|^2 + 2\operatorname{Re}(a\langle v, u \rangle) + |a|^2 \|v\|^2$. Setting $a = -\langle u, v \rangle / \|v\|^2$ we obtain $-(|\langle u, v \rangle| / \|v\|)^2 \geq 0$ so that $\langle u, v \rangle = 0$. $\square$

Two subspaces M and N are said to be *orthogonal* if every element of M is orthogonal to every element of N. Then clearly $M \cap N = \{0\}$ so the direct sum of M and N is defined. In the case at hand this is called the *orthogonal sum* of M and N and denoted by $M \oplus N$. Thus $M \oplus N$ is the set of all sums $u + v$ with $u \in M$, $v \in N$ and $\|u + v\|^2 = \|u\|^2 + \|v\|^2$. If M and N are closed, orthogonal subspaces of $\mathcal{H}$, then their orthogonal sum is also a closed subspace of $\mathcal{H}$; see Exercise 2.1.7.

If A is an arbitrary subset of $\mathcal{H}$ we define

$$A^{\perp} = \{u \in \mathcal{H} : \langle u, v \rangle = 0 \text{ for all } v \in A\} .$$

This is called the *orthogonal complement* of A. It is easy to see that $A^{\perp}$ is a closed linear subspace of $\mathcal{H}$, that $A \subset B$ implies $B^{\perp} \subset A^{\perp}$ and that $A \subset (A^{\perp})^{\perp}$; see Exercise 2.1.8.

When M is a linear subspace of $\mathcal{H}$ an alternative way of writing $M^{\perp}$ is $\mathcal{H} \ominus M$. This makes sense because of the following theorem of central importance.

Theorem 2.1.7. *Suppose M is a closed linear subspace of $\mathcal{H}$. Then $M \oplus M^{\perp} = \mathcal{H}$.*

Proof. $M \oplus M^{\perp}$ is a closed linear subspace of $\mathcal{H}$ so if it is not all of $\mathcal{H}$, then it has a non-trivial normal u by Lemma 2.1.6. But if u is orthogonal to both M and $M^{\perp}$, then $u \in M^{\perp} \cap (M^{\perp})^{\perp}$ which shows that u cannot be different from 0. The theorem follows. $\square$

A nearly obvious consequence of Theorem 2.1.7 is that $M^{\perp\perp} = M$ for any closed linear subspace M of $\mathcal{H}$; see Exercise 2.1.9.

A *linear form* ℓ on $\mathcal{H}$ is a complex-valued linear function on $\mathcal{H}$. Naturally ℓ is said to be continuous if $\ell(u_j) \to \ell(u)$ whenever $u_j \to u$. The set of continuous linear forms on a Banach space $\mathcal{B}$ (or a more general topological vector space) is made into a linear space in an obvious way. This space is called the (topological)[1] *dual* of $\mathcal{B}$, and is denoted by $\mathcal{B}'$.

Proposition 2.1.8. *A linear form ℓ on a normed space $\mathcal{B}$ is continuous if and only if it is* bounded *in the sense that there exists a constant C such that $|\ell(u)| \leq C\|u\|$ for all $u \in \mathcal{B}$.*

Proof. If ℓ is continuous we can find $\delta > 0$ such that $|\ell(u)| < 1$ if $\|u\| < \delta$, so that $|\ell(\delta u / \|u\|)| < 1$ for any $u \neq 0$. Thus $|\ell(u)| < \frac{1}{\delta}\|u\|$ for all u, so ℓ is bounded

Conversely, if ℓ is bounded by C then $|\ell(u_j) - \ell(u)| = |\ell(u_j - u)| \leq C\|u_j - u\| \to 0$ if $u_j \to u$, so a bounded linear form is continuous. $\square$

---

[1] The space of all linear forms on $\mathcal{B}$, continuous or not, is the *algebraic dual* of $\mathcal{B}$, but this is not often a useful notion.

The infimum of all bounds of a bounded linear form ℓ is also a bound, and this smallest possible bound is called the *norm* of ℓ, denoted $\|\ell\|$. It is also easy to see that provided with this norm $\mathcal{B}'$ is complete, so the dual of a normed space is a Banach space (Exercise 2.1.10).

A simple example of a bounded linear form on a Hilbert space $\mathcal{H}$ is $\ell(u) = \langle u, v \rangle$, where v is some fixed element of $\mathcal{H}$. By the Cauchy–Schwarz inequality $|\ell(u)| \le \|v\|\|u\|$ so $\|\ell\| \le \|v\|$. But $\ell(v) = \|v\|^2$ so in fact $\|\ell\| = \|v\|$. The following theorem, which has far-reaching consequences for many applications of analysis, states that this is the only kind of bounded linear form there is on a Hilbert space. In other words, the theorem allows us to identify the dual of a Hilbert space with the space itself.

Theorem 2.1.9 (Riesz' representation theorem). *For any bounded linear form ℓ on $\mathcal{H}$ there is a unique element $v \in \mathcal{H}$ such that $\ell(u) = \langle u, v \rangle$ for all $u \in \mathcal{H}$. The norm of ℓ is then $\|\ell\| = \|v\|$.*

There are other theorems known by the name Riesz' representation theorem, all more or less due to F. Riesz and in effect identifying the duals of various spaces. One example is Theorem B.4.15

Proof. The uniqueness of v is clear, since the difference of two possible choices of v must be orthogonal to all of $\mathcal{H}$ (for example to itself). If $\ell(u) = 0$ for all u then we may take $v = 0$. Otherwise we set $M = \{u \in \mathcal{H} : \ell(u) = 0\}$, which is obviously linear because ℓ is, and closed since ℓ is continuous.

Since M is not all of $\mathcal{H}$ it has a normal $w \ne 0$ by Lemma 2.1.6, so $\ell(w) \ne 0$ and we may assume $\|w\| = 1$. For arbitrary $u \in \mathcal{H}$ we put $u_1 = u - (\ell(u)/\ell(w))w$ so that $\ell(u_1) = \ell(u) - \ell(u) = 0$. Thus $\underline{u_1} \in M$ so that $\langle u_1, w \rangle = 0$. Hence $\langle u, w \rangle = \ell(u)/\ell(w)$ or $\ell(u) = \langle u, v \rangle$, where $v = \overline{\ell(w)}w$. We have already proved that $\|\ell\| = \|v\|$. $\square$

So far we have understood convergence in a Hilbert space to mean convergence in norm, i.e., $u_j \to u$ means $\|u_j - u\| \to 0$. This is called *strong convergence*; one writes s-lim $u_j = u$ or $u_j \to u$. There is also another notion of convergence which is very important.

Definition 2.1.10. We say that u_j tends to u *weakly*, in symbols w-lim $u_j = u$ or $u_j \rightharpoonup u$, if $\langle u_j, v \rangle \to \langle u, v \rangle$ for every $v \in \mathcal{H}$.

It is obvious that strong convergence implies weak convergence to the same limit since the scalar product is continuous in its arguments by the Cauchy–Schwarz inequality; see Exercise 1.3.6. However, a weakly convergent sequence does not necessarily converge strongly; see Exercise 2.1.11. We have the following important theorem.

Theorem 2.1.11. *Every bounded sequence in a Hilbert space $\mathcal{H}$ has a weakly convergent subsequence. Conversely, every weakly convergent sequence is bounded.*

In a finite-dimensional space it is easily seen that weak and strong convergence is the same. In a finite-dimensional space one also has the *Bolzano–Weierstrass theorem* (Corollary 1.3.2), which says that a bounded sequence has a convergent subsequence. In an infinite-dimensional space this is not true for strong convergence (Exercise 2.1.11). But Theorem 2.1.11 states that for *weak* convergence the Bolzano–Weierstrass theorem remains true in any Hilbert space.

Proof (Theorem 2.1.11). The first claim is an immediate consequence of the weak* compactness of the unit ball of the dual of a Banach space. Since we prefer not to assume familiarity with this, we shall give a direct proof, valid in a separable space. The proof is easily extended to a non-separable space; see Exercise 2.1.12.

Suppose that $v_1, v_2, \ldots$ is the given sequence, bounded by C, and let $e_1, e_2, \ldots$ be an orthonormal basis in $\mathcal{H}$. The numerical sequence $\{\langle v_j, e_1 \rangle\}_{j=1}^{\infty}$ is bounded and so by the Bolzano–Weierstrass theorem it has a convergent subsequence, corresponding to a subsequence $\{v_{1j}\}_{j=1}^{\infty}$ of the v's. The numerical sequence $\{\langle v_{1j}, e_2 \rangle\}_{j=1}^{\infty}$ is again bounded, so it has a convergent subsequence, corresponding to a subsequence $\{v_{2j}\}_{j=1}^{\infty}$ of $\{v_{1j}\}_{j=1}^{\infty}$.

Proceeding in this manner we get a sequence of sequences $\{v_{kj}\}_{j=1}^{\infty}, k = 1, 2, \ldots,$ each element of which is a subsequence of those preceding it, and with the property that $\hat{v}_n = \lim_{j \to \infty} \langle v_{nj}, e_n \rangle$ exists. We claim that $\{v_{jj}\}_{j=1}^{\infty}$ converges weakly to $v = \sum \hat{v}_n e_n$.

To see this, first note that $\{\langle v_{jj}, e_n \rangle\}_{j=1}^{\infty}$, a subsequence of $\{\langle v_{nj}, e_n \rangle\}_{j=1}^{\infty}$ from $j = n$ on, converges to $\hat{v}_n$. Furthermore, $\sum_{n=1}^{N} |\hat{v}_n|^2 \le C^2$ for all N since it is the limit of $\sum_{n=1}^{N} |\langle v_{nj}, e_n \rangle|^2$ as $j \to \infty$. By Bessel's inequality (Theorem 1.3.6) this is bounded by $\|v_{nj}\|^2 \le C^2$. It follows that $\sum_{n=1}^{\infty} |\hat{v}_n|^2 \le C^2$ so that v is an element of $\mathcal{H}$.

To show the weak convergence, let $u = \sum \hat{u}_n e_n$ be in $\mathcal{H}$. Suppose $\varepsilon > 0$. Writing $u = u' + u''$ where $u' = \sum_{n=1}^{N} \hat{u}_n e_n$ we may now choose N so large that $\|u''\| < \varepsilon$ so that $|\langle v_{jj}, u'' \rangle| < C\varepsilon$. Furthermore $|\langle v, u'' \rangle| < C\varepsilon$ and

$$\langle v_{jj}, u' \rangle = \sum_{n=1}^{N} \langle v_{jj}, e_n \rangle \overline{\hat{u}_n} \to \sum_{n=1}^{N} \hat{v}_n \overline{\hat{u}_n} = \langle v, u' \rangle$$

so $\overline{\lim}_{j \to \infty} |\langle v_{jj}, u \rangle - \langle v, u \rangle| \le 2C\varepsilon$. Since $\varepsilon > 0$ is arbitrary the weak convergence follows.

The converse is an immediate consequence of the Banach–Steinhaus *principle of uniform boundedness*. This is Theorem A.5. Our application concerns a sequence of bounded linear forms $\ell_1, \ell_2, \ldots$ on a Banach space $\mathcal{B}$ which are pointwise bounded, i.e., such that for each $u \in \mathcal{B}$ the sequence $\ell_1(u), \ell_2(u), \ldots$ is bounded. Then the theorem states that $\ell_1, \ell_2, \ldots$ is *uniformly bounded*, i.e., there is a constant C such that $|\ell_j(u)| \le C\|u\|$ for every $u \in \mathcal{B}$ and $j = 1, 2, \ldots$.

Using Theorem A.5 we can complete the proof of Theorem 2.1.11, since a weakly convergent sequence $v_1, v_2, \ldots$ can be identified with a sequence of linear forms $\ell_1, \ell_2, \ldots$ by setting $\ell_j(u) = \langle u, v_j \rangle$. Since a convergent sequence of numbers is bounded it follows that we have a pointwise bounded sequence of linear functionals.

By Theorem A.5 there is a constant C such that $|\langle u, v_j \rangle| \leq C\|u\|$ for every $u \in \mathcal{H}$ and $j = 1, 2, \ldots$. In particular, setting $u = v_j$ gives $\|v_j\| \leq C$ for every j. □

Exercises

Exercise 2.1.1 Prove the completeness of ℓ^2.

Hint: Given a Cauchy sequence show first that each coordinate converges.

Exercise 2.1.2 Prove that any separable Hilbert space is unitarily equivalent to ℓ^2, i.e., there is a bijective (one-to-one and onto) linear map $\mathcal{H} \ni u \mapsto \hat{u} \in \ell^2$ such that $\langle u, v \rangle = \langle \hat{u}, \hat{v} \rangle$ for any u and v in $\mathcal{H}$.

Exercise 2.1.3 Consider the linear space $C[0, 1]$ of functions continuous on $[0, 1]$, and define on this space $\|u\|_\infty = \sup_{x \in [0,1]} |u(x)|$. Show that this makes $C[0, 1]$ into a Banach space, but that there is no scalar product which generates this norm.

Exercise 2.1.4 Suppose $\mathcal{V}$ is a linear space with norm $\|\cdot\|$ which satisfies the parallelogram identity for all $u, v \in \mathcal{V}$. Show that $\langle u, v \rangle = \frac{1}{4} \sum_{k=0}^{3} i^k \|u + i^k v\|^2$ is a scalar product on $\mathcal{V}$.

Hint: Show first that $\langle u, u \rangle = \|u\|^2$, that $\langle v, u \rangle = \overline{\langle u, v \rangle}$ and that $\langle iu, v \rangle = i\langle u, v \rangle$. Then show that $\langle u + v, w \rangle - \langle u, w \rangle - \langle v, w \rangle = 0$ and from that $\langle \lambda u, v \rangle = \lambda \langle u, v \rangle$ for any rational number λ. Finally use continuity.

Exercise 2.1.5 Suppose W is a subspace of a normed space $\mathcal{V}$ and define a semi-norm on $\mathcal{V}/W$ by $\|u + W\| = \inf_{v \in W} \|u + v\|$.

Show that this is a norm if and only if W is closed, and that if this is the case, then $\mathcal{V}/W$ is a Banach space if and only if $\mathcal{V}$ is.

Exercise 2.1.6 Consider the semi-norm on the space $\mathcal{V}_c$ defined on page 32. Verify that $\mathcal{B} = \mathcal{V}_c/\mathcal{N}$ can be given a norm under which it is complete, that $\mathcal{V}$ may be viewed as isometrically and densely embedded in $\mathcal{B}$, and that $\mathcal{B}$ is a Euclidean space (a space with the norm derived from a scalar product) if $\mathcal{V}$ is.

Also carry out the same procedure to complete a metric space.

Exercise 2.1.7 Show that if M and N are closed, orthogonal subspaces of $\mathcal{H}$, then $M \oplus N$ is also closed.

Exercise 2.1.8 Show that if $A \subset \mathcal{H}$, then $A^\perp$ is a closed linear subspace of $\mathcal{H}$, that $A \subset (A^\perp)^\perp$ and that $A \subset B$ implies $B^\perp \subset A^\perp$.

Exercise 2.1.9 Verify that $M^{\perp\perp} = M$ if M is a closed linear subspace of $\mathcal{H}$. Use this to show that $M^{\perp\perp}$ is the closure of M for any linear subspace M of $\mathcal{H}$, and that for an arbitrary set $A \subset \mathcal{H}$ the smallest closed linear subspace containing A is $A^{\perp\perp}$.

Exercise 2.1.10 Show that a bounded linear form on a Banach space $\mathcal{B}$ has a least bound, which is a norm on $\mathcal{B}'$, and that $\mathcal{B}'$ is complete under this norm.

Exercise 2.1.11 Show that an orthonormal sequence does not have any strongly convergent subsequences but tends weakly to 0. Conclude that if in a Euclidean space every weakly convergent sequence is strongly convergent, then the space is finite-dimensional.

Hint: Show that the distance between two elements in the sequence is $\sqrt{2}$ and use Bessel's inequality to show weak convergence to 0.

Exercise 2.1.12 Prove Theorem 2.1.11 for a non-separable space $\mathcal{H}$.

Hint: The closed linear span of the given sequence is a separable Hilbert space.

2.2 Operators and relations

A *bounded linear operator* from a Banach space $\mathcal{B}_1$ to another Banach space $\mathcal{B}_2$ is a linear mapping $T : \mathcal{B}_1 \to \mathcal{B}_2$ such that for some constant C we have $\|Tu\|_2 \le C \|u\|_1$ for every $u \in \mathcal{B}_1$. The smallest such constant C (cf. Exercise 2.1.10) is the *norm* of the operator T and denoted by $\|T\|$.

Boundedness of T is equivalent to continuity, in the sense that $\|Tu_j - Tu\|_2 \to 0$ if $\|u_j - u\|_1 \to 0$ (Exercise 2.2.1), which follows as in the proof of Proposition 2.1.8. If $\mathcal{B}_1 = \mathcal{B}_2 = \mathcal{B}$ one says that T is an operator *on* $\mathcal{B}$. The operator-norm has the following properties (here $T : \mathcal{B}_1 \to \mathcal{B}_2$ and S are bounded linear operators, and $\mathcal{B}_j$, $j = 1, 2, 3$, are Banach spaces).

(1) $\|T\| \ge 0$, equality only if $T = 0$,

(2) $\|\lambda T\| = |\lambda| \|T\|$ for any $\lambda \in \mathbb{C}$,

(3) $\|S + T\| \le \|S\| + \|T\|$ if $S : \mathcal{B}_1 \to \mathcal{B}_2$,

(4) $\|ST\| \le \|S\|\|T\|$ if $S : \mathcal{B}_2 \to \mathcal{B}_3$.

We leave the proof to the reader (Exercise 2.2.1). Thus we have made the set of bounded operators from $\mathcal{B}_1$ to $\mathcal{B}_2$ into a normed space $\mathcal{L}(\mathcal{B}_1, \mathcal{B}_2)$. In fact, $\mathcal{L}(\mathcal{B}_1, \mathcal{B}_2)$ is a Banach space (Exercise 2.2.2). We write $\mathcal{L}(\mathcal{B})$ for the bounded operators on $\mathcal{B}$. Because of the property (4) for $\mathcal{B}_1 = \mathcal{B}_2 = \mathcal{B}_3$ the space $\mathcal{L}(\mathcal{B})$ is called a *Banach algebra*.

Now let $\mathcal{H}_1$ and $\mathcal{H}_2$ be Hilbert spaces with scalar products $\langle \cdot, \cdot \rangle_1$ and $\langle \cdot, \cdot \rangle_2$ respectively. Then every bounded operator $T : \mathcal{H}_1 \to \mathcal{H}_2$ has an *adjoint*[2] $T^* : \mathcal{H}_2 \to \mathcal{H}_1$ defined as follows. Consider a fixed element $v \in \mathcal{H}_2$ and the linear form $\mathcal{H}_1 \ni u \mapsto \langle Tu, v \rangle_2$, which is obviously bounded by $\|T\|\|v\|_2$. By Riesz' representation theorem there is therefore a unique element $v^* \in \mathcal{H}_1$ such that $\langle Tu, v \rangle_2 = \langle u, v^* \rangle_1$ for all u. By the uniqueness, and since $\langle Tu, v \rangle_2$ depends anti-linearly on v, it follows that $T^* : v \mapsto v^*$ is a linear operator from $\mathcal{H}_2$ to $\mathcal{H}_1$.

---

[2] Also operators between general Banach spaces, or even more general topological vector spaces, have adjoints, but they will not concern us here.

Thus $\langle Tu, v\rangle_2 = \langle u, T^*v\rangle_1$ for all $u \in \mathcal{H}_1$ and $v \in \mathcal{H}_2$. T^* is also bounded, since $\|v^*\|_1^2 = \langle Tv^*, v\rangle_2 \leq \|T\| \|v^*\|_1 \|v\|_2$, so that $\|T^*\| \leq \|T\|$.

Proposition 2.2.1. *The adjoint operation* $\mathcal{L}(\mathcal{H}_1, \mathcal{H}_2) \ni T \mapsto T^* \in \mathcal{L}(\mathcal{H}_2, \mathcal{H}_1)$ *has the properties:*

(1) $(T_1 + T_2)^* = T_1^* + T_2^*$ *if* $T_j \in \mathcal{L}(\mathcal{H}_1, \mathcal{H}_2)$, $j = 1, 2$,

(2) $(\lambda T)^* = \bar{\lambda} T^*$ *for any complex number* λ,

(3) $(T_2 T_1)^* = T_1^* T_2^*$ *if* $T_1 \in \mathcal{L}(\mathcal{H}_1, \mathcal{H}_2)$, $T_2 \in \mathcal{L}(\mathcal{H}_2, \mathcal{H}_3)$,

(4) $T^{**} = T$,

(5) $\|T^*\| = \|T\|$,

(6) $\|T^*T\| = \|T\|^2$.

Note that (3), (4) imply that T^*T and TT^* are both *self-adjoint*, i.e., equal to their adjoint. They are also *positive*, since $\langle T^*Tu, u\rangle = \langle Tu, Tu\rangle \geq 0$ with a similar calculation for TT^*.

Proof. The first four properties are very easy to show and are left as exercises for the reader. To prove (5), note that we have already shown that $\|T^*\| \leq \|T\|$ and combining this with (4) gives the opposite inequality. Use of (5) shows that $\|T^*T\| \leq \|T^*\| \|T\| = \|T\|^2$ and the opposite inequality follows from $\|Tu\|_2^2 = \langle T^*Tu, u\rangle_1 \leq \|T^*Tu\|_1 \|u\|_1 \leq \|T^*T\| \|u\|_1^2$ so (6) follows. The reader is asked to fill in the details missing in the proof; see Exercise 2.2.3. $\square$

Proposition 2.2.2. *A bounded operator is* weakly continuous, *i.e., if T is bounded and $u_j \rightharpoonup u$ weakly, then $Tu_j \rightharpoonup Tu$ weakly.*

Proof. We have $\langle Tu_j, v\rangle = \langle u_j, T^*v\rangle \to \langle u, T^*v\rangle = \langle Tu, v\rangle$. $\square$

If $\mathcal{H}_1 = \mathcal{H}_2 = \mathcal{H}_3$, then the properties (1)–(4) above are the properties required for the star operation to be called an *involution* on the algebra $\mathcal{L}(\mathcal{H}_1)$, and a Banach algebra with an involution, also satisfying (5) and (6), is called a B^* (or C^*) *algebra*.

There are no less than three different useful notions of convergence for operators in $\mathcal{L}(\mathcal{H}_1, \mathcal{H}_2)$. We say that T_j tends to T

- *uniformly* if $\|T_j - T\| \to 0$, denoted by $T_j \rightrightarrows T$,
- *strongly* if $\|T_j u - Tu\|_2 \to 0$ for every $u \in \mathcal{H}_1$, denoted $T_j \to T$,
- *weakly* if $\langle T_j u, v\rangle_2 \to \langle Tu, v\rangle_2$ for all $u \in \mathcal{H}_1$ and $v \in \mathcal{H}_2$, denoted $T_j \rightharpoonup T$.

It is clear that uniform convergence implies strong convergence to the same limit and strong convergence implies weak convergence to the same limit, but neither of these implications can be reversed if the spaces are infinite-dimensional (Exercise 2.2.4).

Of particular interest are so-called *projection operators*. A *projection* on $\mathcal{H}_1$ is a linear operator P defined on $\mathcal{H}_1$ (but not necessarily bounded) for which $P^2 = P$. If P is a projection then so is $I - P$, where I is the identity on $\mathcal{H}_1$, since $(I-P)(I-P) = I - P - P + P^2 = I - P$. Setting $M = P\mathcal{H}_1$ and $N = (I-P)\mathcal{H}_1$ it follows that M is the null-space of $I - P$ since M clearly consists of those elements $u \in \mathcal{H}_1$ for which $Pu = u$. Similarly N is the null-space of P.

It is clear that $M \cap N = \{0\}$ and the direct sum $M \dotplus N$ of M and N is $\mathcal{H}_1$ (recall that this means that any element of $\mathcal{H}_1$ may be written uniquely as $u + v$ with $u \in M$ and $v \in N$). Conversely, if M and N are linear subspaces of $\mathcal{H}_1$, $M \cap N = \{0\}$ and $M \dotplus N = \mathcal{H}_1$, then we may define a projection P by setting $Pw = u$ if $w = u + v$ with $u \in M$ and $v \in N$. This is called the *projection onto M along N*.

If P is bounded, i.e. continuous, so is $I - P$ and the direct sum $M \dotplus N$ is called *topological*. In this case it follows that M and N are closed. The converse is also true, so we have the following theorem.

Theorem 2.2.3. *Suppose $\mathcal{H}_1 = M \dotplus N$, where M and N are subspaces of $\mathcal{H}_1$. Then the projection P onto M along N is bounded if and only if M and N are closed.*

The proof is left to Exercise 2.2.10. If M and N happen to be orthogonal subspaces then P is called an *orthogonal projection*. Since the direct sum of M and N is all of $\mathcal{H}_1$ we then have $N = M^\perp$, $M = N^\perp$ so N and M are closed. Thus an orthogonal projection is bounded by Theorem 2.2.3. We have the following characterization of orthogonal projections; see Exercise 2.2.5 for another characterization.

Proposition 2.2.4. *A projection P is orthogonal if and only if it satisfies $P^* = P$.*

Proof. If $P^* = P$ and $u \in M$, $v \in N$, then $\langle u, v \rangle = \langle Pu, v \rangle = \langle u, P^*v \rangle = \langle u, Pv \rangle = \langle u, 0 \rangle = 0$ so M and N are orthogonal.

Conversely, suppose M and N orthogonal. For arbitrary $u, v \in \mathcal{H}_1$ we then have $\langle Pu, v \rangle = \langle Pu, Pv \rangle + \langle Pu, (I - P)v \rangle = \langle Pu, Pv \rangle$ so that also $\langle u, Pv \rangle = \langle Pu, Pv \rangle$. Hence $\langle Pu, v \rangle = \langle u, Pv \rangle$ for all $u, v \in \mathcal{H}_1$, i.e., $P^* = P$. $\square$

An operator T for which $T^* = T$ is called *self-adjoint*. Hence an orthogonal projection is the same as a self-adjoint projection. We shall have much more to say about self-adjoint operators in a more general context later.

Another important class of operators are the *unitary operators*. These are operators $U : \mathcal{H}_1 \to \mathcal{H}_2$ for which $U^* = U^{-1}$, i.e., U^*U and UU^* are the identities on $\mathcal{H}_1$ and $\mathcal{H}_2$ respectively. Since $\langle Uu, Uv \rangle_2 = \langle U^*Uu, v \rangle_1 = \langle u, v \rangle_1$ the operator U preserves the scalar product; such an operator is called *isometric*. If U is isometric we have $\langle u, v \rangle_1 = \langle Uu, Uv \rangle_2 = \langle U^*Uu, v \rangle_1$, so that U^* is a left-inverse of U for any isometric operator, but to be unitary U must also be surjective (Exercise 2.2.6).

If $\dim \mathcal{H}_1 = \dim \mathcal{H}_2 < \infty$, then a left-inverse of a linear operator is also a right-inverse,[3] so in this case isometric and unitary (called *orthogonal* in the case of a real space) are the same thing. If $\dim \mathcal{H}_1 \neq \dim \mathcal{H}_2$ or both spaces are infinite-dimensional, however, this is not the case. It is clear that an invertible linear operator preserves linear independence, so if the spaces are of different dimensions there are no unitary operators between them. However, if both spaces are separable of infinite dimension some isometric operators are unitary and some are not.

---

[3] Since the criterion for both is a non-zero determinant.

Example 2.2.5. On the space ℓ^2 we may define the operator $S : (x_1, x_2, \dots) \mapsto (0, x_1, x_2, \dots)$, a so-called *shift operator*. Clearly this is an isometric operator, but the vector $(1, 0, 0, \dots)$ is not the image of anything, so the operator is not unitary. Its adjoint is $S^* : (x_1, x_2, \dots) \mapsto (x_2, x_3, \dots)$, which is only a *partial isometry* (an isometry on the orthogonal complement of $(1, 0, 0, \dots)$ while $S^*(1, 0, 0, \dots) = 0$). See also Exercise 2.2.12.

It is not possible to interpret a differential operator as a bounded operator on some Hilbert space of functions.[4] We therefore need to discuss unbounded operators as well. Similarly, we shall need to discuss operators that are not defined on all of $\mathcal{H}_1$. Thus we now consider a linear operator $T : \mathcal{D}(T) \to \mathcal{H}_2$, where the domain $\mathcal{D}(T)$ of T is some, not necessarily closed, subspace of $\mathcal{H}_1$. T is not assumed to be bounded. Another such operator S is said to be an *extension* of T if $\mathcal{D}(T) \subset \mathcal{D}(S)$ and $Su = Tu$ for every $u \in \mathcal{D}(T)$. We then write $T \subset S$.

We must discuss the concept of *adjoint*. The form $u \mapsto \langle Tu, v \rangle_2$ is, for fixed $v \in \mathcal{H}_2$, only defined for $u \in \mathcal{D}(T)$, and though linear not necessarily bounded, so there may not be any $v^* \in \mathcal{H}_1$ such that $\langle Tu, v \rangle_2 = \langle u, v^* \rangle_1$ for all $u \in \mathcal{D}(T)$. Even if there is, it may not be uniquely determined, since if w is orthogonal to $\mathcal{D}(T)$ one may replace v^* by $v^* + w$ with no change in $\langle u, v^* \rangle$. To avoid this problem one may assume that the orthogonal complement of $\mathcal{D}(T)$ equals $\{0\}$, i.e., $\mathcal{D}(T)$ is *dense* in $\mathcal{H}_1$, so that its closure equals $\mathcal{H}_1$.

T is then said to be *densely defined*. In this case $v^* \in \mathcal{H}_1$ is clearly uniquely determined by $v \in \mathcal{H}_2$, if it exists. It is also obvious that v^* depends linearly on v, so we define $\mathcal{D}(T^*)$ to be those $v \in \mathcal{H}_2$ for which we can find a $v^* \in \mathcal{H}_1$, and set $T^* v = v^*$. In general one cannot expect the adjoint T^* to be densely defined and one may even have $\mathcal{D}(T^*) = \{0\}$; see Exercise 2.2.11. Thus T^* may not itself have an adjoint. To understand this rather confusing situation it turns out to be helpful to consider *graphs* of operators or, more generally, *linear relations*.

The *graph* of T is the set $\{(u, Tu) : u \in \mathcal{D}(T)\} \subset \mathcal{H}_1 \times \mathcal{H}_2$. Clearly T is known if and only if its graph is known, so we shall not distinguish the two and denote the graph of T also by T. The graph is a linear subset of the *orthogonal direct sum* $\mathcal{H}_1 \oplus \mathcal{H}_2$, consisting of all pairs (u, f) with $u \in \mathcal{H}_1$ and $f \in \mathcal{H}_2$ with the natural linear operations and provided with the scalar product $\langle (u, f), (v, g) \rangle = \langle u, v \rangle_1 + \langle f, g \rangle_2$. This makes $\mathcal{H}_1 \oplus \mathcal{H}_2$ into a Hilbert space (Exercise 2.2.7). Note that the operator S is an extension of T precisely if the graph of S contains the graph of T. This explains the notation $T \subset S$.

It is now natural to generalize the concept of a linear operator in the following way.

Definition 2.2.6. A linear subspace T of $\mathcal{H}_1 \oplus \mathcal{H}_2$ is called a linear relation from $\mathcal{H}_1$ to $\mathcal{H}_2$. The *domain* of T is the set

$$\mathcal{D}(T) = \{u \in \mathcal{H}_1 : (u, f) \in T \text{ for some } f \in \mathcal{H}_2\},$$

and the *range* of T is the set

---

[4] On more general spaces it is, as in the spaces of test functions and distributions of Appendix C.

$$\mathcal{R}_T = \{f \in \mathcal{H}_2 : (u, f) \in T \text{ for some } u \in \mathcal{H}_1\}\,.$$

Clearly a linear relation T is (the graph of) an operator precisely if the second component of any element of T is determined by the first, which in turn means precisely that $(0, f) \in T$ implies that $f = 0$.

Relations between the same Hilbert spaces may be multiplied by a constant and added; if S and T are linear relations from $\mathcal{H}_1$ to $\mathcal{H}_2$ and $\lambda \in \mathbb{C}$ we define

$$\lambda T = \{(u, \lambda f) : (u, f) \in T\},$$
$$S + T = \{(u, f_S + f_T) : (u, f_S) \in S \text{ and } (u, f_T) \in T\}.$$

Note that $\mathcal{D}(S + T) = \mathcal{D}(S) \cap \mathcal{D}(T)$ may be much smaller than $\mathcal{D}(S)$ and $\mathcal{D}(T)$. While this addition is associative and commutative, and the relation $\mathcal{H}_1 \times \{0\}$ is a neutral element for addition, a relation has no additive inverse unless its domain is the whole space. Thus the set of linear relations from $\mathcal{H}_1$ to $\mathcal{H}_2$ provided with these operations is emphatically *not* a linear space. The same is true of operators that are not everywhere defined. For this reason this notion of addition is in general not of much use unless the domain of one summand is contained in the domain of the other.

The inverse of a linear relation may be defined as $T^{-1} = \{(f, u) : (u, f) \in T\}$ so that $(T^{-1})^{-1} = T$ and $\mathcal{D}(T^{-1}) = \mathcal{R}_T$, $\mathcal{R}_{T^{-1}} = \mathcal{D}(T)$.

In order to define the adjoint of a relation we introduce the *boundary operator*[5] $\mathcal{U} : \mathcal{H}_1 \oplus \mathcal{H}_2 \to \mathcal{H}_2 \oplus \mathcal{H}_1$ by $\mathcal{U}(u_1, u_2) = (-iu_2, iu_1)$. It is clear that $\mathcal{U}$ is unitary, i.e., $\mathcal{U}^*$ is the inverse of $\mathcal{U}$. If $\mathcal{H}_1 = \mathcal{H}_2$ it is clear that $\mathcal{U}$ is also self-adjoint and involutary (i.e., $\mathcal{U}^2$ is the identity). We define the adjoint of a linear relation T by

$$T^* := \mathcal{U}((\mathcal{H}_1 \oplus \mathcal{H}_2) \ominus T) = (\mathcal{H}_2 \oplus \mathcal{H}_1) \ominus \mathcal{U}T\,. \tag{2.2.1}$$

The second equality is left for the reader to verify. For bounded operators the definition coincides with the earlier one; see Exercise 2.2.8.

We say that a linear relation is *closed* if it is closed as a subspace of $\mathcal{H}_1 \oplus \mathcal{H}_2$. By the closed graph theorem an everywhere defined operator is closed if and only if it is bounded (Exercise 2.2.9). It is clear that all adjoints, being orthogonal complements, are closed.

Any linear relation has a closure, which is also a linear relation, but the closure of (the graph of) a linear operator may not be the graph of an operator. Such an operator is said to not be *closeable*. We also use the following terminology.

Definition 2.2.7. Given a relation $T \subset \mathcal{H}_1 \oplus \mathcal{H}_2$ let the space $\mathcal{H}$ be the closure of $\mathcal{D}(T)$ and $\mathcal{H}_\infty = \{g \in \mathcal{H}_1 : (0, g) \in T^*\}$.

If $T_0 \subset T$ are relations from $\mathcal{H}_1$ to $\mathcal{H}_2$ and T is the closure of T_0 we say that T_0 is a *core* of T.

---

[5] The terminology is explained in Section 4.3.

Proposition 2.2.8.

(1) *If* $T \subset S$, *then* $S^* \subset T^*$.

(2) *The closure of* T *is* T^{**}.

(3) $(T^{-1})^* = (T^*)^{-1}$.

(4) $\mathcal{H}_1 = \mathcal{H} \oplus \mathcal{H}_\infty$.

(5) T^* *is an operator if and only if* $\mathcal{D}(T)$ *is dense in* $\mathcal{H}_1$.

(6) *An operator* T *is closeable if and only if* $\mathcal{D}(T^*)$ *is dense in* $\mathcal{H}_2$.

Proof. The first three claims are immediate consequences of defining the adjoint as an orthogonal complement. Next suppose $(u, f) \in T$. Then $\langle u, g \rangle_1 = \langle u, g \rangle_1 - \langle f, 0 \rangle_2 = -i \langle (u, f), \mathcal{U}(0, g) \rangle$, so $(0, g) \in T^*$ if and only if $g \in \mathcal{H}_1 \ominus \mathcal{D}(T)$, which proves the fourth claim. But clearly T^* is an operator precisely if $(0, g) \in T^*$ implies $g = 0$. The fifth claim follows immediately from this, and the sixth from (2) and (5). $\qquad\square$

The advantage of dealing with relations is that any relation is closeable and has an adjoint which is also a relation. This facilitates many calculations; one may for example find adjoints of operators that are not densely defined.

In the rest of this section we specialize to the case $\mathcal{H}_1 = \mathcal{H}_2$. A linear relation T on $\mathcal{H}_1$ is said to be *symmetric* if $T \subset T^*$. If (u, f) and $(v, g) \in T$ this means that $\langle f, v \rangle = \langle u, g \rangle$. Thus $\langle u, f \rangle$ is always real. It therefore makes sense to say that a symmetric relation is *positive* if $\langle u, f \rangle \geq 0$ for all $(u, f) \in T$.

If I is the identity on $\mathcal{H}_1$, then a symmetric relation T is said to be *bounded from below* by $c \in \mathbb{R}$ if $\langle u, f \rangle \geq c \langle u, u \rangle$ for all $(u, f) \in T$, which we may write as $T \geq cI$.

Note that a densely defined symmetric operator is always closeable since T^* is automatically densely defined, being an extension of T. A linear relation T for which $T = T^*$ is said to be *self-adjoint*. The theory of self-adjoint relations is easily reduced to the theory of self-adjoint operators by the following considerations.

If T is a self-adjoint relation then $\mathcal{H}_\infty = \{ f \in \mathcal{H}_1 : (0, f) \in T \}$ may be viewed as the 'eigenspace' of T corresponding[6] to the 'eigenvalue' ∞, and by Proposition 2.2.8 we have $\mathcal{H}_1 = \mathcal{H} \oplus \mathcal{H}_\infty$.

Now put $T_\infty = \{0\} \times \mathcal{H}_\infty$. Then it is clear that $T = (T \cap \mathcal{H}^2) \oplus T_\infty$ so we have split T into its *multi-valued part* T_∞ and $T \cap \mathcal{H}^2$, which is called the *operator part* of T because of the following theorem.

Theorem 2.2.9 (Spectral theorem for self-adjoint relations). *If* T *is a self-adjoint relation and* $\mathcal{H}$ *the closure of its domain, then* $T \cap \mathcal{H}^2$ *is (the graph of) a densely defined self-adjoint operator in* $\mathcal{H}$ *with domain* $\mathcal{D}(T)$.

Proof. $T \cap \mathcal{H}^2$ is the graph of a densely defined operator on $\mathcal{H}$ since $(0, w) \in T \cap \mathcal{H}^2$ implies $w \in \mathcal{H}_\infty \cap \mathcal{H} = \{0\}$. $T \cap \mathcal{H}^2$ is self-adjoint since it equals $T \ominus T_\infty$ so its adjoint (in $\mathcal{H}$) is $(T \cap \mathcal{H}^2)^* = \mathcal{H}^2 \cap (T^* \oplus \mathcal{U}T_\infty) = \mathcal{H}^2 \cap T$ (check this calculation carefully!). $\qquad\square$

---

[6] For, it is the eigenspace at 0 of T^{-1}.

Remark 2.2.10. The most important theorem in this book is the *spectral theorem* for self-adjoint operators, which will be stated and proved in the next chapter. Theorem 2.2.9 allows an easy extension of the spectral theorem for self-adjoint operators to self-adjoint relations.

Since the spectral theorem deals with (possibly unbounded) self-adjoint operators it is important to be able to recognize if an operator or relation is self-adjoint. In practice it is often quite easy to see if a relation is symmetric, but much more difficult to decide whether a symmetric relation is self-adjoint.

When one wants to interpret a differential operator as a Hilbert space operator one has to choose a domain of definition; in many cases it is clear how one may choose a domain so that the operator becomes symmetric. With luck this operator may have a self-adjoint closure, in which case the operator is said to be *essentially self-adjoint*. In this connection we note the following proposition.

Proposition 2.2.11. *A symmetric operator or relation is essentially self-adjoint if and only if its adjoint is symmetric.*

Proof. T^{**} is the closure of T so if $T \subset T^*$ we have $T^{**} \subset T^*$. Thus T^* is symmetric precisely if $T^* = T^{**}$, so if and only if T^* and thus the closure T^{**} of T is self-adjoint. □

If T is symmetric but not essentially self-adjoint one will look for self-adjoint extensions of T, because then in a sense the adjoint T^* is too big, so that T is too small. To be more precise, if S is a symmetric extension of T, then $T \subset S \subset S^* \subset T^*$ so any symmetric extension of T is a restriction of the adjoint T^*. There is now obviously a need for a theory of symmetric extensions of a symmetric relation. We shall postpone the discussion of this to Section 2.4. Right now we shall instead turn to some very simple, but typical, examples.

Example 2.2.12. Consider the differential operator $\frac{d}{dx}$ on some open interval Ω. We want to interpret it as a densely defined operator in the Hilbert space $L^2(\Omega)$ and so must choose a suitable domain. A convenient choice, which would work for any linear differential operator with coefficients which are locally square integrable, is the set $C_0^\infty(\Omega)$ of infinitely differentiable functions with compact support in Ω, i.e., for each function there is a compact subset of Ω outside of which the function vanishes.

The set $C_0^\infty(\Omega)$ is dense in $L^2(\Omega)$ (see Appendix C). Let us denote the corresponding operator T_0; it is usually called the *minimal operator* for $\frac{d}{dx}$. Often it is the closure of this operator which is called the minimal operator, but this will make no difference to the calculations that follow. We must now find the adjoint of the minimal operator.

Let $v \in \mathcal{D}(T_0^*)$. This means that there is an element $v^* \in L^2(\Omega)$ such that $\int_\Omega \varphi' \overline{v} = \int_\Omega \varphi \overline{v^*}$ for all $\varphi \in C_0^\infty(\Omega)$ and that $T_0^* v = v^*$. In the sense of distributions, this means that $v^* = -v'$ (Appendix C). Concretely this means the following.

Integrating by parts we have $\int_\Omega \varphi \overline{v^*} = -\int_\Omega (\varphi' \overline{\int v^*})$ since the boundary terms vanish. Here $\int v^*$ denotes any integral function of v^*. Thus we have $\int_I \varphi' \overline{(v + \int v^*)} = 0$ for all $\varphi \in C_0^\infty(\Omega)$. We need the following lemma.

Lemma 2.2.13 (du Bois-Reymond). *Suppose u is locally square integrable on* $\mathbb{R}$, *i.e.,* $u \in L^2(I)$ *for every bounded real interval I. Also suppose that* $\int u\varphi' = 0$ *for every* $\varphi \in C_0^\infty(\mathbb{R})$. *Then u is (almost everywhere) equal to a constant.*

This is a special case of Theorem C.11. The lemma shows that (choosing the appropriate representative in the equivalence class of v) $v + \int v^*$ is constant. Hence v is locally absolutely continuous (see page 299) with derivative $-v^*$. It follows that $\mathcal{D}(T_0^*)$ consists of functions $v \in L^2(\Omega)$ which are locally absolutely continuous in Ω with derivative in $L^2(\Omega)$, and that $T_0^* v = -v'$. Conversely, all such functions are in $\mathcal{D}(T_0^*)$, as follows immediately by partial integration in $\int_\Omega \varphi' \overline{v} = \int_\Omega \varphi \overline{v^*}$. The operator T_0^* is therefore also a differential operator, generated by $-\frac{d}{dx}$, but with a larger domain than T_0.

The differential operator $-\frac{d}{dx}$ is called the *formal adjoint* of $\frac{d}{dx}$ and the operator T_0^* is called the *maximal operator* associated with $-\frac{d}{dx}$. In the same way any linear differential operator (with sufficiently smooth coefficients) has a formal adjoint which is also a differential operator and obtained by integration by parts.

For ordinary differential operators with sufficiently smooth coefficients one can always calculate adjoints in essentially the way we just did, although we shall often use less regular coefficients and therefore find it convenient to use different methods; for partial differential operators matters are more subtle and one needs to use the language of distribution theory.

For the minimal operator of a differential operator to be symmetric it is clear that the differential operator has to be *formally symmetric*, i.e., the formal adjoint has to coincide with the original operator. In Example 2.2.12 $\mathcal{D}(T_0) \subset \mathcal{D}(T_0^*)$ but there is a minus sign preventing T_0 from being symmetric. However, it is clear that had we started with the differential operator $i\frac{d}{dx}$ instead, then the minimal operator would have been symmetric, but the domains of the minimal and maximal operators unchanged. One may then ask if there are self-adjoint extensions of the minimal operator or, equivalently, self-adjoint restrictions of the maximal operator.

Example 2.2.14. Let T_1 be the maximal operator of $i\frac{d}{dx}$ on the interval Ω. Let u, $v \in \mathcal{D}(T_1)$ and $a, b \in \Omega$. Then $\int_a^b T_1 u \overline{v} - \int_a^b u \overline{T_1 v} = i \int_a^b (u'\overline{v} + u\overline{v'}) = iu(b)\overline{v(b)} - iu(a)\overline{v(a)}$. Since $u, v, T_1 u$ and $T_1 v$ are all in $L^2(\Omega)$ the limit of $u\overline{v}$ exists at both endpoints of Ω.

Consider now the case $\Omega = \mathbb{R}$. Since $|u(x)|^2$ has limits as $x \to \pm\infty$ and is integrable, the limits must both be 0. Hence $\langle T_1 u, v \rangle - \langle u, T_1 v \rangle = 0$ for any u, $v \in \mathcal{D}(T_1)$, so the maximal operator is symmetric and therefore self-adjoint according to Proposition 2.2.11, and the minimal operator essentially self-adjoint.

Example 2.2.15. Consider the operator in Example 2.2.14 for $\Omega = (0, \infty)$. If $u \in \mathcal{D}(T_1)$ we obtain $\langle T_1 u, u \rangle - \langle u, T_1 u \rangle = -i |u(0)|^2$. To have a symmetric restriction of T_1 we must therefore require $u(0) = 0$, and with only this restriction on the domain of T_1 we obtain a *maximal symmetric* operator T.

If now $u \in \mathcal{D}(T)$ and $v \in \mathcal{D}(T_1)$ we obtain

$$\langle Tu, v \rangle - \langle u, T_1 v \rangle = -\mathrm{i}u(0)\overline{v(0)} = 0 ,$$

so that $T^* = T_1$. T is therefore *not* self-adjoint so *no matter how we choose a symmetric restriction of* T_1 we shall not obtain an operator self-adjoint in $L^2(0, \infty)$. One says that $\mathrm{i}\frac{d}{dx}$ has *no self-adjoint realization* in $L^2(0, \infty)$.

Example 2.2.16. Finally consider the operator of Example 2.2.14 for Ω bounded, say $\Omega = (-\pi, \pi)$. We now have

$$\langle T_1 u, v \rangle - \langle u, T_1 v \rangle = \mathrm{i}\big(u(\pi)\overline{v(\pi)} - u(-\pi)\overline{v(-\pi)}\big) . \tag{2.2.2}$$

In particular, for $u = v$ it follows that for u to be in the domain of a symmetric restriction of T_1 we must require $|u(\pi)| = |u(-\pi)|$. If $u(\pm\pi) = 0$ for all u in the domain of the restriction, then its adjoint is T_1 so this operator is not self-adjoint. If on the other hand $u(\pi) \neq 0$ for some u in the restriction then u satisfies the *boundary conditions* $u(\pi) = \mathrm{e}^{\mathrm{i}\theta}u(-\pi)$ for some real θ. From (2.2.2) it then follows that if v is in the domain of the adjoint, then v will have to satisfy the same boundary condition. On the other hand, if we impose this condition on all u, then the resulting operator will be self-adjoint since its adjoint is symmetric. It follows that restricting the domain of T_1 by such a boundary condition is exactly what is required to obtain a self-adjoint restriction. Each θ in $[0, 2\pi)$ gives a different self-adjoint realization, but there are no others.

The above examples show that there may be a unique self-adjoint realization of our formally symmetric differential operator, none at all, or infinitely many depending on circumstances. It can be a very difficult problem to decide which of these possibilities occur in a given case. In particular, much effort has been devoted to decide whether a given differential operator on a given domain has a unique self-adjoint realization.

Exercises

Exercise 2.2.1 Prove that boundedness is equivalent to continuity for a linear operator between normed spaces. Then prove the properties of the operator norm listed at the beginning of the section.

Exercise 2.2.2 Suppose $\mathcal{B}_1$ and $\mathcal{B}_2$ are Banach spaces. Show that so is $\mathcal{L}(\mathcal{B}_1, \mathcal{B}_2)$ (this is true even if $\mathcal{B}_1$ is not complete).

Exercise 2.2.3 Fill in the details of the proof of Proposition 2.2.1.

Exercise 2.2.4 Suppose $e_1, e_2, \ldots$ is an orthonormal basis in $\mathcal{H}$ and that T_j are linear operators with $T_j e_j = e_j$ and $T_j e_k = 0$ for $j \neq k$. Show that T_j tend strongly, but not uniformly, to zero.

Similarly, if $T_j e_k = e_{k+j}$, show that T_j tends weakly, but not strongly, to zero.

Exercise 2.2.5 Show that a non-trivial, i.e., not identically zero, projection is orthogonal if and only if its operator norm is 1.

Exercise 2.2.6 Show that an isometric operator is unitary if and only if it surjective (onto) and also if and only if its adjoint is invertible.

Exercise 2.2.7 Suppose $\mathcal{H}_1$ and $\mathcal{H}_2$ are Hilbert spaces. Show that the orthogonal direct sum $\mathcal{H}_1 \oplus \mathcal{H}_2$ is also a Hilbert space.

Exercise 2.2.8 Show that the definition of adjoint for bounded operators via the Riesz representation theorem is equivalent to the definition as an orthogonal complement in (2.2.1).

Exercise 2.2.9 Show that a bounded, everywhere defined operator is automatically closed. Conversely, that an everywhere defined, closed operator is bounded.

Hint: The closed graph Theorem A.4.

Exercise 2.2.10 Prove Theorem 2.2.3.

Hint: First prove that P is closed, and then use the previous exercise.

Exercise 2.2.11 Show that the adjoint of a densely defined linear operator need not be densely defined. Then strengthen the argument to give an example of a densely defined linear operator with adjoint domain $\{0\}$.

Hint: If $e_1, e_2, \ldots$ is an orthonormal basis define $Te_k = e_1$ for every k. Show that T is densely defined but the domain of T^* is orthogonal to e_1.

Next show that $e_1, e_2, \ldots$ may be split into a sequence of disjoint subsequences, using that $\mathbb{N} \times \mathbb{N}$ is countable. Use the previous argument to show that we may define T so that the domain of T^* is $\{0\}$.

Exercise 2.2.12 An *eigenvalue* of a relation T is a number λ such that $(u, \lambda u) \in T$ for some $u \neq 0$, which is called an eigenvector to λ. Show that if T is a symmetric relation, then all eigenvalues of T are real, and if U is an isometric operator on some Hilbert space, then all eigenvalues of U have absolute value 1.

In both cases also show that if u and v are eigenvectors corresponding to eigenvalues λ respectively μ, then u and v are orthogonal if $\lambda \neq \mu$.

However, note the next exercise!

Hint: See (1.4.3).

Exercise 2.2.13 Show that neither unitary operators nor self-adjoint relations have to have any eigenvalues.

Hint: Consider the Hilbert space $L^2(0, 1)$ and on this an operator $Tu(x) = f(x)u(x)$, where f is continuous. Find f such that T is unitary, respectively self-adjoint, but has no eigenvalues.

2.3 Resolvents

Here we consider a closed relation T on the Hilbert space $\mathcal{H}_1$. In Section 1.4 we defined the spectrum of a boundary value problem as the set of its eigenvalues. We now need to introduce a more general notion of spectrum.

Definition 2.3.1. Suppose T is a closed relation on $\mathcal{H}_1$ and define:

(1) An *eigenvalue* of T is a complex number λ such that $(u, \lambda u) \in T$ for some non-zero $u \in \mathcal{H}_1$. Such a u is called an *eigenvector* of T to λ. The set of all such u, together with 0, is the *eigenspace* corresponding to λ, and the dimension of this space is called the (geometric) *multiplicity* of the eigenvalue.

(2) The *solvable space* of T at $\lambda \in \mathbb{C}$ is

$$S_\lambda = \{u \in \mathcal{H}_1 : (v, \lambda v + u) \in T \text{ for some } v \in \mathcal{D}(T)\} .$$

(3) The *defect space* D_λ of T at $\lambda \in \mathbb{C}$ is the eigenspace at λ of T^*, i.e.,

$$D_\lambda = \{u \in \mathcal{D}(T^*) : (u, \lambda u) \in T^*\} .$$

(4) The set $\rho(T)$ of all $\lambda \in \mathbb{C}$ which are not eigenvalues of T and for which $S_\lambda = \mathcal{H}_1$ is called the *resolvent set* for T.

(5) The complement $\sigma(T)$ of $\rho(T)$ is called the *spectrum* of T.

We make some comments related to these definitions. First note that an eigenspace for a closed relation is automatically closed. In particular, D_λ is always closed. If T is an operator the condition $u \in S_\lambda$ means precisely that the equation $(T - \lambda)v = u$ has a solution v. Similarly, if T^* is an operator then $u \in D_\lambda$ means precisely that u is a solution of $T^*u = \lambda u$.

The resolvent set of a closed relation may well be empty, but as we shall soon see this cannot happen for self-adjoint relations.

If λ is not an eigenvalue of T the condition $(v, \lambda v + u) \in T$ defines an operator $S_\lambda \ni u \mapsto v \in \mathcal{H}_1$ which is closed because T is. Thus, if this operator is bounded S_λ must be closed, and conversely, if S_λ is closed we have a closed operator mapping a Hilbert space into another, so it is bounded by the closed graph theorem on page 271. In particular, if $S_\lambda = \mathcal{H}_1$, so that $\lambda \in \rho(T)$, we have a bounded operator defined everywhere in $\mathcal{H}_1$.

Definition 2.3.2. Let T be a closed relation on $\mathcal{H}_1$. The bounded, everywhere defined operator $R_\lambda : \mathcal{H}_1 \ni u \mapsto v \in \mathcal{H}_1$ which is defined for $\lambda \in \rho(T)$ by $(v, \lambda v + u) \in T$ is called the *resolvent* of T at λ.

If I is the identity on $\mathcal{H}_1$ and $\lambda \in \rho(T)$, then $R_\lambda = (T - \lambda I)^{-1}$. We note the following general properties of the resolvent of a closed relation.

Theorem 2.3.3. *The resolvent of a closed relation T on $\mathcal{H}_1$ has the properties:*

(1) *The nullspace $\{u \in \mathcal{H}_1 : R_\lambda u = 0\}$ is independent of $\lambda \in \rho(T)$ and equals $\{u : (0, u) \in T\}$, so is $\{0\}$ if and only if T is an operator.*

(2) $R_\lambda - R_\mu = (\lambda - \mu)R_\lambda R_\mu = (\lambda - \mu)R_\mu R_\lambda$ *for λ and $\mu \in \rho(T)$.*

Conversely, if $\rho \subset \mathbb{C}$ is an open set and $\{R_\lambda\}_{\lambda \in \rho}$ a family of bounded, everywhere defined operators satisfying (2) for $\lambda, \mu \in \rho$, then R_λ is the resolvent at λ for some closed relation T. If for some $\lambda \in \rho$ also $\bar{\lambda} \in \rho$ and $R_\lambda^* = R_{\bar{\lambda}}$, then T is self-adjoint.

The statement (2) is called the (first) *resolvent relation*.

Proof. If $\lambda \in \rho(T)$, then $v = R_\lambda u$ is equivalent to $(v, \lambda v + u) \in T$, so $R_\lambda u = 0$ is equivalent to $(0, u) \in T$.

Now suppose $R_\mu u = w$, i.e., $(w, \mu w + u) \in T$. Since T is linear it follows that $R_\lambda u = v$, i.e., $(v, \lambda v + u) \in T$, if and only if

$$(v, \lambda v + u) - (w, \mu w + u) = (v - w, \lambda(v - w) + (\lambda - \mu)w) \in T.$$

But this means precisely that the first equality of (2) holds and the second is obtained by interchanging λ and μ in the first.

Conversely, by (2) $R_\lambda = (I + (\lambda - \mu)R_\lambda)R_\mu = R_\mu(I + (\lambda - \mu)R_\lambda)$ where I is the identity, so the nullspace and range of R_μ are contained in those of R_λ. By symmetry all R_λ have the same range $\mathcal{D}$ and nullspace $\mathcal{H}_\infty$. Setting $\mathcal{H} = \mathcal{H}_1 \ominus \mathcal{H}_\infty$ the restriction of R_λ to $\mathcal{H}$ thus has a closed inverse $R_\lambda^{-1} : \mathcal{D} \to \mathcal{H}$, which may be unbounded. From (2) follows that $R_\lambda^{-1} + \lambda I = R_\mu^{-1} + \mu I$ so $T = R_\lambda^{-1} + \lambda I$ is independent of λ. This is the operator part of a closed relation in $\mathcal{H}_1$ with multi-valued part $\{0\} \times \mathcal{H}_\infty$.

If λ and $\bar{\lambda}$ are both in ρ and $R_\lambda^* = R_{\bar{\lambda}}$ we obtain

$$T^* = (R_\lambda^{-1})^* + \bar{\lambda}I = (R_\lambda^*)^{-1} + \bar{\lambda}I = (R_{\bar{\lambda}})^{-1} + \bar{\lambda}I = T. \qquad \square$$

Theorem 2.3.4. *The resolvent set $\rho(T)$ of a closed relation is open, and the function $\rho(T) \ni \lambda \mapsto R_\lambda$ is analytic in the uniform operator topology as a $\mathcal{L}(\mathcal{H}_1)$-valued function.*

This means (by definition) that R_λ can be expanded in a power series with respect to λ around any point in $\rho(T)$ and that the series converges in operator norm in a neighborhood of the point. In fact, if $\mu \in \rho(T)$, then $\lambda \in \rho(T)$ for $|\lambda - \mu| < 1/\|R_\mu\|$ and

$$R_\lambda = \sum_{k=0}^{\infty} (\lambda - \mu)^k R_\mu^{k+1} \text{ for } |\lambda - \mu| < 1/\|R_\mu\|. \tag{2.3.1}$$

Proof. The series is norm convergent if $|\lambda - \mu| < 1/\|R_\mu\|$ since $\|(\lambda - \mu)^k R_\mu^{k+1}\| \leq \|R_\mu\|(|\lambda - \mu|\,\|R_\mu\|)^k$, which is a term in a convergent geometric series. The triangle inequality immediately shows that the partial sums of the series in (2.3.1) yield a Cauchy sequence, and hence that the series converges in operator norm.

Denoting the sum by R we must prove that $R = R_\lambda$, i.e., for any $u \in \mathcal{H}_1$ we have $(Ru, \lambda Ru + u) \in T$. Now $(R_\mu^{k+1}u, \mu R_\mu^{k+1}u + R_\mu^k u) \in T$ for $k = 0, 1, 2 \ldots$. Multiplying by $(\lambda - \mu)^k$ and adding up the first component becomes Ru, so since T is closed it will be enough to show that the second component becomes $(\lambda R + I)u$ where I is the identity. Now

$$\sum_{k=0}^{\infty} (\lambda - \mu)^k (\mu R_\mu^{k+1} + R_\mu^k) = \mu R + I + (\lambda - \mu)R = \lambda R + I .$$

This proves $R = R_\lambda$ and it also follows that $\rho(T)$ is open. □

That $\rho(T)$ is open means of course that the spectrum $\sigma(T)$ is always closed. It is customary to split the spectrum into disjoint parts as follows.

Definition 2.3.5. Suppose T is a closed relation on $\mathcal{H}_1$.

(1) The set of eigenvalues for T is called the *point spectrum* of T and denoted $\sigma_p(T)$.

(2) The set of $\lambda \in \mathbb{C}$ which are not eigenvalues for T and for which S_λ is dense but not closed in $\mathcal{H}_1$ is denoted $\sigma_c(T)$ and called the *continuous spectrum* of T.

(3) The set of $\lambda \in \mathbb{C}$ which are not eigenvalues for T and for which S_λ is not dense in $\mathcal{H}_1$ is denoted $\sigma_r(T)$ and called the *residual spectrum* of T.

It is clear that $\sigma(T)$ is the disjoint union of $\sigma_p(T)$, $\sigma_c(T)$ and $\sigma_r(T)$. We shall now turn our attention to symmetric and self-adjoint relations. The following lemma is basic.

Lemma 2.3.6. *Suppose T is a closed linear relation on $\mathcal{H}_1$. Then*

(1) $D_{\overline{\lambda}} = \mathcal{H}_1 \ominus S_\lambda$.

(2) *If T is symmetric and $\mathrm{Im}\,\lambda \neq 0$, then λ is not an eigenvalue, and if $(v, \lambda v + u) \in T$, then $\|v\| \leq \|u\|/|\mathrm{Im}\,\lambda|$. Furthermore, S_λ is closed and $\mathcal{H}_1 = S_\lambda \oplus D_{\overline{\lambda}}$.*

Note that (1) explains the term *defect space* and shows that $\lambda \in \sigma_r(T)$ precisely if it is not an eigenvalue of T but $\overline{\lambda}$ is an eigenvalue of T^*.

Proof. Any element of T may be written $(v, \lambda v + u)$, where $u \in S_\lambda$, for if $(v, f) \in T$ we simply put $u = f - \lambda v$. If $\mathcal{U}$ is the boundary operator defined on page 42 we have

$$\langle (v, \lambda v + u), \mathcal{U}(w, \overline{\lambda}w) \rangle = -\mathrm{i}\langle u, w \rangle ,$$

so it follows that $(w, \overline{\lambda}w) \in T^*$, i.e., $w \in D_{\overline{\lambda}}$, if and only if w is orthogonal to S_λ. This proves (1).

If T is symmetric and $(v, \lambda v + u) \in T$ it follows that $\langle \lambda v + u, v \rangle = \langle v, \lambda v + u \rangle$, i.e., $\mathrm{Im}\,\lambda \|v\|^2 = \mathrm{Im}\langle v, u \rangle$, the right-hand side of which has absolute value less than $\|v\|\|u\|$ by the Cauchy–Schwarz inequality.

If $\mathrm{Im}\,\lambda \neq 0$ we obtain $\|v\| \leq \frac{1}{|\mathrm{Im}\,\lambda|}\|u\|$, so that $(v, \lambda v + u) \in T$ determines a bounded linear operator $S_\lambda \ni u \mapsto v \in \mathcal{H}_1$; in particular T has no non-real eigenvalues. Since T is closed this operator is also closed, so that S_λ is closed and (2) follows. □

The first point of the lemma implies that a self-adjoint relation has empty residual spectrum; any λ for which S_λ is not dense must be an eigenvalue. We summarize the properties of the resolvent of a self-adjoint relation as follows.

Theorem 2.3.7. *The resolvent R_λ of a self-adjoint relation T has the properties:*

(1) *The nullspace $\{u \in \mathcal{H}_1 : R_\lambda u = 0\}$ is independent of $\lambda \in \rho(T)$ and equals $\mathcal{H}_\infty$, so equals $\{0\}$ if and only if T is an operator.*

(2) *If $\mathrm{Im}\,\lambda \neq 0$, then $\lambda \in \rho(T)$ and $\|R_\lambda\| \leq 1/|\mathrm{Im}\,\lambda|$.*

(3) *$(R_\lambda)^* = R_{\overline{\lambda}}$ for $\lambda \in \rho(T)$.*

(4) *$R_\lambda - R_\mu = (\lambda - \mu)R_\lambda R_\mu = (\lambda - \mu)R_\mu R_\lambda$ for λ and $\mu \in \rho(T)$.*

(5) *The function $\rho(T) \ni \lambda \mapsto \langle R_\lambda u, v \rangle$ is analytic for all $u, v \in \mathcal{H}_1$, and for $u = v$ it maps the upper and lower half-planes into themselves.*

Proof. The first claim follows directly from Theorem 2.3.3 (1) and the definition of $\mathcal{H}_\infty$. If T is self-adjoint, then $T^* = T$ is symmetric so it has no non-real eigenvalues by Lemma 2.3.6. Thus $D_{\overline{\lambda}} = \{0\}$ if $\mathrm{Im}\,\lambda \neq 0$ so by the same lemma $\lambda \in \rho(T)$ and (2) follows.

Now (3) is an immediate consequence of Proposition 2.2.8 (3) and (4) is just a restatement of Theorem 2.3.3 (2).

By (3) and (4) we have $2\mathrm{i}\,\mathrm{Im}\langle R_\lambda u, u\rangle = \langle R_\lambda u, u\rangle - \langle u, R_\lambda u\rangle = \langle (R_\lambda - R_{\overline{\lambda}})u, u\rangle = 2\mathrm{i}\,\mathrm{Im}\,\lambda\langle R_{\overline{\lambda}}R_\lambda u, u\rangle = 2\mathrm{i}\,\mathrm{Im}\,\lambda\|R_\lambda u\|^2$. It follows that $\mathrm{Im}\langle R_\lambda u, u\rangle$ has the same sign as $\mathrm{Im}\,\lambda$. The analyticity of $\langle R_\lambda u, v\rangle$ follows since we have a power series expansion of it around any point $\mu \in \rho(T)$, by the series for R_λ. Alternatively, from (4) it easily follows that $\frac{d}{d\lambda}\langle R_\lambda u, v\rangle = \langle R_\lambda^2 u, v\rangle$ (Exercise 2.3.1). $\qquad\square$

Analytic functions that map the upper half-plane into itself have particularly nice properties and are called *Nevanlinna functions*, or sometimes Herglotz or Pick functions. The main fact about such functions is a representation formula (E.1.1) given in Appendix E. Our proof of the general spectral theorem is based on the use of this formula for $\langle R_\lambda u, u\rangle$.

Exercises

Exercise 2.3.1 Suppose that R_λ is the resolvent of a self-adjoint relation T in a Hilbert space $\mathcal{H}_1$. Show directly from Theorem 2.3.7 (4) that if $u, v \in \mathcal{H}_1$, then $\lambda \mapsto \langle R_\lambda u, v\rangle$ is analytic (has a complex derivative) for $\lambda \in \rho(T)$, and find an expression for the derivative. Also show that if $u \in \mathcal{H}_1$, then $\lambda \mapsto \langle R_\lambda u, u\rangle$ is increasing in every subinterval of $\rho(T) \cap \mathbb{R}$.

Exercise 2.3.2 It is easy to see that the *Möbius* (or *fractional linear*) transform $w = (z + \mathrm{i})/(z - \mathrm{i}) = 1 + 2\mathrm{i}/(z - \mathrm{i})$ maps $\mathbb{R} \cup \{\infty\}$ to the unit circle in a one-to-one fashion, the image of ∞ being 1. Its inverse is $z = \mathrm{i}(w + 1)/(w - 1)$.

Replacing z by T and i by $\mathrm{i}I$ where I is the identity we obtain the expression $(T + \mathrm{i}I)(T - \mathrm{i}I)^{-1}$ which is meaningless since we have not defined (and will not define) composition of relations. The expression $1 + 2\mathrm{i}/(z - \mathrm{i})$, however, may after this replacement be interpreted as $I + 2\mathrm{i}R_\mathrm{i}$, where R_i is the resolvent of T at i.

Show that if T is a self-adjoint relation, then the operator $U = I + 2iR_i$ is unitary. Conversely, show that if U is unitary, then, *properly interpreted*, the formula $T = i(U + I)(U - I)^{-1}$ yields a self-adjoint relation, and that the two transforms are inverse to each other. What property of U corresponds to T being an operator?

The above transform of T, reminiscent of a Möbius transform, is called the *Cayley transform* of T and was the basis for von Neumann's proof of the spectral theorem for unbounded operators.

Exercise 2.3.3 (Hellinger–Töplitz theorem) Suppose B is a symmetric, everywhere defined relation in $\mathcal{H}_1$. Show that B is a self-adjoint, bounded operator.

Hint: Use Proposition 2.2.8 (4) and then the closed graph Theorem A.4.

Exercise 2.3.4 Show that if B is an everywhere defined, bounded operator and T a linear relation from $\mathcal{H}_1$ to $\mathcal{H}_2$, then $(T + B)^*$ has the same domain as T^* and equals $T^* + B^*$.

Exercise 2.3.5 Show that if B is a bounded, self-adjoint operator and T a self-adjoint relation in $\mathcal{H}_1$, then $T + B$ is self-adjoint with the same domain as T and the resolvent of $T + B$ at λ is the sum of the series $\sum_0^\infty (-R_\lambda B)^k R_\lambda$ if this converges, and that it does if $|\operatorname{Im} \lambda|$ is large.

2.4 Extension of symmetric relations

Here we shall complete the discussion on symmetric extensions of a symmetric relation started in Section 2.2. Although our proofs are different, this is a minor extension of results originally due to von Neumann, who based his proofs on the Cayley transform (see Exercise 2.3.2).

We shall find criteria for the existence of self-adjoint extensions of a symmetric relation, which according to the discussion just before Example 2.2.12 on page 44 must be restrictions of the adjoint relation.

Recall that the adjoint of a linear relation T on $\mathcal{H}_1$ is defined by

$$T^* = \mathcal{U}(\mathcal{H}_1^2 \ominus T) = \mathcal{H}_1^2 \ominus \mathcal{U}T$$

according to (2.2.1), where $\mathcal{H}_1^2 = \mathcal{H}_1 \oplus \mathcal{H}_1$ and $\mathcal{U} : \mathcal{H}_1^2 \ni (v, f) \mapsto (-if, iv) \in \mathcal{H}_1^2$ is the boundary operator introduced on page 42. Also recall that $\mathcal{U}$ is self-adjoint, unitary and involutary on $\mathcal{H}_1^2$.

So, assume we have a symmetric relation T. We want to investigate what self-adjoint extensions, if any, T has. Since the closure of a symmetric relation is also symmetric with the same adjoint we may as well assume that T is closed to begin with. Recall that if S is a symmetric extension of T, then it is a restriction of T^* since we then have $T \subset S \subset S^* \subset T^*$. Now put

$$\mathbf{D}_{\pm i} = \{U \in T^* : \mathcal{U}U = \pm U\} .$$

It is immediately seen that $\mathbf{D_i}$ and $\mathbf{D_{-i}}$ consist of the elements of T^* of the form (u, iu) and $(u, -iu)$ respectively. If T^* is an operator this means that u satisfies the equation $T^*u = iu$ respectively $T^*u = -iu$. We may identify these spaces with the defect spaces $D_{\pm i}$ introduced in Section 2.3. Also $\mathbf{D_{\pm i}}$ are therefore called defect spaces.

Theorem 2.4.1 (von Neumann). *If T is a closed and symmetric relation on $\mathcal{H}_1$, then $T^* = T \oplus \mathbf{D_i} \oplus \mathbf{D_{-i}}$. This is an orthogonal sum of closed subspaces of $\mathcal{H}_1^2 = \mathcal{H}_1 \oplus \mathcal{H}_1$.*

Proof. The facts that $\mathbf{D_i}$ and $\mathbf{D_{-i}}$ are eigenspaces of the self-adjoint operator $\mathcal{U}$ for different eigenvalues and $\langle T, \mathcal{U}T^* \rangle = 0$ imply that T, $\mathbf{D_i}$ and $\mathbf{D_{-i}}$ are orthogonal subspaces of T^* (cf. Exercise 2.2.12), but this is also clear since $\langle T, \mathbf{D_{\pm i}} \rangle = \pm \langle T, \mathcal{U}\mathbf{D_{\pm i}} \rangle = 0$ and $\langle (u, iu), (v, -iv) \rangle = \langle u, v \rangle + \langle iu, -iv \rangle = 0$.

It remains to show that $\mathbf{D_i} \oplus \mathbf{D_{-i}}$ contains $T^* \ominus T$. However, if $U \in T^* \ominus T$ then $U \in \mathcal{H}_1^2 \ominus T$ and thus $\mathcal{U}U \in T^*$. Denoting the identity on $\mathcal{H}_1^2$ by I define $U_+ = \frac{1}{2}(I + \mathcal{U})U \in T^*$ and $U_- = \frac{1}{2}(I - \mathcal{U})U \in T^*$ so that $U = U_+ + U_-$. Using $\mathcal{U}^2 = I$ clearly $\mathcal{U}U_\pm = \pm U_\pm$ so $U_\pm \in \mathbf{D_{\pm i}}$ and we are done. $\qquad\square$

We define the *defect indices* of T to be

$$n_+ = \dim \mathbf{D_i} = \dim D_i \quad \text{and} \quad n_- = \dim \mathbf{D_{-i}} = \dim D_{-i}$$

so these are natural numbers or infinity. We may now characterize the symmetric extensions of T.

Theorem 2.4.2. *If S is a closed, symmetric extension of the closed symmetric relation T, then $S = T \oplus D$ where $D \subset \mathbf{D_i} \oplus \mathbf{D_{-i}}$ is a linear space such that*

$$D = \{u + Ju : u \in \mathcal{D}(J) \subset \mathbf{D_i}\}$$

for some linear isometry J of a closed subspace $\mathcal{D}(J)$ of $\mathbf{D_i}$ onto part of $\mathbf{D_{-i}}$. Conversely, every space D gives rise to a closed symmetric extension $S = T \oplus D$ of T.

This is obvious after noting that if $u_+, v_+ \in \mathbf{D_i}$ and $u_-, v_- \in \mathbf{D_{-i}}$, then $\langle u_+, v_+ \rangle = \langle u_-, v_- \rangle$ precisely if $\langle u_+ + u_-, \mathcal{U}(v_+ + v_-) \rangle = 0$, cf. Exercise 2.4.1. Some immediate consequences of Theorem 2.4.2 follow.

Corollary 2.4.3. *The closed symmetric relation T is* maximal symmetric [7] *precisely if one of n_+ and n_- equals zero and* self-adjoint *precisely if $n_+ = n_- = 0$.*

Example 2.4.4. Consider the operator $i\frac{d}{dx}$ of Examples 2.2.14, 2.2.15 and 2.2.16. The space $D_{\pm i}$ consist of the solutions of $iu' = \pm iu$ which are in the relevant space. The solutions are the multiples of $e^{\pm x}$, and these are not in $L^2(\mathbb{R})$. Thus the defect indices of the operator considered on $\mathbb{R}$ are both 0, and the minimal operator is essentially self-adjoint, as we already saw in Example 2.2.14.

---

[7] That is, T has no proper symmetric extension.

Clearly e^{-x} is in $L^2(0, \infty)$ but e^x is not, so considered in this space our operator has defect indices $n_+ = 0$ and $n_- = 1$. The minimal operator is therefore maximal symmetric, but there are no self-adjoint realizations in $L^2(0, \infty)$. This is in agreement with what we deduced in Example 2.2.15.

Finally, $e^{\pm x} \in L^2(-\pi, \pi)$, so in this space our operator has defect indices $n_\pm = 1$ and self-adjoint realizations are given by a unitary map $J : \mathbf{D}_i \to \mathbf{D}_{-i}$. Now $\mathbf{D}_{\pm i}$ are one-dimensional and the norms of (e^x, ie^x) and $(e^{-x}, -ie^{-x})$ are the same, so such a unitary map is obtained by mapping one pair onto the other multiplied by $e^{i\varphi}$ for some $\varphi \in [0, 2\pi)$. The space D of Theorem 2.4.2 is therefore spanned by $(e^x + e^{i\varphi}e^{-x}, ie^x - ie^{i\varphi}e^{-x})$.

The value of the first component at π is $z = e^\pi + e^{i\varphi}e^{-\pi}$ and at $-\pi$ it is $e^{-\pi} + e^{i\varphi}e^\pi = \bar{z}e^{i\varphi}$. The quotient of the value at π by that at $-\pi$ is therefore $e^{i(2\arg z - \varphi)}$ for any element of the domain of the self-adjoint realization. With $\theta = 2\arg z - \varphi$ we obtain the boundary condition $u(\pi) = e^{i\theta}u(-\pi)$. It is not hard to see that choosing $\arg z \in [0, 2\pi)$, the function $[0, 2\pi) \ni \varphi \mapsto \theta \in [0, 2\pi)$ is strictly increasing and onto, so the self-adjoint realizations given by the abstract theory are in a one-to-one correspondence with the boundary conditions derived in Example 2.2.16.

Corollary 2.4.5. *If S is a symmetric extension of the closed symmetric relation T given as in Theorem 2.4.2 by an isometry J with domain $\mathcal{D}(J) \subset \mathbf{D}_i$ and range $\mathcal{R}_J \subset \mathbf{D}_{-i}$, then the defect spaces for S are $\mathbf{D}_i(S) = \mathbf{D}_i \ominus \mathcal{D}(J)$ and $\mathbf{D}_{-i}(S) = \mathbf{D}_{-i} \ominus \mathcal{R}_J$ respectively.*

Proof. If $D \subset \mathbf{D}_i \oplus \mathbf{D}_{-i}$ and $S = T \oplus D$ is symmetric, then $u \in \mathbf{D}_i(S) \subset \mathbf{D}_i$ precisely if $\langle T \oplus D, \mathcal{U}u \rangle = 0$. But $\langle T, \mathcal{U}u \rangle = 0$ and if $u_+ + u_- \in D$ with $u_+ \in \mathbf{D}_i$, $u_- \in \mathbf{D}_{-i}$ then $\langle u_+ + u_-, u \rangle = \langle u_+, u \rangle$, which shows that $\mathbf{D}_i(S) = \mathbf{D}_i \ominus \mathcal{D}(J)$. Similarly the statement about $\mathbf{D}_{-i}(S)$ follows. $\square$

Corollary 2.4.6. *Every symmetric relation has a maximal symmetric extension. If one of n_+ and n_- is finite, then all or none of the maximal symmetric extensions are self-adjoint depending on whether $n_+ = n_-$ or not. If $\mathcal{H}$ is separable and $n_+ = n_- = \infty$, however, some maximal symmetric extensions are self-adjoint and some are not.*

We now give a generalization of Theorem 2.4.1 which is sometimes useful. To do this, we use the notation of Lemma 2.3.6 and also define

$$\mathbf{D}_\lambda = \{(u, \lambda u) \in T^*\} = \{(u, \lambda u) : u \in D_\lambda\} ,$$
$$\mathbf{E}_\lambda = \{(v, \lambda v + u) \in T^* : u \in D_{\bar{\lambda}}\} .$$

We shall prove the following theorem.

Theorem 2.4.7. *Suppose T is closed and symmetric and that S_λ is closed. Then $T^* = \mathrm{Sp}(T, \mathbf{E}_\lambda)$, and if λ is not an eigenvalue for T, then $T^* = T \dotplus \mathbf{E}_\lambda$ as a topological direct sum, i.e., the projection from T^* onto T along $\mathbf{E}_\lambda$ is bounded.*

In particular, this is the case if $\mathrm{Im}\, \lambda \neq 0$, and then we also have $\mathbf{E}_\lambda = \mathbf{D}_\lambda \dotplus \mathbf{D}_{\bar{\lambda}}$ as a topological direct sum.

Proof. Since T^* is closed it follows that $\mathbf{D}_\lambda$ is closed so also $\mathbf{E}_\lambda$ is closed. Since also T is closed it follows from Theorem 2.2.3 that if we can prove the statements about direct sums algebraically they are also true topologically.

Let $(u, f) \in T^*$. By Lemma 2.3.6 (1) $\mathcal{H}_1 = S_\lambda \oplus D_{\overline{\lambda}}$ if S_λ is closed so we may write $f - \lambda u = u_0 + u_{\overline{\lambda}}$ with $u_{\overline{\lambda}} \in D_{\overline{\lambda}}$ and $u_0 \in S_\lambda$. Thus there exists a $v_0 \in \mathcal{H}_1$ such that $(v_0, \lambda v_0 + u_0) \in T$, so that $(u, f) = (v_0, \lambda v_0 + u_0) + (u - v_0, \lambda(u - v_0) + u_{\overline{\lambda}})$. The first term is in T and the second clearly in $\mathbf{E}_\lambda$.

Now, if $(u, f) \in T \cap \mathbf{E}_\lambda$ we have $f - \lambda u \in S_\lambda \cap D_{\overline{\lambda}} = \{0\}$ so that λ is an eigenvalue of T unless $(u, f) = 0$.

If $\operatorname{Im} \lambda \neq 0$, then λ is not an eigenvalue of the symmetric relation T and S_λ is closed by Lemma 2.3.6 (2), so we have $T^* = T \dotplus \mathbf{E}_\lambda$. Furthermore, $\mathbf{E}_\lambda = \mathbf{D}_\lambda \dotplus \mathbf{D}_{\overline{\lambda}}$ since if $a = \frac{1}{2 \operatorname{Im} \lambda}$ we have $(v, \lambda v + u) = a(u, \overline{\lambda} u) + (v - au, \lambda(v - au))$. The first term is in $\mathbf{D}_{\overline{\lambda}}$ so the second is in T^* and thus in $\mathbf{D}_\lambda$. □

Corollary 2.4.8. *If* $\operatorname{Im} \lambda > 0$ *then* $\dim \mathbf{D}_\lambda = n_+$ *and* $\dim \mathbf{D}_{\overline{\lambda}} = n_-$.

Proof. Suppose (u, f) and (v, g) are in T^*. The *boundary form*

$$\langle (u, f), \mathcal{U}(v, g) \rangle = i(\langle u, g \rangle - \langle f, v \rangle)$$

is a bounded Hermitian form on T^*. It is immediately verified that it is positive definite on $\mathbf{D}_\lambda$, negative definite on $\mathbf{D}_{\overline{\lambda}}$, non-negative on $T \dotplus \mathbf{D}_\lambda$ and non-positive on $T \dotplus \mathbf{D}_{\overline{\lambda}}$.

Let μ be a complex number with $\operatorname{Im} \mu > 0$. We get a linear map $P_{\mu\lambda} : \mathbf{D}_\mu \to \mathbf{D}_\lambda$ in the following way. Given $U \in \mathbf{D}_\mu$ we may write $U = U_0 + U_\lambda + U_{\overline{\lambda}}$ uniquely with $U_0 \in T$, $U_\lambda \in \mathbf{D}_\lambda$ and $U_{\overline{\lambda}} \in \mathbf{D}_{\overline{\lambda}}$ according to Theorem 2.4.7. Let $P_{\mu\lambda} U = U_\lambda$. Then U_λ cannot be 0 unless U is since the boundary form is positive definite on $\mathbf{D}_\mu$ but non-positive on $T \dotplus \mathbf{D}_{\overline{\lambda}}$. If $e_1, \ldots, e_n$ are linearly independent in $\mathbf{D}_\mu$ it follows that $P_{\mu\lambda} e_1, \ldots, P_{\mu\lambda} e_n$ are linearly independent in $\mathbf{D}_\lambda$, so $\dim \mathbf{D}_\mu \leq \dim \mathbf{D}_\lambda$. Interchanging μ and λ gives the opposite inequality, so the dimensions of $\mathbf{D}_\lambda$ and $\mathbf{D}_\mu$ are equal. For $\mu = i$ we obtain $\dim \mathbf{D}_\lambda = n_+$. Similarly one shows that $\dim \mathbf{D}_{\overline{\lambda}} = n_-$. □

Note that even if $n_+ = \infty$ the operator $P_{\mu\lambda} : \mathbf{D}_\mu \to \mathbf{D}_\lambda$ is a bounded linear map with bounded inverse (since it is obviously closed and using the closed graph theorem). Theorem 2.4.7 also has interesting implications for real λ.

Theorem 2.4.9. *Suppose* T *is a closed and symmetric relation and that* $\lambda \in \mathbb{R}$ *with* S_λ *closed. Then* $n_+ = n_- = \dim \mathbf{D}_\lambda \ominus (\mathbf{D}_\lambda \cap T)$, T *has self-adjoint extensions, and we may find such extensions for which any closed subspace of* $\mathbf{D}_\lambda$ *containing* $\mathbf{D}_\lambda \cap T$ *is the eigenspace at* λ.

We shall need the following proposition. The simple proof is left to Exercise 2.4.2.

Proposition 2.4.10. *Suppose* $B(u, v)$ *is a sesquilinear form on a complex linear space. Then the* polarization identity

$$B(u, v) = \frac{1}{4} \sum_{k=0}^{3} i^k B(u + i^k v, u + i^k v)$$

holds.

Proof (Theorem 2.4.9). We must prove equality of the defect indices $n_{\pm}$. Let $\mathbf{D}_\lambda(T) = \mathbf{D}_\lambda \cap T$ and $\tilde{\mathbf{D}}_\lambda = \mathbf{D}_\lambda \ominus \mathbf{D}_\lambda(T)$ so that $\mathbf{D}_\lambda(T) \subset \mathbf{D}_\lambda \subset \mathbf{E}_\lambda$. By Theorems 2.4.1 and 2.4.7 we have

$$\dim(\mathbf{E}_\lambda \ominus \mathbf{D}_\lambda(T)) \geq n_+ + n_- \ .$$

The map $\mathbf{E}_\lambda \ominus \mathbf{D}_\lambda(T) \ni (v, \lambda v + u) \mapsto (u, \lambda u) \in \mathbf{D}_\lambda$ has nullspace and range contained in $\tilde{\mathbf{D}}_\lambda$ since $\langle (w, \lambda w), \mathcal{U}(v, \lambda v + u) \rangle = i\langle w, u \rangle$, so if $(w, \lambda w) \in T$, then $(w, \lambda w)$ and $(u, \lambda u)$ are orthogonal. Thus

$$\dim \mathbf{E}_\lambda \ominus \mathbf{D}_\lambda(T) \leq 2 \dim \tilde{\mathbf{D}}_\lambda \ .$$

It is easily verified that $\langle \tilde{\mathbf{D}}_\lambda, \mathcal{U}\tilde{\mathbf{D}}_\lambda \rangle = 0$ so the relation $T \dotplus \tilde{\mathbf{D}}_\lambda$ is symmetric. Thus $\dim \tilde{\mathbf{D}}_\lambda \leq \min(n_+, n_-)$ and it follows that $n_+ + n_- \leq 2 \dim \tilde{\mathbf{D}}_\lambda \leq 2 \min(n_+, n_-)$, so that $n_+ = n_- = \dim \tilde{\mathbf{D}}_\lambda$.

Let D_1 be a closed subspace of $\tilde{\mathbf{D}}_\lambda$ so that $\langle D_1, \mathcal{U}D_1 \rangle = 0$. Furthermore, set $D_2 = \{(v, \lambda v + u) \in \mathbf{E}_\lambda \ominus \mathbf{D}_\lambda : \langle u, w \rangle = 0 \text{ for } (w, \lambda w) \in D_1\}$. For $(v, \lambda v + u) \in D_2$ we obtain $0 = \langle (v, \lambda v + u), (u, \lambda u) \rangle = (1 + \lambda^2)\langle v, u \rangle + \lambda \|u\|^2$ so that $\langle v, u \rangle$ is real.

A previous calculation shows that if $(w, \lambda w) \in D_1$ and $(v, \lambda v + u) \in D_2$, then $\langle (w, \lambda w), \mathcal{U}(v, \lambda v + u) \rangle = i\langle w, u \rangle = 0$ so $\langle D_1, \mathcal{U}D_2 \rangle = 0$. Since $\langle v, u \rangle$ is real we also have $\langle (v, \lambda v + u), \mathcal{U}(v, \lambda v + u) \rangle = i(\langle v, u \rangle - \langle u, v \rangle) = 0$. By the polarization identity (Proposition 2.4.10) it follows that $\langle D_2, \mathcal{U}D_2 \rangle = 0$ so that $T \dotplus (D_1 \oplus D_2)$ is symmetric. If $D = D_1 \oplus D_2$, then for finite $\dim D = n_{\pm}$ this shows that $T \dotplus D$ is self-adjoint. For infinite $n_{\pm}$ an additional argument is needed to show this; see Exercise 2.4.3.   □

An important case to which Theorem 2.4.9 applies is when T is bounded from below. Recall that this means that there is a constant a such that $\langle u, f \rangle \geq a\langle u, u \rangle$ for all $(u, f) \in T$. It is obvious that the closure of a symmetric relation which is bounded from below has an unchanged lower bound, so one may as well assume it closed to begin with.

Corollary 2.4.11. *If T is closed, symmetric and bounded from below by a, then S_λ is closed for every $\lambda < a$, and for any such λ there is a self-adjoint extension of T which is bounded from below by λ.*

Proof. We may write any $(u, f) \in T$ as $(u, \lambda u + v)$, and if T is bounded from below by a this shows that $(a - \lambda)\langle u, u \rangle \leq \langle u, v \rangle \leq \|u\|\|v\|$. If $\lambda < a$ this shows that $\|u\| \leq (a - \lambda)^{-1}\|v\|$ so the map $S_\lambda \ni v \mapsto (u, \lambda u + v) \in T$ is bounded and obviously closed. Thus S_λ is closed. The rest of the proof is left for Exercise 2.4.4.   □

There is an important strengthening of this corollary, given in Corollary 3.4.5.

Exercises

Exercise 2.4.1 Fill in all missing details in the proofs of Theorem 2.4.2 and Corollaries 2.4.3–2.4.6.

Exercise 2.4.2 Prove Proposition 2.4.10. Is there a corresponding formula for bilinear forms on a real linear space? For a symmetric bilinear form?

Exercise 2.4.3 Finish the proof of Theorem 2.4.9. Are there other self-adjoint extensions than the one given which will do?

Exercise 2.4.4 Finish the proof of Corollary 2.4.11. This is a theorem of von Neumann, who also conjectured that there is actually a self-adjoint extension with the *same* lower bound as T. For this, see Corollary 3.4.5.

2.5 Notes and remarks

The basic idea of an infinite-dimensional Euclidean space is due to Hilbert [101] and the notion of an abstract Hilbert space was introduced by von Neumann [171] and fully developed in [172].

Linear relations appear to have been first discussed by Arens [5]. They were used by Orcutt [176] to investigate the spectral theory of systems of ordinary differential equations, although his contributions seems to have been largely overlooked, certainly by the present authors. Our presentation is based on Bennewitz [14] and may also be found in Bennewitz [15, Section 1]. The main purpose of defining relations is to simplify the handling of adjoints. In the literature one sometimes defines composition of relations by

$$ST = \{(u, f) : (u, v) \in T \text{ and } (v, f) \in S \text{ for some } v \in \mathcal{R}_T \cap \mathcal{D}(S)\}.$$

Since neither $T^{-1}T$ nor TT^{-1} is in general an identity it is clear that this notion has very limited use. For example, its use in connection with the Cayley transform discussed in Exercise 2.3.2 would lead to absurd consequences.

The operator version of Theorem 2.4.1 is generally attributed to von Neumann, but appears not to have been explicitly stated by him, whereas Theorem 2.4.2 and its corollaries were all proved in his paper [172].

A complete exposition of the theory of Hilbert spaces may be found in Akhiezer and Glazman [3], Weidmann [216], Lax [140] and many other books.

Chapter 3
Abstract spectral theory

3.1 The spectral theorem

In Section 1.4 we found an orthonormal basis of eigenfunctions for some Sturm–Liouville boundary value problems, using $\langle u, v \rangle = \int_a^b u\bar{v}w$ as the scalar product. Define the operator T in the corresponding Hilbert space $L_w^2(a, b)$ by setting $Tu = f$ if u satisfies the boundary conditions and the equation $-u'' + qu = wf$. Let H be as in Section 1.4 so that integration by parts shows that $\langle Tu, v \rangle = H(u, v)$ if u and v are both in $\mathcal{D}(T)$, so that $\langle Tu, v \rangle = \langle u, Tv \rangle$ since H is Hermitian. Thus T is symmetric, and it is not hard to see that it is in fact self-adjoint (Exercise 3.1.1).

One might perhaps hope that all self-adjoint operators (in a separable Hilbert space) have an orthonormal basis of eigenvectors, but this is not the case. To see this we may look at Exercise 2.2.13 or, if we prefer dealing with differential operators, Example 2.2.14, where the operator is $i\frac{d}{dx}$ in the Hilbert space $L^2(\mathbb{R})$. An eigenfunction for an eigenvalue λ satisfies $iu' = \lambda u$ so it is a multiple of $e^{-i\lambda x}$, but there are no values of λ for which this is in $L^2(\mathbb{R})$. The operator is self-adjoint, but does not have a single eigenvalue!

Nevertheless, there exists something very similar to an eigenfunction expansion for this equation. A clue to what it looks like is obtained by looking at the same equation on a bounded but very long interval $(-a, a)$. Like in Example 2.2.16 one then needs a boundary condition to get a self-adjoint operator T, and we shall consider the periodic condition $u(-a) = u(a)$ though any other of the conditions found in Example 2.2.16 would do as well.

For λ to be an eigenvalue we must then have $e^{i\lambda a} = e^{-i\lambda a}$ or $e^{2i\lambda a} = 1$ so that $\lambda = k\pi/a$, where k is an integer. Since $e^{-i\lambda x}$ is in $L^2(-a, a)$ these are genuine eigenfunctions which multiplied by $(2a)^{-1/2}$ yield an orthonormal basis[1] when k runs through $\mathbb{Z}$ using the scalar product $\langle u, v \rangle = \int_{-a}^a u\bar{v}$. As a increases the distance between eigenvalues decreases, and as $a \to \infty$ they accumulate everywhere on $\mathbb{R}$.

---

[1] By the theory of Section 1.4 since they are also the eigenfunctions for periodic boundary conditions of $-u'' = \lambda^2 u$.

© Springer Nature Switzerland AG 2020

C. Bennewitz et al., *Spectral and Scattering Theory for Ordinary Differential Equations*,
Universitext, https://doi.org/10.1007/978-3-030-59088-8_3

Now consider a bounded interval Δ and the orthogonal projection E_Δ onto the space spanned by all eigenfunctions with eigenvalues in Δ. If there is just one eigenvalue $\lambda_j \in \Delta$ with eigenvector e_j we have $E_\Delta u = \hat{u}_j e_j$, where $\hat{u}_j = \langle u, e_j \rangle$. Note that $E_\Delta u \in \mathcal{D}(T)$ and $TE_\Delta u = \lambda_j \hat{u}_j e_j$.

In general we have $E_\Delta u = \sum_{\lambda_j \in \Delta} \hat{u}_j e_j$ and $TE_\Delta u = \sum_{\lambda_j \in \Delta} \lambda_j \hat{u}_j e_j$. It is clear that the range of E_Δ increases with Δ and that $\langle E_\Delta u, E_\Delta u \rangle = \sum_{\lambda_j \in \Delta} |\hat{u}_j|^2$ and $\langle TE_\Delta u, E_\Delta u \rangle = \sum_{\lambda_j \in \Delta} \lambda_j |\hat{u}_j|^2$. It follows that if $\Delta = (\alpha, \beta)$ we have $\alpha \langle E_\Delta u, E_\Delta u \rangle < \langle TE_\Delta u, E_\Delta u \rangle < \beta \langle E_\Delta u, E_\Delta u \rangle$ or

$$\frac{\langle TE_\Delta u, E_\Delta u \rangle}{\langle E_\Delta u, E_\Delta u \rangle} \in \Delta . \tag{3.1.1}$$

Since the eigenfunctions yield an orthonormal basis it follows that E_Δ converges to the identity as Δ increases to $\mathbb{R}$. These facts are independent of a. In fact, it may be shown that as $a \to \infty$ the projection E_Δ will have a limit (Exercise 3.1.2), even though the number of eigenvalues in Δ tends to infinity with a, and the range of the limiting projection will have infinite dimension. The limiting projections still increase with Δ and converge to the identity as $\Delta \to \mathbb{R}$. Also (3.1.1) remains true in the limit.

So, if we look at projections associated with *intervals* instead of projections associated with individual *eigenvalues*, something important survives the loss of the eigenvalues. These facts give a reasonably satisfactory generalization of the spectral theorem for symmetric $n \times n$ matrices to the case of a self-adjoint operator in Hilbert space.

An even better way of expressing this is the following. Let E_t be the orthogonal projection associated with $(-\infty, t)$, so that we have an increasing family of projections $\{E_t\}_{t \in \mathbb{R}}$. This means that the function $t \mapsto \langle E_t u, u \rangle$ is increasing for every $u \in \mathcal{H}$.

For finite a it is clear that E_t tends to zero as $t \to -\infty$ and to the identity as $t \to \infty$, since the eigenfunctions yield an orthonormal basis in $L^2(-a, a)$. One may hope that these facts survive when $a \to \infty$. Thus, if we can make sense of the integrals, $\int_{-\infty}^{\infty} dE_t$ should equal the identity and $T = \int_{-\infty}^{\infty} t \, dE_t$. This is the general spectral theorem.

It follows that to understand the statement of the spectral theorem one must have some familiarity with Stieltjes integrals, for which we refer to the first two sections of Appendix B.

Theorem 3.1.1 (Spectral theorem). *Suppose T is a densely defined, self-adjoint operator in a Hilbert space $\mathcal{H}$. Then there exists a unique, increasing and left-continuous family $\{E_t\}_{t \in \mathbb{R}}$ of orthogonal projections with the following properties:*

(1) *E_t tends strongly to zero as $t \to -\infty$ and to the identity on $\mathcal{H}$ as $t \to \infty$.*

(2) *$T = \int_{-\infty}^{\infty} t \, dE_t$ in the following sense: $u \in \mathcal{D}(T)$ if and only if $\int_{-\infty}^{\infty} t^2 \, d\langle E_t u, u \rangle < \infty$, and then $\langle Tu, v \rangle = \int_{-\infty}^{\infty} t \, d\langle E_t u, v \rangle$ and $\|Tu\|^2 = \int_{-\infty}^{\infty} t^2 \, d\langle E_t u, u \rangle$, the integrals converging absolutely.*

(3) *E_t commutes with T, in the sense that TE_t is the closure of $E_t T$.*

A family $\{E_t\}_{t\in\mathbb{R}}$ of orthogonal projections is increasing if and only if $E_t E_s = E_{\min(t,s)}$ for all $t, s \in \mathbb{R}$, see Exercise 3.1.3. The family of Theorem 3.1.1 is called the *resolution of the identity* for T, and the operator E_t the *spectral projector* for the interval $(-\infty, t)$. We note that spectral projectors for more general sets are defined in Example 3.1.3.

If T is a self-adjoint relation we may apply the spectral theorem to the restriction of the relation T to the space $\mathcal{H}$ of Theorem 2.2.9, and then extend the definition of the projections E_t to all of $\mathcal{H}_1$ by linearity and setting $E_t \mathcal{H}_\infty = 0$. We define E_∞ as the identity on $\mathcal{H}_1$ so that $E_\infty - E_t$ tends to the orthogonal projection onto $\mathcal{H}_\infty$ as $t \to \infty$, Thus we may consider $\{E_t\}_{t\in\mathbb{R}\cup\{\infty\}}$ as a resolution of the identity for T.

An important consequence of the spectral theorem is that it leads to a well rounded *functional calculus* for self-adjoint operators. However, in this book the functional calculus is only used to define the square root of a positive operator, in Theorem 3.4.4, so the reader may omit Theorem 3.1.2 at a first reading.

In Theorem 3.1.2 we shall use $B(\sigma)$ as a notation for the Borel functions defined on a closed set σ, but the reader unfamiliar with the notion of a Borel function (Definition B.6.5) may interpret $B(\sigma)$ as the functions which are pointwise limits of a sequence of functions continuous on σ. This interpretation will be enough for most purposes. In many cases it would even be enough to think of the functions in $B(\sigma)$ as piecewise continuous on σ.

Given a self-adjoint operator T and $f \in B(\sigma(T))$ we shall define a closed operator $f(T)$ with dense domain by the symbolic formula $f(T) = \int_{-\infty}^{\infty} f(t)\, dE_t$ and give a set of calculation rules for such operators.

Theorem 3.1.2 (Functional calculus). *Suppose T is self-adjoint with resolution of the identity $\{E_t\}_{t\in\mathbb{R}}$ and $f \in B(\sigma(T))$. Define $\mathcal{D}(f(T))$ as the set of $u \in \mathcal{H}$ for which $\int_{-\infty}^{\infty} |f(t)|^2\, d\langle E_t u, u\rangle < \infty$. Then $\mathcal{D}(f(T))$ is dense in $\mathcal{H}$ and*

$$\langle f(T)u, v\rangle = \int_{-\infty}^{\infty} f(t)\, d\langle E_t u, v\rangle$$

defines a closed operator $f(T)$ with domain $\mathcal{D}(f(T))$. The integral is absolutely convergent if $u \in \mathcal{D}(f(T))$ and $v \in \mathcal{H}$. If C is constant and $g \in B(\sigma(T))$ we also have

(1) *$(Cf)(T) = C\, f(T)$ (for $C \neq 0$ if f is unbounded),*

(2) *$\mathcal{D}(f(T)^*) = \mathcal{D}(f(T))$ and $f(T)^* = \overline{f}(T)$,*

(3) *$f(T) + g(T) \subset (f+g)(T)$ with equality if f or g is bounded,*

(4) *$f(T)g(T)$ and $g(T)f(T)$ are both cores of $fg(T)$,*

(5) *$f(T)$ is unitary if $|f| = 1$, self-adjoint if f is real-valued and bounded or semi-bounded if f (restricted to $\sigma(T)$) is, with the same bound.*

Often some version of the *spectral mapping theorem* is considered part of the functional calculus. We have left this to Exercise 3.1.7. Note that $f(T)g(T)$ is closed if g is bounded (Exercise 3.1.5), but may not be otherwise. Clearly $f(T)$ and $g(T)$ commute in the sense that $f(T)g(T)$ and $g(T)f(T)$ have the same closure $fg(T)$.

Example 3.1.3. A particular case of a function of a self-adjoint operator is obtained for $f(t) = 1/(t - \lambda)$ which is bounded and continuous on $\sigma(T)$ if $\lambda \in \rho(T)$ and in this case $f(T)$ is the resolvent of T at λ.

If $M \subset \mathbb{R}$ with characteristic function χ_M it is clear that $\chi_M^2 = \chi_M$ so if $\chi_M \in B(\sigma(T))$ it follows that $E_M = \chi_M(T)$ is a self-adjoint, and thus orthogonal, projection. In particular $E_t = \chi_{(-\infty, t]}(T)$, so if one can create a functional calculus for self-adjoint operators without using the spectral theorem one will obtain the spectral theorem as a consequence, and particularly for bounded operators this is a common method of proof.

The functional calculus also shows that $E_{M_1} E_{M_2} = E_{M_1 \cap M_2}$. In particular, we have $\langle E_{M_1} u, E_{M_2} v \rangle = \langle E_{M_2} E_{M_1} u, v \rangle = 0$ for all $u, v \in \mathcal{H}$ provided $M_1 \cap M_2$ does not intersect $\sigma(T)$, so in this case the ranges of E_{M_2} and E_{M_1} are orthogonal. Note that if T is a self-adjoint relation, then the range of $E_{\mathbb{R}}$ is the orthogonal complement of $\mathcal{H}_\infty$.

To prove the spectral theorem we need three lemmas, the first of which contains the main step in the proof.

Lemma 3.1.4. *For $u, v \in \mathcal{H}$ there is a unique left-continuous function $\eta_{u,v}$ of bounded variation on $\mathbb{R}$, tending to zero at $-\infty$, and*

(1) *$\eta_{u,v}$ is pointwise a Hermitian sesquilinear form in u, v.*

(2) *$\eta_{u,u}$ is non-decreasing.*

(3) *$\langle R_\lambda u, v \rangle = \int_{-\infty}^{\infty} \frac{d\eta_{u,v}(t)}{t - \lambda}$.*

(4) *$\int_{-\infty}^{\infty} h \, d\eta_{u,v}$ is a bounded Hermitian form of $u, v \in \mathcal{H}$ for every bounded Borel function h and the total variation measure of $d\eta_{u,v}$ satisfies*

$$\int\limits_{-\infty}^{\infty} |h| \, |d\eta_{u,v}| \leq \left(\int\limits_{-\infty}^{\infty} |h| \, d\eta_{u,u} \int\limits_{-\infty}^{\infty} |h| \, d\eta_{v,v} \right)^{1/2} \leq \|h\|_\infty \|u\| \|v\|.$$

Proof. The uniqueness of $\eta_{u,v}$ follows from (3) and the Stieltjes inversion formula (Lemma E.1.5), applied to $F(\lambda) = \langle R_\lambda u, v \rangle$. Since $\langle R_\lambda u, v \rangle$ is sesquilinear in u, v and $R_\lambda^* = R_{\bar{\lambda}}$, it then follows that $\eta_{u,v}$ is pointwise Hermitian and sesquilinear if it exists.

To show existence, note that by Theorem 2.3.7 the function $\lambda \mapsto \langle R_\lambda u, u \rangle$ is a Nevanlinna function of λ for any $u \in \mathcal{H}$, so Theorem E.1.1 shows that

$$\langle R_\lambda u, u \rangle = A + B\lambda + \int\limits_{-\infty}^{\infty} \left(\frac{1}{t - \lambda} - \frac{t}{1 + t^2} \right) d\eta_{u,u}(t), \tag{3.1.2}$$

where $\eta_{u,u}$ is non-decreasing and $A \in \mathbb{R}$, $B \geq 0$. Since $\|R_\lambda\| \leq \frac{1}{|\operatorname{Im} \lambda|}$, we find that $\|u\|^2$ is an upper bound for $v \langle R_{iv} u, u \rangle$ when $v > 0$, the imaginary part of which is

$Bv^2 + \int_{-\infty}^{\infty} \frac{v^2 \, d\eta_{u,u}(t)}{t^2 + v^2}$. Hence $B = 0$, and letting $v \to \infty$ Fatou's lemma (or monotone convergence) shows that $\int_{-\infty}^{\infty} d\eta_{u,u} \le \|u\|^2$.

We may now assume $\eta_{u,u}$ normalized so as to be left-continuous and $\eta_{u,u}(-\infty) = 0$. Clearly $\int_{-\infty}^{\infty} \frac{t}{1+t^2} \, d\eta_{u,u}(t)$ is absolutely convergent since $|t|/(1+t^2) \le 1/2$, so this part of the integral in (3.1.2) may be incorporated in the constant A. So, with absolute convergence, we have $\langle R_\lambda u, u \rangle = A' + \int_{-\infty}^{\infty} \frac{d\eta_{u,u}(t)}{t-\lambda}$ for some real A'. However, as $\lambda \to \infty$ along the imaginary axis, both the left-hand side and the integral tend to zero (Exercise 3.1.4), so $A' = 0$. This finishes the proof in the case $u = v$.

By the polarization identity (Proposition 2.4.10)

$$\langle R_\lambda u, v \rangle = \tfrac{1}{4} \sum_{k=0}^{3} i^k \langle R_\lambda(u + i^k v), u + i^k v \rangle,$$

so we obtain $\langle R_\lambda u, v \rangle = \int_{-\infty}^{\infty} \frac{d\eta_{u,v}(t)}{t-\lambda}$ by setting

$$\eta_{u,v} = \tfrac{1}{4} \sum_{k=0}^{3} i^k \eta_{u+i^k v, u+i^k v}.$$

This function $\eta_{u,v}$ is of bounded variation with the correct normalization, so only the bound on the total variation remains to be proved. But if Δ is an interval, then $\int_{\Delta} |h| \, d\eta_{u,v}$ is a semi-scalar product on $\mathcal{H}$, so the Cauchy–Schwarz inequality $\left| \int_{\Delta} |h| \, d\eta_{u,v} \right|^2 \le \int_{\Delta} |h| \, d\eta_{u,u} \int_{\Delta} |h| \, d\eta_{v,v}$ is valid. If $\{\Delta_j\}_1^n$ is a partition of $\mathbb{R}$ into disjoint intervals we obtain

$$\sum_j \left| \int_{\Delta_j} |h| \, d\eta_{u,v} \right| \le \sum_j \left(\int_{\Delta_j} |h| \, d\eta_{u,u} \int_{\Delta_j} |h| \, d\eta_{v,v} \right)^{\frac{1}{2}}$$

$$\le \left(\int_{-\infty}^{\infty} |h| \, d\eta_{u,u} \right)^{1/2} \left(\int_{-\infty}^{\infty} |h| \, d\eta_{v,v} \right)^{1/2} \le \|h\|_\infty \|u\| \|v\|,$$

where the second inequality is the Cauchy–Schwarz inequality in $\mathbb{R}^n$. Taking the least upper bound over all partitions of $\mathbb{R}$ gives the bound on the total variation, and since $\left| \int_{-\infty}^{\infty} h \, d\eta_{u,v} \right| \le \int_{-\infty}^{\infty} |h| \, |d\eta_{u,v}|$ this finishes the proof. □

Lemma 3.1.5. $\int_{-\infty}^{\infty} d\eta_{u,v} = \langle u, v \rangle$ for any $u, v \in \mathcal{H}$.

Proof. First assume that $u \in \mathcal{D}(T)$ and $v > 0$. We then have $u = R_{iv}(f - ivu)$, where $f = Tu$. Thus $\langle u, v \rangle = -iv \langle R_{iv}u, v \rangle + \langle R_{iv}f, v \rangle$. Now $-iv \int_{-\infty}^{\infty} \frac{d\eta_{u,v}(t)}{t-iv} \to \int_{-\infty}^{\infty} d\eta_{u,v}$ as $v \to \infty$ by bounded convergence (Exercise 3.1.4). In particular $\langle R_{iv}f, v \rangle \to 0$, so the lemma is true for $u \in \mathcal{D}(T)$, which is dense in $\mathcal{H}$.

However, $\int_{-\infty}^{\infty} d\eta_{u,v}$ is a bounded Hermitian form on $\mathcal{H}$ since by Lemma 3.1.4 $\left| \int_{-\infty}^{\infty} d\eta_{u,v} \right| \le \int_{-\infty}^{\infty} |d\eta_{u,v}| \le \|u\| \|v\|$, so the general case follows by continuity. □

Lemma 3.1.6. *The integral in the formula* $\langle f(T)u, v \rangle = \int_{-\infty}^{\infty} f \, d\eta_{u,v}$ *converges absolutely for every* $u \in \mathcal{H}$ *such that* $\int_{-\infty}^{\infty} |f|^2 \, d\eta_{u,u} < \infty$ *and* $v \in \mathcal{H}$ *if* $f \in B(\sigma(T))$ *and defines the operator* $f(T)$ *uniquely.*

Proof. Let $f \in B(\sigma(T))$ and define mutually disjoint Borel sets

$$M_k = \{t \in \sigma(T) : 2^{k-1} \leq |f(t)| < 2^k\}, \ k \in \mathbb{Z}.$$

By Lemma 3.1.4 (4) with h the characteristic function of M_k

$$\int_{M_k} |f| \, |d\eta_{u,v}| \leq 2^k \int_{M_k} |d\eta_{u,v}| \leq 2\left(2^{2(k-1)} \int_{M_k} d\eta_{u,u} \int_{M_k} d\eta_{v,v}\right)^{1/2}$$

$$\leq 2\left(\int_{M_k} |f|^2 \, d\eta_{u,u} \int_{M_k} d\eta_{v,v}\right)^{1/2}.$$

Summing over k, using monotone convergence and Cauchy–Schwarz for ℓ^2, we obtain

$$\int_{-\infty}^{\infty} |f| \, |d\eta_{u,v}| \leq 2\left(\int_{-\infty}^{\infty} |f|^2 \, d\eta_{u,u}\right)^{1/2} \|v\|.$$

Thus $\mathcal{H} \ni v \mapsto \int_{-\infty}^{\infty} f \, d\eta_{u,v}$ is a bounded anti-linear form for any $u \in \mathcal{H}$ for which $\int_{-\infty}^{\infty} |f|^2 \, d\eta_{u,u} < \infty$, so by Riesz' representation theorem there is a unique operator $f(T)$ defined for such u so that

$$\langle f(T)u, v \rangle = \int_{-\infty}^{\infty} f(t) \, d\eta_{u,v}. \qquad \square$$

Proof (Theorem 3.1.1). We first show the uniqueness of the resolution of the identity. Assume there exists a resolution of the identity with the claimed properties. Then $E_t E_s = E_{\min(s,t)}$, so for $w \in \mathcal{D}(T)$ and s fixed we obtain

$$\langle E_s T w, v \rangle = \langle T w, E_s v \rangle = \int_{-\infty}^{\infty} t \, d_t \langle E_t w, E_s v \rangle = \int_{(-\infty,s]} t \, d\langle E_t w, v \rangle,$$

so the measures $d\langle E_t T w, v \rangle$ and $t \, d\langle E_t w, v \rangle$ are equal (Theorem B.2.5). Now replace w by $R_\lambda u$. We then get

$$\int_{-\infty}^{\infty} \frac{d\langle E_t u, v \rangle}{t - \lambda} = \int_{-\infty}^{\infty} \frac{d\langle E_t (T - \lambda) R_\lambda u, v \rangle}{t - \lambda} = \int_{-\infty}^{\infty} d\langle E_t R_\lambda u, v \rangle = \langle R_\lambda u, v \rangle.$$

Applying the Stieltjes inversion formula (Lemma E.1.5) to this the uniqueness of the spectral projectors follows.

To prove the existence note that if h is the characteristic function of $(-\infty, t]$ Lemma 3.1.4 (4) shows that the linear form $u \mapsto \eta_{u,v}(t)$ is bounded by $\|v\|$ for each $v \in \mathcal{H}$. Riesz' representation theorem shows that this form equals $\langle u, v_t \rangle$ for a unique $v_t \in \mathcal{H}$ with $\|v_t\| \leq \|v\|$. It is obvious that v_t depends linearly on v so $v_t = E_t v$ where E_t is a linear operator with norm ≤ 1. E_t is self-adjoint since $\eta_{u,v}(t)$ is Hermitian. Furthermore, by the normalization of $\eta_{u,v}$, $E_t u \rightharpoonup 0$ weakly as $t \to -\infty$ and $E_t u \rightharpoonup u$ weakly as $t \to \infty$, using Lemma 3.1.5.

Suppose we knew that E_t is a projection. Since E_t is self-adjoint it is then an orthogonal projection, so $\|E_t u\|^2 = \langle E_t u, E_t u \rangle = \langle E_t^2 u, u \rangle = \langle E_t u, u \rangle \to 0$ as $t \to -\infty$ and similarly $\|u - E_t u\|^2 = \langle u - E_t u, u \rangle \to 0$ as $t \to \infty$. Hence, to show (1) it only remains to show that E_t is a projection increasing with t.

The resolvent relation $R_\lambda - R_\mu = (\lambda - \mu) R_\lambda R_\mu$ shows that

$$\int_{-\infty}^{\infty} \frac{1}{t-\lambda} \frac{d\langle E_t u, v \rangle}{t-\mu} = \frac{1}{\lambda - \mu} \left(\int_{-\infty}^{\infty} \frac{d\langle E_t u, v \rangle}{t-\lambda} - \int_{-\infty}^{\infty} \frac{d\langle E_t u, v \rangle}{t-\mu} \right)$$

$$= \frac{\langle R_\lambda u, v \rangle - \langle R_\mu u, v \rangle}{\lambda - \mu} = \langle R_\lambda R_\mu u, v \rangle = \int_{-\infty}^{\infty} \frac{d\langle E_t R_\mu u, v \rangle}{t-\lambda},$$

so the Stieltjes inversion formula (Lemma E.1.5) shows that

$$\int_{(-\infty, t]} \frac{d\langle E_s u, v \rangle}{s-\mu} = \langle E_t R_\mu u, v \rangle = \langle R_\mu u, E_t v \rangle = \int_{-\infty}^{\infty} \frac{d_s \langle E_s u, E_t v \rangle}{s-\mu}.$$

So, again by uniqueness, $\langle E_t E_s u, v \rangle = \langle E_s u, E_t v \rangle = \langle E_{\min(s,t)} u, v \rangle$. For $s = t$ this shows that E_t is a projection, and if $t > s$ we get $0 \leq (E_t - E_s)^*(E_t - E_s) = (E_t - E_s)^2 = E_t - E_s$ so that $\{E_t\}_{t \in \mathbb{R}}$ is an increasing family of orthogonal projections.

Now suppose $u \in \mathcal{D}(T)$. For any non-real λ we have $u = R_\lambda(Tu - \lambda u)$ or $R_\lambda T u = u + \lambda R_\lambda u$. Since $1 + \lambda/(t-\lambda) = t/(t-\lambda)$ we obtain

$$\int_{-\infty}^{\infty} \frac{d\langle E_t T u, v \rangle}{t-\lambda} = \int_{-\infty}^{\infty} \frac{t \, d\langle E_t u, v \rangle}{t-\lambda}$$

so that $\langle Tu, E_t v \rangle = \langle E_t T u, v \rangle = \int_{(-\infty, t]} s \, d\langle E_s u, v \rangle$. In particular, this shows that $\langle Tu, v \rangle = \int_{-\infty}^{\infty} t \, d\langle E_t u, v \rangle$ so that $\|Tu\|^2 = \int_{-\infty}^{\infty} t \, d\langle E_t u, Tu \rangle = \int_{-\infty}^{\infty} t^2 \, d\langle E_t u, u \rangle$. Thus this integral is finite for every $u \in \mathcal{D}(T)$.

We must also show that if the integral is finite, then $u \in \mathcal{D}(T)$. But Lemma 3.1.6 with $f(t) = t$ gives an operator S defined for $u \in \mathcal{H}$ for which $\int_{-\infty}^{\infty} t^2 \, d\eta_{u,u}(t) < \infty$, so S is an extension of T. It is clear that S is symmetric, so since T is self-adjoint, and thus maximal symmetric, we obtain $S = T$.

Finally, we must show that $T E_t$ is the closure of $E_t T$. But from (2) follows that if $u \in \mathcal{D}(T)$ then $E_t u \in \mathcal{D}(T)$. For $v \in \mathcal{H}$ we have $\langle T E_t u, v \rangle = \int_{-\infty}^{\infty} s \, d_s \langle E_s u, E_t v \rangle =$

$\langle Tu, E_t v \rangle = \langle E_t Tu, v \rangle$ so TE_t is an extension of $E_t T$. Since E_t is bounded and T is closed it follows that TE_t is closed (Exercise 3.1.5).

Now suppose $E_t u \in \mathcal{D}(T)$. We must find $u_j \in \mathcal{D}(T)$ such that $u_j \to u$ and $E_t Tu_j \to TE_t u$. Since $\mathcal{D}(T)$ is dense in $\mathcal{H}$ we can find $v_j \in \mathcal{D}(T)$ so that $v_j \to u$. Now set $u_j = v_j - E_t v_j + E_t u$. Clearly $u_j \in \mathcal{D}(T)$, $u_j \to u$ and $E_t Tu_j = TE_t u_j = TE_t u$ and the proof is complete. $\square$

We now turn to the functional calculus.

Proof (Theorem 3.1.2). The definition of $f(T)$ with domain $\mathcal{D}(f(T))$ was given in Lemma 3.1.6. Consider first the case when f and g are bounded. In this case all involved operators are everywhere defined, closed and bounded, and (1) and (3) are immediate consequences of the definition. Furthermore $f(T)^* = \overline{f}(T)$ since

$$\langle \overline{f}(T)u, v \rangle = \int_{-\infty}^{\infty} \overline{f} \, d\eta_{u,v} = \overline{\int_{-\infty}^{\infty} f \, d\eta_{v,u}} = \langle u, f(T)v \rangle . \tag{3.1.3}$$

Moreover, we have $\langle f(T)g(T)u, v \rangle = \int_{-\infty}^{\infty} f(t) \, d\langle E_t g(T)u, v \rangle$ and $\langle g(T)u, E_t v \rangle = \int_{-\infty}^{\infty} g(s) \, d\langle E_s u, E_t v \rangle = \int_{(-\infty,t]} g(s) \, d\langle E_s u, v \rangle$ so by Corollary B.8.1

$$\langle f(T)g(T)u, v \rangle = \int_{-\infty}^{\infty} fg(t) \, d\langle E_t u, v \rangle = \langle fg(T)u, v \rangle , \tag{3.1.4}$$

and similarly $g(T)f(T) = fg(T)$.

If $g = \overline{f}$ we obtain $f(T)f(T)^* = f(T)^* f(T) = |f|^2 (T)$ so $f(T)$ is unitary precisely if[2] $|f| = 1$ on $\sigma(T)$. If instead f is the characteristic function of a Borel set, so that $f^2 = f$, then f is real-valued so $f(T)$ is self-adjoint, and $f(T)^2 = f(T)$ so $f(T)$ is an orthogonal projection. Because of (3.1.4) all such projections commute.

If f is un-bounded let χ_j be the characteristic function of $\{t \in \sigma(T) : |f(t)| \le j\}$ and $P_j = \chi_j(T)$ so that P_j is an orthogonal projection. It follows that $P_j u \in \mathcal{D}(f(T))$ for all j, for

$$\|f(T)P_j u\|^2 = \int_{-\infty}^{\infty} |f(t)|^2 \, d\langle E_t P_j u, P_j u \rangle = \int_{-\infty}^{\infty} \chi_j(t) |f(t)|^2 \, d\langle E_t u, u \rangle \le j^2 \|u\|^2 ,$$

since $\langle E_t P_j u, P_j u \rangle = \langle P_j E_t u, P_j u \rangle = \langle E_t u, P_j u \rangle$. Now P_j tends strongly to the identity as $j \to \infty$, so $\mathcal{D}(f(T))$ is dense.

We still have (1), at least if $C \ne 0$, and if $v \in \mathcal{D}(f(T))$ and $u \in \mathcal{D}(\overline{f}(T))$ the calculation (3.1.3) shows that $\overline{f}(T) \subset f(T)^*$. Similarly, if $u \in \mathcal{D}(g(T))$ and $g(T)u \in \mathcal{D}(f(T))$, then (3.1.4) is still true. Setting $v = f(T)g(T)u$ we obtain $\|f(T)g(T)u\|^2 = \int_{-\infty}^{\infty} |fg(t)|^2 \, d\langle E_t u, u \rangle$ so that $f(T)g(T) \subset fg(T)$.

---

[2] More exactly, if $|f| = 1$ a.e. with respect to every measure $d\eta_{u,u}$.

To prove (2) we must show that $\int_{-\infty}^{\infty} |f|^2 \, d\eta_{u,u} < \infty$ if $u \in \mathcal{D}(f(T)^*)$, so assume $u \in \mathcal{D}(f(T)^*)$ and $v \in \mathcal{H}$. Then $P_j v \in \mathcal{D}(f(T))$ and

$$\langle P_j f(T)^* u, v \rangle = \langle u, f(T) P_j v \rangle = \overline{\langle f(T) P_j v, u \rangle} = \int_{-\infty}^{\infty} \chi_j \overline{f(t)} \, d\langle E_t u, v \rangle.$$

Now, if $v = P_j f(T)^* u$ this gives $\|P_j f(T)^* u\|^2 = \int_{-\infty}^{\infty} |\chi_j f(t)|^2 \, d\langle E_t u, u \rangle$ so letting $j \to \infty$ we obtain $\int_{-\infty}^{\infty} |f(t)|^2 \, d\langle E_t u, u \rangle = \|f(T)^* u\|^2 < \infty$.

If also g is unbounded we must show that the closure of $g(T)f(T)$ is $fg(T)$. So we must show that if $u \in \mathcal{D}(fg(T))$, then there is a sequence $u_j \to u$ such that $u_j \in \mathcal{D}(f(T))$, $f(T)u_j \in \mathcal{D}(g(T))$ and $g(T)f(T)u_j \to fg(T)u$.

Put $u_j = P_j u$, which certainly is in $\mathcal{D}(f(T))$. Now

$$\int_{-\infty}^{\infty} |g(t)|^2 \, d\langle E_t f(T)u_j, f(T)u_j \rangle = \int_{-\infty}^{\infty} |\chi_j f g(t)|^2 \, d\langle E_t u, u \rangle$$

$$\leq \int_{-\infty}^{\infty} |fg(t)|^2 \, d\langle E_t u, u \rangle < \infty,$$

so that $f(T)u_j \in \mathcal{D}(g(T))$. To show that $fg(T)u_j \to fg(T)u$ consider

$$\|fg(T)(1 - P_j)u\|^2 = \int_{-\infty}^{\infty} |fg(t)|^2 (1 - \chi_j(t))^2 \, d\langle E_t u, u \rangle.$$

The integrand is bounded by $|fg|^2$ and tends pointwise to zero, so by dominated convergence $fg(T)u_j \to fg(T)u$.

To prove (5), we have already dealt with unitary and self-adjoint $f(T)$, and if $f(t) \geq C$ and $u \in \mathcal{D}(f(T))$ we obtain

$$\langle f(T)u, u \rangle = \int_{-\infty}^{\infty} f(t) \, d\langle E_t u, u \rangle \geq C \int_{-\infty}^{\infty} d\langle E_t u, u \rangle = C\|u\|^2.$$

Similar calculations give bounds from above, completing the proof. $\square$

Example 3.1.7. If H is self-adjoint define $U(t) = e^{-itH}$, $t \in \mathbb{R}$, using Theorem 3.1.2 with $f(s) = e^{-its}$. Thus $U(t)$ is unitary, $U(0)$ the identity and $U(t + s) = U(t)U(s)$ for real t, s. Since $|e^{-ist} - 1|^2$ tends boundedly to 0 as $t \to 0$ through real t, for $s \in \mathbb{R}$, it also follows that $U(t)u \to u$ strongly as $t \to 0$, so it follows that $U(t + s) \to U(t)$ strongly as $s \to 0$. Such an operator family is called a *strongly continuous unitary group* of operators.

We also have $|(e^{-its} - 1)/t|^2 = 4 \sin^2(ts/2)/t^2$ so this is bounded by s^2 and tends pointwise to s^2 as $t \to 0$. Thus, if $u_0 \in \mathcal{D}(H)$ so that $\int_{-\infty}^{\infty} s^2 \, d\langle E_s u_0, u_0 \rangle < \infty$, dominated convergence shows that $U(t)u_0$ is (strongly) differentiable at 0 with derivative $-iHu_0$. An important fact is that *any* strongly continuous unitary group of operators has an *infinitesimal generator* H, which is a densely defined self-adjoint operator such that $U(t) = e^{-itH}$; see Stone [206].

Since $U(t+h) - U(t) = (U(h) - 1)U(t)$ and $U(t)$ maps $\mathcal{D}(H)$ to $\mathcal{D}(H)$ (explain why!) it follows that $U(t)u_0$ is strongly differentiable at every t with derivative $-iHU(t)u_0$ so that $u(t) = U(t)u_0$ solves the *Schrödinger equation* $iu'_t = Hu$ with initial value $u(0) = u_0$.

Much more on this circle of ideas, and on the semi-groups of Exercise 3.1.8, may be found in Hille and Phillips [106] or the more contemporary Engel and Nagel [67].

Remark 3.1.8. The standard interpretation of quantum mechanics is that one cannot predict events, only probabilities of events. The state of a quantum mechanical system at time t is supposed to be given by the *wave function* $\Psi(\cdot, t)$, which for each t is a unit vector in an L^2-space on the configuration space. The probability of finding the system in a subset M of the configuration space at time t is then $\int_M |\Psi(\cdot, t)|^2$.

One assumes that the wave function at time t depends linearly on the initial state, so that $\Psi(\cdot, t) = U(t)\Psi(\cdot, 0)$ where $U(t)$ is a linear operator. One also assumes that the state depends only on the initial state and the time passed. This leads to the equation $U(t + s) = U(t)U(s)$. Thus $U(t)$ is invertible with inverse $U(-t)$ and has to transform unit vectors into unit vectors, so it is unitary. We therefore have a group of unitary operators that describes how wave functions evolve in time. Assuming the group to be strongly continuous it will have a self-adjoint infinitesimal generator H (Engel and Nagel [67, Theorem 3.24]) so that $U(t) = e^{-itH}$.

By Example 3.1.7 this leads to the Schrödinger equation $i\Psi_t = H\Psi$, where the operator H is determined by the physics of the system studied. Particularly interesting are eigenvalues of H. If the initial state ψ is a (normalized) eigenvector of H with eigenvalue λ we have $i\Psi_t = H\Psi = e^{-itH}H\psi = \lambda\Psi$ so that $\Psi(\cdot, t) = e^{-i\lambda t}\psi$. Thus $|\Psi(\cdot, t)| = |\psi|$, from which it follows that the probability of finding the system in any subset of the configuration space is independent of time. Such a state is called stationary. To find stationary states we therefore have to find eigenvalues of H. In Section 6.6 we will do this for a basic quantum mechanical system.

Exercises

Exercise 3.1.1 Show that an operator T defined as on page 59 by an eigenvalue problem of Section 1.4 is self-adjoint.

Hint: By Lemma 2.3.6 (1) and Theorem 2.4.7 the operator T is self-adjoint if the equation $Tv - \lambda v = u$ for $\text{Im}\,\lambda \neq 0$ has a solution for any $u \in L^2_w(a, b)$. Show that $u \in \mathcal{D}(T)$ if and only if $\sum |\lambda_j \hat{u}_j|^2 < \infty$ and use this to find such solutions.

Exercise 3.1.2 Let E_Δ be as in (3.1.1) and suppose $u \in C_0(\mathbb{R})$. Show that $E_\Delta u(x)$ converges to $\frac{1}{2\pi} \int_\Delta \hat{u}(\xi) e^{i\xi x} \, d\xi$ as $a \to \infty$.

Exercise 3.1.3 Suppose P and Q are orthogonal projections on a Hilbert space, and that $P \geq Q$, i.e., $P - Q \geq 0$. Show that $PQ = QP = Q$ and that $P - Q$ is also an orthogonal projection.

Conclude that a family $\{E_t\}_{t \in \mathbb{R}}$ of orthogonal projections is increasing if and only if $E_t E_s = E_{\min(t,s)}$ for all $t, s \in \mathbb{R}$.

Exercise 3.1.4 Suppose η is non-decreasing and $\int_{-\infty}^{\infty} d\eta < \infty$. Show that then $-\lambda \int_{-\infty}^{\infty} \frac{d\eta(t)}{t-\lambda} \to \int_{-\infty}^{\infty} d\eta$ as $\lambda \to \infty$ along any non-real ray (half-line) originating in the origin. In particular, $\int_{-\infty}^{\infty} \frac{d\eta(t)}{t-\lambda} \to 0$.

Exercise 3.1.5 Show that if T is a closed operator on $\mathcal{H}$ and S is bounded and everywhere defined, then TS, but not necessarily ST, is closed.

Exercise 3.1.6 In the proof of Lemma 3.1.6 it was shown that

$$\int\limits_{-\infty}^{\infty} |f| \, |d\eta_{u,v}| \leq C \left(\int\limits_{-\infty}^{\infty} |f|^2 \, d\eta_{u,u} \right)^{1/2} \|v\|$$

for $C = 2$. Prove that this is true for any $C > 1$ and then also for $C = 1$.

Exercise 3.1.7 Let T be a self-adjoint operator and $\overline{\sigma}(T)$ the *extended spectrum* of T, which is $\sigma(T)$ if T is bounded and $\sigma(T) \cup \{\infty\}$ if T is unbounded. Suppose f is continuous on $\overline{\sigma}(T)$ and prove that $\overline{\sigma}(f(T)) = f(\overline{\sigma}(T))$. This is one version of the *spectral mapping theorem*.

Exercise 3.1.8 Let T be self-adjoint and non-negative. For $t \geq 0$ define $K(t) = e^{-tT}$ and show that $K(t)$ is bounded, $\|K(t)\| \to 0$ as $t \to \infty$, $K(0)$ equals the identity, $K(t+s) = K(t)K(s)$ for $s, t \geq 0$ and $K(t+s) \to K(t)$ strongly as $s \to 0$. Such a family of operators is called a *strongly continuous semi-group*.

Also prove that $K(t)$ is self-adjoint with norm ≤ 1 and that if $u_0 \in \mathcal{D}(T)$, then $u(t) = K(t)u_0$ solves a *linear evolution equation* $u_t' + Tu = 0$ for $t > 0$ with initial data $u(0) = u_0$. This will be the ordinary, one-dimensional heat equation if $T = S^2$ where S is the (only) self-adjoint realization in $L^2(\mathbb{R})$ of $i\frac{d}{dx}$ (see Example 2.2.14).

Exercise 3.1.9 Let U be a unitary operator on a Hilbert space $\mathcal{H}$ and define a self-adjoint relation $T = i(U + I)(U - I)^{-1}$ by a Cayley transform as in Exercise 2.3.2. Use this to derive a spectral theorem for unitary operators, integrating over the unit circle.

Exercise 3.1.10 Prove that $\lambda R_{\mu+\lambda}$ for real μ tends weakly to $E_{\{\mu\}}$ as $\lambda \to 0$ in a double sector $|\text{Re }\lambda| \leq C \, |\text{Im }\lambda|$. Also show the convergence is in the strong operator sense.

Hint: For the first part, extend Theorem E.1.7. Then use the resolvent relation.

3.2 The spectrum

From Definition 2.3.1 the spectrum of a closed relation is the disjoint union of the point, continuous and residual spectra. However, we noted earlier that Lemma 2.3.6 implies that a self-adjoint relation has no residual spectrum, so in this case $\sigma(T)$ is the disjoint union of $\sigma_p(T)$ and $\sigma_c(T)$. Obviously these components of the spectrum of a self-adjoint relation are determined by the behavior of the spectral projectors. The following theorem makes this connection explicit.

Theorem 3.2.1. *Suppose T is a self-adjoint relation. Then*

(1) $\lambda \in \sigma_p(T)$ *if and only if* E_t *jumps at* λ, *i.e.,* $E_{\{\lambda\}} = E_{[\lambda,\lambda]} \neq 0$.

(2) $\lambda \in \rho(T) \cap \mathbb{R}$ *if and only if* E_t *is constant in a neighborhood of* $t = \lambda$.

It follows that the continuous spectrum consists of those points of increase of E_t which are not jumps.[3]

Proof. If E_t jumps at λ we can find a unit vector e in the range of $E_{\{\lambda\}}$, i.e., such that $E_{\{\lambda\}}e = e$. It follows immediately from the spectral theorem that $e \in \mathcal{D}(T)$ and $(T - \lambda)e = 0$. Conversely, suppose that e is a unit vector in $\mathcal{D}(T)$ with $Te = \lambda e$. Then

$$0 = \|(T - \lambda)e\|^2 = \int_{-\infty}^{\infty} |t - \lambda|^2 \, d\langle E_t e, e\rangle ,$$

so that the support of the non-zero, non-negative measure $d\langle E_t e, e\rangle$ is contained in $\{\lambda\}$. Hence E_t jumps at λ, and the proof of (1) is complete.

Now assume E_t is constant in $(\lambda-\varepsilon, \lambda+\varepsilon)$. Then λ is not an eigenvalue of T so S_λ is dense in $\mathcal{H}$. Thus the inverse of $T - \lambda$ exists as a closed, densely defined operator. We need only show that this inverse is bounded, to see that its domain is all of $\mathcal{H}$ so that $\lambda \in \rho(T)$. But $\|(T - \lambda)u\|^2 = \int_{-\infty}^{\infty} |t - \lambda|^2 \, d\langle E_t u, u\rangle \geq \varepsilon^2 \int_{-\infty}^{\infty} d\langle E_t u, u\rangle = \varepsilon^2 \|u\|^2$ so the inverse of $T - \lambda$ is bounded by $1/\varepsilon$.

Conversely, assume that E_t is not constant near λ. Then there are arbitrarily short intervals Δ containing λ such that $E_\Delta \neq 0$, i.e., there are non-zero vectors u such that $E_\Delta u = u$. But then $\|(T - \lambda)u\| \leq |\Delta| \, \|u\|$, where $|\Delta|$ is the length of Δ. Hence we can find a sequence of unit vectors u_j, $j = 1, 2, \ldots$, for which $(T - \lambda)u_j \to 0$. Consequently either $T - \lambda$ is not injective or else the inverse is unbounded so that $\lambda \notin \rho(T)$.   □

The proof also shows that if $\lambda \in \rho(T)$, then R_λ is bounded by $1/d$, where d is the distance from λ to the spectrum. A small extra effort shows that $1/d$ is in fact the norm of R_λ (Exercise 3.2.1).

There is another way of decomposing the spectrum which is often more relevant.

Definition 3.2.2. Suppose T is a closed relation on $\mathcal{H}$.

---

[3] A point of increase for E_t is a point λ such that $E_\Delta \neq 0$ for every open $\Delta \ni \lambda$.

(1) The set $\sigma_d(T)$ of eigenvalues of T which are *isolated* points of $\sigma(T)$ and have *finite* multiplicity is called the *discrete spectrum* for T.

(2) The set $\sigma_e(T)$ of $\lambda \in \mathbb{C}$ for which S_λ is not closed or λ is an eigenvalue of *infinite* multiplicity, is called the *essential spectrum* for T.

Clearly $\sigma_d(T)$ and $\sigma_e(T)$ are disjoint subsets of $\sigma(T)$, and we shall soon see that if T is self-adjoint, then $\sigma(T)$ is the union of $\sigma_d(T)$ and $\sigma_e(T)$.

It should be noted that a self-adjoint realization of an ordinary differential equation (scalar, or acting on functions with values in a finite-dimensional space) can never have eigenvalues of infinite multiplicity. However, for partial differential equations there are important cases where eigenvalues of infinite multiplicity occur.

The reason for the name essential spectrum is that $\sigma_e(T)$ is rather insensitive to small changes ('perturbations') of the relation T, whereas the discrete spectrum is very sensitive to such changes. It should be noted that there are several other, slightly different definitions of essential spectrum in the literature, but for self-adjoint operators they all coincide; see Edmunds and Evans [66, Theorem 1.6, p. 424]. The following theorem is of particular interest when dealing with ordinary differential equations.

Theorem 3.2.3. *All finite-dimensional extensions of a closed relation are closed and have the same essential spectrum.*

Proof. Clearly an eigenvalue of infinite multiplicity will also be an eigenvalue of infinite multiplicity for any extension. However, making a finite-dimensional extension cannot introduce new eigenvalues of infinite multiplicity.

From Lemma A.6 it follows that a finite-dimensional extension T of a closed relation T_0 is also closed, and it is also clear that such an extension has a solvable space $S_\lambda(T)$ which is at most a finite-dimensional extension of $S_\lambda(T_0)$. We need to show that $S_\lambda(T)$ is closed if and only if $S_\lambda(T_0)$ is.

Again using Lemma A.6 it follows that $S_\lambda(T)$ is closed if $S_\lambda(T_0)$ is. To prove the reverse implication we consider the map $T_0 \ni (v, f) \mapsto f - \lambda v \in S_\lambda(T)$ which has range $S_\lambda(T_0)$, is linear, defined on the Hilbert space T_0 and obviously bounded. Since the codimension of $S_\lambda(T_0)$ in $S_\lambda(T)$ is finite, Lemma A.7 shows that $S_\lambda(T_0)$ is closed if $S_\lambda(T)$ is. □

We have the following immediate consequence of Theorem 3.2.3.

Corollary 3.2.4. *Suppose T is symmetric with one finite defect index. Then all maximal symmetric extensions, in particular any self-adjoint extensions, have the same essential spectrum.*

Theorem 3.2.5. *If T is a self-adjoint relation on $\mathcal{H}$, then $\sigma_e(T)$ is closed, and is the subset of $\sigma(T)$ consisting of all eigenvalues of infinite multiplicity and all non-isolated points of $\sigma(T)$.*

Proof. It will be enough to prove that (1) $\sigma_e(T)$ is closed and (2) S_λ is closed but not all of $\mathcal{H}$ if and only if λ is an isolated eigenvalue. To this end, first note that if λ

is an eigenvalue we may consider the relation $\tilde{T} = T \cap \overline{S}_\lambda^2$, which is self-adjoint on $\overline{S}_\lambda$ and does not have the eigenvalue λ. We denote the resolvent of this relation by $\tilde{R}_\lambda$, and note that $S_\lambda(\tilde{T}) = S_\lambda(T)$.

Now, if λ is isolated in $\sigma(T)$, then $\mu \in \rho(T)$ if $\mu \neq \lambda$ but sufficiently close to λ. Since $\tilde{R}_\mu$ is the restriction of R_μ to $\overline{S}_\lambda(T)$ we also have $\mu \in \rho(\tilde{T})$ for such μ. By Theorem 3.2.1 an isolated point of the spectrum is an eigenvalue, but since λ is not an eigenvalue of $\tilde{T}$ we have $\lambda \in \rho(\tilde{T})$ so that S_λ is closed.

On the other hand, if $S_\lambda(T)$ is closed clearly $\lambda \in \rho(\tilde{T})$, so also all μ close to λ are in $\rho(\tilde{T})$. If $u \in \mathcal{H}$ we have $u = u_1 + u_0$ with $u_1 \in S_\lambda$ and $u_0 \in D_\lambda$ by Lemma 2.3.6. Now $T \ni (u_0, \lambda u_0) = (u_0, \mu u_0 + (\lambda - \mu)u_0)$ so setting $v = \tilde{R}_\mu u_1 + (\lambda - \mu)^{-1} u_0$ we have $(v, \mu v + u) \in T$. Thus $S_\mu(T) = \mathcal{H}$ so $\mu \in \rho(T)$ and λ is isolated in $\sigma(T)$.

Finally, removing an isolated point from a closed set leaves the set closed, so the intersection of all sets obtained by removing an isolated eigenvalue of finite multiplicity from $\sigma(T)$ is closed. Since this set is $\sigma_e(T)$, the essential spectrum is closed.   □

Corollary 3.2.6. *If T is a self-adjoint relation, then $\sigma(T)$ is the disjoint union of $\sigma_d(T)$ and $\sigma_e(T)$.*

Note that while all self-adjoint extensions of a symmetric relation with finite defect indices have the same essential spectrum, the discrete spectrum depends very much on the choice of extension, see for example Theorem 2.4.9.

Naturally we can characterize the essential and discrete spectra by the behavior of the spectral projectors.

Corollary 3.2.7. *If T is a self-adjoint relation, then $\sigma_e(T)$ consists of all λ such that the spectral projector E_Δ has infinite rank[4] for all open intervals Δ containing λ.*

Similarly, $\sigma_d(T)$ are those λ for which $E_\Delta \neq 0$ but with finite rank if $\lambda \in \Delta$ and the length of Δ is sufficiently small.

Proof. By Theorem 3.2.1 an isolated eigenvalue is an isolated point of increase for E_t, so if it has finite multiplicity the rank of E_Δ is finite and non-zero for every sufficiently small open interval containing the point. Conversely, if the rank of E_Δ is finite, then Δ can only contain a finite number of points of increase, all being eigenvalues of finite multiplicity.

Consequently, if $\lambda \in \sigma_e(T)$ the projection E_Δ has infinite rank for every open interval Δ containing λ.   □

It follows from the corollary that $\lambda \in \sigma_e(T)$ if and only if there is a sequence of unit vectors $u_1, u_2, \ldots$ tending weakly to zero such that $Tu_j - \lambda u_j \to 0$ strongly. Such a sequence is called a *singular sequence*. Explicit construction of singular sequences is an often used method to identify points in the essential spectrum for specific operators.

---

[4] i.e., the range of E_Δ has infinite dimension.

Spectral multiplicity. The (geometric) multiplicity of an eigenvalue is simply the dimension of the eigenspace, but multiplicity for points in the essential spectrum of a self-adjoint operator is less elementary. We shall adopt the following definition. Consider a self-adjoint operator T on a Hilbert space $\mathcal{H}$ with associated resolution of the identity E_t, $t \in \mathbb{R}$. Let $u \in \mathcal{H}$ and consider all finite linear combinations of vectors $E_t u$, $t \in \mathbb{R}$. If it is possible to choose u so that these vectors are dense in $\mathcal{H}$ the *spectral multiplicity* of T is said to be one, and the vector u is called *cyclic*. In general, if the minimal number of vectors $u_1, u_2, \ldots$ required for the vectors $E_t u_j$, $t \in \mathbb{R}$, $j = 1, 2, \ldots$, to be dense in $\mathcal{H}$ is n, $1 \le n \le \infty$, then T is said to have spectral multiplicity n.

There is also a more local notion of multiplicity. If $\Delta \subset \mathbb{R}$ is an interval and the minimal number of vectors $u_j \in \mathcal{H}$ required for $E_\Delta E_t u_j$ to be dense in $E_\Delta \mathcal{H}$ is n, then the spectral multiplicity of T is said to be n in Δ. We shall use the notion of spectral multiplicity very rarely.

Exercise

Exercise 3.2.1 Show that $\|R_\lambda\| = 1/d(\lambda)$ where $d(\lambda)$ is the distance from λ to the spectrum; cf. the proof of Theorem 3.2.1.

3.3 Compactness

If a self-adjoint operator T has an orthonormal basis of eigenvectors $e_1, e_2, \ldots$, then for any $f \in \mathcal{H}$ we have $f = \sum \hat{f}_j e_j$, where $\hat{f}_j = \langle f, e_j \rangle$ are the generalized Fourier coefficients; we have a generalized Fourier series. However, $\sigma_p(T)$ can still be very complicated; it may for example be dense in $\mathbb{R}$ (so that $\sigma(T) = \mathbb{R}$), and each eigenvalue can have infinite multiplicity. A considerably simpler situation, more similar to the case of the classical Fourier series or the equations considered in Section 1.4, is obtained if $\sigma(T) = \sigma_d(T)$, which is the case precisely if the resolvent is *compact*.

Definition 3.3.1.

- A subset of a Hilbert space is called *precompact* (or *relatively compact*) if every sequence of points in the set has a strongly convergent subsequence.

- A closed, precompact set is called *compact*. Equivalently, a subset of a Hilbert space is called compact if every sequence of points in the set has a subsequence converging strongly to a point in the set.

- An operator $A : \mathcal{H}_1 \to \mathcal{H}_2$ is called *compact* if it maps bounded sets into precompact sets.

Clearly a precompact set has to be bounded, but note that in an infinite-dimensional space this is *not* enough for it to be precompact. For example, the closed unit sphere in a Hilbert space is closed and bounded, and it contains an orthonormal sequence. But no orthonormal sequence has a strongly convergent subsequence, by Exercise 2.1.11.

Since precompact sets are bounded a compact operator also has to be bounded, but in an infinite-dimensional space this is not enough; the identity is compact only in a finite-dimensional space. The third point means that if $\{u_j\}_1^\infty$ is a bounded sequence in $\mathcal{H}_1$, then $\{Au_j\}_1^\infty$ has a subsequence which converges strongly in $\mathcal{H}_2$.

Theorem 3.3.2.

(1) *A bounded operator with finite-dimensional range is compact.*

(2) *An operator is compact if and only if every weakly convergent sequence is mapped onto a strongly convergent sequence. Equivalently, $u_j \rightharpoonup 0$ implies that $Au_j \to 0$.*

(3) *If $A : \mathcal{H}_1 \to \mathcal{H}_2$ is compact and $B : \mathcal{H}_2 \to \mathcal{H}_3$ bounded, then BA is compact.*

(4) *If $A : \mathcal{H}_1 \to \mathcal{H}_2$ is compact and $B : \mathcal{H}_3 \to \mathcal{H}_1$ bounded, then AB is compact.*

(5) *If $A : \mathcal{H}_1 \to \mathcal{H}_2$ is compact, then so is $A^* : \mathcal{H}_2 \to \mathcal{H}_1$.*

(6) *If $A : \mathcal{H}_1 \to \mathcal{H}_2$ and $B : \mathcal{H}_1 \to \mathcal{H}_2$ are compact and $\lambda, \mu \in \mathbb{C}$, then $\lambda A + \mu B$ is also compact.*

Proof. The first item is an immediate consequence of Theorem 1.3.1 and the Bolzano–Weierstrass theorem 1.3.2. Now $u_j \rightharpoonup u$ is equivalent to $u_j - u \rightharpoonup 0$, and $A(u_j - u) \to 0$ is equivalent to $Au_j \to Au$. Thus the last statement of (2) is obvious. By Theorem 2.1.11 every bounded sequence has a weakly convergent subsequence, so if A maps weakly convergent sequences into strongly convergent ones, then A is compact.

Conversely, suppose $u_j \rightharpoonup u$ and A is compact. Since weakly convergent sequences are bounded (Theorem 2.1.11), any subsequence of $\{Au_j\}_1^\infty$ has a convergent subsequence. Suppose $Au_{j_k} \to v$. Then for any $w \in \mathcal{H}$ we have $\langle v, w \rangle = \lim\langle Au_{j_k}, w \rangle = \lim\langle u_{j_k}, A^*w \rangle = \langle u, A^*w \rangle = \langle Au, w \rangle$, so that $v = Au$. Hence the only point of accumulation of $\{Au_j\}_1^\infty$ is Au, so $Au_j \to Au$. If not, there would be a neighborhood $\mathcal{O}$ of Au and a subsequence of $\{Au_j\}_{j=1}^\infty$ outside $\mathcal{O}$. But we could then find a convergent subsequence which does not converge to Au. This proves (2). We leave the rest of the proof to the reader (Exercise 3.3.1). $\qquad\square$

Theorem 3.3.3. *Suppose T is self-adjoint relation and for some $\mu \in \rho(T)$ its resolvent R_μ is compact. Then R_λ is compact for all $\lambda \in \rho(T)$, and T has discrete spectrum, i.e., $\sigma(T) = \sigma_d(T)$ and $\sigma_e(T) = \emptyset$, so the spectrum consists of isolated eigenvalues of finite multiplicity.*

Proof. By the resolvent relation $R_\lambda = (I + (\lambda - \mu)R_\lambda)R_\mu$ where I is the identity, so the first factor to the right is bounded. Hence R_λ is compact by Theorem 3.3.2(3).

Now let Δ be a bounded interval and $\text{Im}\,\lambda \neq 0$. If $u \in E_\Delta \mathcal{H}$ then $\|R_\lambda u\|^2 = \int_\Delta \frac{d\langle E_t u, u \rangle}{|t - \lambda|^2} \geq K\|u\|^2$, where $K = \inf_{t \in \Delta} |t - \lambda|^{-2} > 0$ (verify this calculation!).

We have $R_\lambda u_j \to 0$ if $u_j \to 0$, so the inequality shows that any weakly convergent sequence in $E_\Delta \mathcal{H}$ is strongly convergent (the identity operator on $E_\Delta \mathcal{H}$ is compact).

This implies that $E_\Delta \mathcal{H}$ has finite dimension (for example since any orthonormal sequence converges weakly to 0 but is not strongly convergent). In particular eigenspaces are finite-dimensional. It also follows that any bounded interval can only contain a finite number of points of increase for E_t, because projections belonging to disjoint intervals have orthogonal ranges (Example 3.1.3). This completes the proof. □

Resolvents for different self-adjoint extensions of a symmetric relation are closely related. In particular, we have the following corollary of Corollary 3.2.4.

Corollary 3.3.4. *Suppose a symmetric relation T has finite defect indices and a self-adjoint extension with compact resolvent. Then every self-adjoint extension of T has compact resolvent.*

A self-adjoint operator has compact resolvent precisely if its essential spectrum is empty so this follows from Corollary 3.2.4. It also follows since if R_λ, $\tilde{R}_\lambda$ are resolvents of self-adjoint extensions of T then the range of $A = R_\lambda - \tilde{R}_\lambda$ is contained in the finite-dimensional space D_λ so A is compact by Theorem 3.3.2. Thus if $\tilde{R}_\lambda$ is compact, so is R_λ.

An important class of compact operators frequently encountered is the class of *Hilbert–Schmidt operators*.

Definition 3.3.5. *If $\mathcal{H}$ is a separable Hilbert space, then $A : \mathcal{H} \to \mathcal{H}$ is called a Hilbert–Schmidt operator if we have $\sum \|Ae_j\|^2 < \infty$ for some orthonormal basis $e_1, e_2, \ldots$. The number $\|A\| = \sqrt{\sum \|Ae_j\|^2}$ is the Hilbert–Schmidt norm of A.*

Theorem 3.3.6. *$\|A\|$ is independent of the particular orthonormal basis used in the definition, it is a norm, $\|A\| = \|A^*\|$, and any Hilbert–Schmidt operator is compact. The set of Hilbert–Schmidt operators on $\mathcal{H}$ is a Hilbert space normed by the Hilbert–Schmidt norm.*

Proof. It is clear that $\|\cdot\|$ is a norm. Now suppose $\{e_j\}_1^\infty$ and $\{f_j\}_1^\infty$ are arbitrary orthonormal bases. Using Parseval's formula twice it follows that

$$\sum_j \|Ae_j\|^2 = \sum_{j,k} |\langle Ae_j, f_k \rangle|^2 = \sum_{j,k} |\langle e_j, A^* f_k \rangle|^2 = \sum_k \|A^* f_k\|^2 .$$

Thus the Hilbert–Schmidt norm has the claimed properties. We have used that in a positive double series one may change the order of summation, a special case of Tonelli's theorem B.9.5.

To see that A is compact, suppose $u_k \to 0$ weakly and let $\varepsilon > 0$. Choose N so large that $\sum_N^\infty \|A^* e_j\|^2 < \varepsilon$ and let C be a bound for the sequence $\{u_k\}_1^\infty$ (Theorem 2.1.11). By Parseval's formula we then have $\|Au_k\|^2 = \sum |\langle Au_k, e_j \rangle|^2 = \sum |\langle u_k, A^* e_j \rangle|^2$. We obtain

$$\|Au_k\|^2 \le \sum_1^N |\langle u_k, A^*e_j \rangle|^2 + C^2\varepsilon \to C^2\varepsilon$$

as $k \to \infty$ since $|\langle u_k, A^*e_j \rangle| \le C\|A^*e_j\|$. It follows that $Au_k \to 0$ strongly so that A is compact. We leave the proof of the last statement to the reader in Exercise 3.3.5.
□

It is common to consider a differential operator defined in some domain $\Omega \subset \mathbb{R}^n$ as an operator in the space $L_w^2(\Omega)$ where w is a positive (Radon) measure and the scalar product in the space is given by $\langle u, v \rangle = \int_\Omega u\bar{v}\, w$. In many cases the resolvent of such an operator can be realized as an integral operator, i.e., an operator of the form

$$Au(x) = \int_\Omega g(x, \cdot)u\, w \text{ for } x \in \Omega. \tag{3.3.1}$$

The function g, defined in $\Omega \times \Omega$, is called the *integral kernel* of the operator A. The integral kernel of the resolvent of a differential operator is usually called *Green's function* for the operator.

Theorem 3.3.7. *Assume $g(x, y)$ is measurable with respect to $w \otimes w$ and that $y \mapsto g(x, y)$ is in $L_w^2(\Omega)$ for a.e.(w) $x \in \Omega$. Then the operator A of (3.3.1) is a Hilbert–Schmidt operator in $L_w^2(\Omega)$ if and only if $g \in L_w^2(\Omega) \otimes L_w^2(\Omega))$, i.e., if and only if*

$$\iint_{\Omega \times \Omega} |g(\cdot, \cdot)|^2\, w \otimes w < \infty.$$

Proof. Let $\{e_j\}_1^\infty$ be an orthonormal basis in the space $L_w^2(\Omega)$. For fixed $x \in \Omega$ we may view $Ae_j(x)$ as the jth Fourier coefficient of $\overline{g(x, \cdot)}$ so Parseval's formula gives

$$\sum |Ae_j(x)|^2 = \int_\Omega |g(x, \cdot)|^2\, w$$

for a.e.(w) $x \in \Omega$. By monotone convergence this function is in $L_w^1(\Omega)$ if and only if the Hilbert–Schmidt norm of A is finite. The theorem now follows by an application of Tonelli's theorem, Theorem B.9.5.   □

Example 3.3.8. Consider a self-adjoint differential operator T in $L^2(\Omega)$, where $\Omega \subset \mathbb{R}$ is an interval. If Green's function $g(x, y, \lambda)$ for this operator exists we have

$$R_\lambda u(x) = \int_\Omega u(y)g(x, y, \lambda)\, dy = \int_\Omega (T - \lambda)R_\lambda u(y)g(x, y, \lambda)\, dy.$$

If T is a realization of $i\frac{d}{dx}$ and u is in the range of the minimal operator, then $R_\lambda u$ has compact support and the formula above means that $(-T - \lambda)g(x, \cdot, \lambda) = \delta_x$ in the sense of distributions, where δ_x is the Dirac measure at x. As a function of x one also expects Green's function to satisfy the boundary conditions determining T, and

$(R_\lambda)^* = R_{\bar\lambda}$ should be reflected in the formula $g(x, y, \lambda) = \overline{g(y, x, \bar\lambda)}$. For fixed y we therefore expect $g(\cdot, y, \lambda)$ to satisfy $Tu - \lambda u = \delta_y$. We shall not prove this here, but simply use this to find Green's function. Once found, one may check that the function constructed is Green's function by calculation.

Consider the operator T in $L^2(-\pi, \pi)$ with domain $\mathcal{D}(T)$, consisting of those absolutely continuous functions u with derivative in $L^2(-\pi, \pi)$ for which $u(\pi) = u(-\pi)$, and given by $Tu = i\frac{du}{dx}$ (cf. Example 2.2.16). This operator is self-adjoint and the solutions of the equation $Tu = \lambda u$ are the multiples of $e^{-i\lambda x}$ so Green's function should equal $a(x)e^{i\lambda y}$ for $x < y$ and $b(x)e^{i\lambda y}$ for $x > y$, where $a(-\pi) = b(\pi)$ and $a(x) = Ae^{-i\lambda x}$, $b(x) = Be^{-i\lambda x}$ with constants A, B. Thus $Ae^{i\lambda\pi} = Be^{-i\lambda\pi}$. A brief calculation now shows that Green's function is given by

$$g(x, y, \lambda) = \begin{cases} -\dfrac{e^{-i\lambda\pi}}{2\sin\lambda\pi}\, e^{i\lambda(y-x)}, & x < y\,, \\[3mm] -\dfrac{e^{i\lambda\pi}}{2\sin\lambda\pi}\, e^{i\lambda(y-x)}, & x > y\,. \end{cases}$$

The reader should verify this. Of course, we have an eigenvalue if $e^{-i\lambda\pi} = e^{i\lambda\pi}$ or $e^{2i\pi\lambda} = 1$, so the set of eigenvalues is $\mathbb{Z}$. For such λ the resolvent does not exist, so it is not surprising that the formula makes no sense then, $\sin\lambda\pi$ vanishing for $\lambda \in \mathbb{Z}$. Except for such λ clearly $g(x, y, \lambda)$ is bounded so $\int_{-\pi}^{\pi}\int_{-\pi}^{\pi} |g(x, y, \lambda)|^2 \, dx\, dy < \infty$. Thus the resolvent is a Hilbert–Schmidt operator, therefore compact, $\sigma(T) = \mathbb{Z}$ and the eigenfunctions are an orthonormal basis in $L^2(-\pi, \pi)$.

Now consider the operator of Example 2.2.14. Similar calculations show that Green's function is now only defined for non-real λ and equals

$$g(x, y, \lambda) = \begin{cases} i\dfrac{\operatorname{Im}\lambda}{|\operatorname{Im}\lambda|}\, e^{i\lambda(y-x)} & \text{if } (y - x)\operatorname{Im}\lambda > 0\,, \\[2mm] 0 & \text{otherwise}\,. \end{cases} \tag{3.3.2}$$

The reader should try to verify this as well. In this case there is no value of λ for which $g(\cdot, \cdot, \lambda) \in L^2(\mathbb{R}^2)$ so the resolvent is *not* a Hilbert–Schmidt operator (verify this) and $\sigma(T) = \mathbb{R}$.

Exercises

Exercise 3.3.1 Prove Theorem 3.3.2 (3)–(6).

Exercise 3.3.2 Show the converse of Theorem 3.3.3, i.e., if the spectrum of a self-adjoint operator consists of isolated eigenvalues of finite multiplicity, then the resolvent is compact.

Hint: Let $\lambda_1, \lambda_2, \dots$ be the eigenvalues ordered by increasing absolute value and repeated according to multiplicity and let the corresponding normalized eigenvectors be $e_1, e_2, \ldots$. Show that $\|R_\lambda u\|^2 = \sum |\langle u, e_j \rangle|^2 / |\lambda - \lambda_j|^2$ and use this to see that $R_\lambda u_k \to 0$ if $u_k \rightharpoonup 0$.

Exercise 3.3.3 Prove that a compact self-adjoint operator has a spectrum consisting of 0 and a finite or infinite sequence of real, non-zero eigenvalues of finite multiplicity, which if infinitely many converge to zero.

Conversely, given an orthonormal sequence $e_1, e_2, \ldots$ and a sequence $\lambda_j \to 0$, the operator $Tu = \sum \lambda_j \langle u, e_j \rangle e_j$ defines a compact operator which is self-adjoint if all λ_j are real. If they are not, at least the operator commutes with its adjoint. Prove this as well.

Hint: The proof of Theorem 3.3.3.

Exercise 3.3.4 A bounded operator T which commutes with its adjoint ($T^*T = TT^*$) is called *normal*.

Define the real part of a bounded operator T as $A = \frac{1}{2}(T + T^*)$ and the imaginary part as $B = \frac{1}{2i}(T - T^*)$. Thus $T = A + iB$ and $T^* = A - iB$. Show that the real and imaginary parts of an operator are self-adjoint and that they commute if and only if the operator is normal.

There exists a general spectral theorem valid for normal operators. In particular, show that if T has an orthonormal basis of eigenvectors, then T is normal. Conversely, if T is normal and compact, then the real and imaginary parts of T are compact, and T has an orthonormal basis of eigenvectors (in a non-separable space the statement is that T has an orthonormal basis of eigenvectors in the orthogonal complement of its nullspace).

Hint: Show that if E is an eigenspace for A, then the restriction of B to the (finite-dimensional) space E is a self-adjoint operator on E.

Exercise 3.3.5 Prove the last statement of Theorem 3.3.6.

Exercise 3.3.6 Verify all claims made in Example 3.3.8.

Exercise 3.3.7 Let T be a self-adjoint operator with discrete spectrum, i.e., the spectrum consists of isolated eigenvalues of finite multiplicity, and let λ_j, $j = 1, 2, \ldots$, be the non-zero eigenvalues of T, repeated according to multiplicity.

Show that the resolvent R_λ of T is a Hilbert–Schmidt operator if and only if $\sum_{j=1}^\infty \lambda_j^{-2} < \infty$.

3.4 Quadratic forms and the Minimax principle

If T is a symmetric operator on some linear space with scalar product we obtain a Hermitian sesquilinear form H defined on $\mathcal{D}(T)$ by setting $H(u, v) = \langle u, Tv \rangle$. Here we shall briefly discuss when this process can be reversed, i.e., given a Hermitian form H defined on some linear space with scalar product, is it possible to find a symmetric operator T such that H is given by the formula above?

Let $\mathcal{V}$ be a linear space with scalar product $\langle \cdot, \cdot \rangle$ and norm $\|\cdot\|$ on which is defined a Hermitian sesquilinear form H. Such a form is commonly called a *quadratic form* with *form domain* $\mathcal{V}$, although the term quadratic is really only appropriate if $\mathcal{V}$ is a real space.

We may of course complete $\mathcal{V}$ with respect to $\|\cdot\|$ to a Hilbert space $\mathcal{H}$, but in general the form domain cannot be extended to all of $\mathcal{H}$ unless H is *bounded*, i.e., $|H(u,v)| \le C\|u\|\|v\|$ for some constant C and all $u, v \in \mathcal{V}$. In this case we may extend the form domain to $\mathcal{H}$ by continuity, with the same bound C. This follows from

$$\left| H(u_j, v_j) - H(u_k, v_k) \right| = \left| H(u_j - u_k, v_j) + H(u_k, v_j - v_k) \right|$$
$$\le C(\|u_j - u_k\|\|v_j\| + \|u_k\|\|v_j - v_k\|),$$

for if $\mathcal{V} \ni u_j \to u \in \mathcal{H}$ and $\mathcal{V} \ni v_j \to v \in \mathcal{H}$ as $j \to \infty$ it follows that $H(u_j, v_j), j = 1, 2, \ldots$, is a Cauchy sequence of numbers and we define $H(u,v) = \lim H(u_j, v_j)$. This definition is easily seen to make the extended form well defined, Hermitian, sesquilinear and bounded by C; see Exercise 3.4.1.

For a fixed $v \in \mathcal{H}$ the linear form $\mathcal{H} \ni u \mapsto H(u,v)$ is bounded by $C\|v\|$, so the Riesz representation theorem shows that there exists a unique $v^* \in \mathcal{H}$ such that $H(u,v) = \langle u, v^* \rangle$. Clearly v^* depends linearly on v and $\|v^*\|^2 = H(v^*, v) \le C\|v^*\|\|v\|$ so that $\|v^*\| \le C\|v\|$. It follows that there is a linear operator T on $\mathcal{H}$, bounded by C and such that $H(u,v) = \langle u, Tv \rangle$ for $u, v \in \mathcal{H}$. T is symmetric since $\langle Tu, v \rangle = \overline{\langle v, Tu \rangle} = \overline{H(v,u)} = H(u,v) = \langle u, Tv \rangle$, so we have proved the following theorem.

Theorem 3.4.1. *Every bounded quadratic form H on a Hilbert space $\mathcal{H}$ with dense form domain can be uniquely continued to have form domain $\mathcal{H}$ with the same bound. H is then given by $H(u,v) = \langle u, Tv \rangle$, where T is a bounded, symmetric operator on $\mathcal{H}$ with the same bound as H.*

If H is not bounded one might hope to represent it in a similar way by an unbounded operator, but this is not always possible. However, if H is semi-bounded, say bounded below by a so that $H(u,u) \ge a\|u\|^2$ for all $u \in \mathcal{V}$, one may proceed as follows.

Choose c such that $a + c > 0$ and put $\langle u, v \rangle_1 = H(u,v) + c\langle u, v \rangle$. Then $\langle \cdot, \cdot \rangle_1$ is a scalar product on $\mathcal{V}$, and it is clear that as long as $a + c > 0$ all such scalar products give equivalent norms $\|\cdot\|_1$. We therefore fix $c = 1 - a$ so that $\|u\|_1 \ge \|u\|$ for all $u \in \mathcal{V}$.

Completing $\mathcal{V}$ with respect to $\|\cdot\|_1$ we obtain a Hilbert space $\mathcal{H}_1$ with scalar product $\langle \cdot, \cdot \rangle_1$ on which, by Theorem 3.4.1, $\langle \cdot, \cdot \rangle$ is a quadratic form bounded by 1. The theorem also shows that there is a symmetric operator R_1 on $\mathcal{H}_1$ bounded by 1 such that $\langle u, v \rangle = \langle u, R_1 v \rangle_1$ for $u, v \in \mathcal{H}_1$. Since $\langle R_1 u, u \rangle_1 = \|u\|^2 \ge 0$ we have $0 \le R_1 \le I$, I being the identity on $\mathcal{H}_1$.

It is important to note that it is possible that $R_1 u = 0$ for some nonzero $u \in \mathcal{H}_1$, or equivalently that $\|\cdot\|$ is only a semi-norm on $\mathcal{H}_1$. A simple example is given in Exercise 3.4.2.

We postpone the discussion of this and for the moment assume that $\|\cdot\|$ is a norm on $\mathcal{H}_1$. We may then complete $\mathcal{H}_1$ (or $\mathcal{V}$) with respect to $\|\cdot\|$ and obtain a Hilbert space $\mathcal{H}$. The quadratic form H then has dense form domain $\mathcal{H}_1$ in $\mathcal{H}$. The form domain $\mathcal{H}_1$ is complete with respect to $H(u, v) + c\langle u, v\rangle$, and such a form is called *closed*. Any form which may be extended to a closed form is called *closeable*. As we have seen, not all semi-bounded quadratic forms are closeable, but for closed forms we have the following theorem.

Theorem 3.4.2. *Given a closed, semi-bounded quadratic form H on a Hilbert space $\mathcal{H}$ with form domain $\mathcal{H}_1$, there is a unique densely defined, self-adjoint operator T on $\mathcal{H}$ such that $H(u, v) = \langle u, Tv\rangle$ for $u \in \mathcal{H}_1$, $v \in \mathcal{D}(T)$.*

If H is bounded below by a and $a + c > 0$ the form domain $\mathcal{H}_1$ equals the domain of $\sqrt{T + c}$, where the square root is given by the functional calculus of Theorem 3.1.2.

Proof. The linear form $\mathcal{H}_1 \ni u \mapsto \langle u, v\rangle$ is bounded by $\|v\|$ for any $v \in \mathcal{H}$ since $|\langle u, v\rangle| \le \|u\|\|v\| \le \|v\|\|u\|_1$, so R_1 extends to an operator $R : \mathcal{H} \to \mathcal{H}_1$ bounded by 1. One may also view R as an operator on $\mathcal{H}$ bounded by 1, symmetric since $\langle Ru, v\rangle = \langle Ru, Rv\rangle_1 = \langle u, Rv\rangle$ and positive since $\langle Ru, u\rangle = \|Ru\|_1^2 \ge 0$. Thus $0 \le R \le I$, where I is the identity on $\mathcal{H}$.

If $Rv = 0$ and $u \in \mathcal{H}_1$, then $\langle u, v\rangle = \langle u, Rv\rangle_1 = 0$ so v is orthogonal to the dense set $\mathcal{H}_1$. Thus $v = 0$ so R has a, possibly unbounded, self-adjoint inverse. Symmetry of R then shows that the range of R is dense in $\mathcal{H}$. Now put $T = R^{-1} - cI$ where I is the identity on $\mathcal{H}$. T is then densely defined and self-adjoint in $\mathcal{H}$. We have $H(u, v) = \langle u, v\rangle_1 - c\langle u, v\rangle = \langle u, R^{-1}v\rangle - c\langle u, v\rangle = \langle u, Tv\rangle$ if $u \in \mathcal{H}_1$ and $v \in \mathcal{D}(T)$.

By the spectral theorem

$$\langle u, u\rangle_1 = \langle u, (T + c)u\rangle = \int_{[a,\infty)} (t + c)\, d\langle E_t u, u\rangle \ge 0$$

for $u \in \mathcal{D}(T)$, where $\{E_t\}_{t \in \mathbb{R}}$ is the resolution of the identity for T. Completing $\mathcal{D}(T)$ in this norm-square gives $\mathcal{H}_1$. By the functional calculus of Theorem 3.1.2 $\mathcal{H}_1$ is therefore the domain of $\sqrt{T + c}$. $\square$

Remark 3.4.3. In the proof are defined operators R and R_1. Here R_1 is clearly a core of R since $\mathcal{H}_1$ is dense in $\mathcal{H}$. It follows that the operator $T_1 = R_1^{-1} - cI$, where I is the identity on $\mathcal{H}_1$, is densely defined and self-adjoint in $\mathcal{H}_1$, and T_1 is a core of T.

Now consider the case when $\|\cdot\|$ is just a semi-norm on $\mathcal{H}_1$. The reason could be that H is not closeable, but also that $\langle \cdot, \cdot\rangle$ is just a semi-scalar product while $\langle \cdot, \cdot\rangle_1$ is a genuine scalar product on $\mathcal{V}$. Let $\mathcal{H}_\infty$ be the nullspace of R_1 and define $\tilde{\mathcal{H}}_1 = \mathcal{H}_1 \ominus \mathcal{H}_\infty$. Then $\langle \cdot, \cdot\rangle$ is a scalar product on $\tilde{\mathcal{H}}_1$ which we may therefore complete with respect to $\|\cdot\|$ and so obtain a Hilbert space $\mathcal{H}$. The restriction of H to $\tilde{\mathcal{H}}_1$ is thus a closed quadratic form on $\mathcal{H}$ so by Theorem 3.4.2 there is a unique self-adjoint operator T densely defined in $\tilde{\mathcal{H}}$ such that $H(u, v) = \langle u, Tv\rangle$ for $u \in \tilde{\mathcal{H}}$ and $v \in \mathcal{D}(T)$. We conclude our discussion with the following criterion for a closeable quadratic form.

Theorem 3.4.4. *A semi-bounded quadratic form H on $\mathcal{H}$ with form domain $\mathcal{V}$ is closeable if and only if there is a symmetric operator T_0 on $\mathcal{H}$ with domain dense* *in $\mathcal{V}$ such that $H(u,v) = \langle u, T_0 v \rangle$ for $u \in \mathcal{V}$ and $v \in \mathcal{D}(T_0)$. If it exists, T_0 has a self-adjoint extension T in $\mathcal{H}$ which is the operator of Theorem 3.4.2. H, T_0 and T all have the same lower bound.*

Proof. Suppose there is an operator T_0 with domain dense in $\mathcal{V}$ and values in $\mathcal{H}$ such that $H(u,v) = \langle u, T_0 v \rangle$ for $u \in \mathcal{V}$ and $v \in \mathcal{D}(T_0)$, and consider a Cauchy sequence $\{u_j\}_1^\infty$ in $\mathcal{V}$ with respect to $\|\cdot\|_1$. We must show that if $\|u_j\| \to 0$, then also $\|u_j\|_1 \to 0$. To see this, suppose $u_j \to u$ in $\mathcal{H}_1$ and let $v \in \mathcal{D}(T_0)$. Then $\langle u_j, v \rangle_1 = \langle u_j, T_0 v + cv \rangle \to 0$. The left-hand side converges to $\langle u, v \rangle_1$, so u is orthogonal to the dense set $\mathcal{D}(T_0)$, i.e., $u = 0$. Thus H is closeable, and clearly T_0 is a restriction of T. Clearly the lower bound of H equals the lower bound of T_0 and that of the closure of H equals the lower bound of T, and H and its closure have the same lower bound.

The converse follows from Theorem 3.4.2 by picking T_0 equal to the restriction of T to $\mathcal{V}$. $\qquad\square$

The following important result by K. O. Friedrichs [76] is an immediate corollary. By Exercise 2.4.4 a symmetric operator T bounded from below by a has self-adjoint extensions with lower bound arbitrarily close to but strictly less than a. Friedrichs showed that there is a self-adjoint extension with the *same* lower bound as T.

Corollary 3.4.5 (Friedrichs extension). *Suppose T is a symmetric operator densely defined on $\mathcal{H}$ and bounded from below by a. Then there exists a self-adjoint extension of T with lower bound a.*

Proof. Setting $H(u,v) = \langle u, Tv \rangle$ we obtain a quadratic form on $\mathcal{H}$ with form domain $\mathcal{D}(T)$. By Theorem 3.4.4 H is closeable, the closure given by a self-adjoint operator with the same lower bound a as H. $\qquad\square$

The theorem is clearly also true, with essentially the same proof, if T is a symmetric relation on $\mathcal{H}$, for example a not densely defined symmetric operator.

The minimax principle. This is a method for calculating eigenvalues below the essential spectrum of a self-adjoint operator bounded from below. Suppose T is self-adjoint with greatest lower bound $a > -\infty$, so that $\inf \sigma(T) = a$. We have $a \leq \inf \sigma_e(T)$, and if $a < \inf \sigma_e(T)$ the spectrum is discrete in $[a, \inf \sigma_e(T))$ and a is the smallest eigenvalue of T. Let $H(u,v) = \langle u, Tv \rangle$ for $u, v \in \mathcal{D}(T)$ and let S_n be the set of all subspaces of $\mathcal{H}_1$ with dimension n, where $\mathcal{H}_1$ is the domain of the closure of H. Since $\mathcal{D}(T)$ is dense in $\mathcal{H}_1$ it will turn out that we may equivalently let S_n be the set of n-dimensional subspaces of $\mathcal{D}(T)$.

Theorem 3.4.6 (Minimax principle). *Suppose there are (finitely or infinitely many) eigenvalues $\lambda_1 \leq \lambda_2 \leq \lambda_3 \leq \ldots$ in $[a, \inf \sigma_e(T))$, counted with multiplicity, i.e., any eigenvalue in the list is repeated as many times as given by the multiplicity. Then*

$$\lambda_n = \sup_{\mathcal{V} \in S_{n-1}} \inf_{0 \neq u \perp \mathcal{V}} H(u,u)/\|u\|^2 \tag{3.4.1}$$

if the number of eigenvalues $< \inf \sigma_e(T)$ is at least n. Otherwise the right-hand side of (3.4.1) equals $\inf \sigma_e(T)$.

Proof. If $\mathcal{V} \in S_{n-1}$ some non-trivial linear combination of the eigenvectors associated with $\lambda_1, \ldots, \lambda_n$ is orthogonal to $\mathcal{V}$ since this condition gives a homogeneous linear system of equations with fewer equations than unknowns. Choosing u as this linear combination we have $H(u, u)/\|u\|^2 \leq \lambda_n$, while if $\mathcal{V}$ is the span of the eigenvectors to $\lambda_1, \ldots, \lambda_{n-1}$ and $u \perp \mathcal{V}$ the spectral theorem shows that $H(u, u)/\|u\|^2 \geq \lambda_n$ with equality if u is the eigenvector to λ_n.

If $\{E_t\}_{t \in \mathbb{R}}$ is the resolution of the identity for T and $a = \inf \sigma_e(T)$, then $E_{[a,a+\varepsilon)}$ has infinite-dimensional range for every $\varepsilon > 0$. Thus we can always choose u in the range of $E_{[a,a+\varepsilon)}$ and orthogonal to any given finite-dimensional space $\mathcal{V}$ so it follows that

$$\inf_{0 \neq u \perp \mathcal{V}} H(u, u)/\|u\|^2 \leq \inf \sigma_e(T) + \varepsilon$$

for every $\mathcal{V}$ and $\varepsilon > 0$. But if the number of eigenvalues $< \inf \sigma_e(T)$ is n, then choosing $\mathcal{V}$ to be the span of all n eigenvectors to the eigenvalues $< \inf \sigma_e(T)$ the spectral theorem shows this expression to be $\geq \inf \sigma_e(T)$, finishing the proof. $\square$

When calculating eigenvalues numerically the minimax principle will yield upper bounds for the eigenvalues, since for numerical purposes $\mathcal{H}_1$ needs to be replaced by a finite-dimensional subspace. This is commonly known as the Rayleigh—Ritz method, see Trefethen and Bau [213]. Thus the minimax principle must be supplemented by some method giving lower bounds . There are several such methods available; see the discussion on this and related questions in Davies [50]. However, the minimax principle is at least as important for theoretical purposes. We give some examples based on the equations discussed in Section 1.4.

Example 3.4.7. Consider the form H of Section 1.4 defined on the space $\mathcal{D}_0$. The corresponding operator has an orthonormal basis of eigenvectors, so its essential spectrum is empty and all eigenvalues are given by (3.4.1). Now replace the coefficients q and w by $\tilde{q} \geq q$ and $0 < \tilde{w} \leq w$. This will increase the quotient $H(u, u)/\langle u, u \rangle$ for any $u \in \mathcal{D}_0$. If the new eigenvalues are $\lambda_1', \lambda_2', \ldots$ the minimax principle tells us that $\lambda_n \leq \lambda_n'$ for all n. Note that the theory of Section 1.4 will only give this for $n = 1$.

We can also compare the eigenvalues for different boundary conditions but the same equation. This may mean changing the form domain $\mathcal{D}_0$ or the form H by adding boundary terms. Let us first look at the latter. Suppose we replace H by $H + Q$, where Q is the Hermitian form $Q(u, v) = Au(a)\overline{v(a)} + Bu(b)\overline{v(b)} - Cu(b)\overline{v(a)} - u(a)\overline{Cv(b)}$ with real constants A, B and a complex constant C. If this form is positive, i.e., $A > 0$ and $AB \geq |C|^2$, then clearly $\lambda_n \leq \lambda_n'$ for every n.

Now suppose $\mathcal{D}_0 = C^1(a, b)$, $\mathcal{D}_0' = \{u \in C^1(a, b) : u(a) = Ku(b)\}$ and $\mathcal{D}_0'' = \{u \in C^1(a, b) : u(a) = u(b) = 0\}$, where K is a fixed complex number, and denote the corresponding eigenvalues λ_n, λ_n' and λ_n''. Since $\mathcal{D}_0'' \subset \mathcal{D}_0' \subset \mathcal{D}_0$ we obtain $\lambda_n \leq \lambda_n' \leq \lambda_n''$ for all n. This means that Dirichlet boundary conditions give the largest eigenvalues and the boundary conditions of (1.4.15) (often called the *free* or *natural* boundary conditions for the corresponding form), the smallest.

One may also change the interval. For example, if $(c, d) \subset (a, b)$ then $C_0^1(c, d) \subset \mathcal{D}_0$ so the Dirichlet eigenvalues of a smaller interval are all larger than the corresponding eigenvalues for any boundary condition on the full interval (Exercise 3.4.4).

Exercises

Exercise 3.4.1 Show that a bounded quadratic form in a Hilbert space $\mathcal{H}$ with dense form domain can be uniquely extended to a quadratic form with form domain $\mathcal{H}$ and the same bound.

Exercise 3.4.2 Let V be the set of trigonometric polynomials $\sum_{-n}^{n} c_k e^{ikx}$ on $[-\pi, \pi]$ with $\langle u, v \rangle = \int_0^\pi u\bar{v}$ and $H(u, v) = \int_{-\pi}^0 u\bar{v}$.

Show that the space $\mathcal{H}_1$ constructed before Theorem 3.4.2 is $\mathcal{H}_1 = L^2(-\pi, \pi)$ and $\|u\| = 0$ precisely if $u = 0$ a.e. in $(0, \pi)$.

Hint: By Example 3.3.8 the exponentials e^{ikx}, $k = 0, \pm 1, \pm 2, \dots$, yield an orthonormal basis in $L^2(-\pi, \pi)$.

Exercise 3.4.3 Suppose T is self-adjoint, not semi-bounded, but with a *gap* in the essential spectrum, i.e., there is some real open interval Ω which does not intersect the essential spectrum. Show that applying the minimax principle to the resolvent R_λ of T for an appropriately chosen $\lambda \in \mathbb{R}$ one can find all eigenvalues in Ω.

Exercise 3.4.4 Show that $\mathcal{D}_0 = C_0^1(c, d)$ give the same eigenvalues as the choice $\mathcal{D}_0 = \{u \in C^1(c, d) : u(c) = u(d) = 0\}$ in Example 3.4.7.

3.5 Notes and remarks

The first version of the spectral theorem general enough to include operators with continuous spectra was by Hilbert [103], who dealt with bounded quadratic forms in ℓ^2. Carleman [44] considered singular integral operators which are not bounded and therefore required a careful analysis of the difference between what we now call symmetric and self-adjoint operators.

The general concept of an unbounded self-adjoint operator and the spectral theorem for such operators, as well as the extension theory for symmetric operators given in Section 2.4 is due to von Neumann [172]. Another early source is the book [207] by Stone.

Our proof of the spectral theorem was sketched in 1934 by Doob and Koopman [56]. There are many other proofs, some of which may be found in Lax [140], but few lead so directly to the goal. Reed and Simon [185] present three different versions of the spectral theorem, viz. the functional calculus form, the multiplication operator form, and the projection valued measure form.

The formulas $T = \int_{-\infty}^{\infty} t\, dE_t$ and $Tu = \int_{-\infty}^{\infty} t\, dE_t u$ can be made sense of directly, at least if T is bounded, by introducing Stieltjes integrals with respect to operator-valued or Hilbert space-valued functions. This is a simple generalization of the scalar-valued case; see F. Riesz and E. R. Lorch [190]. Since this does not appear to be any more useful than the slightly weaker statement used in our version of the spectral theorem we have not done this.

The modern definition of a compact operator was introduced by F. Riesz [188]. Although the von Neumann extension theory shows how to construct the self-adjoint extensions of any non-negative symmetric operator, it does not identify which of these extensions are non-negative. This problem was solved by Kreĭn in [133] where he showed that if T is a non-negative symmetric operator in a Hilbert space H, then T has two distinguished non-negative self-adjoint extensions, the Friedrichs extension T_F and the Kreĭn extension T_K, which will coincide if and only if T is essentially self-adjoint. Any non-negative self-adjoint extension S of T satisfies $T_K \le S \le T_F$.

One of the first papers dealing with perturbations of a self-adjoint operator, before the concept of a Hilbert space operator was defined, is by Weyl [221] in 1908. Here Weyl introduces the essential spectrum, without naming it. This result is also quoted in Weyl [222, Satz 8] and in modern language states that if T is self-adjoint, possibly unbounded, and P is a compact operator on the same space, then $\sigma_e(T + P) = \sigma_e(T)$. A similar result is obtained also for unbounded perturbations P with $\mathcal{D}(P) \supset \mathcal{D}(T)$ if P is *relatively compact* with respect to T, which means that $P(T + \mathrm{i})^{-1}$ is compact. This gives a much more generally applicable result but is easily proved by considering both operators defined on $\mathcal{D}(T)$ provided with the graph norm. Much more concerning this circle of ideas may be found in the classical book by Tosio Kato [122].

In 1912 Hermann Weyl wrote a ground-breaking paper [224] about the asymptotic distribution of eigenvalues of the Dirichlet Laplacian in a bounded domain. The question was raised by the physicist H. A. Lorenz in 1910 in connection with black-body (actually, cavity) radiation.

The physicists believed that in the first approximation the distribution of the high frequency spectral lines depended only on the volume of the cavity. Hilbert is quoted as saying that they were probably right, but he did not expect to see a proof in his lifetime as he was not aware of any possible way to attack the question.

Weyl applied the technique now called the minimax principle together with what is now called *Dirichlet–Neumann bracketing* to prove the conjecture. There are earlier, similar ideas, but Weyl was the first to make substantial use of the minimax principle. Later the idea was picked up by Richard Courant who used and developed it to such an extent that it is today often known as *Courant's minimax principle*. See also the 1950 paper by Hermann Weyl [225], which is required reading for anyone interested in the development of spectral theory in the 20th century.

Chapter 4
Sturm–Liouville equations

4.1 Introduction

A *formal differential operator* on a real interval (a, b) is an expression $M[u] = \sum_{j,k=0}^{n}(p_{jk}u^{(j)})^{(k)}$ where the functions p_{jk}, defined on (a, b), are the coefficients of M. M is applied to u, which belongs to some class $\mathcal{C}$ of functions for which the expression $M[u]$ makes sense. Roughly, a formal differential operator is said to be *formally symmetric* in an interval (a, b) if, integrating by parts, we may write

$$\int_I M[u]\bar{v} = \int_I u\overline{M[v]} + \text{boundary terms}$$

for every compact subinterval $I \subset (a, b)$. A differential equation $M[u] = N[f]$ is said to be formally symmetric if M and N are formally symmetric formal differential operators.

A simple example of a formally symmetric differential equation, corresponding to M being real of order two and N order zero, is given by the general Sturm–Liouville equation

$$-(pu')' + qu = wf \, . \tag{4.1.1}$$

Here the coefficients p, q and w are given real-valued functions in a given interval (a, b). There are (at least) two Hermitian forms naturally associated with this equation, namely $\int_a^b (pu'\bar{v}' + qu\bar{v})$ and $\int_a^b u\bar{v}w$. Under appropriate positivity conditions either of these forms is a suitable choice of scalar product for a Hilbert space in which to study (4.1.1). The corresponding problems are then called *left-definite* and *right-definite* respectively. We shall discuss left-definite problems in Chapter 5, but here we shall study the right-definite case and therefore assume $w \geq 0$.

If p is not differentiable and u a solution of (4.1.1), then u' is not differentiable, but the product pu', called the *quasi-derivative* of u, is. In this case it is often convenient to interpret (4.1.1) as a first order system

C. Bennewitz et al., *Spectral and Scattering Theory for Ordinary Differential Equations*, Universitext, https://doi.org/10.1007/978-3-030-59088-8_4

$$\begin{pmatrix} 0 & -1 \\ 1 & 0 \end{pmatrix} U' + \begin{pmatrix} q & 0 \\ 0 & -1/p \end{pmatrix} U = \begin{pmatrix} w & 0 \\ 0 & 0 \end{pmatrix} V \tag{4.1.2}$$

when discussing existence and uniqueness of solutions. The system (4.1.2) becomes equivalent to (4.1.1) on setting $U = \binom{u}{pu'}$ and letting the first component of V be f. It is a special case of a fairly general first order system

$$JU' + QU = HV , \tag{4.1.3}$$

where J is a constant $n \times n$ matrix which is invertible and skew-Hermitian i.e., $J^* = -J$, the coefficients Q and H are $n \times n$ Hermitian matrix-valued functions locally integrable on (a, b), and U, V are $n \times 1$ matrix-valued functions. We shall study such systems in the second volume of this book.

In this chapter we always assume that (a, b) is a real interval, that the coefficients p, q and w are real-valued and that $1/p$, q and w are *locally integrable* in (a, b), i.e., Lebesgue integrable on every compact subinterval of (a, b). Basic for what follows is the following existence and uniqueness theorem, where $AC_{\text{loc}}(a, b)$ is the set of functions locally absolutely continuous in (a, b); see Section B.8.

Theorem 4.1.1. *Suppose $1/p$, q and w are locally integrable in an interval (a, b) and that $c \in (a, b)$. Then, for any locally integrable function f and arbitrary complex constants A, B and λ the initial value problem*[1]

$$\begin{cases} -(pu')' + qu = \lambda wu + f \text{ in } (a, b) , \\ u(c) = A, \quad pu'(c) = B \end{cases}$$

has a unique solution in $AC_{\text{loc}}(a, b)$ with quasi-derivative $pu' \in AC_{\text{loc}}(a, b)$. If A, B are independent of λ the solution $u(\cdot, \lambda)$ and its quasi-derivative $pu'(\cdot, \lambda)$ will be entire functions of λ, locally uniformly in (a, b).

The theorem has the following immediate consequence.

Corollary 4.1.2. *Let (a, b), p, q, w and λ be as in Theorem 4.1.1. Then the set of solutions of $-(pu')' + qu = \lambda wu$ in (a, b) is a two-dimensional linear space.*

If one rewrites $-(pu')' + (q - \lambda w)u = f$ as a first order system as in (4.1.2), then Theorem 4.1.1 and Corollary 4.1.2 become special cases of Theorem D.6 and Corollary D.4.

In this chapter we shall assume w non-negative and not a.e. zero and study (4.1.1) in the Hilbert space $L_w^2(a, b)$ with scalar product

$$\langle u, v \rangle = \int_a^b u\bar{v}w . \tag{4.1.4}$$

---

[1] Here differentiation is traditionally differentiation a.e. of functions in AC_{loc}, but we may always instead think of differentiation in the sense of distributions.

We may look at $L^2_w(a, b)$ as the completion of $C_0(a, b)$ in the corresponding semi-norm; see Appendix B. In most texts it is assumed that $p > 0$ and $w > 0$ almost everywhere, but it is enough to assume that $1/p$ and w are locally integrable and that w is non-negative almost everywhere. The following technical lemma is crucial.

Lemma 4.1.3. *Suppose* $-(pu')' + ru = 0$ *in* (a, b) *for some locally integrable* r, *that the closure of* $M \subset (a, b)$ *has positive Lebesgue measure and that* $u = 0$ *in* M. *Then* u *is identically zero.*

Proof. This is nearly obvious if the closure of M has an interior point, since if u is zero in an open set, so is its quasi-derivative, and thus u is identically zero by Theorem 4.1.1. If the closure of M has no interior points we may proceed as follows.

Since u is continuous we may as well assume M closed. We also discard any isolated points of M since these make up a nullset anyway. Thus every point of M is a point of accumulation of zeros for u. But almost every point in M is also a Lebesgue point (see Theorem B.8.8) for $1/p$, and since $1/p \neq 0$ a.e. we may assume $a \in M$ is a Lebesgue point for $1/p$ with $1/p(a) \neq 0$. If φ is continuous it follows that a is a Lebesgue point for φ/p as well (Exercise B.8.2).

Now suppose $M \ni a_j \to a$ with all $a_j \neq a$, so that $u(a_j) = 0$ for all j. Then since pu' is continuous $\frac{1}{a_j - a} \int_a^{a_j} \frac{pu'}{p} \to pu'(a)/p(a)$. However, the integral equals $u(a_j) - u(a) = 0$, so not only u, but also pu' vanishes at a. By Theorem 4.1.1 it follows that u is the zero solution. $\qquad\square$

In order to get a spectral theory for (4.1.1) we shall define a corresponding minimal operator, show that it is densely defined and symmetric and calculate its adjoint. In the next section we shall then find all self-adjoint restrictions of the adjoint.

Definition 4.1.4. Let T_c be the relation in $L^2_w(a, b)$ with domain consisting of those $u \in AC_{\text{loc}}$ which have *compact support*[2] for which $pu' \in AC_{\text{loc}}$ and $-(pu')' + qu = wf$ for some $f \in L^2_w(a, b)$ with compact support. We then write $(u, f) \in T_c$.

To see that T_c is an operator and not just a relation it is enough to show that it is symmetric and that $\mathcal{D}(T_c)$ is dense, for by Proposition 2.2.8 (5) T_c^*, and thus $T_c \subset T_c^*$, is then an operator.

To show this and calculate the adjoint T_c^* we need some preparation.

Definition 4.1.5. Suppose u, pu', v and pv' are all in AC_{loc} and define the p-*Wronskian* of u and v as $\mathcal{W}_p(u, v) = pu' v - u pv'$. If $p \equiv 1$ we write the Wronskian as $\mathcal{W}(u, v)$.

It is clear that $\mathcal{W}_p$ is in AC_{loc} and pointwise a bilinear form, and that it is skew-symmetric, i.e., $\mathcal{W}_p(u, v) = -\mathcal{W}_p(v, u)$, in particular $\mathcal{W}_p(u, u) = 0$. The following elementary fact is crucial.

Proposition 4.1.6. *Suppose* u *and* v *satisfy* $-(pu')' + f = 0$ *and* $-(pv')' + g = 0$ *respectively, where* f *and* g *are locally integrable functions. Then the derivative* $\mathcal{W}_p(u, v)' = fv - ug$.

---

[2] That is, for each u there is a compact subset of (a, b) outside of which $u = 0$.

In particular, if u and v are solutions of $-(pu')' + qu = \lambda wu$ in (a, b), then the Wronskian $\mathcal{W}_p(u, v)$ is constant in (a, b). This constant is non-zero precisely if u and v are linearly independent.

Proof. Differentiating we obtain $\mathcal{W}_p(u, v)' = (pu')'v - u(pv')' = fv - ug$ since the two additional terms cancel. If $fv = gu$, in particular if $f = (\lambda w - q)u$ and $g = (\lambda w - q)v$, then the Wronskian is constant.

For any point $c \in (a, b)$ the vectors $(u(c), pu'(c))$ and $(v(c), pv'(c))$ are proportional if and only if the constant is zero. Since the initial value problem has a unique solution this implies that u and v are linearly dependent precisely if $\mathcal{W}_p(u, v) = 0$. □

Now let v_1 and v_2 be solutions of $-(pv')' + qv = \lambda wv$ in (a, b) such that $\mathcal{W}_p(v_1, v_2) = 1$ in (a, b). There are certainly such solutions, for if v_1 and v_2 are linearly independent, then $\mathcal{W}_p(v_1, v_2)$ is constant $\neq 0$ so the value becomes 1 if we replace v_1 by an appropriate multiple. The following lemma is a version of the classical method known as *variation of constants* for solving the inhomogeneous equation in terms of the solutions of the homogeneous equation.

Lemma 4.1.7. *Let v_1, v_2 be solutions of $-(pv')' + qv = \lambda wv$ with $\mathcal{W}_p(v_1, v_2) = 1$, let $c \in (a, b)$ and suppose f locally integrable in (a, b). The solution u of $-(pu')' + qu = \lambda wu + f$ with initial data $u(c) = pu'(c) = 0$ is then given by*

$$u(x) = v_2(x) \int_c^x v_1(y)f(y)\,dy - v_1(x) \int_c^x v_2(y)f(y)\,dy\,. \qquad (4.1.5)$$

Proof. With u given by (4.1.5) clearly $u(c) = 0$. Differentiating we obtain

$$pu'(x) = pv_2'(x) \int_c^x v_1 f - pv_1'(x) \int_c^x v_2 f\,,$$

since the two other terms obtained cancel. Thus $pu'(c) = 0$. Differentiating again we obtain

$$(pu')'(x) = (pv_2')'(x) \int_c^x v_1 f - (pv_1')'(x) \int_c^x v_2 f - \mathcal{W}_p(v_1, v_2)f(x)$$

$$= (q(x) - \lambda w)u(x) - f(x)\,,$$

which was to be proved. □

Corollary 4.1.8. *Suppose $f \in L_w^2(a, b)$ with compact support in (a, b). Then the equation $-(pu')' + qu = wf$ has a solution u with compact support in (a, b) if and only if $\int_a^b vfw = 0$ for all solutions v of the homogeneous equation $-(pv')' + qv = 0$.*

Proof. The function wf is locally integrable in (a, b) since w is, so that 1 is locally in $L^2_w(a, b)$, as is f. If $c \in (a, b)$ is to the left of the support of f, then the function u given by (4.1.5) for $\lambda = 0$ and f replaced by wf is the only solution of the equation $-(pu')' + qu = wf$ which is identically zero near the left endpoint of (a, b).

Since v_1 and v_2 are linearly independent, in particular to the right of the support of f, the equation has a solution of compact support if and only if $\int_a^b v_1 fw = \int_a^b v_2 fw = 0$. Now v_1, v_2 is a basis for the solutions of the homogeneous equation so the corollary follows. □

Theorem 4.1.9. *The relation T_c is a densely defined and symmetric operator. Furthermore, if $u \in \mathcal{D}(T_c^*)$ and $f = T_c^* u$, then $u \in AC_{loc}$, $pu' \in AC_{loc}$, and u satisfies*[3] $-(pu')' + qu = wf$.

Conversely, if $u, f \in L^2_w(a, b)$ satisfy this equation, then $u \in \mathcal{D}(T_c^*)$ and $T_c^* u = f$.

Proof. Let u_1 be a solution of $-(pu_1')' + qu_1 = wf$ and assume $(u_0, f_0) \in T_c$. Integrating by parts twice, using that u_0 has compact support, we get

$$\langle u_0, f \rangle = \int_a^b u_0 \bar{f} w = \int_a^b u_0 (\overline{-(pu_1')' + qu_1})$$

$$= \int_a^b (pu_0' \bar{u_1'} + qu_0 \bar{u_1}) = \int_a^b (-(pu_0')' + qu_0) \bar{u_1} = \int_a^b f_0 \bar{u_1} w . \quad (4.1.6)$$

So, if f is orthogonal to the domain of T_c, then the last integral is zero for all compactly supported elements $f_0 \in L^2_w(a, b)$ for which there is a solution u_0 of $-(pu_0')' + qu_0 = wf_0$ with compact support. By Corollary 4.1.8 it follows that $\bar{u_1}$, and thus u_1, solves[4] $-(pv')' + qv = 0$ so that $wf = 0$. But this means that $f = 0$ as an element of $L^2_w(a, b)$ so T_c is densely defined and T_c^* an operator by Proposition 2.2.8.

The calculation (4.1.6) also proves the converse part of the theorem, taking $u_1 = u$, so that $T_{c'} \subset T_c^*$. Furthermore, if u is in the domain of T_c^* with $T_c^* u = f$ we obtain $0 = \langle u_0, f \rangle - \langle f_0, u \rangle = \int_a^b f_0 (\overline{u_1 - u}) w$. Just as before it follows that $u_1 - u$ satisfies $-(p(u_1 - u)')' + q(u_1 - u) = 0$. It follows that u satisfies the equation $-(pu')' + qu = wf$. The proof is complete. □

Being symmetric and densely defined T_c is closeable, and we define the *minimal operator* T_0 as the closure of T_c and denote the domain of T_0 (the *minimal domain*) by $\mathcal{D}(T_0)$. Similarly, the *maximal operator* T_1 is $T_1 := T_c^*$ with domain $\mathcal{D}(T_1) \supset \mathcal{D}(T_0)$. Thus the maximal domain $\mathcal{D}(T_1)$ consists of all $u \in AC_{loc}$ which are in $L^2_w(a, b)$ and such that $pu' \in AC_{loc}$ and $-(pu')' + qu = wf$ for some $f \in L^2_w(a, b)$. For such u we have $T_1 u = f$.

---

[3] i.e., there is a unique such element in the equivalence class of u.

[4] More precisely, there is a function u_2 satisfying $w(u_1 - u_2) = 0$ a.e. and $-(pu_2')' + qu_2 = 0$. But u_1 and u_2 are both continuous, so $u_1 = u_2$ in supp w. It follows that $wf = 0$ a.e. in supp w, and outside the support $wf = 0$ automatically.

4.2 Boundary conditions

Our next task is to identify the self-adjoint extensions of T_0, and here the extension theory of Section 2.4 comes into play. The defect indices of T_0 are accordingly the number of linearly independent solutions of the equations $-(pu')' + qu = iwu$ and $-(pu')' + qu = -iwu$ respectively which are in $L^2_w(a, b)$. There are only 2 (pointwise) linearly independent solutions for each of these equations so both defect indices are at most 2.

For the equation (4.1.1) the defect indices are always equal, since if $u \in L^2_w(a, b)$ solves $-(pu')' + qu = \lambda wu$, then $\bar{u} \in L^2_w(a, b)$ and solves the equation with λ replaced by $\bar{\lambda}$ (recall that p, q, w are all real-valued), and linear independence (as elements of $L^2_w(a, b)$) is preserved when conjugating functions. Thus, for our equation there are only three possibilities: The defect indices may both be 2, both may be 1, or both may be 0. We shall see later that all three cases can occur, depending on the choice of (a, b), p, q and w.

We shall now take a closer look at how self-adjoint realizations are determined as restrictions of the maximal operator. Suppose u and $v \in \mathcal{D}(T_1)$. Then the boundary form (cf. Section 2.4) is

$$\langle (u, T_1u), \mathcal{U}(v, T_1v) \rangle = i \int_a^b (u\overline{T_1v} - T_1u\,\bar{v})w$$

$$= i \int_a^b (-u\overline{(pv')'} + (pu')'\bar{v}) = i \lim_{K \to (a,b)} (pu'\bar{v} - u\overline{pv'})\big|_K , \quad (4.2.1)$$

the limit being taken over compact subintervals K of (a, b). We must restrict T_1 so that this vanishes. This means that the restriction of T_1 to a self-adjoint operator T is obtained by *boundary conditions* since the limit clearly only depends on the values of u and v in arbitrarily small neighborhoods of the endpoints of (a, b). This is of course the motivation behind the terms boundary operator and boundary form. If one imposes $n_+ = n_-$ linearly independent boundary conditions such that the restricted operator is symmetric, it will automatically be self-adjoint.

The situation is particularly simple near a finite endpoint of (a, b) such that $1/p$, q and w are integrable near the endpoint. Such an endpoint is called *regular*;[5] otherwise the endpoint is *singular*. If both endpoints are regular, we are dealing with a *regular problem*. This is what we did in Section 1.4 (for the case $p \equiv 1$ and q, w continuous with $w > 0$). The problem is *singular* if at least one of the endpoints is infinite, or if not all of $1/p$, q and w are in $L^1(a, b)$.

Regular problems. Clearly the defect indices are both 2 in the regular case, since all solutions of $-(pu')' + qu = \pm iwu$ are continuous on the compact closure of (a, b)

---

[5] It is also possible to regard an infinite endpoint near which $1/p, q, w$ are integrable as a regular endpoint, but traditionally one assumes a finite endpoint.

and thus in $L_w^2(a, b)$, and two pointwise linearly independent solutions cannot belong to the same equivalence class in $L_w^2(a, b)$ by Lemma 4.1.3. We shall determine all boundary conditions which yield self-adjoint restrictions of T_1.

The boundary form depends only on the boundary values $u(a)$, $pu'(a)$, $u(b)$ and $pu'(b)$ of u, so the possible boundary values constitute a linear subspace of $\mathbb{C}^4$. On the other hand, since $\mathbf{D}_{\pm i}$ is the eigenspace of $\mathcal{U}$ to the eigenvalue ± 1 the boundary form is positive definite on $\mathbf{D}_i$ and negative definite on $\mathbf{D}_{-i}$, both of which are two-dimensional spaces.

The boundary values for the defect spaces therefore span two two-dimensional spaces which do not overlap. It follows that as u ranges through $\mathcal{D}(T_1)$ the boundary values range through all of $\mathbb{C}^4$. It also follows that the boundary values for elements of $\mathcal{D}(T_0)$ all vanish, in fact that $\mathcal{D}(T_0)$ consists of all elements of $\mathcal{D}(T_1)$ with vanishing boundary values.

The boundary conditions need to restrict the four-dimensional space $\mathbf{D}_i \oplus \mathbf{D}_{-i}$ to a two-dimensional space D as in Theorem 2.4.2, so two independent linear, homogeneous conditions are needed. This means that there are 2×2 matrices A and B such that the boundary conditions are given by $AU(a) + BU(b) = 0$, where $U = \left({u \atop pu'} \right)$. Linear independence of the conditions means that the 2×4 matrix (A, B) must have linearly independent rows.

Consider first the case when A is invertible. Then the condition is of the form $U(a) = SU(b)$, where $S = -A^{-1}B$. Now, if $J = \left({0 \atop 1} {-1 \atop 0} \right)$ the boundary form is $i\{(U_2(a))^*JU_1(a) - (U_2(b))^*JU_1(b)\}$, so symmetry requires this to vanish. Inserting $U(a) = SU(b)$ the condition becomes $(U_2(b))^*(S^*JS - J)U_1(b) = 0$, where $U_1(b)$ and $U_2(b)$ are arbitrary 2×1 matrices. It follows that the condition $U(a) = SU(b)$ gives a self-adjoint restriction of T_1 precisely if S satisfies $S^*JS = J$. Such a matrix S is called *symplectic*.

Important special cases are when S is plus or minus the unit matrix. These cases are called *periodic* and *semi-periodic* boundary conditions respectively. Another valid choice is $S = J$. Since $\det J = 1 \neq 0$ and $\det(S^*JS) = |\det S|^2 \det J$, clearly any symplectic matrix S satisfies $|\det S| = 1$ (see also Exercise 4.2.2). In particular, it is invertible. Clearly the inverse of a symplectic matrix is also symplectic, so it follows that assuming the matrix B to be invertible again leads to boundary conditions of the form $U(a) = SU(b)$ with a symplectic S.

It remains to consider the case when neither A nor B is invertible. Neither A nor B can then be zero, since the rows of (A, B) are linearly independent. Thus A and B both have linearly dependent rows, one of which has to be non-zero. We may assume the first row in A to be non-zero, if necessary interchanging the rows of (A, B), and since adding a multiple of the first row to the second in (A, B) gives an equivalent set of conditions, we may assume the second row of A to be zero.

The second row of B will then be non-zero since the rows of (A, B) are linearly independent, and then adding an appropriate multiple of the second row to the first we may cancel the first row of B. At this point the first row gives a condition on $U(a)$ and the second a condition on $U(b)$. Such boundary conditions are called *separated*. We must find which separated conditions give a symmetric restriction of T_1.

Separated boundary conditions require $\left(\begin{smallmatrix} u \\ pu' \end{smallmatrix} \right)$ to be multiples of fixed non-zero column vectors at the endpoints of (a, b), and the boundary values at the endpoints are independent. If $\mathcal{A}$ and $\mathcal{B}$ are the fixed vectors the conditions for symmetry are therefore $\mathcal{A}^* J \mathcal{A} = \mathcal{B}^* J \mathcal{B} = 0$.

If $\mathcal{A} = \left(\begin{smallmatrix} a_1 \\ a_2 \end{smallmatrix} \right)$ the condition $\mathcal{A}^* J \mathcal{A} = 0$ reads $\operatorname{Im} a_2 \overline{a_1} = 0$. This means the arguments of a_1 and a_2 differ by a multiple of π, so we may assume that $\mathcal{A}$ is a real unit vector. Thus $\mathcal{A}^* J = (\cos \alpha, \sin \alpha)$ for some α, and we may assume $0 \le \alpha < \pi$. A separated condition at a is therefore of the form

$$u(a) \cos \alpha + pu'(a) \sin \alpha = 0 . \tag{4.2.2}$$

We may use the same arguments in dealing with the other endpoint, and so obtain the following theorem.

Theorem 4.2.1. *Separated, symmetric boundary conditions at a regular endpoint a are of the form (4.2.2) for some $\alpha \in [0, \pi)$. Important special cases are $\alpha = 0$, which gives the condition $u(a) = 0$, called a* Dirichlet *condition, and $\alpha = \pi/2$, which gives the condition $pu'(a) = 0$, called a* Neumann *condition.*

To obtain a self-adjoint realization of a regular problem via separated boundary conditions one condition of the form (4.2.2) is needed at each interval endpoint, perhaps for different values of α.

Every self-adjoint realization of a regular problem not given by symmetric, separated boundary conditions is given by coupled boundary conditions $U(a) = SU(b)$ where S is a symplectic but otherwise arbitrary matrix. Important special cases are periodic *and* semi-periodic *boundary conditions.*

Note that by Exercise 4.2.2 the boundary conditions found in Section 1.4 are precisely the boundary conditions that yield a self-adjoint realization of the Sturm–Liouville equation.

Singular problems. We need the following definition.

Definition 4.2.2. The *rank* of a Hermitian form B on a linear space is the maximal dimension of a subspace $\mathcal{M}$ such that $B(u, \mathcal{M}) = 0$ for some $u \in \mathcal{M}$ only if $u = 0$.

By Theorem 2.4.1 the rank of the boundary form of (4.2.1) on T_1 is thus $n_+ + n_- = 2n_+$. The boundary form $i \lim_{K \to (a,b)} (pu'\overline{v} - u\overline{pv'})\big|_K$ splits into the difference of the boundary form $i \lim_b (pu'\overline{v} - u\overline{pv'})$ at b and the boundary form at a, which is the similar limit at a. The boundary forms at a and b are *independent*, i.e., the rank of the full boundary form is the sum of the ranks of the two forms. To prove this we use the following technical lemma.

Lemma 4.2.3. *Let v_1, v_2 be independent solutions of $-(pv')' + qv = \lambda wv$ and let Ω be a non-trivial open, bounded interval with $\overline{\Omega} \subset (a, b)$ such that w is not zero a.e. in Ω. Then one can find $f \in C^1(\overline{\Omega})$ such that f and f' take arbitrarily prescribed values at the endpoints of Ω and such that $\int_\Omega fv_j w$, $j = 1, 2$, also take arbitrarily prescribed values.*

Proof. Write $f = f_1 + f_2$ where f_1 and f_1' take the prescribed values and $f_2 \in C_0^1(\Omega)$. It is then enough to show that f_2 may be chosen so that $\int_\Omega f_2 v_j w$ assume arbitrarily prescribed values.

But if the claim is not true the linear forms $C_0^1(\Omega) \ni f_2 \mapsto \int_\Omega f_2 v_j w$ are linearly dependent, so for some non-trivial linear combination $v = \alpha v_1 + \beta v_2$ we have $\int_\Omega f_2 v w = 0$ for all $f_2 \in C_0^1(\Omega)$. Thus $vw = 0$ on Ω as a distribution (see Appendix C) so that $v = 0$ a.e. in $\operatorname{supp} w \cap \Omega$. This is excluded by Lemma 4.1.3.   □

We may now prove the independence of the boundary forms at the endpoints.

Lemma 4.2.4. *Any $u \in \mathcal{D}(T_1)$ may be written as $u = u_1 + u_2$ where $u_j \in \mathcal{D}(T_1)$, $u_1 = 0$ near b and $u_2 = 0$ near a.*

Proof. Let $c < d$ be in the interior of (a, b) such that w is not a.e. zero in $[c, d]$ and let f_1 equal $T_1 u$ to the left of c and 0 to the right of d. It remains to define f_1 in $[c, d]$. Now put

$$u_1(x) = v_2(x)\left(u(c) + \int_c^x f_1 v_1 w\right) + v_1(x)\left(pu'(c) - \int_c^x f_1 v_2 w\right),$$

where v_1, v_2 are solutions of $-(pv_j')' + qv_j = 0$ with $v_2(c) = pv_1'(c) = 1$ and $pv_2'(c) = v_1(c) = 0$. By the variation of constants formula $u_1 = u$ to the left of c. If $\int_c^d f_1 v_1 w = -u(c)$, $\int_c^d f_1 v_2 w = pu'(c)$ we have $u_1 = 0$ to the right of d. We can always find such $f_1 \in C^1(c, d)$ by Lemma 4.2.3. Thus $u_1 \in \mathcal{D}(T_1)$. Setting $u_2 = u - u_1$ finishes the proof.   □

Since the rank of the boundary form at a regular endpoint clearly is two and the rank of the full boundary form is even and at most 4, the rank of the boundary form at a singular endpoint is 0 or 2.

Definition 4.2.5. An endpoint at which the boundary form has rank 2 is said to be in the *limit-circle* case, and an endpoint where the boundary form has rank 0 is said to be in the *limit-point* case. Briefly we say that the endpoint is limit-circle respectively limit-point.

A regular endpoint is clearly in the limit-circle case. The terminology derives from the methods used by Weyl [222] to construct the resolvent of a self-adjoint Sturm–Liouville problem. Lemma 4.2.4 shows that whether an endpoint is limit-circle or limit-point only depends on the behavior of the coefficients p, q, w in an arbitrarily small neighborhood of the endpoint. Now recall that the rank of the full boundary form is $\dim D_i \oplus D_{-i}$. The following theorem is therefore an immediate consequence of our analysis.

Theorem 4.2.6. *If $n_\pm = 0$, then both endpoints of (a, b) are limit-point, if $n_\pm = 1$ then one of them is limit-point and the other limit-circle, and if $n_\pm = 2$ both of them are limit-circle.*

The boundary form at a limit-point endpoint is identically zero, so no boundary condition is to be applied there, but if the other endpoint is limit-circle a symmetric, separated boundary condition must be imposed there to obtain a self-adjoint realization.

If both endpoints are limit-circle, then two boundary conditions must be imposed. For a regular problem either a symmetric, separated condition at each end or two linearly independent coupled boundary conditions given by a symplectic matrix.

In a singular limit-circle endpoint in general only $\operatorname{Im} pu'\overline{u}$ and not u and pu' individually will have a limit. We shall discuss how to express boundary conditions in such a situation in Section 4.4. If both endpoints are limit-point clearly T_1 is self-adjoint and T_c essentially self-adjoint.

It is clearly an important problem to find conditions on the interval and the coefficients of the equation which place an endpoint in limit-point or limit-circle cases. By the general theory an endpoint is limit-point if and only if there is a solution u of $-(pu')' + qu = \lambda wu$ for some *non-real* λ for which $|u|^2 w$ is not integrable near the endpoint.

A large number of different sufficient conditions for limit-point are known today. Here is a simple criterion for limit-point, known already to Weyl [222].

Theorem 4.2.7. *Suppose $c \in (a, b)$. Assume $w \notin L^1(c, b)$ and that in (c, b) we have $p > 0$ and $q - Cw \geq 0$ for some constant C. Then* (4.1.1) *is limit-point at b.*

A similar criterion is of course valid for the left endpoint of (a, b).

Proof. The general theory implies that if we can find a solution of $-(pu')'+qu = \lambda wu$ for some non-real λ such that $|u|^2 w$ is not integrable near b, then b is in the limit-point case (as we shall see in Corollary 4.2.11 the same is actually true for real λ). Let therefore u solve $-(pu')' + qu = \lambda wu$ with $u(c) = 1$, $pu'(c) = 0$, where $\operatorname{Re} \lambda = C$. Then

$$\big(\operatorname{Re}(pu'\overline{u})\big)' = \operatorname{Re}\big(p\,|u'|^2 + (pu')'\overline{u}\big) = p\,|u'|^2 + (q - Cw)\,|u|^2 \geq 0 .$$

Thus $\operatorname{Re}(pu'\overline{u})$ is non-decreasing and 0 at c so $\operatorname{Re}(u'\overline{u}) \geq 0$ in $[c, b)$. But $(|u|^2)' = 2\operatorname{Re}(u'\overline{u})$ so $|u|^2$ is non-decreasing. It follows that $|u|^2 \geq 1$ in (c, b) so $\int_c^b |u|^2 w = \infty$. Thus b is in the limit-point condition. $\qquad\square$

Weyl stated this for the interval $[0, \infty)$ and $w \equiv 1$. The conditions are then simply $p > 0$ and that q is bounded from below. A slightly more elaborate proof gives the following well known and useful criterion, due to Levinson [146], see also [46, Chapter 9, Theorem 2.4].

Theorem 4.2.8. *Suppose $c \in (a, b)$. Assume $w \notin L^1(c, b)$ and that $p > 0$ in (c, b). Also suppose there exists a function $M > 0$ which is locally absolutely continuous in (c, b), where $q \geq -Mw$, $\sqrt{w/pM} \notin L^1(c, b)$ and $p\,|M'|^2 \leq AM^3w$ for some positive constant A. Then* (4.1.1) *is limit-point at b.*

This is not quite a generalization of Theorem 4.2.7, since the choice of a constant M requires that $\sqrt{w/p}$ is not integrable near b, which is not required in Theorem 4.2.7. See also Exercise 4.2.3.

Proof. If $-(pu')' + qu = \lambda wu$ with $\mathrm{Re}\,\lambda = 0$ and $|u|^2 w$ is integrable near b the arithmetic-geometric inequality shows that

$$(2\,\mathrm{Re}(pu'\overline{u})/M)' = 2\,\mathrm{Re}(p\,|u'|^2 + (q - \lambda w)\,|u|^2)/M - 2M'\,\mathrm{Re}(pu'\overline{u})M^{-2}$$
$$\geq 2(p\,|u'|^2 + q\,|u|^2)/M - p\,|u'|^2/M - |u|^2\,p\,|M'|^2\,M^{-3}$$
$$\geq p\,|u'|^2/M - (2 + A)w\,|u|^2 \ .$$

Integrating we obtain $2\,\mathrm{Re}(pu'\overline{u})/M \geq \int_c^x p\,|u'|^2/M - C$ for some constant C. This shows that $p\,|u'|^2/M$ is integrable near b, since otherwise $(|u|^2)' = 2\,\mathrm{Re}(u'\overline{u})$ would eventually be positive so that $|u|^2$ would increase strictly and $|u|^2 w$ would not be integrable near b.

If v is another solution with $|v|^2 w$ integrable near b and the Wronskian $pu'v - u\,pv' = 1$ we then obtain, multiplying the Wronskian by $\sqrt{w/(pM)}$,

$$\sqrt{w/(pM)} = (\sqrt{p/M}u'\sqrt{w}v - \sqrt{w}u\sqrt{p/M}v')$$
$$\leq \tfrac{1}{2}(p\,|u'|^2/M + |v|^2 w + |u|^2 w + p\,|v'|^2/M) \ ,$$

again using the arithmetic-geometric inequality. Since the right-hand side is integrable near b this contradicts the assumption. $\qquad\square$

Example 4.2.9. Consider the interval $(0, \infty)$ and let $p \equiv w \equiv 1$ and $q(x) \geq -Cx^\alpha$, $\alpha \in \mathbb{R}$. If $\alpha \leq 0$ or $C \leq 0$, then q is bounded from below near ∞ so by Theorem 4.2.7 infinity is in the limit-point case.

If $\alpha > 0$ and $C > 0$ we take $M = Cx^\alpha$ in Theorem 4.2.8. Then $p\,|M'|^2\,M^{-3} = \alpha^2 x^{-(\alpha+2)}/C$ is bounded above near ∞. Now $\sqrt{w/(pM)} = 1/\sqrt{M} = x^{-\alpha/2}/\sqrt{C}$, which fails to be integrable near ∞ if $\alpha \leq 2$. The equation is therefore limit-point at ∞ if $\alpha \leq 2$. It may be shown that we have limit-circle at ∞ if $q(x) = -Cx^\alpha, C > 0$ and $\alpha > 2$ (Hille [104, Theorem 10.1.3]).

No explicit necessary and sufficient conditions for limit-point are known, but the following theorem due to Atkinson [8, Theorem 8] comes close.

Theorem 4.2.10. *If $\lambda \in \mathbb{C}$ the equation $-(pu')' + qu = \lambda wu$ on (a, b) has a solution for which $|u|^2 w$ is not integrable near b if and only if there exists functions y and f satisfying $-(py')' + qy = wf$ with $|f|^2 w$ locally integrable, and for some $c \in (a, b)$*

$$F(x) = |y(x)|^2\,w(x)\Big(1 + \int_c^x |f|^2\,w\Big)^{-1}$$

is not integrable near b. In particular, the equation is then limit-point at b.

Proof. Suppose there exists a solution of $-(pu')' + qu = \lambda wu$ with $\int_c^b |u|^2\,w = \infty$. Choosing $y = u$ and $f = \lambda u$ we have $\int_c^x F$ equal to $|\lambda|^{-2} \log(1 + |\lambda|^2 \int_c^x |u|^2\,w)$ if

$\lambda \neq 0$ and $\int_c^x |u|^2 w$ if $\lambda = 0$. In either case this tends to infinity as $x \to b$ proving one direction of the theorem.

To prove the other direction, let v_1, v_2 solve $-(pv')' + qv = \lambda wv$ with Wronskian $\mathcal{W}_p(v_1, v_2) = 1$ and put $g = (|v_1|^2 + |v_2|^2)w$. We must show that if we can choose y and f so that $F \notin L^1(c, b)$, then $g \notin L^1(c, b)$.

The variation of constants formula shows that

$$y(x) = v_2(x)\Big(A + \int_c^x v_1 w(f - \lambda y)\Big) - v_1(x)\Big(B + \int_c^x v_2 w(f - \lambda y)\Big)$$

for some constants A, B. By Cauchy–Schwarz both integrals have absolute value less than $(\int_c^x g \int_c^x |f - \lambda y|^2 w)^{1/2}$, which is less than $\frac{1}{2}(\int_c^x g + \int_c^x |f - \lambda y|^2 w)$, so

$$|y(x)|^2 w \leq Cg(x)\Big(2 + \int_c^x g + \int_c^x |f - \lambda y|^2 w + \int_c^x g \int_c^x |f - \lambda y|^2 w\Big)$$

$$\leq Cg(x)\Big(1 + \int_c^x g\Big)\Big(2 + \int_c^x |f - \lambda y|^2 w\Big)$$

for some constant C. Now $|f - \lambda y|^2 \leq 2|f|^2 + 2|\lambda y|^2$ so the last factor above is easily seen to be less than $2(1 + \int_c^x |f|^2 w)(1 + |\lambda|^2 \int_c^x F)$. Thus

$$F(x)\Big(1 + |\lambda|^2 \int_c^x F\Big)^{-1} \leq 2Cg\Big(1 + \int_c^x g\Big).$$

Integrating we obtain $|\lambda|^{-2} \log(1 + |\lambda|^2 \int_c^x F)$ to the left if $\lambda \neq 0$ and $\int_c^x F$ if $\lambda = 0$. In either case this tends to infinity as $x \to b$, and since the integral of the right member is $C(2 + \int_c^x g)\int_c^x g$ this forces $\int_c^x g \to \infty$ as $x \to b$, which finishes the proof.   □

An immediate consequence is the following.

Corollary 4.2.11. *If for some $\lambda \in \mathbb{C}$, real or not, there is a solution of $-(pu')' + qu = \lambda wu$ with $|u|^2 w$ not integrable near b, then this is true for all other values of $\lambda \in \mathbb{C}$.*

It follows that given any coefficients p and q it is possible to choose w so that we have the endpoints in the limit-point or limit-circle cases as desired. This is clear, since if v_1, v_2 are linearly independent solutions of $-(pv')' + qv = 0$ we simply have to choose w so that $(|v_1|^2 + |v_2|^2)w$ is integrable near a desired limit-circle endpoint, but not near a desired limit-point endpoint.

We give two limit-point criteria as corollaries of Theorem 4.2.10.

Corollary 4.2.12. *Suppose $w \equiv 1$ in $[0, \infty)$. Then the equation (4.1.1) is limit-point at infinity if $q \in L^2(0, \infty)$ or even just $\int_c^x |q|^2 = \mathcal{O}(x \prod_1^k \log_j x)$ as $x \to \infty$ for some k and $c > 0$, where $\log_j$ is a j times iterated logarithm.*

Proof. If we pick $y \equiv 1$ and $f = q$, Theorem 4.2.10 shows that we have limit-point at infinity if $q \in L^2(c, \infty)$.

Since $\log_{k+1} x \to \infty$ as $x \to \infty$ and is a primitive of $\left(x \prod_1^k \log_j x\right)^{-1}$ it is even enough if $\int_c^x |q|^2 = \mathcal{O}\left(x \prod_1^k \log_j x\right)$ as $x \to \infty$. □

Note that there are no sign restrictions on p in this corollary, and neither is there in the next.

Corollary 4.2.13. *Consider* (4.1.1) *on* (a, b) *where* $|q| \le Cw$ *near* b *for some constant* C *and let* P *be a primitive of* $1/p$. *Then* b *is in the limit-point condition if* $(1 + P^2)w$ *is not integrable near* b. *In particular, we have limit-point at* b *unless* w *is integrable near* b.

Proof. We use Theorem 4.2.10 with either $y = 1$ or $y = P$. In either case $(py')' = 0$ so we must pick $f = q/w$ respectively $f = qP/w$, where we set $f = 0$ wherever $w = 0$.

Using $|q| \le Cw$ we obtain $\int_c^x |f|^2 w \le C^2 \int_c^x w$ if $y = 1$. It follows that $\int_c^x F \ge C^{-2} \log(1 + C^2 \int_c^x w)$. Thus we have limit-point at b if w is not integrable near b. Similarly, $\int_c^x |f|^2 w \le C^2 \int_c^x P^2 w$ if $y = P$ so in this case $\int_c^x F \ge C^{-2} \log(1 + C^2 \int_c^x P^2 w)$. We therefore also have limit-point at b if $P^2 w$ is not integrable near b. □

There are very few conditions for limit-point known which do not assume a fixed sign for p near the endpoint. Note that the bound from below required of q in Theorem 4.2.7 where p is positive corresponds to a bound from above if p is negative. The last corollary therefore gives an intuitively reasonable extension of the condition of Theorem 4.2.7 to the case when p changes sign arbitrarily.

Exercises

Exercise 4.2.1 A more general equation of Sturm–Liouville type is

$$-(pu')' + i(su)' + i\overline{s}u' + qu = wf .$$

Show that this equation is formally symmetric if the coefficients p, q, w are real-valued and that an existence and uniqueness theorem for the initial value problem is valid if $1/p$, $|s|^2/p$, q and w are locally integrable.

Assuming this, also show that if s is locally absolutely continuous one may find a locally absolutely continuous function h with $|h| = 1$ such that setting $v = hu$, $g = hf$ the equation $-(pv')' + \tilde{q}v = wg$ is satisfied, where $\tilde{q}$ does not depend on u or f and is real-valued and locally integrable. Conclusions?

Exercise 4.2.2 Show that a symplectic 2×2 matrix S is of the form λP where $|\lambda| = 1$ and P is a real 2×2 matrix with $\det P = 1$.

Exercise 4.2.3 If we specialize Theorem 4.2.7 to the case when M is a positive constant the statement is that we have limit-point if $p > 0$, $q \geq -Mw$ and neither w nor $\sqrt{w/p}$ are integrable near the endpoint, whereas Theorem 4.2.6 only requires non-integrability of w but is otherwise the same.

Show that if $p > 0$ and $q \geq -Mw$ for a constant M we also have limit-point if $\sqrt{w/p}$ is not integrable near the endpoint, with no further assumptions.

Hint: First show that the equation $-(pu')' + (q + Mw)u = 0$ has a solution for which $1/(pu^2)$ is integrable near the endpoint, and then show that if $w\,|u|^2$ is integrable near the endpoint so is $\sqrt{w/p}$. Then use Corollary 4.2.11.

To show the existence of u, first show that for c near the endpoint the solution v with $v(c) = 1$, $pv'(c) = 0$ is positive up to the endpoint. Then show that $u(x) = v(x) \int_c^x 1/(pv^2)$ has the desired properties (see also the discussion in Section 5.5).

4.3 Expansion in eigenfunctions

The spectral theorem we proved in Chapter 3.1 is very powerful, but sometimes its abstract nature is a drawback, and one needs a more explicit expansion, analogous to a Fourier series or a Fourier transform. We shall prove such an expansion here. As in our proof of the spectral theorem, we shall deduce our results from properties of the resolvent, but now need a more explicit description of the resolvent operator. The first step is to prove that the resolvent is an integral operator.

Theorem 4.3.1. *Suppose T is a self-adjoint realization of (4.1.1) in $L_w^2(a, b)$ and that R_λ is the resolvent of T.*

If $\lambda \in \rho(T)$ there exists Green's function $g(x, y, \lambda)$ for T, which is in $L_w^2(a, b)$ as a function of y for every fixed $x \in (a, b)$ and such that

$$R_\lambda u(x) = \langle u, \overline{g(x, \cdot, \lambda)} \rangle$$

for any $u \in L_w^2(a, b)$. There is also a kernel $g_1(x, \cdot, \lambda)$ in $L_w^2(a, b)$ for every $x \in (a, b)$ such that $p(R_\lambda u)'(x) = \langle u, \overline{g_1(x, \cdot, \lambda)} \rangle$, and both $\|g(x, \cdot, \lambda)\|$ and $\|g_1(x, \cdot, \lambda)\|$ are locally uniformly bounded in (a, b).

Proof. By the variation of constants formula we have

$$R_\lambda u(x) = v_2(x)\Big(A(u) + \int_c^x uv_1 w\Big) - v_1(x)\Big(B(u) + \int_c^x uv_2 w\Big), \qquad (4.3.1)$$

where $c \in (a, b)$, v_1, v_2 solve $-(pv')' + qv = \lambda wv$ with $\mathcal{W}_p(v_1, v_2) = 1$ and A, B depend on u and λ but not on x. If K is a compact subinterval of (a, b) containing c and x the integrals are bounded by $(\int_K |v_j|^2 w)^{1/2}\|u\|$, $j = 1, 2$, by the Cauchy–Schwarz inequality.

With K so large that $K \cap \operatorname{supp} w$ is not a nullset, Lemma 4.2.3 guarantees that we may find $\varphi \in C_0(K)$ with $\int_a^b \overline{\varphi} v_1 w = 0$ and $\int_a^b \overline{\varphi} v_2 w = 1$. Then the scalar product of (4.3.1) with φ gives

$$A(u) = \langle R_\lambda u, \varphi \rangle - \left\langle v_2(x) \int_c^x u v_1 w, \varphi(x) \right\rangle + \left\langle v_1(x) \int_c^x u v_2 w, \varphi(x) \right\rangle .$$

The last two terms are each bounded by $(\int_K |v_1|^2 \, w \int_K |v_2|^2 \, w)^{1/2} \|\varphi\| \|u\|$, so since R_λ is a bounded operator on $L_w^2(a, b)$ it follows that $u \mapsto A$ is a bounded linear form on $L_w^2(a, b)$. Similarly one shows that B is a bounded linear form. Thus there are v_A and $v_B \in L_w^2(a, b)$ such that $A(u) = \langle u, \overline{v_A} \rangle$ and $B(u) = \langle u, \overline{v_B} \rangle$, and Green's function is

$$g(x, y, \lambda) = v_2(x)(v_A(y) + \chi(y)v_1(y)) - v_1(x)(v_B(y) + \chi(y)v_2(y)) . \qquad (4.3.2)$$

Here χ is the characteristic function of $[c, x]$ if $c \le x$, i.e., χ equals 1 in $[c, x]$ and 0 outside. If instead $c > x$, then $-\chi$ is the characteristic function of $[x, c]$. The solutions v_j are continuous so they are bounded in K. Thus $u \mapsto R_\lambda u(x)$ are linear forms uniformly bounded for $x \in K$, which means that $x \mapsto \|g(x, \cdot, \lambda)\|$ is locally uniformly bounded. Differentiating (4.3.1) we obtain

$$p(R_\lambda u)'(x) = p v_2'(x) \left(A + \int_c^x u v_1 w \right) - p v_1'(x) \left(B + \int_c^x u v_2 w \right) ,$$

so in the same way it follows that $p(R_\lambda u)'(x)$ are locally uniformly bounded linear forms on $L_w^2(a, b)$. $\qquad \square$

We can now prove the following important corollary.

Corollary 4.3.2. *Suppose both endpoints of (a, b) are limit-circle for (4.1.1). Then the resolvent of any self-adjoint restriction of the maximal operator is a Hilbert–Schmidt operator in $L_w^2(a, b)$.*

Thus all such operators have discrete spectrum and an orthonormal basis of eigenfunctions in $L_w^2(a, b)$. In particular this holds for all regular problems.

Proof. By (4.3.2) $(x, y) \mapsto g(x, y, \lambda)$ is measurable and

$$|g(x, y, \lambda)| \le |v_2(x)| (|v_A(y)| + |v_1(y)|) + |v_1(x)| (|v_B(y)| + |v_2(y)|) .$$

If the equation is limit-circle at both a and b the functions v_1, v_2 are both in $L_w^2(a, b)$, so then the right member is a sum of functions which are a product of a function of x and one of y, both of which are in $L_w^2(a, b)$. The corollary now follows from Theorem 3.3.7. $\qquad \square$

It should be pointed out that having limit-circle at both endpoints is by no means a *necessary* condition for the resolvent to be compact, or even Hilbert–Schmidt, as we shall see in Theorem 4.5.8 and Example 6.1.13. It follows from Theorem 4.3.1 that $R_\lambda u_j \to R_\lambda u$ and $p(R_\lambda u_j)' \to p(R_\lambda u)'$ locally uniformly in (a, b) if $u_j \to u$ in $L_w^2(a, b)$ and pointwise and locally boundedly if $u_j \rightharpoonup u$ weakly. Since $R_\lambda u_j(x) = R_\lambda u_j(c) + \int_c^x p(R_\lambda u_j)'/p$ it follows by dominated convergence that $R_\lambda u_j$ converges locally uniformly also if $u_j \rightharpoonup u$ only weakly.

If $\{e_j\}_1^\infty$ is the sequence of orthonormal eigenfunctions for a problem with discrete spectrum and $s_N u = \sum_1^N \hat u_j e_j$ a partial sum of the generalized Fourier series of $u \in L_w^2(a, b)$ we know that $\|u - s_N u\| \to 0$ as $N \to \infty$. We obtain much better convergence if $u \in \mathcal{D}(T)$.

Corollary 4.3.3. *Suppose a self-adjoint realization T of* (4.1.1) *in $L_w^2(a, b)$ has an orthonormal basis of eigenfunctions. If $u \in \mathcal{D}(T)$, then the generalized Fourier series of u, as well as the series of quasi-derivatives, converge locally uniformly in (a, b).*

If both endpoints of (a, b) are limit-circle and v_1, v_2 are as in (4.3.1), *then both $s_N u/(|v_1| + |v_2|)$ and $p(s_N u)'/(|pv_1'| + |pv_2'|)$ converge uniformly in (a, b).*

In particular, if v_1, v_2 are bounded, then $s_N u$ converges uniformly in (a, b) and if pv_1', pv_2' are bounded, then $p(s_N u)'$ converges uniformly in (a, b). All these assumptions are satisfied if both endpoints are regular.

Proof. Suppose $u \in \mathcal{D}(T)$, i.e., $Tu = g$ for some $g \in L_w^2(a, b)$, and let $f = g - \lambda u$, so that $u = R_\lambda f$. If e_j is an eigenfunction of T with eigenvalue λ_j we have $Te_j = \lambda_j e_j$ or $(T - \lambda)e_j = (\lambda_j - \lambda)e_j$ so $R_\lambda e_j = e_j/(\lambda_j - \lambda)$. It follows that $\hat u_j = \langle R_\lambda f, e_j \rangle = \langle f, R_{\bar\lambda} e_j \rangle = \hat f_j/(\lambda_j - \lambda)$, so that $\langle u, e_j \rangle e_j = \langle f, e_j \rangle R_\lambda e_j$.

Thus $s_N u = R_\lambda(s_N f)$, where $s_N f$ is the Nth partial Fourier sum for f. Since $s_N f \to f$ in $L_w^2(a, b)$, it follows from Theorem 4.3.1 and the remark before Theorem 4.3.3 that $s_N u \to u$ and $p(s_N u)' \to pu'$ locally uniformly.

Now, if v_1, v_2 are in $L_w^2(a, b)$, then the integrals in (4.3.1) with u replaced by $f - s_N f$ may be estimated by $\|v_j\| \|f - s_N f\| \to 0$ so they converge uniformly to zero. The second part of the corollary follows. $\square$

The convergence is even better than the corollary shows, since it is *absolute* and uniform; see Exercise 4.3.7.

Expansions with one regular endpoint. We now have a satisfactory eigenfunction expansion theory for Sturm–Liouville boundary value problems that are regular, or at least have both endpoints in the limit-circle condition, so we turn next to singular problems which may have at least one endpoint in the limit-point condition. We then need to take a closer look at Green's function.

In this section we shall only consider the case of separated boundary conditions for (a, b) where a is a regular endpoint and b possibly singular, leaving all other singular problems to the next section. With these assumptions Green's function has a particularly simple structure.

Let φ, θ be solutions of $-(pu')' + qu = \lambda wu$ with initial data

$$
\begin{cases} \varphi(a, \lambda) = -\sin\alpha \,, \\ p\varphi'(a, \lambda) = \cos\alpha \,, \end{cases}
\qquad
\begin{cases} \theta(a, \lambda) = \cos\alpha \,, \\ p\theta'(a, \lambda) = \sin\alpha \,. \end{cases}
\tag{4.3.3}
$$

These solutions and their quasi-derivatives are entire functions of λ, locally uniformly with respect to x, according to Theorem 4.1.1. The solution φ satisfies the boundary condition (4.2.2) and θ another similar condition such that the Wronskian $\mathcal{W}_p(\varphi, \theta) = 1$.

Theorem 4.3.4. *Suppose T is a self-adjoint realization of (4.1.1) on (a, b) with a regular, given by the separated condition (4.2.2) at a, and another separated, self-adjoint condition at b if needed, i.e., if b is regular or limit-circle.*

There is then a function m defined in $\rho(T)$ such that Green's function equals $g(x, y, \lambda) = \varphi(\min(x, y), \lambda)\psi(\max(x, y), \lambda)$ where the Weyl *solution ψ is given by $\psi(x, \lambda) = \theta(x, \lambda) + m(\lambda)\varphi(x, \lambda)$.*

The coefficient $m(\lambda)$, called the Weyl–Titchmarsh *m-function, is a Nevanlinna function in the sense of Appendix E. The kernel g_1 is $g_1(x, y, \lambda) = p(x)\frac{\partial}{\partial x}g(x, y, \lambda)$, which equals $p\varphi'(x, \lambda)\psi(y, \lambda)$ if $x < y$ and $\varphi(y, \lambda)p\psi'(x, \lambda)$ if $x > y$.*

Proof. First note that the minimal operator cannot have an eigenvalue, for an eigenfunction would solve $-(pu')' + qu = \lambda wu$ with $u(a) = pu'(a) = 0$ so it cannot be non-trivial. According to Theorem 2.4.9 we then have dim $D_\lambda = n_\pm$ for all $\lambda \notin \sigma_e(T)$.

The solution φ satisfies the boundary condition at a and can therefore only satisfy the boundary condition at b if λ is an eigenvalue and thus in $\sigma(T)$. On the other hand, if $\lambda \in \rho(T)$ there will be a solution in $L_w^2(a, b)$ satisfying the boundary condition at b, since if defect indices are 1 there is no condition at b, and if defect indices are 2, then the condition at b is a linear, homogeneous condition on a two-dimensional space, which leaves a one-dimensional space and thus a non-trivial solution in $L_w^2(a, b)$. We may therefore find a unique $m(\lambda)$ so that $\psi = \theta + m\varphi$ satisfies the boundary condition at b, and we have $\mathcal{W}_p(\varphi, \psi) = \mathcal{W}_p(\varphi, \theta) + m\mathcal{W}_p(\varphi, \varphi) = 1$.

Now assume $u \in L_w^2(a, b)$ with supp u a compact subset of $[a, b)$ and define

$$v(x) = \psi(x, \lambda) \int_a^x u\varphi(\cdot, \lambda)w + \varphi(x, \lambda) \int_x^b u\psi(\cdot, \lambda)w ,$$

so that $v(a) = -\sin\alpha \int_a^b u\psi(\cdot, \lambda)w$. Differentiating we obtain

$$pv'(x) = p\psi'(x, \lambda) \int_a^x u\varphi(\cdot, \lambda)w + p\varphi'(x, \lambda) \int_x^b u\psi(\cdot, \lambda)w . \qquad (4.3.4)$$

Thus $pv'(a) = \cos\alpha \int_a^b u\psi(\cdot, \lambda)w$ so v satisfies the boundary condition at a. If x is to the right of the support of u we obtain $v(x) = \psi(x, \lambda) \int_a^b u\varphi(\cdot, \lambda)w$ so v also satisfies the boundary condition at b, being a multiple of ψ near b. Differentiating again we obtain

$$-(pv')'(x) + (q(x) - \lambda w)v(x) = \mathcal{W}_p(\varphi, \psi)wu(x) = wu(x) .$$

It follows that $v = R_\lambda u$ for compactly supported u, and since such functions are dense in $L_w^2(a, b)$ and R_λ bounded, this formula is correct for every $u \in L_w^2(a, b)$. Thus Green's function for our operator is as stated, and from (4.3.4) it follows that the kernel g_1 is also as stated.

It remains to show that m is a Nevanlinna function. If u and v both have compact supports in (a, b) and $\lambda \in \rho(T)$ we have

$$\langle R_\lambda u, v \rangle = \iint g(x, y, \lambda) u(y) w(y) \overline{v(x)} w(x) \, dx \, dy \, ,$$

the double integral being absolutely convergent. Similarly

$$\langle u, R_{\overline{\lambda}} v \rangle = \iint \overline{g(y, x, \overline{\lambda})} u(y) w(y) \overline{v(x)} w(x) \, dx \, dy \, ,$$

and since the integrals are equal for all u, v by Theorem 2.3.7 (3) we obtain $g(x, y, \lambda) = \overline{g(y, x, \overline{\lambda})}$ for $x, y \in \operatorname{supp} w$ or, if also $x < y$,

$$\varphi(x, \lambda)\theta(y, \lambda) + \varphi(x, \lambda)\varphi(y, \lambda)m(\lambda) = \overline{\varphi(x, \lambda)\theta(y, \lambda) + \varphi(x, \lambda)\varphi(y, \lambda)m(\overline{\lambda})} \, ,$$

since $\overline{\varphi(\cdot, \overline{\lambda})} = \varphi(\cdot, \lambda)$ and similarly for θ. Now $\varphi(x, \lambda) \ne 0$ for almost all $x \in \operatorname{supp} w$ by Lemma 4.1.3 so $\overline{m(\overline{\lambda})} = m(\lambda)$. Also, $\lambda \mapsto R_\lambda u(x)$ is analytic for $\lambda \in \rho(T)$ and for compactly supported u

$$R_\lambda u(x) = \theta(x, \lambda) \int_a^x u\varphi(\cdot, \lambda)w + \varphi(x, \lambda) \int_x^b u\theta(\cdot, \lambda)w$$

$$+ m(\lambda)\varphi(x, \lambda) \int_a^b u\varphi(\cdot, \lambda)w \, .$$

The first two terms on the right are obviously entire functions according to Theorem 4.1.1, as is the coefficient of $m(\lambda)$, and since by choice of u we may always assume that this coefficient is non-zero in a neighborhood of any given λ it follows that $m(\lambda)$ is analytic in $\rho(T)$.

Finally, integration by parts shows that if $\operatorname{Im} \lambda \ne 0$

$$\lambda \int_a^x |\psi|^2 w = \int_a^x (-(p\psi')' + q\psi)\overline{\psi} = \left[-p\psi'\overline{\psi}\right]_a^x + \int_a^x (p|\psi'|^2 + q|\psi|^2) \, .$$

Taking the imaginary part of this and using the fact that ψ satisfies the boundary condition at b so that $\operatorname{Im}(p\psi'\overline{\psi}) \to 0$ at b we obtain

$$0 < \int\limits_a^b |\psi(\cdot,\lambda)|^2\, w = \frac{\operatorname{Im} m(\lambda)}{\operatorname{Im}\lambda}\,, \tag{4.3.5}$$

since a simple calculation shows that $\operatorname{Im}(p\psi'(a,\lambda)\overline{\psi(a,\lambda)}) = \operatorname{Im} m(\lambda)$. Thus m has all the properties required of a Nevanlinna function. □

Remark 4.3.5. By Equation (4.3.5) $\psi \in L^2_w(a,b)$ for $\operatorname{Im}\lambda \neq 0$. But if $n_\pm = 1$ there can be no linearly independent solution in $L^2_w(a,b)$ so viewing (4.3.5) as an equation for m there is just one solution in this case. On the other hand, if $n_\pm = 2$ every solution is in $L^2_w(a,b)$. In this case we may write (4.3.5) as

$$|m|^2\,\|\varphi\|^2 + \operatorname{Re}(m(2\langle\varphi,\theta\rangle + i/\operatorname{Im}\lambda)) + \|\theta\|^2 = 0\,.$$

Viewing m as an unknown and φ, θ and λ as known, such an equation describes a circle in the complex plane, a single point or it may have no solution at all. Since we know there is a solution for each boundary condition at b, in this case it describes a circle. This is the origin of the terms limit-point and limit-circle. See also Exercise 4.3.2.

Remark 4.3.6. If the regular endpoint a is the right endpoint of our interval we may of course carry out the construction of the resolvent analogously. However, in this case it is convenient to define the m-function so that $\psi = \theta - m\varphi$, to make m a Nevanlinna function also in this case.

Theorem E.1.1 now shows that there is a unique increasing and left-continuous function η with $\eta(0) = 0$ and unique real numbers A and $B \geq 0$ such that

$$m(\lambda) = A + B\lambda + \int\limits_{-\infty}^{\infty}\left(\frac{1}{t-\lambda} - \frac{t}{t^2+1}\right)d\eta(t)\,. \tag{4.3.6}$$

We call η the *spectral function* and $d\eta$ the *spectral measure* for T for reasons which will become clear presently. The spectral measure gives rise to a Hilbert space L^2_η, which consists of (equivalence classes of) those functions $\hat{u}$ which are measurable with respect to $d\eta$ and for which $\|\hat{u}\|^2_\eta = \int_{-\infty}^{\infty}|\hat{u}|^2\,d\eta$ is finite. Alternatively, we may think of L^2_η as the completion in this (semi-)norm of compactly supported, continuous functions. By Theorem B.7.5 these alternative definitions give the same space. We denote the scalar product in L^2_η by $\langle\cdot,\cdot\rangle_\eta$. The main result of this section is the following.

Theorem 4.3.7.

(1) If $u \in L^2_w(a,b)$ the integral $\int_a^x u\varphi(\cdot,t)w$ converges in L^2_η as $x \to b$. The limit is called the *generalized Fourier transform of u and is denoted by* $\mathcal{F}(u)$ *or* $\hat{u}$. We write this as $\hat{u}(t) = \langle u, \varphi(\cdot,t)\rangle$, although the integral may not converge pointwise.

(2) *The mapping $u \mapsto \hat{u}$ is unitary between $L_w^2(a, b)$ and L_η^2 so that Parseval's formula $\langle u, v \rangle = \langle \hat{u}, \hat{v} \rangle_\eta$ is valid if $u, v \in L_w^2(a, b)$.*

(3) *The integral $\int_K \hat{u}(t) \varphi(x, t) \, d\eta(t)$ converges in $L_w^2(a, b)$ as $K \to \mathbb{R}$ through compact intervals. If $\hat{u} = \mathcal{F}(u)$ the limit is u, so the integral is the inverse of the generalized Fourier transform. Again, we write $u(x) = \langle \hat{u}, \varphi(x, \cdot) \rangle_\eta$, although the integral may not converge pointwise.*

(4) *Let E_Δ denote the spectral projector of T for the interval (or Borel set) Δ. Then $E_\Delta u(x) = \int_\Delta \hat{u} \varphi(x, \cdot) \, d\eta$.*

(5) *If $u \in \mathcal{D}(T)$ then $\mathcal{F}(Tu)(t) = t\hat{u}(t)$. Conversely, if $\hat{u}$ and $t\hat{u}(t)$ are in L_η^2, then $\mathcal{F}^{-1}(\hat{u}) \in \mathcal{D}(T)$.*

Before we prove this theorem, let us interpret it in terms of the spectral theorem. If the interval Δ shrinks to a point t, then E_Δ tends to zero, unless t is an eigenvalue, in which case we obtain the projection on the eigenspace. By (4) this means that eigenvalues are precisely those points at which the function η has a (jump) discontinuity; the continuous spectrum corresponds to points where η is continuous, but which are still points of increase for η, i.e., there is no neighborhood of the point where η is constant.

In terms of measure theory, this means that the *atomic part* of the measure $d\eta$ determines the eigenvalues and the spectrum equals supp $d\eta$. Note that T has only simple eigenvalues, since an eigenfunction has to be a multiple of φ. Therefore, the discrete spectrum consists of the isolated points in supp $d\eta$ and the essential spectrum of all non-isolated points of supp $d\eta$. In terms of the m-function the discrete spectrum is the set of poles of m and the essential spectrum consists of all other singularities of m.

We shall prove Theorem 4.3.7 through a long (but finite!) sequence of lemmas. First note that for $u \in L_w^2(a, b)$ with compact support in $[a, b]$ the function $\hat{u}(\lambda) = \langle u, \varphi(\cdot, \overline{\lambda}) \rangle$ is an entire function of λ since $\varphi(x, \lambda)$ is entire, locally uniformly in x, according to Theorem 4.1.1.

Lemma 4.3.8. *The function $\langle R_\lambda u, v \rangle - m(\lambda) \hat{u}(\lambda) \hat{v}(\overline{\lambda})$ is entire for all $u, v \in L_w^2(a, b)$ with compact supports in $[a, b]$.*

Proof. If the supports are inside $[a, c]$, direct calculation shows that the function is

$$\int\limits_a^c \left(\theta(x, \lambda) \int\limits_a^x u \varphi(\cdot, \lambda) w + \varphi(x, \lambda) \int\limits_x^c u \theta(\cdot, \lambda) w \right) \overline{v(x)} w(x) \, dx \, .$$

This is obviously an entire function of λ.   □

In order to be able to integrate $\lambda \mapsto \langle R_\lambda u, v \rangle$ around a contour intersecting $\mathbb{R}$ we need the following lemma.

Lemma 4.3.9. *Let η be increasing in $[-1, 1]$ and differentiable at 0. Then the integral $\int_{-1}^1 \int_{-1}^1 (t^2 + s^2)^{-1/2} \, ds \, d\eta(t)$ converges absolutely.*

Proof. We have

$$\int\limits_{-1}^{1} \frac{ds}{\sqrt{t^2 + s^2}} = \log(\sqrt{t^2 + s^2} + s)\Big|_{s=-1}^{s=1} = 2\log(\sqrt{t^2 + 1} + 1) - \log t^2$$

for $t \neq 0$. The first term is positive, continuous, and bounded and thus absolutely integrable on $(-1, 1)$ with respect to $d\eta$. Integrating by parts

$$\int\limits_{-1}^{1} \log t^2 \, d\eta = -2 \int\limits_{-1}^{1} \frac{\eta(t) - \eta(0)}{t} \, dt \ .$$

The integrand is bounded since η is differentiable at 0. By Tonelli's theorem B.9.5 the double integral is absolutely convergent. □

As in the spectral theorem we denote the resolution of the identity for T by $\{E_t\}_{t \in \mathbb{R}}$. The following lemma shows that $E_t u$ may be calculated by integration of $\hat{u}$ for compactly supported u.

Lemma 4.3.10. *Let $u \in L_w^2(a, b)$ have compact support in $[a, b]$ and assume $c < d$, where c, d are points of differentiability for both $\langle E_t u, u \rangle$ and $\eta(t)$. Then*

$$\langle E_d u, u \rangle - \langle E_c u, u \rangle = \int\limits_{c}^{d} |\hat{u}(t)|^2 \, d\eta(t) \ . \tag{4.3.7}$$

Proof. Let γ be the positively oriented rectangle with corners in $c \pm i$, $d \pm i$. According to Lemma 4.3.8

$$\oint\limits_{\gamma} \langle R_\lambda u, u \rangle \, d\lambda = \oint\limits_{\gamma} \hat{u}(\lambda)\overline{\hat{u}(\overline{\lambda})} m(\lambda) \, d\lambda$$

if either of these integrals exist. However, by (4.3.6),

$$\oint\limits_{\gamma} \hat{u}(\lambda)\overline{\hat{u}(\overline{\lambda})} m(\lambda) \, d\lambda = \oint\limits_{\gamma} \hat{u}(\lambda)\overline{\hat{u}(\overline{\lambda})} \int\limits_{-\infty}^{\infty} \left(\frac{1}{t - \lambda} - \frac{t}{t^2 + 1} \right) d\eta(t) \, d\lambda \ .$$

The double integral is absolutely convergent except perhaps where $t = \lambda$. The difficulty is thus caused by

$$\int\limits_{-1}^{1} ds \int\limits_{\mu-1}^{\mu+1} \frac{\hat{u}(\mu + is)\overline{\hat{u}(\mu - is)} \, d\eta(t)}{t - \mu - is}$$

for $\mu = c, d$. However, Lemma 4.3.9 ensures the absolute convergence of these integrals. Changing the order of integration gives

$$\oint_\gamma \hat{u}(\lambda)\overline{\hat{u}(\overline{\lambda})}\,m(\lambda)\,d\lambda = \int_{-\infty}^\infty \oint_\gamma \hat{u}(\lambda)\overline{\hat{u}(\overline{\lambda})}\Big(\frac{1}{t-\lambda}-\frac{t}{t^2+1}\Big)\,d\lambda\,d\eta(t)$$

$$= -2\pi \mathrm{i}\int_c^d |\hat{u}(t)|^2\,d\eta(t)$$

since for $c < t < d$ the residue of the inner integral is $-|\hat{u}(t)|^2$ whereas $t = c, d$ do not carry any mass and the inner integrand is regular for $t < c$ and $t > d$.

Similarly we have

$$\oint_\gamma \langle R_\lambda u, u\rangle\,d\lambda = \int_{-\infty}^\infty d\langle E_t u, u\rangle \oint_\gamma \frac{d\lambda}{t-\lambda} = -2\pi \mathrm{i}\int_c^d d\langle E_t u, u\rangle\,,$$

which completes the proof.   □

The next lemma extends the result of the previous lemma to general $u \in L^2_w(a, b)$, which gives a general form of Parseval's formula.

Lemma 4.3.11. *If $u \in L^2_w(a, b)$ the generalized Fourier transform $\hat{u} \in L^2_\eta$ exists as the L^2_η-limit of $\int_a^x u\varphi(\cdot, t)w$ as $x \to b$. Furthermore,*

$$\langle E_t u, v\rangle = \int_{-\infty}^t \hat{u}\overline{\hat{v}}\,d\eta\,.$$

In particular, $\langle u, v\rangle = \langle \hat{u}, \hat{v}\rangle_\eta$ if u and $v \in L^2_w(a, b)$.

Proof. If u has compact support Lemma 4.3.10 shows that (4.3.7) holds for a dense set of values c, d since functions of bounded variation are a.e. differentiable (Corollary B.8.10). Since E_t is left-continuous we obtain, by letting $d \uparrow t, c \to -\infty$ through such values,

$$\langle E_t u, v\rangle = \int_{(-\infty, t)} \hat{u}\overline{\hat{v}}(t)\,d\eta(t)$$

with absolutely convergent integral when u, v have compact supports. This is clear if $u = v$ and then in general by polarization (Proposition 2.4.10). As $t \to \infty$ we also obtain that $\langle u, v\rangle = \langle \hat{u}, \hat{v}\rangle_\eta$ when u and v have compact supports.

For arbitrary $u \in L^2_w(a, b)$ we set, for $c \in (a, b)$,

$$u_c(x) = \begin{cases} u(x) & \text{for } x < c \\ 0 & \text{otherwise} \end{cases} \tag{4.3.8}$$

and obtain a transform $\hat{u}_c$. If also $d \in (a, b)$ it follows that $\|\hat{u}_c - \hat{u}_d\|_\eta = \|u_c - u_d\|$, and since $u_c \to u$ in $L^2_w(a, b)$ as $c \to b$, Cauchy's convergence principle shows that $\hat{u}_c$ converges to an element $\hat{u} \in L^2_\eta$ as $c \to b$. The lemma now follows in full generality by continuity. $\square$

Note that we have proved that $\mathcal{F} : L^2_w(a,b) \to L^2_\eta$ is an isometry. At this point we have proved Theorem 4.3.7 (1) and (4). We need to prove (2), i.e., the transform is unitary, but first turn to (3), i.e., we calculate the adjoint of the transform.

Lemma 4.3.12.

(1) *The integral $\int_K \hat{u}\varphi(x,\cdot)\,d\eta$ is in $L^2_w(a,b)$ if K is a compact interval and $\hat{u} \in L^2_\eta$, and as $K \to \mathbb{R}$ the integral converges in $L^2_w(a,b)$. The limit is denoted $\mathcal{F}^{-1}(\hat{u})$.*

(2) *$\mathcal{F}^{-1}$ is the adjoint of $\mathcal{F}$ and $\mathcal{F}^{-1}(\hat{u})$ is called the* inverse transform *of $\hat{u}$. If $u \in L^2_w(a,b)$ then $\mathcal{F}^{-1}(\mathcal{F}(u)) = u$.*

Proof. If $\hat{u} \in L^2_\eta$ has compact support, then $u(x) = \langle \hat{u}, \varphi(x,\cdot) \rangle_\eta$ is continuous, so u_c, defined as in (4.3.8), is in $L^2_w(a,b)$ for $c \in (a,b)$, and has a transform $\hat{u}_c$. We have

$$\|u_c\|^2 = \int\limits_a^c \left(\int\limits_{-\infty}^\infty \hat{u}\varphi(x,\cdot)\,d\eta \right)\overline{u(x)}w(x)\,dx \ .$$

As an integral with respect to the product measure $d\eta(t) \otimes w(x)dx$ this is absolutely convergent, so by Tonelli's Theorem B.9.5 we may change the order of integration to obtain

$$\|u_c\|^2 = \int\limits_{-\infty}^\infty \left(\int\limits_a^c \overline{u}\varphi(\cdot,t)w \right)\hat{u}(t)\,d\eta(t) = \langle \hat{u}, \hat{u}_c \rangle_\eta \le \|\hat{u}\|_\eta \|\hat{u}_c\|_\eta = \|\hat{u}\|_\eta \|u_c\| \ ,$$

according to Lemma 4.3.11. Hence $\|u_c\| \le \|\hat{u}\|_\eta$, so $u \in L^2_w(a,b)$, and $\|u\| \le \|\hat{u}\|_\eta$. If now $\hat{u} \in L^2_\eta$ is arbitrary, this inequality shows (like in the proof of Lemma 4.3.11) that $\int_K \hat{u}(t)\varphi(x,t)\,d\eta(t)$ converges to a limit $u_1 \in L^2_w(a,b)$ as $K \to \mathbb{R}$ through compact intervals. If $v \in L^2_w(a,b)$, $\hat{v}$ is its generalized Fourier transform, K is a compact interval, and $c \in (a,b)$, we have

$$\int\limits_K \left(\int\limits_a^c \overline{v}\varphi(\cdot,t)w \right)\hat{u}(t)\,d\eta(t) = \int\limits_a^c \left(\int\limits_K \hat{u}\varphi(x,\cdot)\,d\eta \right)\overline{v(x)}w(x)\,dx$$

by absolute convergence. Letting $c \to b$, using Lemma 4.3.11, and then letting $K \to \mathbb{R}$ we obtain $\langle \hat{u}, \hat{v} \rangle_\eta = \langle u_1, v \rangle$. If $\hat{u}$ is the transform of u, then by Lemma 4.3.11 $u_1 - u$ is orthogonal to $L^2_w(a,b)$, so $u_1 = u$. $\square$

It now only remains to prove Theorem 4.3.7 (2) and (5). We have shown the inverse transform to be the adjoint of the transform as an operator from $L^2_w(a,b)$ into L^2_η. To prove (2) it remains to prove that the transform is surjective. The following lemma will enable us to do this.

Lemma 4.3.13. *The transform of* $R_\lambda u$ *is* $\hat{u}(t)/(t - \lambda)$.

Proof. We have $\langle E_t u, v \rangle = \int_{(-\infty,t)} \hat{u}(t) \overline{\hat{v}(t)} \, d\eta(t)$ by Lemma 4.3.11 so by Corollary B.8.1

$$\langle R_\lambda u, v \rangle = \int_{-\infty}^{\infty} \frac{d\langle E_t u, v \rangle}{t - \lambda} = \int_{-\infty}^{\infty} \frac{\hat{u}(t) \overline{\hat{v}(t)} \, d\eta(t)}{t - \lambda} = \langle \hat{u}(t)/(t - \lambda), \hat{v}(t) \rangle_\eta .$$

By Theorem 2.3.7 (3) and (4) on page 51

$$\|R_\lambda u\|^2 = \frac{1}{2i \operatorname{Im} \lambda} \langle R_\lambda u - R_{\overline{\lambda}} u, u \rangle = \int_{-\infty}^{\infty} \frac{d\langle E_t u, u \rangle}{|t - \lambda|^2} = \|\hat{u}(t)/(t - \lambda)\|_\eta^2 .$$

Setting $v = R_\lambda u$ and using Lemma 4.3.11, it then follows that $\|\hat{u}(t)/(t - \lambda)\|_\eta^2 = \langle \hat{u}(t)/(t - \lambda), \mathcal{F}(R_\lambda u) \rangle_\eta = \|\mathcal{F}(R_\lambda u)\|_\eta^2$. Thus $\|\hat{u}(t)/(t - \lambda) - \mathcal{F}(R_\lambda u)\|_\eta = 0$. □

Lemma 4.3.14. *The generalized Fourier transform* $\mathcal{F} : L_w^2(a, b) \to L_\eta^2$ *is surjective.*

Proof. To show that $\mathcal{F}$ is surjective we must show that if $\hat{u} \in L_\eta^2$ is orthogonal to all transforms, then $\hat{u} = 0$ as an element of L_η^2, i.e., $\hat{u} \, d\eta$ is the zero measure. But if $\hat{v}$ is a transform, then by Lemma 4.3.13 so is $\hat{v}(t)/(t - \overline{\lambda})$ for all non-real λ. Thus $\int_{-\infty}^{\infty} \frac{1}{t - \lambda} \hat{u}(t) \overline{\hat{v}(t)} \, d\eta(t) = 0$ for all non-real λ. By the Stieltjes inversion formula (Lemma E.1.5) it follows that $\hat{u}\overline{\hat{v}} \, d\eta$ is the zero measure.

Now, the Fourier transform of a function in $C_0(a, b)$ is entire, so to prove that t is outside the support of $\hat{u} \, d\eta$ it is enough to show that there exists a $v \in C_0(a, b)$ for which $\hat{v}(t) \neq 0$. But $\hat{v}(t) = \int_a^b v\varphi(\cdot, t)w$, so if this is zero for all $v \in C_0(a, b)$, then $\varphi(\cdot, t)w$ is the zero measure, which contradicts Lemma 4.1.3. The lemma follows. □

To prove Theorem 4.3.7 only (5) now remains.

Lemma 4.3.15. *If* $u \in \mathcal{D}(T)$, *then* $\mathcal{F}(Tu)(t) = t\hat{u}(t)$. *Conversely, if* $\hat{u}$ *and* $t\hat{u}(t)$ *are in* L_η^2, *then* $\mathcal{F}^{-1}(\hat{u}) \in \mathcal{D}(T)$.

Proof. We have $u \in \mathcal{D}(T)$ if and only if $u = R_\lambda(Tu - \lambda u)$, which holds if and only if $\hat{u}(t) = (\mathcal{F}(Tu)(t) - \lambda\hat{u}(t))/(t - \lambda)$, i.e., $\mathcal{F}(Tu)(t) = t\hat{u}(t)$, according to Lemma 4.3.13. □

This completes the proof of Theorem 4.3.7. We also have the following generalization of Corollary 4.3.3.

Theorem 4.3.16. *Suppose* $u \in \mathcal{D}(T)$. *Then the inverse transform* $\langle \hat{u}, \varphi(x, \cdot) \rangle_\eta$ *converges locally uniformly to* $u(x)$ *and* $\langle \hat{u}, p\varphi'(x, \cdot) \rangle_\eta$ *converges locally uniformly to* $pu'(x)$.

Proof. The proof is very similar to that of Corollary 4.3.3. Let $\lambda \in \rho(T)$ and put $f = (T - \lambda)u$ so that $u = R_\lambda f$. Let K be a compact interval, and put $u_K(x) = \int_K \hat{u}(t)\varphi(x,t)\,d\eta(t) = \mathcal{F}^{-1}(\chi\hat{u})(x)$, where χ is the characteristic function for K. Define f_K similarly. Then by Lemma 4.3.13

$$R_\lambda f_K = \mathcal{F}^{-1}\left(\frac{\chi(t)\hat{f}(t)}{t - \lambda}\right) = \mathcal{F}^{-1}(\chi\hat{u}) = u_K .$$

Since $f_K \to f$ in $L_w^2(a,b)$ as $K \to \mathbb{R}$, it follows from Theorem 4.3.1 that $u_K \to u$ and $pu'_K \to pu'$ locally uniformly as $K \to \mathbb{R}$. The proof is finished once we prove that $p(x)\frac{d}{dx}\int_K \hat{u}\varphi(x,t)\,d\eta(t) = \int_K \hat{u}\,p\varphi'(x,\cdot)\,d\eta$. For this, see Exercise 4.3.6. $\qquad\square$

We shall prove one more theorem which will be useful in later chapters, but also has intrinsic interest.

Theorem 4.3.17. $\mathcal{F}(\psi(\cdot,\lambda))(t) = 1/(t - \lambda)$ *and* $\psi(\cdot,\lambda) \to 0$ *in* $L_w^2(a,b)$ *as* $\lambda \to \infty$ *in a double sector* $|\mathrm{Re}\,\lambda| \leq C\,|\mathrm{Im}\,\lambda|$. *In the representation formula* (4.3.6) *for the m-function we always have* $B = 0$.

Proof. By Theorem 4.3.13 $\mathcal{F}(R_\lambda u)(t) = \hat{u}(t)/(t - \lambda)$. It therefore follows that $\int_{-\infty}^{\infty}\hat{u}(t)\varphi(x,t)/(t-\lambda)\,d\eta(t)$ converges in $L^2(0,b)$ to $\langle u, \overline{g(x,\cdot,\lambda)}\rangle$. Thus the Fourier transform of $g(x,\cdot,\lambda)$ is $\varphi(x,t)/(t - \lambda)$, so the integral is pointwise and absolutely convergent. In particular, $g(0,\cdot,\lambda) = -\psi(\cdot,\lambda)\sin\alpha$ has Fourier transform $\varphi(0,t)/(t-\lambda) = -\sin\alpha/(t - \lambda)$. This proves the first claim for $\alpha \neq 0$.

Otherwise similar arguments involving $p(R_\lambda u)'(x)$ and the kernel $g_1(x,y,\lambda)$ of Theorem 4.3.1 gives the desired result; see Exercise 4.3.8.

By Parseval's formula $\|\psi(\cdot,\lambda)\|^2 = \int_{-\infty}^{\infty}|t - \lambda|^{-2}\,d\eta(t)$. Since $(1+t^2)/|t-\lambda|^2$ is easily seen to be bounded and tend to 0 as $\lambda \to \infty$ in any double sector $|\mathrm{Re}\,\lambda| \leq C\,|\mathrm{Im}\,\lambda|$ and $\int_{-\infty}^{\infty}\frac{d\eta(t)}{t^2+1} < \infty$ the second claim follows by dominated convergence.

Finally, $B = 0$ since by Parseval and (4.3.5) we have

$$\int_{-\infty}^{\infty}\frac{d\eta(t)}{|t - \lambda|^2} = \|\psi(\cdot,\lambda)\|^2 = \frac{\mathrm{Im}\,m(\lambda)}{\mathrm{Im}\,\lambda} = B + \int_{-\infty}^{\infty}\frac{d\eta(t)}{|t - \lambda|^2} . \qquad\square$$

Example 4.3.18 (Sine and cosine transforms). Let us interpret Theorem 4.3.7 for the case of the equation $-u'' = \lambda u$ on the interval $[0,\infty)$. We shall look at the cases when the boundary condition at 0 is either a Dirichlet condition ($\alpha = 0$ in (4.2.2)) or a Neumann condition ($\alpha = \pi/2$). The general solution of the equation is $u(x) = Ae^{\sqrt{-\lambda}x} + Be^{-\sqrt{-\lambda}x}$. Let the root be the principal branch, i.e., the branch where the real part is ≥ 0. Then the only solutions in $L^2(0,\infty)$ are, unless $\lambda \geq 0$, the multiples of $e^{-\sqrt{-\lambda}x} = \cos(i\sqrt{-\lambda}x) + i\sin(i\sqrt{-\lambda}x)$. It follows that the equation is in the limit-point condition at infinity (this is also a consequence of Theorem 4.2.7).

With a Dirichlet condition at 0 we have $\theta(x,\lambda) = \cos(i\sqrt{-\lambda}x)$ and $\varphi(x,\lambda) = -i\sin(i\sqrt{-\lambda}x)/\sqrt{-\lambda}$. It follows that the *m*-function is $m_0(\lambda) = -\sqrt{-\lambda}$. Similarly, the *m*-function in the case of a Neumann condition at 0 is $m_{\pi/2}(\lambda) = 1/\sqrt{-\lambda}$, using again the principal branch of the root.

Using the Stieltjes inversion formula Lemma E.1.5 we see that the corresponding spectral measures are given by $d\eta_0(t) = \frac{1}{\pi}\sqrt{t}\,dt$ for $t \geq 0$, $d\eta_0 = 0$ in $(-\infty, 0)$, respectively $d\eta_{\pi/2}(t) = dt/(\pi\sqrt{t})$ for $t \geq 0$, $d\eta_{\pi/2} = 0$ in $(-\infty, 0)$. If $u \in L^2(0, \infty)$ and we define $\hat{u}(t) = \int_0^\infty u(x)\frac{\sin(\sqrt{t}x)}{\sqrt{t}}\,dx$, as a generalized integral converging in $L^2_{\eta_0}$, then the inversion formula reads $u(x) = \frac{1}{\pi}\int_0^\infty \hat{u}(t)\sin(\sqrt{t}x)\,dt$.

In this case one usually changes variable in the transform and defines the *sine transform* $S(u)(\xi) = \int_0^\infty u(x)\sin(\xi x)\,dx = \xi\hat{u}(\xi^2)$. Changing variable to $\xi = \sqrt{t}$ in the inversion formula above then shows that $u(x) = \frac{2}{\pi}\int_0^\infty S(u)(\xi)\sin(\xi x)\,d\xi$.

Similarly, if we set $\hat{u}(t) = \int_0^\infty u(x)\cos(\sqrt{t}x)\,dx$ the inversion formula obtained is $u(x) = \frac{1}{\pi}\int_0^\infty \hat{u}(t)\frac{\cos(\sqrt{t}x)}{\sqrt{t}}\,dt$. In this case it is again common to use $\xi = \sqrt{t}$ as the transform variable to define the *cosine transform* $C(u)(\xi) = \int_0^\infty u(x)\cos(\xi x)\,dx$. Changing variables in the inversion formula above then gives the inversion formula $u(x) = \frac{2}{\pi}\int_0^\infty C(u)(\xi)\cos(\xi x)\,d\xi$ for the cosine transform.

Note that there are no eigenvalues in either of these cases; the spectrum is *purely continuous* and consists of the interval $[0, \infty)$. In these cases the spectrum is even purely *absolutely continuous*, meaning that the spectral measures are locally absolutely continuous with respect to Lebesgue measure.

Exercises

Exercise 4.3.1 In Theorem 4.1.9 we showed that if u is in the domain of the maximal operator, then u and pu' are continuous. This allows a more functional analytic proof of the existence of Green's function and the kernel $g_1(x, y, \lambda)$ for the quasi-derivative, as follows.

Use the graph norm on T and let K be a compact subset of (a, b) and consider the map taking $(u, f) \in T$ to the restriction of (u, pu') to K where we consider (u, pu') an element of $C(K) \times C(K)$, normed by $\|(v_1, v_2)\| = \sup_K |v_1| + \sup_K |v_2|$. Show that this map is closed and conclude the existence of the two kernels by use of the closed graph theorem and the Riesz representation theorem.

Exercise 4.3.2 If b is a regular endpoint another way of writing (4.3.5) is $\mathrm{Im}(\psi(b, \lambda)\overline{p\psi'(b, \lambda)}) = 0$ or $p\psi'(b, \lambda)/\psi(b, \lambda) \in \mathbb{R}$. Viewing the left-hand side as a function of m, show that it is a fractional linear transformation (a Möbius transformation). Such transformations map a circle to another circle or an extended line (a line plus ∞). Find the center and radius of the circle determining m, expressed in the values of, or integrals involving, φ and θ.

Exercise 4.3.3 Show that the following functions are Nevanlinna functions and find their Nevanlinna representations.

(1) $m(\lambda) = \alpha + i\beta$ where $\alpha \in \mathbb{R}$ and $\beta > 0$.
(2) $m(\lambda) = \log(\lambda)$.

(3) $m(\lambda) = \log(-1/\lambda)$.
(4) $m(\lambda) = \lambda^r$ where $0 < r < 1$.
(5) $m(\lambda) = -\lambda^{-r}$ where $0 < r < 1$.
(6) $m(\lambda) = \tan(\lambda)$.
(7) $m(\lambda) = -\cot(\lambda)$.
(8) $m(\lambda) = (z - \beta)/(z - \alpha)$ where $\alpha < \beta$.
(9) $m(\lambda) = \log((z - \beta)/(z - \alpha))$ where $\alpha < \beta$.

Exercise 4.3.4 Suppose (a, b) is a gap in the spectrum of T, i.e., (a, b) is a nullset with respect to $d\eta$. Show that the m-function, which is defined for $\lambda \in (a, b)$, is a strictly increasing, real-valued function in (a, b).

Exercise 4.3.5 Prove Theorem 4.3.10 integrating only along two horizontal line segments, similar to the proof of the Stieltjes inversion formula Lemma E.1.5. This proof is more elementary in that it does not use the a.e. differentiability of a monotone function, but the details become somewhat messier.

Exercise 4.3.6 Show that if $\hat{u} \in L_\eta^2$ has compact support and $u(x) = \int_{-\infty}^{\infty} \hat{u}\varphi(x, \cdot)\, d\eta$, then $pu'(x) = \int_{-\infty}^{\infty} \hat{u}p\varphi'(x, \cdot)\, d\eta$.

Hint: Write $\varphi(x, t) = \varphi(c, t) + \int_c^x p\varphi'/p$ and note that $p\varphi'$ is continuous. Now use Tonelli's theorem to show that $\int_c^x \frac{1}{p}\int_{-\infty}^{\infty} \hat{u}p\varphi'\, d\eta = u(x) - u(c)$.

Exercise 4.3.7 As a stronger version of Theorem 4.3.16 one may prove that the inverse transform of $\mathcal{F}u$ for $u \in \mathcal{D}(T)$ converges *absolutely* and locally uniformly to u. Show this.

Hint: Recall that $x \mapsto \|g(x, \cdot, \lambda)\|$ is bounded on any compact subinterval of (a, b). Now use Lemma 4.3.13 to show that $\mathcal{F}(g(x, \cdot, \lambda))(t) = \varphi(x, t)/(t - \lambda)$.

Combine these facts with Parseval's formula to show that as $K \to \mathbb{R}$ through compact intervals $\int_{\mathbb{R}\setminus K} |\hat{u}\varphi(\cdot, x)|\, d\eta$ tends to zero locally uniformly in x.

Exercise 4.3.8 Use Exercise 4.3.6 to show that the kernel $\overline{g_1(x, \cdot, \lambda)}$ of Theorem 4.3.1 has transform $p\varphi'(x, t)/(t - \overline{\lambda})$ and then finish the proof of Lemma 4.3.17.

4.4 Two singular endpoints

In Section 4.3 we singled out, from the general singular case, the case when one endpoint is regular and provided with a separated boundary condition. The reason was that assuming this simplifies the statement and proof of the expansion theorem.

If the singular endpoint is in the limit-circle condition, the spectrum is not necessarily simple if we use coupled boundary conditions, but we saw in Corollary 4.3.2 that it will be discrete so the expansion theorem will be a generalized Fourier series and no complications arise. The corollary also shows that the case of two singular endpoints which are both in the limit-circle condition again gives a discrete spectrum.

If the initial point is not regular but limit-circle one still obtains an expansion theorem much like Theorem 4.3.7. The only problem is that one may not be able to state a separated boundary condition at a in terms of the values of u and pu' at a, because these values may not be well defined. To show how to express boundary conditions in this case we need the following lemma.

Lemma 4.4.1. *Let v_1, v_2 be solutions of $-(pv')'+qv = 0$ with $\mathcal{W}_p(v_1, v_2) = 1$. If $f\sqrt{w}$ is in L^2 near a limit-circle endpoint of (a, b) near which u satisfies $-(pu')'+qu = wf$, then the Wronskians $\mathcal{W}_p(u, v_j)$, $j = 1, 2$, have finite limits at this endpoint. There is a unique solution u for arbitrarily specified values of the limits.*

If $-(pu')' + qu = \lambda w u$ and λ-independent limits for the Wronskians are specified, the solution $u(x, \lambda)$ is an entire function of λ, locally uniformly in x.

Proof. Let $V = \begin{pmatrix} v_2 & v_1 \\ pv_2' & pv_1' \end{pmatrix}$, let $U = \begin{pmatrix} u \\ pu' \end{pmatrix}$ and put $Z = V^{-1}U$. It is easily verified that $Z' = \begin{pmatrix} v_1 f w \\ -v_2 f w \end{pmatrix}$, where the right-hand side is integrable near the limit-circle endpoint, so Z has finite limits at it which may be arbitrarily specified.

If instead $f = \lambda u$, then since $U = VZ$ we obtain $u = \begin{pmatrix} v_2 & v_1 \end{pmatrix} Z$ so that Z satisfies $Z' = \lambda \begin{pmatrix} v_1 v_2 w & v_1^2 w \\ -v_2^2 w & -v_1 v_2 w \end{pmatrix} Z$. Near the endpoint the coefficient matrix for Z is integrable so Theorem 4.1.1 ensures the existence of a unique solution for arbitrarily specified initial values there, and if the initial value is independent of λ the solution is an entire function of λ locally uniformly near the endpoint.

Now $Z = V^{-1}U$ has rows $-\mathcal{W}_p(u, v_1)$ and $\mathcal{W}_p(u, v_2)$, proving the lemma. $\square$

Theorem 4.4.2. *Symmetric boundary conditions at a limit-circle endpoint are precisely the same as for a regular endpoint, provided the values of u and pu' at the endpoint are replaced by the values of $\mathcal{W}_p(u, v_1)$ and $\mathcal{W}_p(u, v_2)$, where v_1, v_2 are as in Lemma 4.4.1. Elements of the domain of the minimal operator are precisely those elements of the maximal domain for which these Wronskians vanish.*

Proof. First note that if u_1, u_2, v_1, v_2 are functions in AC_{loc} with quasi-derivatives in AC_{loc}, then by an elementary computation[6]

$$\mathcal{W}_p(u_1, v_1)\mathcal{W}_p(u_2, v_2) - \mathcal{W}_p(u_1, v_2)\mathcal{W}_p(u_2, v_1) = \mathcal{W}_p(u_1, u_2)\mathcal{W}_p(v_1, v_2) .$$

The boundary form for T_1 is $\langle (u_1, f_1), \mathcal{U}(u_2, f_2) \rangle = i\mathcal{W}_p(u_1, u_2)\big|_a^b$, so if we pick v_j so that $\mathcal{W}_p(v_1, v_2) = 1$ the boundary form at a is

$$i\left(\lim_a \mathcal{W}_p(u_1, v_1) \lim_a \mathcal{W}_p(u_2, v_2) - \lim_a \mathcal{W}_p(u_1, v_2) \lim_a \mathcal{W}_p(u_2, v_1) \right) .$$

Picking v_1 and v_2 as in Lemma 4.4.1 the lemma tells us that the limits at a of $\mathcal{W}_p(u_j, v_k)$ exist finitely. We may now extend the analysis of Section 4.2 for a regular endpoint to the case of a limit-circle endpoint a by replacing $u(a)$ and $pu'(a)$ by $\lim_a \mathcal{W}_p(u, v_1)$ and $\lim_a \mathcal{W}_p(u, v_2)$. All of Section 4.3 now extends to

---

[6] If u_1, u_2 are linearly dependent clearly both sides vanish, and if not we may write $v_2 = \alpha u_1 + \beta u_2$ so $\mathcal{W}(v_2, u_1) = \beta \mathcal{W}(u_2, u_1)$ and $\mathcal{W}(v_2, u_2) = \alpha \mathcal{W}(u_1, u_2)$. Insertion of this gives the identity.

this case with no other change. In particular the minimal operator is the restriction of T_1 for which these limits vanish at any limit-circle endpoint.   □

Limit-point endpoints. It remains to discuss the case of two singular endpoints which are both in the limit-point case. This requires a bit more effort, since in this case we may in general have a continuous spectrum of multiplicity two, as defined on page 73.

We shall deal with a self-adjoint realization of (4.1.1) with both endpoints singular and determined by separated boundary conditions. The only case not already covered is when both endpoints are in the limit-point condition, when no boundary conditions at all are required, but we need not assume that this is the case.

Let $c \in (a, b)$ be such that neither in (a, c) nor in (c, b) is w a.e. zero. We call c the *anchor point*. If we consider the equation on (c, b) with the original (separated) boundary condition at b and some other separated, symmetric boundary condition at c we may therefore find a corresponding Weyl solution $\psi_+(x, \lambda)$. Similarly we may find a Weyl solution $\psi_-(x, \lambda)$ for the interval (a, c). The solutions $\psi_\pm$ have of course unique extensions to all of (a, b) by Theorem 4.1.1.

Theorem 4.4.3. *Green's function for the interval (a, b) with the given separated boundary conditions is given by*

$$g(x, y, \lambda) = \frac{\psi_-(\min(x, y), \lambda)\psi_+(\max(x, y), \lambda)}{\mathcal{W}_p(\psi_-(\cdot, \lambda), \psi_+(\cdot, \lambda))} . \tag{4.4.1}$$

Similarly $p(R_\lambda u)'(x) = \langle u, \overline{g_1(x, \cdot, \lambda)} \rangle$, where the kernel g_1 is the quasi-derivative of $g(x, y, \lambda)$ with respect to x.

Proof. Suppose $u \in L_w^2(a, b)$ has compact support in (a, b) and define

$$v(x) = \psi_+(x, \lambda) \int_a^x u\psi_-(\cdot, \lambda)w + \psi_-(x, \lambda) \int_x^b u\psi_+(\cdot, \lambda)w .$$

To the left of supp u the function v is a multiple of $\psi_-(\cdot, \lambda)$ and to the right of supp u it is a multiple of $\psi_+(\cdot, \lambda)$, so v satisfies the required boundary conditions. Differentiating, multiplying by p and differentiating again we obtain $-(pv')' + qv = \lambda wv + \mathcal{W}_p(\psi_-, \psi_+)wu$ so that $R_\lambda u(x) = \langle u, \overline{g(x, \cdot, \lambda)} \rangle$ and $p(R_\lambda u)'(x) = \langle u, \overline{g_1(x, \cdot, \lambda)} \rangle$ for compactly supported u. Since compactly supported u are dense in $L_w^2(a, b)$, R_λ is bounded and the integrals in the scalar products converge for $u \in L_w^2(a, b)$, the formulas for $R_\lambda u$ and $p(R_\lambda u)'$ are correct in general.   □

Green's function will of course be independent of the anchor point c and the boundary conditions we choose at c for the auxiliary problems on $(a, c]$ and $[c, b)$, since the Weyl solutions are determined up to multiples by the boundary condition at the singular endpoint. Multiplying $\psi_\pm$ by a non-zero factor, such a factor appears in both the numerator and denominator and may be cancelled.

However, for definiteness we fix c and the same boundary conditions at c for both auxiliary problems, and we may as well settle for a Dirichlet condition. Thus we introduce solutions $\varphi(\cdot, \lambda)$ and $\theta(\cdot, \lambda)$ satisfying the initial conditions

$$
\begin{cases} \varphi(c, \lambda) = 0 , \\ p\varphi'(c, \lambda) = 1 , \end{cases} \qquad \begin{cases} \theta(c, \lambda) = 1 , \\ p\theta'(c, \lambda) = 0 . \end{cases} \tag{4.4.2}
$$

There are then m-coefficients $m_\pm(\lambda)$ such that

$$
\psi_+(\cdot, \lambda) = \theta(\cdot, \lambda) + m_+(\lambda)\varphi(\cdot, \lambda) ,
$$
$$
\psi_-(\cdot, \lambda) = \theta(\cdot, \lambda) - m_-(\lambda)\varphi(\cdot, \lambda) .
$$

Note the minus sign in front of m_- which makes m_- a Nevanlinna function also in the case when the right endpoint is the regular (initial) point, as mentioned in Remark 4.3.6.

Since $\mathcal{W}_p(\varphi, \theta) = 1$ the denominator of (4.4.1) is $-m_+ - m_-$, so with $\Phi(x, \lambda)$ the 2×1 matrix with components $\varphi(x, \lambda)$ and $\theta(x, \lambda)$ we may write

$$
g(x, y, \lambda) = \Phi^*(\min(x, y), \overline{\lambda})(M(\lambda) - \tfrac{1}{2}J)\Phi(\max(x, y), \lambda) ,
$$

where $M = \frac{1}{m_+ + m_-} \begin{pmatrix} m_+ m_- & (m_- - m_+)/2 \\ (m_- - m_+)/2 & -1 \end{pmatrix}$ and $J = \begin{pmatrix} 0 & -1 \\ 1 & 0 \end{pmatrix}$. It is easily verified that $\frac{1}{2i}(M(\lambda) - M^*(\lambda))$ is a positive matrix for all λ in the upper half-plane, i.e., the function $M(\lambda)$ is a 2×2 matrix-valued Nevanlinna function. This means that there is an increasing 2×2 matrix-valued function Ξ on $\mathbb{R}$ such that

$$
M(\lambda) = A + B\lambda + \int_{-\infty}^{\infty} \left(\frac{1}{t - \lambda} - \frac{t}{1 + t^2} \right) d\Xi(t) ,
$$

where A is a self-adjoint and B a positive, uniquely determined 2×2 matrix. To see this, let U be a 2×1 matrix and apply Theorem E.1.1 to the scalar Nevanlinna function $U^* M(\lambda) U$, noting that the uniqueness of the constants and measure in this representation implies that they are quadratic forms in U; cf. the proof of Lemma 3.1.4.

We now introduce the Hilbert space L_Ξ^2 of 2×1 matrix-valued functions $\hat{u}$ for which $\|\hat{u}\|_\Xi^2 = \int_{-\infty}^{\infty} \hat{u}^* \, d\Xi \hat{u} < \infty$. The corresponding scalar product is denoted $\langle \cdot, \cdot \rangle_\Xi$. The reader is asked to supply the missing measure-theoretic and other details in Exercise 4.4.1.

For a compactly supported function $u \in L_w^2(a, b)$ we define a 2×1 generalized Fourier transform $\hat{u}$ with components $\hat{u}_\varphi(t) = \langle u, \overline{\varphi(\cdot, t)} \rangle$ and $\hat{u}_\theta(t) = \langle u, \overline{\theta(\cdot, t)} \rangle$ so that $\hat{u}(t) = \int_a^b u\Phi(\cdot, t)w$. We may now state a result analogous to Theorem 4.3.7.

Theorem 4.4.4.

(1) *If $u \in L_w^2(a, b)$ the column matrix $\int_K u\Phi(\cdot, t)w$ converges in L_Ξ^2 as the compact interval K increases to (a, b). The limit is called the* generalized Fourier *transform of u and is denoted by $\mathcal{F}(u)$ or $\hat{u}$.*

(2) *The mapping $u \mapsto \hat{u}$ is unitary between $L_w^2(a, b)$ and L_Ξ^2 so that the Parseval formula $\langle u, v \rangle = \langle \hat{u}, \hat{v} \rangle_\Xi$ is valid if $u, v \in L_w^2(a, b)$.*

(3) *If $\hat{u} \in L_\Xi^2$ the integral $\int_K \Phi^*(x, \cdot)\, d\Xi\, \hat{u}$ converges in $L_w^2(a, b)$ as $K \to \mathbb{R}$ through compact intervals. If $\hat{u} = \mathcal{F}(u)$ the limit is u, so the integral is the inverse of the generalized Fourier transform.*

(4) *Let E_Δ denote the spectral projector of T for the interval (or Borel set) Δ. Then $E_\Delta u(x) = \int_\Delta \Phi^*(x, \cdot)\, d\Xi\, \hat{u}$.*

(5) *If $u \in \mathcal{D}(T)$ then $\mathcal{F}(Tu)(t) = t\hat{u}(t)$. Conversely, if $\hat{u}$ and $t\hat{u}(t)$ are in L_Ξ^2, then $\mathcal{F}^{-1}(\hat{u}) \in \mathcal{D}(T)$.*

The proof of Theorem 4.4.4 is very similar to the proof of Theorem 4.3.7 so we shall be brief. First note that for $u \in L_w^2(a, b)$ with compact support in (a, b) the function $\hat{u}(\lambda) = \langle u, \Phi(\cdot, \bar{\lambda}) \rangle$ is an entire function of λ since the components of $\Phi(x, \lambda)$ are entire, locally uniformly in x, by Theorem 4.1.1.

Lemma 4.4.5. *The function $\langle R_\lambda u, v \rangle - \hat{v}^*(\bar{\lambda})M(\lambda)\hat{u}(\lambda)$ is entire for all u, $v \in L_w^2(a, b)$ with compact supports in (a, b).*

Proof. If the supports are contained in $[c, d] \subset (a, b)$, direct calculation shows that the function is

$$\frac{1}{2} \int_c^d \left(\int_c^x wu\, \Phi^*(\cdot, \bar{\lambda}) - \int_x^d wu\, \Phi^*(\cdot, \bar{\lambda}) \right) J\Phi(x, \lambda)\overline{v(x)}w(x)\, dx .$$

This is obviously an entire function of λ. □

As in Chapter 3 we denote the resolution of the identity for T by $\{E_t\}_{t \in \mathbb{R}}$. The following lemma shows that $E_t u$ may be calculated by integration of $\hat{u}$ for compactly supported u.

Lemma 4.4.6. *Let $u \in L_w^2(a, b)$ have compact support in (a, b) and assume $c < d$, where c, d are points of differentiability for both $\langle E_t u, u \rangle$ and $\Xi(t)$. Then*

$$\langle E_d u, u \rangle - \langle E_c u, u \rangle = \int_c^d \hat{u}^*\, d\Xi\, \hat{u} .$$

Proof. Let Γ be the positively oriented rectangle with corners in $c \pm i$, $d \pm i$. According to Lemma 4.4.5

$$\oint_\Gamma \langle R_\lambda u, u \rangle\, d\lambda = \oint_\Gamma \hat{u}^*(\bar{\lambda})M(\lambda)\hat{u}(\lambda)\, d\lambda$$

if either of these integrals exist. The proof may now be completed along the same lines as the proof of Theorem 4.3.10. □

The next lemma extends the result of the previous lemma to general $u \in L_w^2(a, b)$, which gives a general form of Parseval's formula.

Lemma 4.4.7. *If $u \in L_w^2(a, b)$ the generalized Fourier transform $\hat{u} \in L_\Xi^2$ exists as the L_Ξ^2-limit of $\int_a^x u\varphi(\cdot, t)w$ as $x \to b$. Furthermore,*

$$\langle E_t u, v \rangle = \int\limits_{-\infty}^{t} \hat{v}^* \, d\Xi\hat{u} \ .$$

In particular, $\langle u, v \rangle = \langle \hat{u}, \hat{v} \rangle_\Xi$ if u and $v \in L_w^2(a, b)$.

The proof of this runs entirely parallel to the proof of Lemma 4.3.11 and will be omitted. We have now verified Theorem 4.4.4 (1) and (4) and next turn to (3).

Lemma 4.4.8.

(1) *The integral $\int_K \Phi^*(x, \cdot) \, d\Xi\hat{u}$ is in $L_w^2(a, b)$ if K is a compact interval and $\hat{u} \in L_\Xi^2$, and as $K \to \mathbb{R}$ the integral converges in $L_w^2(a, b)$. The limit is denoted $\mathcal{F}^{-1}(\hat{u})$.*

(2) *$\mathcal{F}^{-1}$ is the adjoint of $\mathcal{F}$ and $\mathcal{F}^{-1}(\hat{u})$ is called the inverse transform of $\hat{u}$. If $u \in L_w^2(a, b)$ then $\mathcal{F}^{-1}(\mathcal{F}(u)) = u$.*

Again, the proof runs entirely parallel to the proof of Lemma 4.3.12 and will be omitted, as will the proof of the following analogue of Lemma 4.3.13, needed to prove that the generalized Fourier transform is surjective.

Lemma 4.4.9. *The transform of $R_\lambda u$ is $\hat{u}(t)/(t - \lambda)$.*

Lemma 4.4.10. *The generalized Fourier transform is unitary from $L_w^2(a, b)$ to L_Ξ^2 and the inverse transform is the inverse of this map.*

Proof. The proof runs mostly like the proof of Lemma 4.3.14. Thus we assume $\hat{u} \in L_\Xi^2$ is orthogonal to all Fourier transforms $\hat{v}$ and conclude, using Lemma 4.4.9, that $\hat{v}^* \, d\Xi \, \hat{u}$ is the zero measure for all Fourier transforms $\hat{v}$.

The components of the Fourier transform of a function in $C_0(a, b)$ are entire, so to prove that t is outside the support of $d\Xi\hat{u}$ it is enough to show that there exist $v_1, v_2 \in C_0(a, b)$ for which $\hat{v}_1(t)$ and $\hat{v}_2(t)$ are linearly independent. But if not there is a fixed, non-zero 1×2 matrix A such that $A\hat{v}(t) = 0$ for all $v \in C_0(\omega)$. Thus $\int_a^b vA\Phi(\cdot, t)w = 0$ for all $v \in C_0(a, b)$. This means that $A\Phi(\cdot, t)$, which is a non-trivial solution of $-u'' + qu = twu$, is zero on supp w, which contradicts Lemma 4.1.3. □

To prove Theorem 4.4.4 only (5) now remains, but this is proved precisely like in Lemma 4.3.15. This completes the proof of Theorem 4.4.4. We also have the following generalization of Corollary 4.3.3, proved in essentially the same way.

Theorem 4.4.11. *Suppose $u \in \mathcal{D}(T)$. Then the inverse transform $\langle \hat{u}, \Phi(x, \cdot) \rangle_\Xi$ converges locally uniformly to $u(x)$ and $\langle \hat{u}, p\Phi'(x, \cdot) \rangle_\Xi$ converges locally uniformly to $pu'(x)$.*

The weakness of the results in this section is the dependence of the Fourier transform and spectral measure on the anchor point c and the auxiliary boundary conditions at c, since these are not part of the original boundary value problem. We therefore end the section by examining this dependence.

Proposition 4.4.12. *Changing the anchor point c to $\tilde{c}$ and the initial data (4.4.2) at $\tilde{c}$ to the general (4.3.3) changes the Fourier transform $\mathcal{F}u$ to $\tilde{\mathcal{F}}u$, where $\mathcal{F}u(\lambda) = K(\lambda)\tilde{\mathcal{F}}u(\lambda)$ and*

$$
K(\lambda) = \begin{pmatrix} \varphi(\tilde{c}, \lambda) & p\varphi'(\tilde{c}, \lambda) \\ \theta(\tilde{c}, \lambda) & p\theta'(\tilde{c}, \lambda) \end{pmatrix} \begin{pmatrix} -\sin\alpha & \cos\alpha \\ \cos\alpha & \sin\alpha \end{pmatrix}.
$$

Similarly M is replaced by $\tilde{M}(\lambda) = K^*(\bar{\lambda})M(\lambda)K(\lambda)$ and the spectral measure $d\Xi$ by $d\tilde{\Xi}(t) = K^*(t)d\Xi(t)K(t)$.

Since $K(\lambda)$ is entire the measures $d\tilde{\Xi}$ and $d\Xi$ are equivalent, i.e., each is absolutely continuous with respect to the other. The proof of the theorem is an elementary computation, based on expressing φ, θ and their derivatives in terms of $\tilde{\varphi}$, $\tilde{\theta}$ and their derivatives by comparing the values at $\tilde{c}$. We leave this to Exercise 4.4.2.

Exercises

Exercise 4.4.1 Explain in detail how the representation formula for matrix-valued Nevanlinna functions is derived and give a detailed definition of the space L_Ξ^2.

Exercise 4.4.2 Prove Proposition 4.4.12.

4.5 The spectrum

If the coefficient p in a regular right-definite problem for (4.1.1) is positive the corresponding self-adjoint operators are all bounded from below (Exercise 4.5.1), but for a singular problem they may not be; see Exercise 4.5.2. On the other hand, if p takes both signs on sets of positive Lebesgue measure the spectrum of a self-adjoint realization is never bounded, neither above nor below. To prove this we use the following well-known lemma.

Lemma 4.5.1 (Riemann–Lebesgue). *Suppose $u \in L^1(a, b)$ and $\lambda \in \mathbb{R}$. If $(c, d) \subset (a, b)$, then $\int_c^d u(x)e^{i\lambda x}\, dx \to 0$ uniformly with respect to (c, d) as $\lambda \to \pm\infty$.*

Proof. If $u \in C_0^1(a, b)$ an integration by parts shows that

$$\int\limits_c^d u(x) e^{i\lambda x}\, dx = \left(-iu(x) e^{i\lambda x} \big|_c^d + i \int\limits_c^d u'(x) e^{i\lambda x}\, dx \right) \lambda^{-1} .$$

The absolute value of this is less than $(2 \sup_{(a,b)} |u| + \int_a^b |u'|)/|\lambda|$, which tends to zero independently of c, d as $\lambda \to \pm\infty$.

$C_0^1(a, b)$ is dense in $L^1(a, b)$ by Theorem B.7.3 and Lemma C.4, so given $u \in L^1(a, b)$ and $\varepsilon > 0$ we may find $\tilde{u} \in C_0^1(a, b)$ so that $\int_a^b |u - \tilde{u}| < \varepsilon$. Now $\int_c^d \tilde{u}(x) e^{i\lambda x}\, dx$ tends to 0 uniformly with respect to $c, d \in (a, b)$ and

$$\left| \int\limits_c^d u(x) e^{i\lambda x}\, dx - \int\limits_c^d \tilde{u}(x) e^{i\lambda x}\, dx \right| \le \int\limits_c^d |u - \tilde{u}| < \varepsilon ,$$

so $\left| \int_c^d u(x) e^{i\lambda x}\, dx \right| < 2\varepsilon$ if $|\lambda|$ is sufficiently large, independently of c and d. This proves the lemma.   □

Theorem 4.5.2. *Suppose* $1/p$ *and* q *are locally integrable and that the integral* $\int_a^b (p |u'|^2 + q |u|^2) \ge 0$ *for all* $u \in C_0(a, b)$ *with* $pu' \in C_0(a, b)$. *Then* $p > 0$ *a.e.*

Proof. If $\varphi \in C_0(a, b)$ and ≥ 0, then the same is true for $\sqrt{\varphi}$. If $k \in \mathbb{Z}_+$, then by the Riemann–Lebesgue lemma $\int_{-\infty}^x \frac{\sqrt{\varphi(t)}}{p(t)} e^{ikt}\, dt \to 0$ uniformly as $k \to \infty$ and is 0 to the left and equal to a constant c_k to the right of $\operatorname{supp} \varphi$. Suppose $\psi \in C_0(a, b)$ with $\int_a^b \psi/p = 1$ and define

$$u_k(x) = \int\limits_{-\infty}^x \frac{\sqrt{\varphi(t)}\, e^{ikt}}{p(t)}\, dt - c_k \int\limits_{-\infty}^x \frac{\psi(t)}{p(t)}\, dt .$$

Then u_k is continuous, zero outside a k-independent compact subinterval of (a, b), and tends uniformly to 0 as $k \to \infty$, so $\int_a^b q |u_k|^2 \to 0$. Furthermore, $pu_k'(x) = \sqrt{\varphi(x)}\, e^{ikx} - c_k \psi(x)$ where the second term has support contained in $\operatorname{supp} \psi$ and tends uniformly to 0. It follows that

$$0 \le \int\limits_a^b (p |u_k'|^2 + q |u_k|^2) \to \int\limits_a^b \varphi/p \quad \text{as } k \to \infty .$$

This is non-negative for any non-negative $\varphi \in C_0(a, b)$ by assumption, so it follows that $dx/p(x)$ is a positive measure, i.e., $p > 0$ a.e., as follows from Theorem B.8.3, page 299.   □

Corollary 4.5.3. *Suppose p takes both signs on sets of positive measure. Then the spectrum of any self-adjoint realization of* (4.1.1) *is neither bounded above nor below.*

Proof. If T is a self-adjoint realization bounded below, then for some constant C we have $\langle Tu, u \rangle \geq C \langle u, u \rangle$ for $u \in \mathcal{D}(T)$. In particular this is true if u is in the minimal domain, in which case $\langle Tu, u \rangle = \int_a^b (p |u'|^2 + q |u|^2)$. Hence we have $\int_a^b (p |u'|^2 + (q - Cw) |u|^2) \geq 0$ for $u \in C_0(a, b)$ for which $pu' \in C_0(a, b)$. Thus $p > 0$ by Theorem 4.5.2. Similarly, if T is bounded above, then $p < 0$. □

Glazman decomposition. Since regular and limit-circle problems have a discrete spectrum the existence and location of the essential spectrum depends entirely on the behavior of the coefficients of the equation in an arbitrary neighborhood of a limit-point endpoint. This is the idea behind a very effective method of finding properties of the essential spectrum for singular problems discovered by Glazman [96] in 1949. Glazman called his method the decomposition method (at least in English translation), and we shall present it in the case of a Sturm–Liouville operator and apply it in a few typical cases.

Let us first discuss the direct sum of relations T_1 and T_2 acting in Hilbert spaces $\mathcal{H}_1$ and $\mathcal{H}_2$ respectively. The direct sum $T_1 \oplus T_2$ is then a relation on the orthogonal direct sum $\mathcal{H}_1 \oplus \mathcal{H}_2$ so that $(u, f) \in T_1 \oplus T_2$ if $u = (u_1, u_2)$ and $f = (f_1, f_2)$ with $(u_j, f_j) \in T_j$. It is clear that $\sigma(T_1 \oplus T_2) = \sigma(T_1) \cup \sigma(T_2)$ and $\sigma_e(T_1 \oplus T_2) = \sigma_e(T_1) \cup \sigma_e(T_2)$. These facts are the reason for the utility of the decomposition method, which we will now describe.

Suppose we have a Sturm–Liouville operator defined in $L_w^2(a, b)$ and $c \in (a, b)$. We suppose T is a self-adjoint realization in $L_w^2(a, b)$, so it is a finite-dimensional extension of the corresponding minimal operator. Now note that this minimal operator is an extension of the direct sum of the corresponding minimal operators on the intervals (a, c) and (c, b). Also this extension is finite-dimensional, since to obtain the minimal operator for (a, b) we have simply to replace the two conditions $u(c) = pu'(c) = 0$ satisfied by functions in the domain of the minimal operators for (a, c) and (c, b) by continuity of u, pu' at c. The essential spectrum of T is therefore the same as that of the direct sum by Corollary 3.2.4. We obtain the following proposition.

Proposition 4.5.4. *Suppose T is a self-adjoint realization of a Sturm–Liouville equation in $L_w^2(a, b)$. If $c \in (a, b)$ and T_- and T_+ are self-adjoint realizations of the same equation in $L_w^2(a, c)$ respectively $L_w^2(c, b)$, then $\sigma_e(T) = \sigma_e(T_-) \cup \sigma_e(T_+)$.*

This is of course also clear from expressing the corresponding m-matrix M in terms of the m-functions $m_\pm$ as in Section 4.4.

Theorem 4.5.5. *Suppose q and $\tilde{q}$ are real-valued, locally integrable functions on (a, b) such that $-u'' + qu = \lambda u$ and $-u'' + \tilde{q}u = \lambda u$ are both limit-circle at a. Let T and $\tilde{T}$ be corresponding self-adjoint realizations in $L^2(a, b)$ and suppose that $\lim_b (q - \tilde{q}) = 0$. Then $\sigma_e(T) = \sigma_e(\tilde{T})$.*

Corollary 4.5.6. *Suppose q is real-valued and locally integrable in* $(0, \infty)$, *that* 0 *is limit-circle for* $-u'' + qu = \lambda u$ *and that* $q(x) \to A$ *as* $x \to \infty$. *If T is a self-adjoint realization of* $-u'' + qu$ *in* $L^2(0, \infty)$, *then* $\sigma_e(T) = [A, \infty)$.

Proof. The only solutions of $-u'' + Au = \lambda u$ in $L^2(0, \infty)$ for $\lambda \notin \mathbb{R}$ are multiples of $e^{-x\sqrt{A-\lambda}}$ where the square root is the one with a positive real part. It easily follows that the m-function for a Dirichlet condition at 0 is $m(\lambda) = -\sqrt{A - \lambda}$, which is analytic as long as $A - \lambda$ is not negative, i.e. unless $\lambda \in [A, \infty)$, so $[A, \infty)$ is the essential spectrum of the corresponding operator. By Theorem 4.5.5 $\sigma_e(T) = [A, \infty)$. □

The proof of Theorem 4.5.5 requires the following abstract lemma.

Lemma 4.5.7. *Suppose T and B are self-adjoint operators in a Hilbert space* $\mathcal{H}$ *and that* $\|B\| < \varepsilon$. *If* $[\lambda - \varepsilon, \lambda + \varepsilon] \cap \sigma_e(T) = \varnothing$, *then* $\lambda \notin \sigma_e(T + B)$.

Proof. By the spectral theorem $\|(T - \lambda)u\|^2 = \int_{-\infty}^{\infty} |t - \lambda|^2 \, d\langle E_t u, u \rangle$. Suppose $u \in \mathcal{H} \ominus E$ where E is the span of all eigenvectors to eigenvalues in $(\lambda - \varepsilon, \lambda + \varepsilon)$, a finite-dimensional space. We then have

$$\|(T - \lambda)u\|^2 = \int_{|t-\lambda| \geq \varepsilon} |t - \lambda|^2 \, d\langle E_t u, u \rangle \geq \varepsilon^2 \int_{-\infty}^{\infty} d\langle E_t u, u \rangle = \varepsilon^2 \|u\|^2 \, .$$

By Exercise 2.3.4 $T + B$ is self-adjoint and $\|(T + B - \lambda)u\| \geq \|(T - \lambda)u\| - \|Bu\| \geq (\varepsilon - \|B\|)\|u\|$, so since dim $E < \infty$ it follows that λ is at worst an isolated eigenvalue of $T + B$ of finite multiplicity. □

Proof (Theorem 4.5.5). Given $\varepsilon > 0$ choose $c \in (a, b)$ so that $|q - \tilde{q}| < \varepsilon$ in (c, b). By Lemma 4.5.7 any self-adjoint realization of $-u'' + \tilde{q}u$ in $L^2(c, b)$ has its essential spectrum in an ε-neighborhood of the essential spectrum of any self-adjoint realization of $-u'' + qu$ in $L^2(c, b)$. By the decomposition principle the same is true of the essential spectra of $\tilde{T}$ and T, and since these are closed and $\varepsilon > 0$ is arbitrary we have $\sigma_e(\tilde{T}) \subset \sigma_e(T)$. Interchanging the roles of T and $\tilde{T}$ the opposite inclusion is also true so that $\sigma_e(\tilde{T}) = \sigma_e(T)$. □

The idea of proof of the corollary also shows the following theorem.

Theorem 4.5.8. *Suppose* 0 *is limit-circle for* $-u'' + qu = \lambda u$ *on* $(0, \infty)$ *and that* $\underline{\lim} \, q(x) \geq Q$ *at* ∞. *Then* $\sigma_e(T) \subset [Q, \infty)$ *for any self-adjoint realization T of* $-u'' + qu$ *in* $L^2(0, \infty)$. *In particular, if* $Q = \infty$, *then T has compact resolvent, i.e.,* $\sigma_e(T) = \varnothing$.

We shall leave the proof to the reader in Exercise 4.5.3. We finally prove a well known criterion for discrete spectrum by Molchanov [169].

Theorem 4.5.9. *Suppose q is locally integrable and bounded from below in* $(0, \infty)$ *and that* $-u'' + qu = f$ *is limit-circle at* 0. *Then a necessary and sufficient condition for discrete spectrum for any self-adjoint realization T on* $(0, \infty)$ *is that* $\int_x^{x+\varepsilon} q \to \infty$ *as* $x \to \infty$ *for any* $\varepsilon > 0$.

We need the following lemma.

Lemma 4.5.10. *Suppose K is a compact interval on which $q \geq 0$ is integrable with $\int_K q > 0$. If $u' \in L^2(K)$ and $x \in K$, then*

$$|u(x)|^2 \leq C \int_K (|u'|^2 + q\,|u|^2),\qquad (4.5.1)$$

where $C = |K| + 1/\int_K q$ and $|K|$ is the length of K.

This is essentially inequality (1.4.5) (see also Exercise 1.4.4) but we repeat the proof.

Proof. If $x, y \in K$ the fundamental theorem of calculus shows that $|u(x)| \leq |u(y)| + |y - x|^{1/2} \left(\int_K |u'|^2 \right)^{1/2}$ by use of Cauchy–Schwarz. Multiplication by $q(y)$ and integrating with respect to y gives

$$|u(x)| \int_K q \leq \int_K q\,|u| + |K|^{1/2} \int_K q \left(\int_K |u'|^2 \right)^{1/2}.$$

Now $\int_K q\,|u| \leq \left(\int_K q \int_K q\,|u|^2 \right)^{1/2}$ so dividing by $\int_K q$ and applying Cauchy–Schwarz for $\mathbb{R}^2$ we obtain (4.5.1). $\qquad\square$

Proof (Theorem 4.5.9). Rewriting $-u'' + qu = \lambda u$ for a constant C as $-u'' + (q + C)u = (\lambda + C)u$ we may with no loss of generality assume that $q \geq 0$. Given $\varepsilon > 0$, suppose we can choose A so large that $\int_x^{x+\varepsilon} q \geq 1$ for $x \geq A$. Thus if $K \subset (A, \infty)$ is an interval with length ε the constant in (4.5.1) is $C \leq \varepsilon + 1$ so that

$$\int_K |u|^2 \leq \varepsilon(\varepsilon + 1) \int_K (|u'|^2 + q\,|u|^2).$$

We may split (A, ∞) into intervals of length ε and add up the corresponding inequalities, so the inequality is also true for $K = (A, \infty)$. An integration by parts shows that the minimal operator for $-u'' + qu = f$ in $L^2(A, \infty)$ is bounded below by $(\varepsilon(\varepsilon+1))^{-1}$ so by the decomposition principle $\sigma_e(T)$ is bounded below by arbitrarily large constants and is thus empty.

To prove the converse, assume that for some constants $\varepsilon > 0$ and B there is a positive sequence $x_k \to \infty$ such that $\int_{x_k}^{x_k+\varepsilon} q \leq B$ for $k = 1, 2, \ldots$. Suppose $\varphi \in C_0^2(0, \varepsilon)$ equals 1 on $[\varepsilon/3, 2\varepsilon/3]$ and is bounded by 1. Thinning out the sequence $\{x_k\}_1^\infty$ if necessary we may assume $x_k - x_{k-1} > \varepsilon$ for all k and put $\varphi_k(x) = \varphi(x - x_k)$ so that the supports of all φ_k are disjoint. We have $\int_0^\infty |\varphi_k|^2 \geq \varepsilon/3$ while $\int_0^\infty (|\varphi_k'|^2 + q\,|\varphi_k|^2) \leq C + \int_{x_k}^{x_k+\varepsilon} q \leq C + B$ where $C = \int_0^\varepsilon |\varphi'|^2$. Thus

$$\int_0^\infty (Tu)\bar{u} = \int_0^\infty (|u'|^2 + q\,|u|^2) \leq \frac{3(B+C)}{\varepsilon} \int_0^\infty |u|^2$$

if u is any finite linear combination of φ_k, $k = 1, 2 \ldots$, so if $\{E_t\}_{t \in \mathbb{R}}$ is the resolution of the identity for T the range of E_μ for $\mu > 3(B + C)/\varepsilon$ has infinite dimension. Since the operator is bounded below the interval $(-\infty, \mu)$ intersects $\sigma_e(T)$.   □

Exercises

Exercise 4.5.1 Show that a self-adjoint realization T for a regular problem of (4.1.1) with $p > 0$ is always bounded from below.

Hint: First show that if $\operatorname{supp} u \subset I$, then $\int_I q_- |u|^2 \le \int_I q_- \int_I 1/p \int_I p |u'|^2$ where $q_- = \max(0, -q)$. Then show that $\int_I (-(pu')' + qu)\overline{u} \ge 0$ if I is a sufficiently short interval.

Next, consider the restriction of T obtained by requiring functions in the domain to vanish in a finite number of judiciously chosen points, and show this operator to be positive. Finally show that T is a finite-dimensional extension of this operator.

Exercise 4.5.2 Show that a self-adjoint realization for a singular problem (4.1.1) does not have to be bounded from below even if $p > 0$.

Hint: Consider the Schrödinger equation on $[0, \infty)$ with q tending to $-\infty$ at ∞. Show the minimal operator cannot be bounded below by translating a fixed function in $C_0^2(0, \infty)$ to infinity.

Exercise 4.5.3 Give a complete proof of Theorem 4.5.8.

Exercise 4.5.4 Prove a result analogous to Theorem 4.5.9 but with $(0, \infty)$ replaced by $\mathbb{R}$.

4.6 Notes and remarks

Separated boundary conditions for regular problems leading to orthonormal bases of eigenfunctions were known to Sturm and Liouville, but the systematic derivation of all self-adjoint boundary conditions for Sturm–Liouville equations was carried out much later. This may be found in many books from the 1950s onwards, for example Naĭmark [170].

There are a large number of limit-point criteria in the literature in addition to those given at the end of Section 4.2. See for example Brinck [36], Atkinson and Evans [10], Atkinson [8] and references given there. The limit-point criterion in Exercise 4.2.3 may be found in Zettl [228, Theorem 7.4.4].

Many proofs of expansion theorems use a limiting process from regular problems on smaller intervals, see for example [212, 46, 104]. Our approach to the expansion theorems may be found in Bennewitz and Everitt [22]. An alternative approach to the eigenfunction expansion problem is provided by the concept of *direction functionals*.

These were introduced by M. G. Kreĭn in papers from 1946 and 1948. The method is discussed in detail in [170, page 106, Vol 2] together with references to the original papers and in [2]. A more general setting of Kreĭn's work can be found in [78] which also contains applications to contemporary problems.

A large number of results may be found in the literature analogous to the expansions in Theorems 4.3.7 and 4.4.4 but applicable to other situations. We shall give some such results in Chapters 5 and 8. There are also similar results for abstract operators. A general theorem of this nature was given by von Neumann [173] in 1949, but is of a fairly abstract nature. It can be applied to elliptic partial differential equations (Gårding [79]), but gives more satisfactory results when applied to ordinary differential equations, as described by Gårding in an appendix to John, Bers and Schechter [23]. A more general situation, including left-definite equations, was treated in Bennewitz [15]. We have chosen an approach to expansion theorems more in line with traditional methods for ordinary differential equations.

The device of using the case of one regular endpoint to deal with two singular endpoints used in Section 4.4 is due to Weyl [222] and was also used by Titchmarsh [212]. The complications caused by two singular endpoints stem from the possibility of an essential spectrum with multiplicity two. If this can be avoided an expansion theorem derived from a scalar m-function and thus a scalar spectral measure is still feasible. A necessary and sufficient condition for this was given by Kostenko, Sakhnovich and Teschl [130]. A similar, earlier result was given by Gesztesy and Zinchenko [92].

A very important method for investigating the spectrum uses *subordinacy*. This is a concept introduced by Gilbert [94] in 1984. Consider the equation $-u'' + qu = \lambda u$ on an interval (a, b) where b is limit-point and λ real. A solution *subordinate at* b is a solution u of the equation such that, if $c \in (a, b)$, $\int_c^x |u|^2 / \int_c^x |v|^2 \to 0$ as $x \to b$, for any solution v linearly independent of u. If this holds only for a sequence $x_j \to b$ one speaks of *sequential subordinacy*. See also Gilbert and Pearson [95] and Pearson [179, Chapter 7]. These notions immediately extend to Sturm–Liouville equations of general form. They are very useful for identifying intervals in which the spectral measure is locally absolutely continuous, i.e., of the form $d\eta(t) = f(t)\, dt$ with a locally integrable function f.

To give the flavor of the results that may be obtained we quote just one result from Pearson [179, Theorem 7.8].

Theorem 4.6.1. *Let T be a self-adjoint realization of $-u'' + qu = \lambda u$ in $L^2(a, b)$, where a is a regular endpoint and b limit-point. Then*

(i) *Suppose for a.a. $\lambda \in (\lambda_1, \lambda_2)$ the equation has a solution subordinate at b. Then the spectrum of T is purely singular in (λ_1, λ_2).*

(ii) *Suppose for $\lambda \in (\lambda_1, \lambda_2)$ there is no solution sequentially subordinate at b. Then the spectrum of T is absolutely continuous in (λ_1, λ_2).*

The assumptions in this theorem are in many cases much easier to verify than the detailed information on the m-function required to obtain such conclusions by applying the Stieltjes inversion formula to the m-function.

In [90] and [91] Gesztesy, Zinchenko, and Weikard investigated the spectral theory of Schrödinger equations on a half-line with potentials taking values among the bounded operators on a given Hilbert space. The former paper places emphasis on existence and uniqueness of solutions of initial value problems and their consequences culminating in the introduction of the operator-valued Weyl–Titchmarsh function and Green's function. The latter paper establishes expansions in eigenfunctions both for the half-line and the full-line case. These results, of course, generalize earlier work (referenced in [90] and [91]) on Schrödinger equations with matrix-valued potentials.

Chapter 5
Left-definite Sturm–Liouville equations

Spectral theory is concerned with the simultaneous diagonalization of Hermitian forms or, equivalently, a corresponding eigenvalue equation $Su = \lambda T u$. For this to be possible one needs some positivity in the problem,[1] usually that some linear combination of the Hermitian forms is positive. When dealing with differential operators S and T one usually writes the eigenvalue equation such that S has higher order, or somehow dominates T.

In the simplest cases T may be multiplication by a positive function w, and then one traditionally considers the problem in the L^2-space with weight w. This would be a *right-definite* equation, whereas if one requires positivity of a Hermitian form associated with the operator S one talks of a *left-definite* equation.

If we restrict our attention to a Sturm–Liouville equation

$$-(pu')' + qu = \lambda w u$$

on an open interval (a, b), with the standard assumptions that $1/p$, q and w are real-valued, locally integrable functions, the most obvious Hermitian forms associated with the two sides of the equation are $\int_a^b (pu'\overline{v'} + qu\overline{v})$ and $\int_a^b u\overline{v}w$. The problem is then called right-definite if $\int_a^b u\overline{v}w$ is positive, which obviously requires $w \geq 0$, and we have discussed this case in some detail in the previous chapter.

The problem is called left-definite if the form $\int_a^b (pu'\overline{v'} + qu\overline{v})$, called the *Dirichlet integral* for the equation, is positive. We shall study this situation in the present chapter.

5.1 Left-definite equations

The equation. Positivity of the Dirichlet integral requires $p > 0$ a.e., which is an immediate consequence of Theorem 4.5.2. In view of this, if we pick

---

[1] This is so even if S and T are 2×2 matrices!

© Springer Nature Switzerland AG 2020
C. Bennewitz et al., *Spectral and Scattering Theory for Ordinary Differential Equations*,
Universitext, https://doi.org/10.1007/978-3-030-59088-8_5

$c \in (a, b)$ and change variables in the equation by setting $t(x) = \int_c^x 1/p$ the equation is transformed[2] into one where $p \equiv 1$, and q, w are replaced by $\tilde{q}(t) = q(x(t))p(x(t))$ respectively $\tilde{w}(t) = w(x(t))p(x(t))$. We also have $\int_a^b (p |u'|^2 + q |u|^2) = \int_{a'}^{b'} (|du/dt|^2 + \tilde{q} |u|^2)$ and $\int_{a'}^{b'} |u|^2 \tilde{w} = \int_a^b |u|^2 w$ where (a', b') is the image of (a, b) under $x \mapsto t(x)$. The new coefficients $\tilde{q}$ and $\tilde{w}$ are locally integrable as functions of t so in this chapter we shall only deal with equations of the form[3]

$$-u'' + qu = wf . \tag{5.1.1}$$

As in Chapter 4 we assume that q, w are real-valued and locally integrable in (a, b). We still assume w is not a.e. zero, but no longer assume positivity, allowing arbitrary sign changes in w. Instead we assume $q \geq 0$ and not a.e. zero. This assumption ensures that $\int_a^b (|u'|^2 + q |u|^2)$ is positive definite, but is by no means necessary for this to be true. We shall explore the ramifications of this fact in Sections 5.5–5.8.

A Hilbert space. Let $\mathcal{H}_1$ be the set of locally absolutely continuous functions u defined in (a, b) and such that $u' \in L^2(a, b)$ and $q |u|^2 \in L^1(a, b)$. As we shall soon see $\mathcal{H}_1$ is a Hilbert space with scalar product

$$\langle u, v \rangle = \int_a^b (u'\bar{v}' + qu\bar{v})$$

and norm $\|u\| = \sqrt{\langle u, u \rangle}$. In order to show completeness of $\mathcal{H}_1$ and discuss how to find self-adjoint realizations of (5.1.1) in $\mathcal{H}_1$ we first note that Lemma 4.5.10 shows that for any compact $K \subset (a, b)$ there exists a constant C_K such that

$$|u(x)| \leq C_K \|u\| \tag{5.1.2}$$

for $u \in \mathcal{H}_1$ and $x \in K$.

Proposition 5.1.1. *The space $\mathcal{H}_1$ is complete.*

Proof. By (5.1.2) a Cauchy sequence $u_1, u_2, \ldots$ in $\mathcal{H}_1$ converges locally uniformly to a continuous function u. Furthermore, $\sqrt{q}u_j$ and u'_j converge in $L^2(a, b)$ to $\sqrt{q}u$ and, say, v, respectively.

Since the characteristic function of a compact subinterval of (a, b) is in $L^2(a, b)$ we obtain from $u_j(x) = u_j(c) + \int_c^x u'_j$ that $u(x) = u(c) + \int_c^x v$. Thus u is locally absolutely continuous with derivative v so that u_j converges to u in $\mathcal{H}_1$. $\square$

Proposition 5.1.2. *There exists a unique real-valued function g_0 defined in $(a, b) \times (a, b)$ such that $g_0(x, \cdot) \in \mathcal{H}_1$ for every fixed $x \in (a, b)$ and $u(x) = \langle u, g_0(x, \cdot) \rangle$ for every $u \in \mathcal{H}_1$ and $x \in (a, b)$.*

---

[2] For more details about such transformations see Section 7.2

[3] Similarly, if $w > 0$ a.e. one may transform the equation into one with weight 1 by setting $t = \int_c^x w$. Since we wanted to cover the case when w vanishes on an open set in Chapter 4 we did not do this.

Furthermore $g_0(x, y) = g_0(y, x)$, $g_0(x, x) > 0$ *and the norm of the linear form* $\mathcal{H}_1 \ni u \mapsto u(x)$ *is* $\sqrt{g_0(x, x)}$.

Proof. By Riesz' representation theorem and (5.1.2) there exists for every $x \in (a, b)$ a unique $g_0(x, \cdot) \in \mathcal{H}_1$ such that the linear form $\ell_x : \mathcal{H}_1 \ni u \mapsto u(x)$ is given by $u(x) = \langle u, g_0(x, \cdot)\rangle$. If $u \in \mathcal{H}_1$ is real-valued we also obtain $u(x) = \langle u, \operatorname{Re} g_0(x, \cdot)\rangle$ by taking the real part of $\langle u, g_0(x, \cdot)\rangle$. Splitting u into its real and imaginary parts the same formula is true for a general u. Thus $\operatorname{Im} g_0(x, \cdot)$ is orthogonal to $\mathcal{H}_1$ and therefore zero.

Now $|u(x)| \le \|g_0(x, \cdot)\|\|u\|$ so $\|\ell_x\| \le \|g_0(x, \cdot)\|$. However, for $u = g_0(x, \cdot)$ we have $g_0(x, y) = \langle g_0(x, \cdot), g_0(y, \cdot)\rangle$ so conjugating this we obtain $g_0(x, y) = g_0(y, x)$, and for $y = x$ we obtain $g_0(x, x) = \|g_0(x, \cdot)\|^2 > 0$ so that $\|\ell_x\| = \|g_0(x, \cdot)\| = \sqrt{g_0(x, x)}$. $\qquad\square$

It follows that the smallest value of C_K for which (5.1.1) holds is $\sup_{x \in K} \sqrt{g_0(x, x)}$.

Denote the set of integrable functions with compact support in (a, b) by L_c. Then, if $u \in \mathcal{H}_1$ and $v \in L_c$, it follows that $|\int u\bar{v}| \le C_K \int |v|\, \|u\|$ if $\operatorname{supp} v \subset K$, so that the linear form $\mathcal{H}_1 \ni u \mapsto \int u\bar{v}$ is bounded. By Riesz' representation theorem we may therefore find a unique $v^* \in \mathcal{H}_1$ so that $\int u\bar{v} = \langle u, v^*\rangle$. Clearly v^* depends linearly on v, so we obtain an operator[4] $G_0 : L_c \to \mathcal{H}_1$ such that

$$\langle u, G_0 v\rangle = \int_a^b u\bar{v} \text{ for } u \in \mathcal{H}_1,\ v \in L_c. \tag{5.1.3}$$

The operator G_0 is central for our approach to the left-definite spectral theory of (5.1.1).

Proposition 5.1.3. *The operator* $G_0 : L_c \to \mathcal{H}_1$ *is an injective integral operator*

$$G_0 u(x) = \int_a^b u\, g_0(x, \cdot),$$

and its restriction to $\mathcal{H}_c := L_c \cap \mathcal{H}_1$ *is symmetric with range dense in* $\mathcal{H}_1$.

Proof. We have $G_0 v(x) = \langle G_0 v, g_0(x, \cdot)\rangle = \int_a^b v\, g_0(x, \cdot)$ for any $v \in L_c$. Thus G_0 is an integral operator with kernel $g_0(x, y)$. If u and $v \in L_c \cap \mathcal{H}_1$, then

$$\langle G_0 u, v\rangle = \overline{\langle v, G_0 u\rangle} = \int_a^b u\bar{v} = \langle u, G_0 v\rangle,$$

so the restriction of G_0 to $\mathcal{H}_c$ is symmetric.

---

[4] A sequence $\{u_j\}_1^\infty$ is said to converge in L_c if it converges in $L^1(a, b)$ and all $\operatorname{supp} u_j$ are in a fixed compact $\subset (a, b)$. With this understanding L_c is complete and G_0 continuous.

This calculation also shows, for $v \in C_0^\infty(a, b) \subset \mathcal{H}_c$, that if $G_0 u = 0$, then $u = 0$ as a distribution and thus a.e., and also that the range of the restriction of G_0 to $C_0^\infty(a, b)$ is dense in $\mathcal{H}_1$. $\qquad\qquad\qquad\qquad\qquad\qquad\qquad\qquad\qquad\qquad\qquad\qquad\qquad$ $\square$

Boundary conditions. Now define a relation on $\mathcal{H}_1$ by setting

$$T_c = \{(G_0(wv), v) : v \in L_c \cap \mathcal{H}_1\} \ .$$

Since w is real-valued, T_c is a symmetric relation on $\mathcal{H}_1$, for if $u, v \in L_c \cap \mathcal{H}_1$, then

$$\langle G_0(wu), v \rangle = \overline{\langle v, G_0(wu) \rangle} = \int_a^b wu\bar{v} = \langle u, G_0(wv) \rangle \ .$$

Proposition 5.1.3 implies that T_c is the graph of a densely defined symmetric operator in $\mathcal{H}_1$ if $\operatorname{supp} w = (a, b)$, but if w vanishes on a non-empty open set it will just be a symmetric relation. We define the *minimal relation* T_0 associated with (5.1.1) on (a, b) as the closure of T_c, and the *maximal relation* T_1 as the adjoint of this, as defined in (2.2.1), *i.e.*,

$$T_1 = \mathcal{H}_1^2 \ominus \mathcal{U} T_c = \{(u, f) \in \mathcal{H}_1^2 : \langle u, v \rangle = \langle f, G_0(wv) \rangle \text{ for all } v \in L_c \cap \mathcal{H}_1\} \ ,$$

where $\mathcal{H}_1^2 = \mathcal{H}_1 \oplus \mathcal{H}_1$.

We next show that T_1 is the maximal realization of (5.1.1) in $\mathcal{H}_1$.

Proposition 5.1.4. *We have $(u, f) \in T_1$ if and only if u and $f \in \mathcal{H}_1$, u' is locally absolutely continuous, and $-u'' + qu = wf$.*

Proof. First note that if u and $f \in \mathcal{H}_1$, then the definition of G_0 shows that

$$\langle u, v \rangle - \langle f, G_0(wv) \rangle = \int_a^b (u'\bar{v}' + qu\bar{v} - wf\bar{v}) \qquad\qquad (5.1.4)$$

for any $v \in L_c \cap \mathcal{H}_1$. This therefore vanishes for all $v \in C_0^\infty(a, b)$ if and only if $-u'' + qu = wf$ in a distributional sense; see Appendix C. Since $wf - qu$ is locally integrable it follows that so is u'', so that u' is locally absolutely continuous and $u \in C^1(a, b)$. Thus (5.1.4) equals zero also if we only have $v \in L_c \cap \mathcal{H}_1$. $\qquad$ $\square$

One may give a proof without the use of distribution theory by use of Theorem 1.4.5 and arguing as in the subsequent lines.

Let $D_\lambda = \{u \in \mathcal{H}_1 : (u, \lambda u) \in T_1\}$ be the defect space of T_0 at λ. Thus $u \in D_\lambda$ precisely if $u \in \mathcal{H}_1$ satisfies $-u'' + qu = \lambda wu$, and according to Corollary 2.4.8 the defect indices of T_0 are $n_+ = \dim D_\lambda$ and $n_- = \dim D_{\bar{\lambda}}$ if $\operatorname{Im} \lambda > 0$.

It is clear that $\dim D_\lambda \le 2$, and that $n_+ = n_-$ since $\bar{u} \in D_{\bar{\lambda}}$ if and only if $u \in D_\lambda$. Thus there are always self-adjoint extensions of T_0. These are at the same time restrictions of T_1, and therefore realizations of (5.1.1). As in the right-definite case this gives only three possibilities: The defect indices may both be 2, both may be 1, or both may be 0. We shall see later that all three cases can occur, depending on the choice of (a, b), q and w.

Suppose (u, f) and $(v, g) \in T_1$. Then the boundary form (cf. Section 2.4) is

$$\langle (u, f), \mathcal{U}(v, g) \rangle = i \int_a^b (u'\overline{g'} + qu\overline{g}) - i \int_a^b (f'\overline{v'} + qf\overline{v}) = i \lim_{K \to (a,b)} (u'\overline{g} - f\overline{v'})\big|_K ,$$

the limit being taken over compact subintervals K of (a, b). To obtain a self-adjoint restriction of T_1 it must be restricted by $n_+ = n_-$ linear, homogeneous conditions so that this vanishes. Since the limit clearly only depends on the values of (u, f) and (v, g) in arbitrarily small neighborhoods of the endpoints of (a, b) this means that restrictions of T_1 to self-adjoint relations are again obtained by *boundary conditions*.

Comparing the condition for symmetry $(u'\overline{g} - f\overline{v'})\big|_a^b = 0$ with the corresponding condition $(u'\overline{v} - u\overline{v'})\big|_a^b = 0$ for the right-definite case, we see that only exceptionally would a self-adjoint boundary condition in the left-definite case also be a self-adjoint boundary condition in the right-definite case. On the other hand, self-adjoint boundary conditions for the left-definite case look the same as those for the right-definite case, *except* that they are imposed on the values of f and u' at the interval endpoints instead of those of u and pu'.

As in the right-definite case we call an endpoint of (a, b) *regular* if it is finite and q and w are integrable near the endpoint. Otherwise the endpoint is singular. If both endpoints are regular, we are dealing with a *regular problem*. The problem is *singular* if at least one of the endpoints is infinite, or if q and w are not both in $L^1(a, b)$. It is clear that a regular problem has defect indices 2. If a is a regular endpoint both f and u' have well-defined values at a by (5.1.2) and Theorem 4.1.1, so reasoning as in the right-definite case a separated, symmetric boundary condition at a is of the form

$$f(a) \cos \alpha + u'(a) \sin \alpha = 0 , \tag{5.1.5}$$

for some $\alpha \in [0, \pi)$. Comparing this with a separated right-definite condition $u(a) \cos \beta + u'(a) \sin \beta = 0$, they coincide only if $\alpha = \beta = \pi/2$, the Neumann boundary condition. However, for eigenfunctions, where $f = \lambda u$, it is clear that the conditions for $\alpha = \beta = 0$, the Dirichlet boundary condition, also coincide as long as $\lambda \neq 0$.

Remark 5.1.5. By changing the scalar product, similar to the changes of the form H made in Section 1.4, other boundary conditions may yield self-adjoint realizations (with respect to the new scalar product). A similar device has long been used in the right-definite case to deal with boundary conditions which contain the eigenvalue parameter, e.g. in Walter [214]. We shall briefly consider such modifications in Section 5.6.

If both endpoints a and b are regular, self-adjoint coupled boundary conditions are of the form $\begin{pmatrix} f(b) \\ u'(b) \end{pmatrix} = S \begin{pmatrix} f(a) \\ u'(a) \end{pmatrix}$ with a symplectic matrix S. Such a condition will never coincide with a similar condition for the right-definite case, but applying the conditions to eigenfunctions for non-zero eigenvalues, when $f = \lambda u$ and $\lambda \neq 0$, the conditions are the same precisely if S is diagonal. This includes periodic and semi-periodic conditions.

Let us now consider the singular case. We shall need the following counterpart to Lemma 4.2.4

Lemma 5.1.6. *Any* $(u, f) \in T_1$ *may be written as* $(u, f) = (u_1, f_1) + (u_2, f_2)$ *where* $(u_j, f_j) \in T_1$, $(u_1, f_1) = 0$ *near the right and* $(u_2, f_2) = 0$ *near the left endpoint of* (a, b).

This is proved in exactly the same way as Lemma 4.2.4, if one keeps in mind to select f_1 smooth enough to be in $\mathcal{H}_1$.

Clearly the boundary form $\langle (u, f), \mathcal{U}(v, g) \rangle = \mathrm{i}(\langle u, g \rangle - \langle f, v \rangle)$ is a bounded Hermitian form on T_1, and in view of Lemma 5.1.6, so are the boundary forms $u'\overline{g} - f\overline{v'}$, evaluated at the endpoints of (a, b). By Lemma 5.1.6 these boundary forms are independent, so that the rank of the full boundary form is the sum of the ranks at the endpoints. Since the rank at a regular endpoint is two it follows that the rank at a singular endpoint is zero or two. As in the right-definite case the endpoint is said to be *limit-circle* if the rank is two and *limit-point* if the rank is zero. On the other hand, by the abstract theory the rank of the full boundary form is dim $\mathbf{D}_i \oplus \mathbf{D}_{-i}$, so we have the following theorem, which is completely analogous to Theorem 4.2.6.

Theorem 5.1.7. *If* $n_\pm = 0$, *then both endpoints of* (a, b) *are limit-point, if* $n_\pm = 1$ *one of the endpoints is limit-point and the other limit-circle, and if* $n_\pm = 2$ *both endpoints are limit-circle.*

The boundary form at a limit-point endpoint is identically zero, so no boundary condition is to be applied there, but if the other endpoint is limit-circle a symmetric, separated boundary condition must be imposed there to obtain a self-adjoint relation.

If both endpoints are limit-circle, then two boundary conditions must be imposed. For a regular problem this means either a symmetric, separated condition at each end or coupled boundary conditions, given by a symplectic matrix.

At regular endpoints self-adjoint boundary conditions are expressed in terms of the values of f and u' at the endpoint, but at a singular limit-circle endpoint in general only Im $u'\overline{f}$ and f but not u' has a limit. We will discuss this case in Theorem 5.1.12 and the subsequent paragraph. Clearly T_1 is self-adjoint and T_c essentially self-adjoint if and only if both endpoints are limit-point.

As in the right-definite case it is of great interest to have criteria in terms of the interval (a, b) and the coefficients q and w for different values of the defect indices. In contrast to the right-definite case, we have the following simple and explicit necessary and sufficient criteria.

Theorem 5.1.8. *Let* W *be a primitive of* w, *i.e.,* $W' = w$ *in the sense of distributions. Then* $n_\pm = 2$ *if and only if* (a, b) *is bounded,* $q \in L^1(a, b)$ *and* $W \in L^2(a, b)$. *In this case* dim $D_\lambda = 2$ *also for all* $\lambda \in \mathbb{R}$.

If (a, b) *is not bounded,* $q \notin L^1(a, b)$ *or* $W \notin L^2(a, b)$, *then* $n_\pm = 1$ *if* (a, b) *has a finite endpoint near which* $q \in L^1$ *and* $W \in L^2$. *Otherwise* $n_\pm = 0$.

Proof. If $c \in (a, b)$, consider for $\mathrm{Re}\,\lambda = 0$ a solution of $-u'' + qu = \lambda wu$. Integration by parts and taking real parts shows that

$$\int\limits_c^x (|u'|^2 + q\,|u|^2) = \mathrm{Re}(u\overline{u'})\Big|_c^x \,,$$

so that $\mathrm{Re}(u\overline{u'})$ is increasing.

Now, if $\int_c^b (|u'|^2 + q\,|u|^2) < \infty$ and $\lim_b \mathrm{Re}(u\overline{u'}) \neq 0$, then $1/(u\overline{u'})$ is bounded near b, so $u'/u = |u'|^2 /(u\overline{u'})$ is integrable near b. This implies that u has a non-zero limit at b since $u \exp(-\int_c^x u'/u)$ is seen to be constant by differentiation. It follows that q is integrable near b, and also that $|u|^{-2} = |u'|^2 /|u'\overline{u}|^2$ is integrable near b. Thus b must be finite. Furthermore, if d is sufficiently close to b,

$$\lambda \int\limits_d^x w = \int\limits_d^x (-u'' + qu)/u = -u'/u\Big|_d^x + \int\limits_d^x \left(q - (u'/u)^2\right) .$$

As $x \to b$ the last integral is bounded since q and $|u'|^2$ are integrable and u has a non-zero limit, and since b is finite it follows that $\lambda \int_c^x w$, and thus any primitive of w, is square integrable near b if $\lambda \neq 0$. In all other cases finiteness of $\int_c^b (|u'|^2 + q\,|u|^2)$ implies that $\lim_b \mathrm{Re}(u\overline{u'}) = 0$. This implies that no solution for which $\mathrm{Re}(u\overline{u'}) > 0$ at c can have $\int_c^b (|u'|^2 + q\,|u|^2) < \infty$, so b is limit-point. Similar arguments apply at the other endpoint a.

Conversely, setting $U = \left(\begin{smallmatrix} u \\ u' + \lambda W u \end{smallmatrix}\right)$ and $A = \left(\begin{smallmatrix} -\lambda W & 1 \\ q - (\lambda W)^2 & \lambda W \end{smallmatrix}\right)$ we obtain $U' = AU$. Assuming $W \in L^2$ near a finite endpoint of (a, b) we have $|W| \leq \frac{1}{2}(|W|^2 + 1)$ so W is also integrable near this endpoint. If also q is integrable there, all entries in A are integrable near the endpoint, so the initial value problem at this point has a unique solution. Thus any solution U has a finite value there. It follows that $q\,|u|^2$ is integrable and $u' = (u' + \lambda W u) - \lambda W u$ square integrable near this endpoint. If in fact (a, b) is bounded and $q + W^2 \in L^1(a, b)$, then for any $\lambda \in \mathbb{C}$ all solutions are in $\mathcal{H}_1$. We have now verified all claims of the theorem. □

The space D_0 of solutions in $\mathcal{H}_1$ for $\lambda = 0$ has a special significance for left-definite equations, which is brought out in the following theorem.

Theorem 5.1.9.

(1) *The set D_0 is the orthogonal complement in $\mathcal{H}_1$ of $L_c \cap \mathcal{H}_1$.*

(2) *If (a, b) has a finite endpoint near which q is integrable, then for all solutions of $-v'' + qv = 0$ there are finite limits of v and v' at such an endpoint, and these limits uniquely determine v.*

(3) *If (a, b) has a finite endpoint near which q is integrable, then every $u \in \mathcal{H}_1$ has a limit at this endpoint which is a bounded linear form on $\mathcal{H}_1$.*

(4) $\dim D_0 = 2$ *if and only if* (a, b) *is bounded and* $q \in L^1(a, b)$ *and* $\dim D_0 = 0$ *if and only if* (a, b) *has no finite endpoint near which* q *is integrable. In other cases* $\dim D_0 = 1$.

(5) *If* b *is not finite or* q *not integrable near* b, *then* $u(x)v'(x) \to 0$ *as* $x \to b$ *for any* $u \in \mathcal{H}_1$ *and any solution* v *of* $-v'' + qv = 0$ *for which* $|v'|^2 + q\,|v|^2$ *is integrable near* b.

Proof. We have $u \in D_0$ precisely if $(u, 0) \in T_1$, which holds precisely if $\langle u, f \rangle = \langle u, f \rangle - \langle 0, G_0(wf) \rangle = 0$ for all $f \in L_c \cap \mathcal{H}_1$, which proves (1).

If b is finite and q integrable near it, Theorem 4.1.1 shows that all solutions of $-v'' + qv = 0$ are continuously differentiable with absolutely continuous derivative near b. Now v is uniquely determined by the values of v and v' at b, so (2) follows.

In this case the inequality (5.1.2) clearly also works for $x = b$. Together with similar arguments for a this proves (3). It follows that elements of the closure of $L_c \cap \mathcal{H}_1$ vanish at a finite endpoint near which q is integrable. Not all $u \in \mathcal{H}_1$ vanish at such an endpoint, so it follows that $\dim D_0 \geq 1$ if (a, b) has such an endpoint. If (a, b) has two such endpoints the values of $u \in \mathcal{H}_1$ at the endpoints are clearly independent, so in this case $\dim D_0 = 2$.

If $-v'' + qv = 0$ and $u, v \in \mathcal{H}_1$ and $c \in (a, b)$, then $\lim_b uv' = \int_c^b (u'v' + quv) + u(c)v'(c)$ is finite. If the limit is not zero $1/(uv')$ is bounded and $u'/u = u'v'/(uv')$ integrable near b, so as in the proof of Theorem 5.1.8 u has a non-zero limit at b. Since $q\,|u|^2$ is integrable it follows that near b so is q. Similarly, $v''/v' = qv/v' = quv/(uv')$ is integrable near b, so v' has a nonzero limit there. Since $|v'|^2$ is integrable it follows that b is finite.

Thus, if these conditions do not apply we have $\lim_b uv' = 0$. Similar arguments apply for a. In particular $\int_a^b (|v'|^2 + q\,|v|^2) = v'\overline{v}|_a^b$, so for $v \not\equiv 0$ the limits cannot be zero at both endpoints. Thus one endpoint has to be finite with q integrable near it if $\dim D_0 = 1$, so this proves (4) and (5). $\square$

Proposition 5.1.10. *For fixed* $x \in (a, b)$ *the kernel* $g_0(x, y)$ *satisfies* $-v'' + qv = \delta_x$ *in* (a, b), *where* δ_x *is the Dirac measure at* x. *There are strictly positive solutions* F_+ *and* F_- *of the equation* $-v'' + qv = 0$ *with Wronskian* $\mathcal{W}(F_-, F_+) = 1$, *and such that if* $u \in \mathcal{H}_1$, *then*

(1) F_- *has finite Dirichlet integral near the left, and* F_+ *near the right endpoint of* (a, b).

(2) $F'_- u$ *has limit 0 at the left endpoint and* $F'_+ u$ *at the right.*

(3) $g_0(x, y) = F_-(\min(x, y))F_+(\max(x, y))$.

Proof. If $v = g_0(x, \cdot)$ and $u \in C_0^\infty(a, b)$ we have $\int_a^b (u'v' + quv) = u(x)$ so that in a distributional sense $-v'' + qv = \delta_x$.

Now, if there are non-trivial solutions $F_\pm$ with properties (1) and (2) they must have Wronskian $\neq 0$ since otherwise they would be proportional and $\|F_+\|^2 = F'_+\overline{F_+}|_a^b = 0$. Thus multiplying F_+ by a non-zero number we may also assume $\mathcal{W}(F_-, F_+) = 1$. A calculation then verifies (3).

The existence of F_+ is clear if the right endpoint is finite and q integrable near it, since we may then require F_+' to be 0 and F_+ non-zero at the endpoint. If the endpoint is not finite or q not integrable near it there is still a non-trivial solution F_+ with the desired properties, by Theorem 5.1.9 (5), and replacing this solution by its real or imaginary part we may assume it real-valued.

Similarly one shows the existence of a non-trivial and real-valued solution F_- with properties (1) and (2). Since $g_0(x, x) > 0$ neither of $F_\pm$ can have a zero, so we may assume both positive. $\qquad\square$

Assuming now that we have a self-adjoint realization T in $\mathcal{H}_1$ of (5.1.1), Theorem 2.2.9 shows that the restriction of T to $\mathcal{H} = \mathcal{H}_1 \ominus \mathcal{H}_\infty$ is a self-adjoint operator and that $\mathcal{H}_\infty = \{u \in \mathcal{H}_1 \mid (0, u) \in T\}$. In the present case one may give a rather complete description of the spaces $\mathcal{H}_\infty$ and $\mathcal{H}$. First note that the complement of supp w is a countable union of disjoint open intervals. We shall call any such interval a *gap* in supp w.

Theorem 5.1.11.

(1) *The space $\mathcal{H}_\infty$ consists of those elements $g \in \mathcal{H}_1$ for which $wg = 0$ a.e., and for which $(0, g)$ satisfies the boundary conditions that define T. In particular, if $wg = 0$ a.e. and $g \in L_c \cap \mathcal{H}_1$, then $g \in \mathcal{H}_\infty$.*

(2) *The projection of $v \in \mathcal{H}_1$ onto $\mathcal{H}$ equals v in supp w, and if (a, b) is a gap in the support of w the projection is determined in the gap as the solution of $-u'' + qu = 0$ which equals v in the endpoints a and b if these are finite.*

    If $a = -\infty$ the restriction of the projection to the gap is the multiple of F_- which equals v in b, and if $b = \infty$ it is the multiple of F_+ which equals v in a.

(3) *The support of an element of $\mathcal{H}$ cannot begin or end inside a gap in the support of w.*

(4) *The reproducing kernel $g_0(x, \cdot) \in \mathcal{H}$ if and only if $x \in$ supp w.*

Proof. That $g \in \mathcal{H}_\infty$ means that $(0, g) \in T \subset T_1$, so that $(0, g)$ satisfies the boundary conditions determining T as well as $wg = 0$ as a distribution, i.e., almost everywhere.

Conversely, if $(0, g)$ satisfies the boundary conditions and $gw = 0$ a.e., then if $(u, f) \in T$, an integration by parts gives

$$\langle u, g \rangle - \langle f, 0 \rangle = (u'\overline{g} - f \cdot 0)\big|_a^b + \int_a^b f\overline{g}w = 0 \, ,$$

i.e., $(0, g) \in T$, which proves the first claim.

The difference between v and its projection onto $\mathcal{H}$ can be non-zero only in gaps of supp w. Clearly $\varphi w = 0$ for any $\varphi \in C_0^\infty(a, b)$ so that $C_0^\infty(a, b)$ is orthogonal to $\mathcal{H}$. It follows that an element of $\mathcal{H}$ satisfies the equation $-u'' + qu = 0$ in any gap of the support of w. The next two items are immediate consequences of this, that non-trivial solutions of $-u'' + qu = 0$ cannot vanish on a subinterval by Theorem D.2, and of the fact that elements of $\mathcal{H}$ are continuous.

The last item is a consequence of the previous two and Proposition 5.1.10. $\qquad\square$

Limit-circle endpoints. In limit-circle endpoints, like in regular endpoints, boundary conditions must be imposed to obtain a self-adjoint realization of (5.1.1). In contrast to the case of a regular endpoint, and similar to the right-definite case, the boundary conditions that $(u, f) \in T$ must satisfy may not always be expressed in terms of the values of f and u', as in (5.1.5), since u' may not be well defined at the endpoint. However, the remedy is now somewhat simpler than in the right-definite case.

Theorem 5.1.12. *Let W be a real-valued primitive of w and $(u, f) \in T_1$. At a limit-circle endpoint the functions f and $u' + Wf$ have well defined finite values.*

Proof. Define $A = \begin{pmatrix} 0 & 1 \\ q & 0 \end{pmatrix}$ and $U = \begin{pmatrix} u \\ u'+Wf \end{pmatrix}$, where W is a real-valued primitive of w. The equation $-u'' + qu = wf$ is then equivalent to $U' = AU + F$, where $F = \begin{pmatrix} -Wf \\ Wf' \end{pmatrix}$. It follows from Theorem 5.1.8 that elements $\mathcal{H}_1$ have well defined values at the endpoint and if $f \in \mathcal{H}_1$ the components of F and A are integrable near a limit-circle endpoint. Thus the initial value problem for $U' = AU + F$ at such an endpoint has a unique solution. The theorem follows. $\qquad\square$

Since evaluation at a limit-circle endpoint is a bounded linear form on $\mathcal{H}_1$ the theorem shows that f has a limit at such an endpoint, and if this is not zero, then u' has a limit there precisely if W has. Since all we know is that $W \in L^2$ near the endpoint, u' does not in general have a limit there. However, the boundary form at a limit-circle endpoint evaluated at (u, f) and (v, g) is $\lim(u'\overline{g} - f\overline{v'})$ which equals $\lim\left((u' + Wf)\overline{g} - f(\overline{v' + Wg})\right)$. The boundary conditions needed at a limit-circle endpoint to obtain a self-adjoint realization of (5.1.1) are therefore the same as in the case of a regular endpoint, except that they must be imposed on the values of f and $u' + Wf$ at the endpoint instead of on those of f and u'. A separated boundary condition thus looks like

$$f(a)\cos\alpha + (u' + Wf)(a)\sin\alpha = 0, \qquad (5.1.6)$$

and coupled boundary conditions are as before given by a symplectic matrix, but now coupling the values of f and $u' + Wf$ at the endpoints. If a is a regular endpoint we may of course choose $W(x) = \int_a^x w$, and then (5.1.6) reverts to (5.1.5).

Note that in a limit-circle endpoint, the boundary condition implied by $f \in \mathcal{H}_\infty$ is in most cases the vanishing of f in that endpoint. This is automatic if the endpoint is in supp w. If not, the endpoint is regular by Theorem 5.1.8. The boundary condition (5.1.5) is now $f(a)\cos\alpha = 0$ so this requires $f(a) = 0$ unless $\alpha = \pi/2$. If none of the endpoints are in supp w and $n_\pm = 2$ some coupled boundary conditions also do not require that f vanishes at the endpoints.

If u is a solution of $-u'' + qu = 0$ the boundary conditions (5.1.5) and (5.1.6) both read $u'(a) = 0$ unless $\alpha = 0$, when no condition at all is imposed at a. We shall call the boundary condition a *Dirichlet condition* when $\alpha = 0$. Any solution for which $u'(a) = 0$ satisfies all separated, symmetric boundary conditions at a. Thus the solution F_- of Theorem 5.1.10 satisfies all separated symmetric boundary conditions at the endpoint a, whether this is limit-circle or limit-point. Similarly for the solution F_+ at the right endpoint b. See also Exercise 5.1.2.

Eigenspace at 0. As we have seen, solutions for $\lambda = 0$ which are in $\mathcal{H}_1$, i.e., elements of D_0, play a special role for the left-definite spectral theory. It is therefore of interest to determine all cases when 0 is an eigenvalue for some self-adjoint realization.

Theorem 5.1.13.

(1) *If both endpoints are limit-point, then 0 is an eigenvalue if* $\dim D_0 > 0$, *and the eigenspace is* D_0.

(2) *If one endpoint is limit-point and the other limit-circle with a Dirichlet condition then the eigenspace at 0 is* D_0. *If the boundary condition is not Dirichlet and* $\dim D_0 = 1$, *then 0 is not an eigenvalue and if* $\dim D_0 = 2$ *the eigenspace is one-dimensional.*

(3) *If both endpoints are limit-circle with Dirichlet conditions, then the eigenspace at 0 is two-dimensional. If only one of the boundary conditions is Dirichlet the eigenspace is one-dimensional and if none of the boundary conditions are Dirichlet 0 is not an eigenvalue.*

(4) *If both endpoints are limit-circle and* $\varphi \in D_0$ *not proportional to* F_+ *or* F_- *then* φ *is an eigenfunction for some coupled boundary conditions. A symplectic matrix* $S = (s_{jk})$ *gives rise to a self-adjoint realization with 0 as an eigenvalue precisely if* $s_{12} = 0$. *Neither* F_- *nor* F_+ *are eigenfunctions, so the eigenvalue is simple.*

Proof. Any element of D_0 satisfies the boundary conditions at a limit-point endpoint as well as at a limit-circle endpoint provided with a Dirichlet condition. By Proposition 5.1.10 F_+ satisfies any separated boundary condition at the right endpoint and F_- at the left endpoint. Together with Theorem 5.1.9 this proves everything except the statements about coupled boundary conditions, which follow from an elementary computation. $\qquad\square$

Exercises

Exercise 5.1.1 Show that if the equation $-(pu')' + qu = wf$ is left-definite, i.e., $p > 0$ and $q \geq 0$, $q \not\equiv 0$, almost everywhere, then the necessary and sufficient conditions for limit-circle of Theorem 5.1.8 translates into integrability of $1/p$, q and W^2/p.

Exercise 5.1.2 Suppose both endpoints of (a, b) are regular or limit-circle and in $\operatorname{supp} w$, and consider an operator given by coupled boundary conditions. Find the implied boundary conditions for $u \in \mathcal{H}_\infty$. Also show that F_- satisfies all separated, symmetric boundary conditions at a, F_+ at b, and that $y \mapsto g_0(x, y)$ for fixed $x \in (a, b)$ satisfies all self-adjoint boundary conditions for (a, b), separated or not.

Exercise 5.1.3 Consider the equation $-u'' = \lambda w u$ on $[a, b)$, a being a regular endpoint, and define $\langle u, v \rangle = u(a)\overline{v(a)} + \int_a^b u'\overline{v'}$. Show that this is a scalar product on the appropriate set of functions and derive a theory like the one in this section for this equation, in particular finding all self-adjoint boundary conditions.

5.2 Expansion in eigenfunctions

If a self-adjoint relation T has discrete spectrum (compact resolvent) with eigenvalues $\lambda_1, \lambda_2, \ldots$ and corresponding orthonormal eigenfunctions $e_1, e_2, \ldots$ the expansion of an element u of the Hilbert space is $u = \sum_1^\infty \hat{u}_j e_j$, where $\hat{u}_j = \langle u, e_j \rangle$ and the series converges in norm. We have the following sufficient criteria for compact resolvent.

Theorem 5.2.1. *Suppose that near each endpoint of (a, b) either $x \mapsto w(x)g_0(x, x)$ is integrable or else there is a primitive W of w such that $W^2(x)g_0(x, x)$ is integrable near the endpoint. Then all self-adjoint realizations of (5.1.1) in $\mathcal{H}_1$ have compact resolvent.*

In particular, if both endpoints of (a, b) are in the limit-circle case, then the resolvent of any self-adjoint realization in $\mathcal{H}_1$ is compact.

Proof. If $c \in (a, b)$, $u \in \mathcal{H}_1$ and $v \in L_c \cap \mathcal{H}_1$ we have $\langle u, G_0(wv) \rangle = \int_a^c u\overline{v}w + \int_c^b u\overline{v}w$, and integrating by parts we obtain

$$\int_a^c u\overline{v}w = W(c)u(c)\overline{v(c)} - \int_a^c W(u'\overline{v} + u\overline{v'}) ,   \tag{5.2.1}$$

and similarly integrating over (c, b). We shall use this and our assumptions to show that the form $\langle u, G_0(wv) \rangle$, defined if $u, v \in \mathcal{H}_1$ with supp v compact, extends to a bounded Hermitian form $B(u, v)$ on $\mathcal{H}_1$. Thus $\mathcal{H}_1 \ni u \mapsto B(u, v)$ is a bounded linear form for every $v \in \mathcal{H}_1$. By Theorem 3.4.1 we therefore obtain a bounded, symmetric operator R_0 on $\mathcal{H}_1$ such that $B(u, v) = \langle u, R_0 v \rangle$. We shall then show that R_0 is compact and is the resolvent at 0 of a self-adjoint realization of (5.1.1).

Now $\left| u(x)\overline{v(x)} \right| \leq g_0(x, x) \|u\| \|v\|$ so $W(c)u(c)\overline{v(c)}$ is a bounded Hermitian form on $\mathcal{H}_1$, as is $\int_a^c u\overline{v}w$ if $w(x)g_0(x, x) \in L^1(a, c)$. Using that $\int_a^c |u'|^2 \leq \|u\|^2$ and the Cauchy–Schwarz inequality

$$\left| -\int_a^c W(u'\overline{v} + u\overline{v'}) \right| \leq \left(\int_a^c W^2 |v|^2 \right)^{1/2} \|u\| + \left(\int_a^c W^2 |u|^2 \right)^{1/2} \|v\| ,$$

which may be estimated by $2 \left(\int_a^c W^2 g_0(\cdot, \cdot) \right)^{1/2} \|u\| \|v\|$, so (5.2.1) extends to a bounded Hermitian form on $\mathcal{H}_1$ also if $W^2(x)g_0(x, x) \in L^1(a, c)$. Arguing the same way for the interval (c, b) we find that our assumptions imply that the form $\langle u, G_0(wv) \rangle$ extends to a bounded, Hermitian form on $\mathcal{H}_1$, proving the existence of the operator R_0; the various alternative assumptions may give rise to different operators, see Exercise 5.2.1.

To show that R_0 is compact, suppose $u_j \rightharpoonup 0$ weakly in $\mathcal{H}_1$. R_0 is bounded so also $R_0 u_j \rightharpoonup 0$ weakly by Proposition 2.2.2, and $\|u_j\|$, $j = 1, 2, \ldots$, is a bounded sequence as follows from Theorem 2.1.11. Furthermore, by Proposition 5.1.2 u_j and $R_0 u_j$ tend pointwise to zero.

Now $\|R_0 u_j\|^2 = B(R_0 u_j, u_j)$ and B is a sum of terms we shall examine separately. Assume first that $w(x) g_0(x, x) \in L^1(a, c)$ and consider $\int_a^c R_0 u_j \overline{u_j} w$. Here the integrand is bounded by $|w(x)| g_0(x, x) \|R_0\| \|u_j\|^2$ and tends pointwise to zero, so the integral tends to zero by dominated convergence.

If instead $W^2(x) g_0(x, x) \in L^1(a, c)$ then $W(c) R_0 u_j(c) u_j(c)$ tends to 0 and the Cauchy–Schwarz inequality shows that

$$\left| \int\limits_a^c W(R_0 u_j)' \overline{u_j} \right| \leq \left(\int\limits_a^c |W^2 u_j|^2 \right)^{1/2} \|R_0\| \|u_j\| .$$

The integrand to the right is bounded by $W^2(x) g_0(x, x) \|u_j\|^2$ and tends pointwise to zero so the integral tends to zero by dominated convergence. Similarly $\int_a^c W R_0 u_j \overline{u_j'} \to 0$. Arguing similarly for the interval (c, b) it follows that $\|R_0 u_j\|^2 = B(R_0 u_j, u_j) \to 0$, so $R_0 u_j \to 0$ strongly, proving that R_0 is compact.

If $v \in C_0^\infty(a, b)$, then $\langle R_0 u, v \rangle = \langle u, R_0 v \rangle = \langle u, G_0(wv) \rangle = \int_a^b u \overline{v} w$ so

$$\int\limits_a^b \left((R_0 u)' \overline{v}' + (q R_0 u - uw) \right) \overline{v} = 0 .$$

Thus, in a distributional sense $-(R_0 u)'' + q R_0 u = wu$, so that R_0 is the resolvent at 0 of a self-adjoint realization of (5.1.1). By Corollary 3.3.4, every other self-adjoint realization also has compact resolvent.

If both endpoints of (a, b) are in the limit-circle case, then (a, b) is bounded, $q \in L^1(a, b)$ and $W \in L^2(a, b)$ by Theorem 5.1.8. Thus $F_\pm$ are continuous in the closure of (a, b) so $g_0(x, x) = F_+(x) F_-(x)$ is bounded in (a, b). The conditions assumed in the second case of the theorem are thus automatically satisfied.   □

It is easy to see that in the case of a finite endpoint near which q is integrable the condition on $w(x) g_0(x, x)$ is the stronger of the two. In general neither condition implies the other; see Exercise 5.2.2.

We have the following result on pointwise convergence when the resolvent is compact.

Theorem 5.2.2. *Suppose T is a self-adjoint realization of (5.1.1) which has an orthonormal basis of eigenfunctions.*

If $u \in \mathcal{H}$, then the partial sums $s_N u$ of the generalized Fourier series for u converge locally uniformly in (a, b) to u and $s_N u(x)(g_0(x, x))^{-1/2}$ converges uniformly. At a finite endpoint near which q is integrable the series converges uniformly up to the endpoint.

Proof. The partial sum $s_N u$ converges to u in norm, so by Proposition 5.1.2

$$|u(x) - s_N u(x)| \, (g_0(x, x))^{-1/2} \le \|u - s_N u\| \to 0 \text{ as } N \to \infty \, .$$

At a finite endpoint near which q is integrable, the value of $u \in \mathcal{H}_1$ at that endpoint is a bounded linear form, so $g_0(x, x)$ has a non-zero limit at such a point, but if q is not integrable $g_0(x, x)$ may tend to zero at such an endpoint. In either case the convergence is uniform up to the endpoint.   □

Near an endpoint the function $g_0(x, x)$ cannot grow faster than a multiple of $1 + x^2$, but can tend to zero arbitrarily fast; see Exercise 5.2.3.

We now have a satisfactory expansion theorem as soon as the resolvent is compact, in particular if no endpoint of (a, b) is in the limit-point case. In other cases we shall base our derivation of the expansion theorem for the relation T on a detailed description of its resolvent R_λ.

Using the kernel g_0 for the evaluation operator on $\mathcal{H}_1$, we have $R_\lambda u(x) = \langle R_\lambda u, g_0(x, \cdot) \rangle = \langle u, R_{\overline{\lambda}} g_0(x, \cdot) \rangle$, since $R_\lambda^* = R_{\overline{\lambda}}$. Thus we may view $G(x, \cdot, \lambda) = \overline{R_{\overline{\lambda}} g_0(x, \cdot)}$ as Green's function for our relation; note, however, that the scalar product is not of L^2 type, so $R_\lambda u(x) = \langle u, \overline{G(x, \cdot, \lambda)} \rangle$ is not a standard integral operator. It will turn out to be convenient to introduce the kernel $g(x, y, \lambda) = G(x, y, \lambda) + g_0(x, y)/\lambda$, so that we obtain

$$R_\lambda u(x) = \langle u, \overline{g(x, \cdot, \lambda)} \rangle - u(x)/\lambda \, . \tag{5.2.2}$$

Note that $G(x, \cdot, \lambda) \in \mathcal{H}$ but this is not true of $g(x, \cdot, \lambda)$ if $x \notin \text{supp } w$.

Proposition 5.2.3. *For all $x, y \in (a, b)$ we have both $\overline{G(x, y, \lambda)} = G(y, x, \overline{\lambda})$ and $g(x, y, \lambda) = g(y, x, \overline{\lambda})$.*

Proof. By Proposition 5.1.2 $g_0(x, y) = g_0(y, x)$ so the second claim follows from the first. However,

$$\overline{G(x, y, \lambda)} = R_{\overline{\lambda}} g_0(x, \cdot)(y) = \langle R_{\overline{\lambda}} g_0(x, \cdot), g_0(y, \cdot) \rangle$$

$$= \langle g_0(x, \cdot), R_\lambda g_0(y, \cdot) \rangle = \overline{R_\lambda g_0(y, \cdot)(x)} = G(y, x, \overline{\lambda}) \, .$$

     □

One regular or limit-circle endpoint. In order to obtain an eigenfunction expansion in cases where the spectrum may not be discrete we shall need a precise description of $g(x, y, \lambda)$, and first consider the case of one limit-circle endpoint. We are primarily interested in the case when the other endpoint is limit point, but do not assume this. For definiteness, assume that the left endpoint a is limit-circle for (5.1.1). We only consider separated boundary conditions since coupled boundary conditions always lead to a discrete spectrum. Assume the boundary condition (5.1.6) at a, and, if needed, another separated boundary condition at b.

When a is limit-circle the functions $F_{\pm}$ of Proposition 5.1.10 and their derivatives have well-defined values at a, so in the rest of this section it will be convenient to fix these functions uniquely by requiring $F_-(a) = 1$. Thus F_- is the solution of $-v'' + qv = 0$ with initial data $F_-(a) = 1$, $F_-'(a) = 0$. Since $\mathcal{W}(F_-, F_+) = 1$ this means that F_+ is fixed by $F_+'(a) = -1$.

To describe the kernel $g(x, y, \lambda)$ we must introduce solutions of (5.1.1) satisfying initial conditions at a, so let $\varphi(x, \lambda)$, $\theta(x, \lambda)$ be solutions of (5.1.1) for $\lambda \neq 0$ satisfying

$$\begin{cases} \lambda\varphi(a, \lambda) = -\sin\alpha \,, \\ (\varphi'(\cdot, \lambda) + W\lambda\varphi(\cdot, \lambda))(a) = \cos\alpha \,, \end{cases} \tag{5.2.3}$$

$$\begin{cases} \lambda\theta(a, \lambda) = \cos\alpha \,, \\ (\theta'(\cdot, \lambda) + W\lambda\theta(\cdot, \lambda))(a) = \sin\alpha \,. \end{cases} \tag{5.2.4}$$

This means that $(\varphi, \lambda\varphi)$ satisfies the boundary condition (5.1.6) and $(\theta, \lambda\theta)$ another similar boundary condition at a, and that $\lambda\mathcal{W}(\varphi, \theta) = 1$. If a is regular it is convenient to let $W(x) = \int_a^x w$ so that $W(a) = 0$ and the boundary conditions look like (5.1.5) while the initial data become

$$\begin{cases} \lambda\varphi(a, \lambda) & = -\sin\alpha \,, \\ \varphi'(a, \lambda) & = \cos\alpha \,, \end{cases} \qquad \begin{cases} \lambda\theta(a, \lambda) & = \cos\alpha \,, \\ \theta'(a, \lambda) & = \sin\alpha \,. \end{cases} \tag{5.2.5}$$

Note that $\lambda\varphi(x, \lambda)$, $\lambda\varphi'(x, \lambda)$, $\lambda\theta(x, \lambda)$ and $\lambda\theta'(x, \lambda)$ are entire functions of λ by Theorem D.4, locally uniformly in x, but that $\varphi(x, \lambda)$ and $\varphi'(x, \lambda)$ have a pole at 0 unless $\alpha = 0$ while $\theta(x, \lambda)$ and $\theta'(x, \lambda)$ have one unless $\alpha = \pi/2$.

The expansion theorem is based on the following description of the resolvent.

Theorem 5.2.4. *Let T be a self-adjoint realization of (5.1.1) given by (5.1.6) and a similar condition at b if b is in the limit-circle case.*

Then there exists a unique function $\rho(T) \setminus \{0\} \ni \lambda \mapsto m(\lambda)$ such that $\psi(x, \lambda) = \theta(x, \lambda) + m(\lambda)\varphi(x, \lambda)$ is in $\mathcal{H}_1$ and satisfies the boundary condition at b if this is in the limit-circle case. For $\lambda \in \rho(T) \setminus \{0\}$ we have

$$g(x, y, \lambda) = \varphi(\min(x, y), \lambda)\psi(\max(x, y), \lambda) \,.$$

The function m is the (left-definite) *m-function* for T and $\psi(x, \lambda)$ is the *Weyl solution* corresponding to the given boundary conditions.

Proof. For $0 \neq \lambda \in \rho(T)$ the pair $(\varphi(\cdot, \lambda), \lambda\varphi(\cdot, \lambda))$ which satisfies the boundary condition at a cannot satisfy the boundary condition at b, since that would make λ an eigenvalue and so in $\sigma(T)$. Thus there is a solution $\psi(x, \lambda) = \theta(x, \lambda) + m(\lambda)\varphi(x, \lambda)$ in $\mathcal{H}_1$ such that $(\psi(\cdot, \lambda), \lambda\psi(\cdot, \lambda))$ satisfies the boundary condition at b. For, if

$\dim D_\lambda = 2$ one linear, homogeneous condition still leaves a one-dimensional space, whereas if $\dim D_\lambda = 1$, no boundary condition is to be imposed at b. Define, for fixed y and $\lambda \in \rho(T) \setminus \{0\}$, the function

$$F(x) = \varphi(\min(x, y), \lambda)\psi(\max(x, y), \lambda) - \lambda^{-1}g_0(x, y) \ .$$

Since $\psi(\cdot, \lambda)$ and F_+ are in $\mathcal{H}_1$ we have $(F, \lambda F + g_0(\cdot, y)) \in \mathcal{H}_1^2$. In fact, it belongs to T_1 since one easily checks that F' is locally absolutely continuous and that F satisfies $-F'' + qF = \lambda wF + wg_0(\cdot, y)$. It is also easy to check that $(F, \lambda F + g_0(\cdot, y))$ satisfies the boundary condition (5.1.6).

Finally, we claim that $(F, \lambda F + g_0(\cdot, y))$ also satisfies the boundary condition at b so that it is in T. In fact, for $x > y$ it is a linear combination of $(\psi(x, \lambda), \lambda\psi(x, \lambda))$ and $(F_+(x), 0)$. The former satisfies the boundary condition at b by construction, and $(F_+, 0)$ satisfies the boundary condition at b by Theorem 5.1.9 (5), since if $(u, f) \in T$, then $F'_+\overline{f} - 0\,\overline{u'} = F'_+\overline{f} \to 0$ at b. All this means that $F = R_\lambda g_0(\cdot, y) = \overline{G(\cdot, y, \overline{\lambda})} = G(y, \cdot, \lambda)$ by Proposition 5.2.3. However, the formula for F is symmetric in x, y so $F(x) = G(x, y, \lambda)$. $\qquad\square$

Theorem 5.2.5. *The function m is analytic in $\rho(T) \setminus \{0\}$, and is a Nevanlinna function in the sense of Appendix E.*

Proof. R_λ is analytic in $\rho(T)$ in the uniform operator topology, so by Proposition 5.1.2 $R_\lambda u(x)$ is analytic locally uniformly in x. It follows that $g(x, \cdot, \lambda)$ is weakly analytic in $\rho(T) \setminus \{0\}$, locally uniformly in x, and thus, again by Proposition 5.1.2, $g(x, y, \lambda)$ is analytic in $\rho(T) \setminus \{0\}$, locally uniformly in x and y.

Since $\varphi(x, \lambda)$ and $\theta(x, \lambda)$ are analytic outside 0, locally uniformly in x, it follows that m is analytic in $\rho(T) \setminus \{0\}$.

By Proposition 5.2.3 $\overline{g(x, y, \lambda)} = g(y, x, \overline{\lambda})$, and since $\overline{\varphi(\cdot, \lambda)} = \varphi(\cdot, \overline{\lambda})$ and $\overline{\theta(\cdot, \lambda)} = \theta(\cdot, \overline{\lambda})$ it follows that $\overline{\psi(\cdot, \lambda)} = \psi(\cdot, \overline{\lambda})$ and thus $\overline{m(\lambda)} = m(\overline{\lambda})$. Integrating by parts for $\operatorname{Im}\lambda \neq 0$ we have

$$\operatorname{Im}\lambda \int_a^x (|\psi'(\cdot, \lambda)|^2 + q\,|\psi(\cdot, \lambda)|^2) = \operatorname{Im}(\overline{\psi'(\cdot, \lambda)}\lambda\psi(\cdot, \lambda))\big|_a^x \ .$$

Since ψ satisfies a separated, symmetric boundary condition at b, the integrated term at x vanishes as $x \to b$. At a the integrated term evaluates to $-\operatorname{Im} m(\lambda)$, so we obtain

$$\|\psi(\cdot, \lambda)\|^2 = \operatorname{Im} m(\lambda)/\operatorname{Im}\lambda \ .$$

Thus $\operatorname{Im} m(\lambda) > 0$ if $\operatorname{Im}\lambda > 0$ so m is a Nevanlinna function. $\qquad\square$

Remark 5.2.6. If b is a regular endpoint of (a, b) we may similarly define solutions $\varphi(\cdot, \lambda)$ and $\theta(\cdot, \lambda)$ by (5.2.3), (5.2.4) with a replaced by b and a Weyl solution $\psi(\cdot, \lambda)$ where it is now convenient to write $\psi(\cdot, \lambda) = \theta(\cdot, \lambda) - m(\lambda)\varphi(\cdot, \lambda)$ since the change of sign in front of m will make it a Nevanlinna function also in this case (Exercise 5.2.4).

By Theorem E.1.1 the m-function has a unique representation

$$m(\lambda) = A + B\lambda + \int\limits_{-\infty}^{\infty} \left(\frac{1}{t - \lambda} - \frac{t}{t^2 + 1} \right) d\eta , \qquad (5.2.6)$$

where $A \in \mathbb{R}$, $B \geq 0$, and $d\eta$ is a positive measure with $\int_{-\infty}^{\infty} \frac{d\eta(t)}{1+t^2} < \infty$. As in the right-definite case we shall call the measure $d\eta$ the *spectral measure* for T, and the left-continuous, non-decreasing distribution function η of the spectral measure with $\eta(0) = 0$ is called the *spectral function* for T.

An expansion theorem, very similar to the expansion theorem in the right-definite case (Theorem 4.3.7) may now be proved, but there are some technical complications due to the special significance of $\lambda = 0$ in the left-definite case.

We shall call functions that are identically zero near the endpoint b *finite*. If $\int_a^b (u'\overline{v'} + qu\overline{v}) = 0$ for all finite $v \in \mathcal{H}_1$, Theorem 5.1.9 and Proposition 5.1.10 show that u is a multiple of F_-. Since F_+ is in $\mathcal{H}_1$ as soon as a is limit-circle it follows that finite functions are dense in $\mathcal{H}_1$ unless $\dim D_0 = 2$.

Definition 5.2.7. Let $\mathcal{H}_0 = \mathcal{H}_1$ unless F_- is an eigenfunction of T. In this case, let $\mathcal{H}_0$ be the orthogonal complement of F_- in $\mathcal{H}_1$. Thus $\mathcal{H}_0$ is the closure of the finite functions in $\mathcal{H}_1$ unless b is limit-circle with a boundary condition which is not Dirichlet.

This means that $T \cap \mathcal{H}_0^2$ is a self-adjoint relation in $\mathcal{H}_0$ (Exercise 5.2.5) for which 0 is an eigenvalue precisely if $\alpha = 0$. If F_- is an eigenfunction it is in $\mathcal{D}(T)$, so in this case F_- is orthogonal to $\mathcal{H}_\infty$. Thus, in all cases $\mathcal{H}_\infty \subset \mathcal{H}_0$.

Definition 5.2.8. For $\alpha = 0$ define $\varphi_0(x) = \varphi(x, 0) - F_+(a)F_-(x)$ if F_- is not an eigenfunction and $\varphi_0(x) = \varphi(x, 0) - \varphi(b, 0)F_-(x)/F_-(b)$ if F_- is an eigenfunction.

Remark 5.2.9. If F_- is not an eigenfunction, then $\varphi_0(a) = -F_+(a)$ and $\varphi_0'(a) = 1 = -\varphi_+'(a)$ so $\varphi_0 = -F_+$ is an eigenfunction to the eigenvalue 0. On the other hand, if F_- is an eigenfunction, then $\dim D_0 = 2$, all elements of D_0 are eigenfunctions and together with their derivatives they have well-defined values at b. Thus $\varphi_0(b) = 0$ and $\langle \varphi_0, F_- \rangle = \varphi_0(b)\varphi_-'(b) = 0$ so φ_0 is an eigenfunction in $\mathcal{H}_0$. Thus an eigenfunction expansion of $u \in \mathcal{H}_0$ should have a contribution at 0 which is a multiple of φ_0, but since $\langle u, F_- \rangle = 0$ for finite $u \in \mathcal{H}_1$ the coefficient in such an expansion could equally well be calculated using $\varphi(\cdot, 0)$ instead of φ_0, whereas φ_0 must be used in the inverse transform.

The spectral measure introduced in (5.2.6) gives rise to a Hilbert space L_η^2 with scalar product $\langle \hat{u}, \hat{v} \rangle_\eta = \int_{-\infty}^{\infty} \hat{u}\overline{\hat{v}}\, d\eta$. We obtain the following expansion theorem.

Theorem 5.2.10.

(1) *The map* $u \mapsto \int_a^b (u'\overline{\varphi'}(\cdot, t) + qu\overline{\varphi}(\cdot, t))$, *defined at least for finite* $u \in \mathcal{H}_1$, *extends to a map* $\mathcal{F} : \mathcal{H}_0 \to L_\eta^2$ *called the* generalized Fourier transform

associated with T. The image of $u \in \mathcal{H}_0$ is denoted by $\mathcal{F}(u)$ or $\hat{u}$. We write this as $\hat{u}(t) = \langle u, \varphi(\cdot, t) \rangle$ although the integral in general does not converge pointwise.

(2) *The map $\mathcal{F} : \mathcal{H}_0 \to L_\eta^2$ has nullspace $\mathcal{H}_\infty$ and is unitary between the space $\mathcal{H} = \mathcal{H}_0 \ominus \mathcal{H}_\infty$ and L_η^2 so that* Parseval's formula $\langle u, v \rangle = \langle \hat{u}, \hat{v} \rangle_\eta$ *holds if at least one of u and v is in $\mathcal{H}$.*

(3) *If $(u, f) \in T \cap \mathcal{H}_0^2$, then $\mathcal{F}(f)(t) = t\hat{u}(t)$. Conversely, if $\hat{u}$ and $t\hat{u}(t)$ are in L_η^2, then $\hat{u} = \mathcal{F}u$ for some $u \in \mathcal{D}(T) \cap \mathcal{H}_0$.*

(4) *Suppose $\alpha \neq 0$ in (5.1.6). Then $\varphi(x, \cdot) \in L_\eta^2$ for each x and $\int_{-\infty}^{\infty} \hat{u}\varphi(x, \cdot)\, d\eta = \langle \hat{u}, \varphi(x, \cdot) \rangle_\eta$ converges in $\mathcal{H}$, and hence locally uniformly in x, for $\hat{u} \in L_\eta^2$. The operator $\hat{u} \mapsto \langle \hat{u}, \varphi(x, \cdot) \rangle$ is the adjoint of $\mathcal{F} : \mathcal{H}_0 \to L_\eta^2$ and thus the inverse of $\mathcal{F}$ restricted to $\mathcal{H}$. If M is a Borel set in $\mathbb{R}$, then*

$$E_M u(x) = \int_M \hat{u}\varphi(x, \cdot)\, d\eta \, .$$

If $\alpha = 0$, the same is true, except that we must replace $\varphi(\cdot, 0)$ by the eigenfunction φ_0 of Definition 5.2.8.

In order to prove Theorem 5.2.10, first note that if u is finite, integration by parts shows that

$$\hat{u}(\lambda) = \langle u, \overline{\varphi(\cdot, \lambda)} \rangle = \int_a^b u\,\lambda\varphi(\cdot, \lambda)w - u(a)\cos\alpha \, .$$

By (5.2.3) $\lambda\varphi(x, \lambda)$ is entire in λ, locally uniformly in x, so Fourier transforms of finite functions are entire.

If b is limit-circle the spectrum of T is discrete by Theorem 5.2.1, and the function $\lambda \mapsto \varphi(\cdot, \lambda) \in \mathcal{H}_1$ is entire if we have a Dirichlet boundary condition at a but otherwise has a simple pole at 0. Thus defining the Fourier transform of $u \in \mathcal{H}_1$ by $\hat{u}(\lambda) = \langle u, \overline{\varphi(\cdot, \lambda)} \rangle$, $\hat{u}$ is analytic except perhaps at 0, where it may have a simple pole unless we have a Dirichlet condition at a.

Lemma 5.2.11. *If u and v are in $\mathcal{H}_0$, then $\hat{u}$ and $\hat{v}$ are in L_η^2. If E_Δ is the spectral projection for T associated with an interval Δ, then $\langle E_\Delta u, v \rangle = \int_\Delta \hat{u}\overline{\hat{v}}\, d\eta$.*

Proof. An elementary (but somewhat tedious) calculation shows that if u and v are finite functions and $\lambda \in \rho(T) \setminus \{0\}$, then (see Exercise 5.2.6) $\langle R_\lambda u, v \rangle = \hat{u}(\lambda)\hat{v}(\overline{\lambda})m(\lambda) + r(\lambda)$, if b is limit-circle, even for arbitrary functions in $\mathcal{H}_0$, where r is entire. The case when b is limit-circle and the boundary condition at a not Dirichlet is an exception, for then $\hat{u}$, $\hat{v}$ and r may have a pole at 0 while $\langle R_\lambda u, v \rangle$ is still analytic at 0. Integrating around a rectangle γ with corners at $c \pm i$ and $d \pm i$ we therefore have

$$\int_{\gamma} \langle R_{\lambda}u, v \rangle \, d\lambda = \int_{\gamma} \hat{u}(\lambda)\hat{v}(\overline{\lambda})m(\lambda) \, d\lambda + 2\pi i \operatorname{Res}_{\lambda=0} r(\lambda)$$

whenever one of the integrals exists and the last term should be dropped unless γ encloses 0.

In the exceptional case $\hat{u}(\lambda)\hat{v}(\overline{\lambda})m(\lambda)$ and $r(\lambda)$ may have poles at 0, but since $\langle R_{\lambda}u, v \rangle$ is then analytic at 0 the residues at 0 of these functions will cancel and may be ignored when calculating the right-hand side.

The rest of the proof may now be carried out precisely as the proof of Theorem 4.3.10 except when b is limit-point. In this case we have proved the lemma only for finite functions. But the map $\mathcal{F} : \mathcal{H}_0 \to L^2_{\eta}$ is then densely defined and bounded by 1 so the theorem follows by continuity. $\square$

Since $E_{\mathbb{R}}$ has nullspace $\mathcal{H}_{\infty}$ (Example 3.1.3, page 62), we now obtain Theorem 5.2.10 (1) and (2) except for the surjectivity of $\mathcal{F}$. To prove this we need a lemma analogous to Theorem 4.3.13.

Lemma 5.2.12. *If $u \in \mathcal{H}_0$, then so is $R_{\lambda}u$, and the transform of $R_{\lambda}u$ for $u \in \mathcal{H}_0$ is $\hat{u}(t)/(t - \lambda)$.*

Proof. Note that if F_- is an eigenfunction, then $(F_-, 0) \in T$ so $\langle \lambda R_{\lambda}u + u, F_- \rangle = \langle R_{\lambda}u, 0 \rangle = 0$, so if $u \in \mathcal{H}_0$ also $R_{\lambda}u \in \mathcal{H}_0$. The rest of the proof is identical to that of Theorem 4.3.13 and left to the reader in Exercise 5.2.7. $\square$

It is now easy to prove that $\mathcal{F}$ is surjective.

Lemma 5.2.13. *The Fourier transform $\mathcal{H}_0 \to L^2_{\eta}$ is surjective.*

Proof. Suppose that $\hat{u} \in L^2_{\eta}$ is orthogonal to all Fourier transforms $\hat{v} = \mathcal{F}v$. Since $\hat{v}(t)/(t - \lambda)$ is the transform of $R_{\lambda}v$ for nonreal λ, we have $\int \frac{1}{t-\lambda} \hat{u}(t)\overline{\hat{v}(t)} \, d\eta(t) = 0$ for all nonreal λ. By the Stieltjes inversion formula it follows that the measure $\hat{u}(t)\overline{\hat{v}(t)} \, d\eta(t) = 0$.

The Fourier transform of a finite function is entire, so to prove that t is outside the support of $\hat{u} \, d\eta$ it is enough to show that there is a finite v for which $\hat{v}(t) \neq 0$. Clearly $\hat{v}(t) = \langle v, \varphi(\cdot, t) \rangle = \int_a^b v \, t\varphi(\cdot, t)w - \cos \alpha \, v(a)$, and if this is zero for all $v \in C_0^{\infty}(a, b)$, then $t\varphi(\cdot, t)w$ is the zero measure.

Now, $t\varphi(\cdot, t) \to -\sin \alpha \, F_-$ as $t \to 0$ so unless $t = 0$ and $\alpha = 0$ Lemma 4.1.3 contradicts this. If $\alpha = 0$ we find that $\hat{u} \, d\eta$ is supported at the origin. But in this case $\hat{v}(0) = -v(a)$, so picking a finite v with $v(a) \neq 0$ we obtain $\hat{v}(0) \neq 0$ so that $\hat{u} \, d\eta$ is the zero measure. $\square$

We next turn to Theorem 5.2.10 (3).

Lemma 5.2.14. *If $(u, f) \in T$, then $\hat{f}(t) = t\hat{u}(t)$. Conversely, if $\hat{u}$ and $t\hat{u}(t)$ are in L^2_{η}, then $\hat{u} = \mathcal{F}u$ where $u \in \mathcal{D}(T) \cap \mathcal{H}_0$.*

This is proved in exactly the same way as Lemma 4.3.15. To finish the proof of Theorem 5.2.10 it only remains to consider the inverse transform. We will need the following lemma.

Lemma 5.2.15. *The spectral measure $d\eta$ has mass at 0 if and only if $\alpha = 0$. If $\alpha = 0$, then $\hat{u}(0) = -u(a)$ for any $u \in \mathcal{H}_0$, and the Fourier transform of φ_0 is $\hat{\varphi}_0 := \mathcal{F}(\varphi_0) = \chi_0/\eta\{0\}$, where χ_0 is the characteristic function of $\{0\}$ and $\eta\{0\}$ the mass of $d\eta$ at 0.*

Proof. If $\alpha = 0$, then by Theorem 5.1.13 $\varphi_0 \in \mathcal{H}_0$ is an eigenfunction to the eigenvalue 0, and by Lemma 5.2.14 we have $t\hat{\varphi}_0(t) = 0$ so that $\hat{\varphi}_0$ is supported at 0. In view of Parseval's identity $\{0\}$ cannot therefore be a nullset for $d\eta$.

For any finite u we have $\hat{u}(0) = \langle u, \varphi(\cdot, 0)\rangle = u\varphi'(\cdot, 0)|_a^b = -u(a)$. This formula remains true for general $u \in \mathcal{H}_0$, since $u \mapsto u(a)$ is a bounded linear form, as is $u \mapsto \hat{u}(0)$ since $\eta\{0\} > 0$. Thus

$$\hat{\varphi}_0(0) = -\varphi_0(a) = \|\varphi_0\|^2 = \|\hat{\varphi}_0\|_\eta^2 = |\hat{\varphi}_0(0)|^2 \, \eta\{0\} \,.$$

It follows that $\hat{\varphi}_0 = \chi_0/\eta\{0\}$.

To complete the proof, by Theorem E.1.7 $\lambda m_\alpha(\lambda)$ has a non-zero limit as $\lambda \to 0$ along the imaginary axis precisely if the corresponding spectral measure has mass at 0. Now, Weyl solutions for the same boundary condition at b but different α in (5.1.6) are proportional so it follows that

$$m_0(\lambda) = \frac{(\psi'(\cdot, \lambda) + W\lambda\psi(\cdot, \lambda))(a)}{\lambda\psi(a, \lambda)} = \frac{\sin\alpha + m_\alpha(\lambda)\cos\alpha}{\cos\alpha - m_\alpha(\lambda)\sin\alpha}, \tag{5.2.7}$$

where m_α denotes the m-function associated with the boundary condition parameter α. Now $m_0 \to \infty$ so (5.2.7) shows that $m_\alpha \to \cot\alpha$ if $\alpha \neq 0$, taking limits at 0 along the imaginary axis. Thus, if $\alpha \neq 0$, the spectral measure for m_α has no mass at 0. $\square$

Theorem 5.2.16. *If $\alpha \neq 0$, then $\varphi(x, \cdot) \in L_\eta^2$ for every x and the integral $\langle \hat{u}, \varphi(x, \cdot)\rangle_\eta$ converges in $\mathcal{H}$ and locally uniformly for every $\hat{u} \in L_\eta^2$. If $\hat{u} = \mathcal{F}(u)$ for some $u \in \mathcal{H}_0$, then the integral is the orthogonal projection of u onto $\mathcal{H}$.*

If $\alpha = 0$, the same statement is true if one replaces $\varphi(\cdot, 0)$ in the integral by the function φ_0 of Definition 5.2.8.

Proof. Let $e(x, t) = \mathcal{F}(Pg_0(x, \cdot))(t)$ where P is the orthogonal projection of $\mathcal{H}_1$ onto $\mathcal{H}_0$ (Exercise 5.2.8), and thus the identity if $\mathcal{H}_0 = \mathcal{H}_1$. If $u \in \mathcal{H}_0$ we obtain $E_\mathbb{R}u(x) = \langle E_\mathbb{R}u, g_0(x, \cdot)\rangle = \langle \hat{u}, e(x, \cdot)\rangle_\eta$. Thus the integral operator $\hat{u} \mapsto \langle \hat{u}, e(x, \cdot)\rangle_\eta$ is the adjoint of $\mathcal{F}$.

We must prove that $e(x, t) = \varphi(x, t)$ for $t \neq 0$, so suppose $\hat{u} \in L_\eta^2$ has compact support and consider $\tilde{u}(x) = \langle \hat{u}, \varphi(x, \cdot)\rangle_\eta$ which satisfies the equation $-\tilde{u}'' + q\tilde{u} = w(x)\langle \hat{u}, t\varphi(x, \cdot)\rangle_\eta$, differentiating under the integral sign. Since $\hat{u}$ has compact support $u = \mathcal{F}^{-1}\hat{u} \in D_T$, so that $-u'' + qu = w(x)\langle t\hat{u}(t), e(x, t)\rangle_\eta$. Thus $u_1 = u - \tilde{u}$ satisfies $-u_1'' + qu_1 = w(x)\langle t\hat{u}(t), e(x, t) - \varphi(x, t)\rangle_\eta$. Now, if v is finite, then

$$\langle \tilde{u}, v \rangle = \iint \hat{u}(t)(\varphi'(\cdot, t)\overline{v'} + q\varphi(\cdot, t)\overline{v})\, d\eta(t) = \langle \hat{u}, \hat{v}\rangle_\eta = \langle u, v\rangle \,, \tag{5.2.8}$$

since the double integral is absolutely convergent. Hence u_1 is orthogonal to all finite v so it is a multiple of F_-. It follows that $w(x)\langle t\hat{u}(t), e(x,t) - \varphi(x,t)\rangle_\eta = 0$ a.e., so that $\langle t\hat{u}(t), e(x,t) - \varphi(x,t)\rangle_\eta = 0$ on a set of positive measure. But this function also satisfies $-v'' + qv = 0$, as is seen by replacing $\hat{u}$ by $t\hat{u}(t)$ in (5.2.8) and the definitions of $\tilde{u}$ and u. From Lemma 4.1.3 it follows that $t(e(x,t) - \varphi(x,t))$ is the zero element of L_η^2 for all x, so that $e(x,t) = \varphi(x,t)$ except possibly if $t = 0$ and $\alpha = 0$.

If $\alpha = 0$, then by Lemma 5.2.15 $\varphi_0(x) = \langle \hat{\varphi}_0, e(x, \cdot)\rangle_\eta = \hat{\varphi}_0(0)\overline{e(x,0)}\eta\{0\} = \overline{e(x,0)}$ so that, φ_0 being real-valued, $e(x,0) = \varphi_0(x)$. $\qquad\square$

The proof of Theorem 5.2.10 is now complete. We finally have the counterpart to Theorem 4.3.17. Here $F_\pm$ are as in Proposition 5.1.10, normalized as on page 139.

Theorem 5.2.17. *If $\alpha \neq 0$ and P the orthogonal projection of $\mathcal{H}_1$ onto $\mathcal{H}_0$, then $\mathcal{F}(PF_+)(t) = -\sin\alpha/t$.*

Also, $\psi(\cdot, \lambda) \in \mathcal{H}_0$, $\mathcal{F}(\psi(\cdot, \lambda))(t) = 1/(t - \lambda)$ and $\psi(\cdot, \lambda) \to 0$ in $\mathcal{H}_0$, and thus locally uniformly, as $\lambda \to \infty$ in a double sector $|\mathrm{Re}\,\lambda| \le C\,|\mathrm{Im}\,\lambda|$. In the representation formula (5.2.6) for the m-function we always have $B = 0$.

Proof. If u is finite, integration by parts shows that $\langle u, F_+\rangle = u(0)$, which equals $\langle \hat{u}, \varphi(0, \cdot)\rangle_\eta$. Thus $\mathcal{F}(PF_+)(t) = \varphi(0, t) = -\sin\alpha/t$.

If F_- is an eigenfunction, then since $\psi(\cdot, \lambda)$ satisfies the boundary condition at b we obtain

$$\overline{\varphi'_-(x)\overline{\psi(x, \lambda)}} = \frac{1}{\lambda}(\varphi'_-(x)\overline{\lambda\psi(x, \lambda)} - 0 \cdot \overline{\psi'(x, \lambda)}) \to 0 \text{ as } x \to b \,,$$

so that $\langle F_-, \psi(\cdot, \lambda)\rangle = -\varphi'_-(0)\overline{\psi(0, \lambda)} = 0$. Thus $\psi(\cdot, \lambda) \in \mathcal{H}_0$.

By Theorem 5.2.12 $\mathcal{F}(R_\lambda u)(t) = \hat{u}(t)/(t - \lambda)$. It follows that

$$\int_{-\infty}^{\infty} \hat{u}(t)\varphi(x,t)/(t - \lambda)\, d\eta(t)$$

converges locally uniformly to $\langle u, \overline{g(x, \cdot, \lambda)}\rangle - u(x)/\lambda$. Thus the Fourier transform of $g(x, \cdot, \lambda)$ is

$$\mathcal{F}(g(x, \cdot, \lambda))(t) = \varphi(x,t)(1/(t - \lambda) + 1/\lambda) = t\varphi(x,t)/(\lambda(t - \lambda)) \,.$$

In particular, the Fourier transform of $\lambda g(0, \cdot, \lambda) = -\psi(\cdot, \lambda)\sin\alpha$ is equal to $t\varphi(0, t)/(t - \lambda) = -\sin\alpha/(t - \lambda)$. This gives $\mathcal{F}(\psi(\cdot, \lambda))$ for $\alpha \neq 0$.

If $\alpha = 0$ similar arguments involving $p(R_\lambda u)'(x)$ gives the desired result; see Exercise 5.2.9.

The rest of the proof may be carried out word for word as in the proof of Theorem 4.3.17. $\qquad\square$

Exercises

Exercise 5.2.1 Show that at a limit-circle endpoint a of (a,b) the operator R_0 defined in the proof of Theorem 5.2.1 corresponds to a realization of (5.1.1) given by the Dirichlet condition $f(a) = 0$ if $w(x)g_0(x,x)$ is integrable and by the free condition $(u' + Wf)(a) = 0$ in the other case, cf. Exercise 5.1.2.

Exercise 5.2.2 In Theorem 5.2.1 two conditions for a compact resolvent are given. Show that at a finite endpoint near which q is integrable the first condition implies the second. Also show that in general neither condition implies the other.

Exercise 5.2.3 Show that, if q in $\int_a^b (|u'|^2 + q\,|u|^2)$ is replaced by an everywhere larger q, then $g_0(x,x)$ is replaced by an everywhere smaller function. Then show that near a finite endpoint $g_0(x,x)$ is bounded, while it may grow at most like a multiple of $1 + x^2$ near an infinite endpoint.

Hint: Use the fundamental theorem of calculus to show that at an infinite endpoint $\overline{\lim} |u(x)|\,(1 + |x|)^{-1/2} \le \|u\|$.

Exercise 5.2.4 Suppose the right endpoint b of (a,b) is in the limit-circle condition. Let φ and θ be the solutions of $-v'' + qv = \lambda w v$ determined for $\lambda \ne 0$ by (5.2.3), (5.2.4) but with a replaced by b. Show that there is a solution satisfying a given symmetric boundary condition at a (if a is limit-point this means a solution in $\mathcal{H}_1$) of the form $\psi(\cdot,\lambda) = \theta(\cdot,\lambda) - m(\lambda)\varphi(\cdot,\lambda)$ where m is a Nevanlinna function.

Exercise 5.2.5 Prove that if φ is an eigenfunction for the self-adjoint relation T in $\mathcal{H}_1$ and $\mathcal{H}_\varphi = \{\varphi\}^\perp$, then $T \cap \mathcal{H}_\varphi^2$ is a self-adjoint relation in $\mathcal{H}_\varphi$.

Exercise 5.2.6 Show the formula $\langle R_\lambda u, v \rangle = \hat{u}(\lambda)\hat{v}(\overline{\lambda})m(\lambda) + r(\lambda)$ used in the proof of Lemma 5.2.11.

Exercise 5.2.7 Carry out the details of the proof of Lemma 5.2.12.

Exercise 5.2.8 Using the notation of the proof of Theorem 5.2.16 show that if $\mathcal{H}_0 \ne \mathcal{H}_1$, then

$$P(g_0(x,\cdot))(y) = -F_-(\min(x,y))\varphi_0(\max(x,y)).$$

Exercise 5.2.9 Fill in the missing details of the proof of Theorem 5.2.17.

Hint: Differentiating the formula for $R_\lambda u(x)$ gives $p(R_\lambda u)'(x) = \langle u, \overline{g_1(x,\cdot,\lambda)} \rangle$. Find expressions for $g_1(x,\cdot,\lambda)$ and its Fourier transform.

5.3 Two limit-point endpoints

As in the right-definite case we may handle the case of two limit-point endpoints by considering two auxiliary problems with a regular endpoint. If our base interval is

(a, b) we select $c \in (a, b)$ so that $q \not\equiv 0$ both in $(a, c]$ and $[c, b)$, fix a separated boundary condition (5.1.6) at c and denote the (left-definite) Weyl solutions associated with the intervals $[c, b)$ and $(a, c]$ by ψ_+ respectively ψ_-. The corresponding m-functions are denoted $m_\pm$. Introducing the kernel

$$g(x, y, \lambda) = \frac{\psi_-(\min(x, y), \lambda)\psi_+(\max(x, y), \lambda)}{\lambda \mathcal{W}(\psi_-, \psi_+)} \tag{5.3.1}$$

it is seen, much as in earlier cases, see Exercise 5.3.1, that

$$R_\lambda u(x) = \langle u, \overline{g(x, \cdot, \lambda)} \rangle - u(x)/\lambda \tag{5.3.2}$$

for $\lambda \in \rho(T) \setminus \{0\}$. Let $\varphi(\cdot, \lambda), \theta(\cdot, \lambda)$ be the solutions corresponding to the solutions satisfying (5.2.5) but with initial data given in c and with $\alpha = 0$. We may then write

$$g(x, y, \lambda) = \Phi^*(\min(x, y), \overline{\lambda})(M(\lambda) - \tfrac{1}{2}J)\Phi(\max(x, y), \lambda),$$

where $\Phi(x, \lambda) = \begin{pmatrix} \varphi(x,\lambda) \\ \theta(x,\lambda) \end{pmatrix}$, $J = \begin{pmatrix} 0 & -1 \\ 1 & 0 \end{pmatrix}$ and

$$M = \frac{1}{m_+ + m_-} \begin{pmatrix} m_+ m_- & (m_- - m_+)/2 \\ (m_- - m_+)/2 & -1 \end{pmatrix}.$$

This is a matrix-valued Nevanlinna function, just like in Section 4.4, so there is a representation formula with a 2×2 matrix-valued measure $d\Xi$. We therefore introduce the Hilbert space L_Ξ^2 of 2×1 matrix-valued functions $\hat{u}$ for which $\|\hat{u}\|_\Xi^2 = \int_{-\infty}^\infty \hat{u}^* \, d\Xi \hat{u} < \infty$. For compactly supported elements of $\mathcal{H}_1$ we also define the Fourier transform $\hat{u}(t) = \mathcal{F}u(t) = \langle u, \overline{\Phi(\cdot, t)} \rangle$ with components $\hat{u}_\varphi(t) = \langle u, \overline{\varphi(\cdot, t)} \rangle$ and $\hat{u}_\theta(t) = \langle u, \overline{\theta(\cdot, t)} \rangle$.

We have the following counterpart to Proposition 4.4.12.

Proposition 5.3.1. *Changing the anchor point c to $\tilde{c}$ and the boundary condition parameter of (5.2.5) at $\tilde{c}$ to α changes the Fourier transform $\mathcal{F}u$ to $\tilde{\mathcal{F}}u$, where $\mathcal{F}u(\lambda) = K(\lambda)\tilde{\mathcal{F}}u(\lambda)$ and*

$$K(\lambda) = \begin{pmatrix} \lambda\varphi(\tilde{c}, \lambda) & \varphi'(\tilde{c}, \lambda) \\ \lambda\theta(\tilde{c}, \lambda) & \theta'(\tilde{c}, \lambda) \end{pmatrix} \begin{pmatrix} -\sin\alpha & \cos\alpha \\ \cos\alpha & \sin\alpha \end{pmatrix}.$$

Similarly M is replaced by $\tilde{M}(\lambda) = K^*(\overline{\lambda})M(\lambda)K(\lambda)$ and the spectral measure $d\Xi$ by $d\tilde{\Xi}(t) = K^*(t)d\Xi(t)K(t)$.

As in Proposition 4.4.12 this follows from an elementary calculation. With initial conditions (5.2.5) for $\alpha = 0$ the spectral measures of $m_\pm$ have masses at 0 by Lemma 5.2.15. It follows that the spectral measure of $m_+ m_-/(m_+ + m_-)$ also has mass at 0, while the spectral measures of the other entries in M do not. There is therefore a one-dimensional subspace of elements of L_Ξ^2 supported at 0. Using Proposition 5.3.1 with $\tilde{c} = c$ it follows that this fact is independent of α and clearly also of c.

However, assuming limit-point at both a and b the eigenspace at 0 is D_0, and $\dim D_0$ can be 0, 1 or 2. There is therefore no connection between the multiplicity of 0 as an eigenspace and the dimension of the subspace of L_Ξ^2 of elements supported at 0. In order to obtain an eigenfunction expansion theorem with a simple statement it is therefore best to remove D_0 from $\mathcal{H}_1$ and the elements supported at 0 from L_Ξ^2.

Definition 5.3.2. Let $\mathcal{H}_0$ be the orthogonal complement of D_0 in $\mathcal{H}_1$ and $\mathcal{H}_\Xi$ the subspace of L_Ξ^2 with codimension 1 of elements vanishing at 0.

Note that $\mathcal{H}_0$ is the closure of the compactly supported elements of $\mathcal{H}_1$ by Theorem 5.1.9. As before we write $\mathcal{H} = \mathcal{H}_0 \ominus \mathcal{H}_\infty$.

We can now state the following analogue of Theorem 4.4.4.

Theorem 5.3.3. *Suppose both endpoints are in the limit-point condition. We then have:*

(1) *The map $u \mapsto \mathcal{F}u$ defined at least for compactly supported elements of $\mathcal{H}_0$ extends to a map $\mathcal{F} : \mathcal{H}_0 \to \mathcal{H}_\Xi$ called the generalized Fourier transform.*

(2) *The restriction $\mathcal{F} : \mathcal{H} \to \mathcal{H}_\Xi$ to $\mathcal{H}$ is unitary and that to $\mathcal{H}_\infty$ vanishes, so that the Parseval formula $\langle u, v \rangle = \langle \hat{u}, \hat{v} \rangle_\Xi$ is valid if at least one of u, v is in $\mathcal{H}$.*

(3) *If $(u, f) \in T$, then $\mathcal{F}(f)(t) = t\hat{u}(t)$. Conversely, if $\hat{u}$ and $t\hat{u}(t)$ are in $\mathcal{H}_\Xi$, then $\hat{u} = \mathcal{F}u$ for some $u \in \mathcal{D}(T)$.*

(4) *The integral $\int_K \Phi^*(x, \cdot)\, d\Xi\, \hat{u}$ converges in $\mathcal{H}$ as $K \to \mathbb{R}$ through compact intervals. If $\hat{u} = \mathcal{F}(u)$ with $u \in \mathcal{H}$ the limit is u, so this defines the inverse, or adjoint, of $\mathcal{F} : \mathcal{H} \to \mathcal{H}_\Xi$.*

(5) *Let E_Δ denote the spectral projector of T for the interval (or Borel set) Δ. Then $E_\Delta u(x) = \int_\Delta \Phi^*(x, \cdot)\, d\Xi\, \hat{u}$.*

The proof of the theorem is very similar to the proof of Theorem 5.2.10, with modifications suggested by the proof of Theorem 4.4.4, so it will be left to the reader as Exercise 5.3.2.

Exercises

Exercise 5.3.1 Prove the formula (5.3.2).

Hint: First consider the special cases of $u \in \mathcal{D}_0$ and supp u compact.

Exercise 5.3.2 Prove Theorem 5.3.3.

5.4 Equations that are both left- and right-definite

We start with the companion result to Corollary 4.5.3.

Theorem 5.4.1. *On (a, b) consider a self-adjoint realization T of the left-definite equation $-u'' + qu = \lambda w u$. Then $\sigma(T)$ is bounded from below if and only if $w \geq 0$ a.e. In this case $\sigma(T)$ contains at most two negative points, and is non-negative if the equation is essentially self-adjoint on (a, b).*

A similar statement is true for the opposite sign.

Proof. If $\varphi \in \mathcal{H}_1$ has compact support, then $\langle G_0(w\varphi), \varphi \rangle = \int_a^b |\varphi|^2 w$ so it is non-negative if $w \geq 0$ a.e. This implies that the minimal relation associated with the equation is non-negative. Since a self-adjoint realization T is an at most two-dimensional extension it must be non-negative except for at most two negative eigenvalues.

If w is not a.e. non-negative we need to find a sequence $(u_k, v_k) \in T, j = 1, 2, \ldots$ with $u_k \neq 0$ such that $\langle v_k, u_k \rangle / \langle u_k, u_k \rangle \to -\infty$. We shall find such a sequence in the minimal relation, so that $u_k = G_0(w v_k)$ where $v_k \in \mathcal{H}_1$ has compact support.

We can find $\varphi \in C_0(a, b)$ with $\int_a^b |\varphi|^2 w < 0$ since $w(x)\, dx$ is not a positive measure. Put $v_k(x) = e^{ikx}\varphi(x)$ so that $\langle v_k, u_k \rangle = \langle v_k, G_0(w v_k)\rangle = \int_a^b |v_k|^2 w = \int_a^b |\varphi|^2 w$, which is strictly negative and independent of k. In particular $u_k \neq 0$. It remains to show that $\|u_k\| = \|G_0(w v_k)\| \to 0$.

However, $\|G_0(w v_k)\|^2 = \int_a^b G_0(w v_k) \overline{v_k} w$ and $|\overline{v_k} w| = |\varphi w|$ is integrable, so by dominated convergence (Theorem B.5.7) we are done if we can show that $G_0(w v_k)$ is bounded independent of k on supp φ and tends pointwise to zero.

Now $G_0(w v_k)(x) = \langle G_0(w v_k), g_0(x, \cdot)\rangle = \int_a^b e^{iky}\varphi(y)g_0(x, y)w(y)\, dy$. This tends pointwise to zero by the Riemann–Lebesgue lemma (Theorem 4.5.1) and is bounded by $\int_a^b |\varphi w| g_0(x, \cdot) = G_0(|\varphi w|)(x)$. This function is continuous and thus bounded on supp φ. □

It is of some interest to compare the spectral theories obtained in a case when the equation is both right- and left-definite. We consider the equation $-u'' + qu = \lambda w u$ on an interval (a, b) on which both q and w are non-negative and not identically zero. We first show that all combinations of limit-point and limit-circle in the right- and left-definite theories are possible.

Theorem 5.4.2. *We have:*

- *The point a is limit-circle for both the left- and right-definite theories precisely if a is regular, i.e., a is finite and both q and w are integrable near a.*

- *The point a is limit-circle for the left-definite and limit-point for the right-definite theory precisely if a is finite and q integrable near a but w is not, while a primitive of w is square integrable near a. An example is $w(x) = |x - a|^{-1}$.*

- *The point a is limit-point for the left-definite and limit-circle for the right-definite theory precisely if a is infinite or q not integrable near a, while w is so small that $|u|^2 w$ is integrable near a for every solution u of $-u'' + qu = 0$.*

- *In all other cases the point a is limit-point for both the left- and right-definite theories.*

Proof. By Theorem 5.1.8 the left-definite equation is limit-circle at a precisely if a is finite, q is integrable near a, and a primitive of w is square integrable near a. Since $q \geq 0$ Theorem 4.2.7 shows that w must be integrable near a if the right-definite equation is limit-circle at a. Altogether this means that a is a regular endpoint.

If instead the right-definite equation is limit-point at a, we cannot have w integrable near a, while a primitive of w still has to be square integrable near a.

The left-definite equation can be limit-point at a for one or more of three reasons. If a is finite and q integrable near a the reason must be that w has a primitive which is not square integrable near a. But if a is finite this means that w cannot be integrable near a so the right-definite equation will automatically be limit-point.

On the other hand, if a is infinite or q not integrable near a, let u_1 and u_2 be a basis for the solutions of $-u'' + qu = 0$ and define $F = |u_1|^2 + |u_2|^2 > 0$. By Corollary 4.2.11 the right-definite equation is then limit-circle at a if and only if wF is integrable near a. Clearly there are $w > 0$ for which this is true, for any locally integrable F.   □

Next we show, in essence, that there is always at least one realization of the equation which is self-adjoint in both the left- and right-definite theories.

Theorem 5.4.3. *Given a right- and left-definite Sturm–Liouville equation on an interval (a, b) there is at least one self-adjoint realization in the right-definite theory which has a core which is also a core of a self-adjoint realization in the left-definite theory.*

Proof. Let $\langle u, v \rangle = \int_a^b u\bar{v}w$, $\langle u, v \rangle_1 = \int_a^b (u'\bar{v}' + qu\bar{v})$ and $\langle u, v \rangle_2 = \langle u, v \rangle_1 + \langle u, v \rangle$ with corresponding (semi-)norms $\|\cdot\|$, $\|\cdot\|_1$ and $\|\cdot\|_2$. These forms are all defined on the Hilbert space $\mathcal{H}$ of locally absolutely continuous functions u for which $\|u\|_2$ is finite. Since q and w are both non-negative and not identically zero we have $\|u\|_2 > 0$ and $\|u\|_1 > 0$ unless u is identically zero, but if w vanishes on an open subset of (a, b) we have $\|u\| = 0$ for any $u \in \mathcal{H}$ with support in this set.

Let $\mathcal{H}_\infty = \{u \in \mathcal{H} : \|u\| = 0\}$, $\mathcal{H}_2 = \mathcal{H} \ominus \mathcal{H}_\infty$, $\mathcal{H}_1$ be the completion of $\mathcal{H}_2$ in $\|\cdot\|_1$ and $L_w^2(a, b)$ be the completion of $\mathcal{H}_2$ in $\|\cdot\|$. Note that $\mathcal{H}_1$ is orthogonal to $\mathcal{H}_\infty$ also with respect to $\langle \cdot, \cdot \rangle_1$. We may think of $\mathcal{H}_2$ as a dense subset both of $\mathcal{H}_1$ and $L_w^2(a, b)$. In particular $\langle \cdot, \cdot \rangle_2$ is a closed quadratic form in $L_w^2(a, b)$ with form domain $\mathcal{H}_2$; see Section 3.4.

By Remark 3.4.3 on page 80 we may find symmetric operators R on $L_w^2(a, b)$ and R_2 on $\mathcal{H}_2$ such that R_2 is a core in R and $\langle u, v \rangle = \langle u, Rv \rangle_2$ if $u \in \mathcal{H}_2$ and $v \in L_w^2(a, b)$. We also have $0 \leq R_2 \leq I$ and $0 \leq R \leq I$ where I is the identity on $\mathcal{H}_2$ respectively $L_w^2(a, b)$. Furthermore, if $u, v \in \mathcal{H}_2$ then

$$\langle R_2 u, v \rangle_1 = \langle R_2 u, v \rangle_2 - \langle Ru, v \rangle = \langle u, R_2 u \rangle_2 - \langle u, Ru \rangle = \langle u, R_2 u \rangle_1$$

so R_2 is a densely defined and symmetric operator on $\mathcal{H}_1$. With $B = I - R_2$ we also have

$$\langle R_2 u, u \rangle_1 = \langle R_2 u, u \rangle_2 - \langle Ru, u \rangle = \langle u, u \rangle - \langle Ru, u \rangle \geq 0,$$
$$\langle Bu, u \rangle_1 = \langle Bu, u \rangle_2 - \langle Bu, u \rangle = \langle Bu, Bu \rangle_2 \geq 0,$$

so as an operator in $\mathcal{H}_1$ we also have $0 \leq R_2 \leq I$. The closure R_1 of this operator is therefore a self-adjoint operator in $\mathcal{H}_1$ with $0 \leq R_1 \leq I$ of which R_2 is a core.

Now, if $u \in L_w^2(a, b)$ and $\varphi \in C_0^\infty$ we have $\langle Ru, \varphi \rangle_2 = \langle u, \varphi \rangle$ so in a distributional sense $-(Ru)'' + (q + w)Ru = wu$. Thus R is the resolvent at -1 of a self-adjoint realization T of $-u'' + qu = wf$ in $L_w^2(a, b)$. Similarly R_1 and R_2 are the resolvents at -1 of self-adjoint realizations T_1 and T_2 in $\mathcal{H}_1$ respectively $\mathcal{H}_2$, and T_2 is a core of both T and T_1. If $v \in \mathcal{H}_2$ then $\langle Ru, v \rangle_2 = \langle u, v \rangle$ so $Ru = 0$ only if $u = 0$. Thus T and T_2 are operators and not relations, and the range of R_2 is dense in $\mathcal{H}_2$. It follows that the range of R_1 is dense in $\mathcal{H}_1$ and thus R_1 is injective, which means that T_1 is also an operator. $\qquad\qquad\qquad\qquad\qquad\qquad\qquad\qquad\qquad\qquad\qquad\qquad\qquad\qquad$ $\square$

Note that it is the introduction of the restricted space $\mathcal{H}_2 = \mathcal{H} \ominus \mathcal{H}_\infty$ that ensure T_1 and T_2 are operators and not relations.

Using the notation of the proof the equation $\langle f, v \rangle = \langle Rf, v \rangle_2$, valid for $f \in L_w^2(a, b)$ and $v \in \mathcal{H}_2$, is equivalent to $\langle Tu, v \rangle = \langle u, v \rangle_1$, where $u = Rf$. Integrating by parts and using the differential equation u satisfies, we obtain $\langle u, v \rangle_1 = u'\overline{v}|_a^b + \langle Tu, v \rangle$, so the integrated term vanishes. By use of Lemma 4.2.4 it follows that $u'\overline{v} \to 0$ at both endpoints for any $u \in \mathcal{D}(T)$ and $v \in \mathcal{H}_2$.

If each endpoint is limit-point in either the left- or right-definite theory the operator T_2 is the only self-adjoint realization[5] of our equation in $\mathcal{H}_2$, and this fixes T and T_1. However, if at least one of the endpoints is limit-circle in both theories, i.e., a regular endpoint, we must find the boundary conditions associated with T and T_1. Now, at a regular endpoint an element of $\mathcal{H}_2$ has a well defined but arbitrary value, so at such an endpoint we must have $u' \to 0$ for any u in $\mathcal{D}(T)$ or $\mathcal{D}(T_1)$.

However, taking a cue from Section 1.4, we have a choice of the space $\mathcal{H}$ we started with. Instead of letting $\mathcal{H}$ be all locally absolutely continuous functions u for which $\|u\|_2 < \infty$ we could consider the completion with respect to $\|\cdot\|_2$ of all locally absolutely continuous functions which are identically zero near one or the other, or perhaps both, endpoints. According to Theorem 5.1.9 this will change the space $\mathcal{H}_2$ precisely if the endpoint in question is regular.

At a regular endpoint the original choice of $\mathcal{H}$ therefore gives a Neumann boundary condition, while the alternative gives a Dirichlet condition. If both endpoints are regular, there is yet another way to define $\mathcal{H}$, since we may then define $\mathcal{H}$ as those absolutely continuous functions u defined in (a, b) for which $u(b) = Ku(a)$, where K is some fixed, complex number. Just as in Section 1.4 this will give boundary conditions $u(b) = Ku(a)$, $u'(a) = \overline{K}u'(b)$, which are coupled if $K \neq 0$.

We next want to compare the m-functions and spectral measures obtained in the various cases. We first assume a is a regular endpoint while b may be regular or singular. At b we select a separated boundary condition which is symmetric for both the left- and right-definite theories. As we have seen, if b is regular there are two such choices, while there is a unique choice if b is singular. At a we choose either a Dirichlet condition ($\alpha = 0$ in (4.2.2) and (5.1.5)) or a Neumann condition ($\alpha = \pi/2$ in (4.2.2) and (5.1.5)). We then have the following theorem.

---

[5] We are considering the left-definite theory for the equation $-u'' + (q + w)u = wf$, which is essentially self-adjoint, and the corresponding operator is $T_2 + I$.

Theorem 5.4.4. *Consider a right- and left-definite equation on an interval $[a, b)$ with the left endpoint regular, and a realization by separated boundary conditions which is self-adjoint in both theories in the sense of Theorem 5.4.3. Let m be the m-function in the right-definite and $\tilde{m}$ that in the left-definite theory, and η respectively $\tilde{\eta}$ the corresponding spectral functions.*

Then $\tilde{m}(\lambda) = \lambda m(\lambda)$ and $d\tilde{\eta}(t) = t \, d\eta(t)$ in the case of Neumann conditions at a. Given Dirichlet conditions at a we instead have $m(\lambda) = \lambda \tilde{m}(\lambda)$ and $d\eta(t) = t \, d\tilde{\eta}(t)$.

Clearly the same formulas apply when the regular endpoint is the right endpoint of the base interval. The spectral measures are positive, so from the formulas it follows that the spectrum is positive, which is also clear from the bounds on R.

Proof. Let φ and θ be the solutions of $-u'' + qu = \lambda wu$ with initial data (4.3.3) for $\alpha = 0$ or $\alpha = \pi/2$ and m the corresponding right-definite m-function so that $\psi(\cdot, \lambda) = \theta(\cdot, \lambda) + m(\lambda)\varphi(\cdot, \lambda)$ is the corresponding Weyl solution.

Let $\tilde{\varphi}$ and $\tilde{\theta}$ be the solutions with initial data (5.2.5) and $\tilde{m}$ the m-function for the same α and the same boundary condition at b as in the right-definite theory. Then the left-definite Weyl solution $\tilde{\psi} = \tilde{\theta} + \tilde{m}\tilde{\varphi}$ is a multiple of ψ since both must be in $\mathcal{H}_2$. If $\alpha = 0$ we have $\tilde{\varphi} = \varphi$ and $\lambda\tilde{\theta}(\cdot, \lambda) = \theta(\cdot, \lambda)$. It follows that $\lambda\tilde{\psi}(\cdot, \lambda) = \psi(\cdot, \lambda)$ and $m(\lambda) = \lambda\tilde{m}(\lambda)$. For $\alpha = \pi/2$ we instead obtain $\lambda\tilde{\varphi}(\cdot, \lambda) = \varphi(\cdot, \lambda)$ and $\tilde{\theta} = \theta$ so that $\tilde{\psi} = \psi$ and $\tilde{m}(\lambda) = \lambda m(\lambda)$. Integrating the formulas for the m-functions around an appropriate rectangle we obtain the formulas for the spectral measures as in the proof of Lemma 4.3.10.   □

We finally consider the case of two singular endpoints. The construction of Section 4.4 gives an M-matrix $M = (m_{jk})$ for the right-definite theory on the form $M = \frac{1}{m_+ + m_-} \begin{pmatrix} m_+ m_- & (m_- - m_+)/2 \\ (m_- - m_+)/2 & -1 \end{pmatrix}$ where $m_\pm$ are right-definite Dirichlet m-functions for intervals $(a, c]$ and $[c, a)$. Similarly Section 5.3 gives an M-matrix $\tilde{M} = (\tilde{m}_{jk})$ for the left-definite theory on the form $\tilde{M} = \frac{1}{\tilde{m}_+ + \tilde{m}_-} \begin{pmatrix} \tilde{m}_+ \tilde{m}_- & (\tilde{m}_- - \tilde{m}_+)/2 \\ (\tilde{m}_- - \tilde{m}_+)/2 & -1 \end{pmatrix}$ where $\tilde{m}_\pm$ are left-definite Dirichlet m-functions for intervals $(a, c]$ and $[c, a)$. There are corresponding matrix-valued spectral measures $d\Xi = (d\eta_{jk})$ and $d\tilde{\Xi} = (d\tilde{\eta}_{jk})$. We obtain the following corollary.

Corollary 5.4.5. *We have*

$$\tilde{M} = \begin{pmatrix} \tilde{m}_{11}(\lambda) & \tilde{m}_{12}(\lambda) \\ \tilde{m}_{21}(\lambda) & \tilde{m}_{22}(\lambda) \end{pmatrix} = \begin{pmatrix} \lambda m_{11}(\lambda) & m_{12}(\lambda) \\ m_{21}(\lambda) & m_{22}(\lambda)/\lambda \end{pmatrix}$$

and

$$d\tilde{\Xi}(t) = \begin{pmatrix} d\tilde{\eta}_{11}(t) & d\tilde{\eta}_{12}(t) \\ d\tilde{\eta}_{21}(t) & d\tilde{\eta}_{22}(t) \end{pmatrix} = \begin{pmatrix} t \, d\eta_{11}(t) & d\eta_{12}(t) \\ d\eta_{21}(t) & d\eta_{22}(t)/t \end{pmatrix}.$$

This is an immediate consequence of Theorem 5.4.4.

Exercise

Exercise 5.4.1 Consider a regular Sturm–Liouville equation which is both left-
and right-definite and impose either coupled boundary conditions using a diagonal,
symplectic matrix, or a Dirichlet or Neumann condition at each end. Show that

- In the left-definite theory 0 is a double eigenvalue if both boundary conditions
  are Dirichlet.

- In the left-definite theory 0 is a simple eigenvalue if one boundary condition
  is Dirichlet and the other Neumann, or if the boundary conditions are coupled
  using a diagonal, symplectic matrix.

- In the left-definite theory 0 is not an eigenvalue if both boundary conditions are
  Neumann.

- In the right-definite theory 0 is never an eigenvalue for any of these boundary
  conditions.

5.5 Generalized left-definite Sturm–Liouville equations

Even if we require positivity of the form $\int_a^b (u'\overline{v'} + q u\overline{v})$ for all sufficiently smooth
functions with compact support in (a, b) it is by no means necessary that $q \geq 0$. We
may for example add a sufficiently rapidly oscillating term with arbitrarily negative
lower bound to a positive q without destroying the positivity of the form. It is also
possible to deal with cases of low coefficient regularity. We may for example consider
$q \in H_{\mathrm{loc}}^{-1}(a, b)$, in which case positivity of q makes no sense unless q is a measure.
Nevertheless the corresponding form may have a positive character.

It does not seem possible to give explicit necessary and sufficient conditions on q
to ensure positivity of the form. Instead we shall, inspired by E. B. Davies' [51] treatment of the right-definite one-dimensional Schrödinger equation, express positivity
in terms of the existence of positive solutions of $-u'' + qu = 0$.

To show that the assumptions we are about to make are reasonable, we shall first
prove the following proposition.

Proposition 5.5.1. *Assume that q is locally integrable in (a, b) and*

$$\int_a^b (|u'|^2 + q\,|u|^2) > 0$$

*for non-zero, compactly supported and absolutely continuous functions with deriva-
tives in $L^2(a, b)$. Then non-trivial solutions of $-u'' + qu = 0$ have at most one zero
in (a, b), and there is at least one real-valued solution without zeros.*

Proof. Suppose u is a nontrivial solution with $u(c) = u(d) = 0$, $a < c < d < b$. If
we modify u outside $[c, d]$ by setting it to zero there we obtain a function satisfying

the assumptions of the proposition, so we obtain

$$0 < \int\limits_a^b (|u'|^2 + q\,|u|^2) = [u'\overline{u}]_c^d + \int\limits_c^d (-u'' + qu)\overline{u} = 0 \,.$$

This contradiction proves the first statement.

Now suppose u and v are real-valued solutions with zeros in c and d respectively, $a < c < d < b$. We may assume u and v to be negative to the left, and positive to the right of c and d respectively. Consider v/u, which has a singularity at c and derivative $(v/u)' = \mathcal{W}(v,u)/u^2$, where $\mathcal{W}(v,u) = v'u - vu'$ is the Wronskian of u and v, and consequently constant (Proposition 4.1.6). If we evaluate the Wronskian in d we see that it is positive. Therefore v/u is positive and strictly increasing to the left of c, so that it has a finite limit $A \geq 0$ at a. At b there is again a strictly positive limit $0 < B \leq \infty$.

If $B > A$, then v/u would take any value in (A, B) twice, so that any solution $v - Cu$ with $A < C < B$ would have two zeros. It follows that $B \leq A$ and that the solution $v - Cu$ has no zeros if $B \leq C \leq A$. If $B < A$ there are therefore linearly independent, real-valued solutions which do not have zeros. If $A = B$ the only zero-free real-valued solutions are multiples of the solution $v - Au$. $\qquad\square$

It is also easy to see that any solution except $v - Au$ has a reciprocal in L^2 near a, and any solution except $v - Bu$ a reciprocal in L^2 near b. Thus, the solutions $v - Au$ and $v - Bu$ are 'small' at a respectively b. We may therefore consider the case $A = B$ somewhat degenerate in that the solution $u - Av$ is then small at both ends of (a, b). On the other hand, if $B < C < A$ the solution $v - Cu$ is positive and has a reciprocal in $L^2(a, b)$.

Below we shall discuss Hilbert spaces associated with a left-definite Sturm–Liouville equation $-u'' + qu = \lambda wu$ without reference to a positive q, assuming only the existence of a solution with the properties mentioned. It will be convenient to use the language of distribution theory and the spaces $H_{\mathrm{loc}}^n(a, b)$, for which we refer to Theorem C.16.

Definition 5.5.2. Let $F \in H_{\mathrm{loc}}^1(a, b)$ be strictly positive and satisfy $1/F \in L^2(a, b)$. Also assume F normalized so that $\int_a^b F^{-2} = 1$.

By Theorem C.16 there are distributional derivatives $F' \in L_{\mathrm{loc}}^2(a, b)$ and $F'' \in H_{\mathrm{loc}}^{-1}(a, b)$, so that $1/F \in H_{\mathrm{loc}}^1(a, b)$ and the product $q = F''/F$ is defined as a distribution in $H_{\mathrm{loc}}^{-1}(a, b)$. Thus F satisfies the equation $-F'' + qF = 0$ in (a, b). Now let $h = F'/F \in L_{\mathrm{loc}}^2$ and define

$$\langle u, v \rangle = \int\limits_a^b (u/F)'\overline{(v/F)'}F^2 = \int\limits_a^b (u' - hu)\overline{(v' - hv)} \,, \qquad (5.5.1)$$

a semi-scalar product defined at least on compactly supported, sufficiently smooth functions. Clearly $\langle u, u \rangle = 0$ precisely if u is a multiple of F. Note that we have

$q = h' + h^2$ and[6] 'integrating by parts' (we are really using the definition of a distributional derivative) we have

$$\langle u, v \rangle = \int_a^b (u' - hu)\overline{(v' - hv)} = \int_a^b (u'\overline{v'} + qu\overline{v})$$

for smooth, compactly supported u, v, where $\int_a^b qu\overline{v}$ means $q(u\overline{v})$, the distribution q evaluated on $u\overline{v}$.

Example 5.5.3. We give a few examples of possible functions F and begin with $F(x) = A \cosh x$ where $A > 0$ is a normalization constant which depends on the interval (a, b) considered. This function is smooth with reciprocal in $L^2(a, b)$, and satisfies the equation $-F'' + F = 0$, so here $q = 1 > 0$. This case is of course covered by our earlier theory.

Another example is $F(x) = 2 + |x|$, where we have ignored the normalization constant. We obtain $-F'' + \delta F = 0$, where δ is the Dirac measure at 0. Since q is now a measure this is a case not covered by our earlier theory, although q is positive and it would be easy to extend that theory to cover the case when q is a positive measure.

Given any interval $(a, b) \neq \mathbb{R}$ we can find a positive linear function F with $1/F \in L^2(a, b)$, for if a is finite we may take $F(x) = 1 - a + x$, and if b is finite $F(x) = 1 + b - x$ will do. These functions satisfy $-F'' = 0$. However, for $(a, b) = \mathbb{R}$ the only positive solutions are constants, which do not have a reciprocal in $L^2(\mathbb{R})$. This is a case of $A = B$ in the proof of Proposition 5.5.1.

All these examples correspond to equations with a non-negative potential q. However, consider now $F(x) = x^\alpha + x^{1-\alpha}$ on $(0, \infty)$. Then $q = F''/F = \alpha(\alpha - 1)x^{-2}$, which is negative in $(0, \infty)$ if $0 < \alpha < 1$, while $1/F \in L^2(0, \infty)$ unless $\alpha = 1/2$. Thus we may find F satisfying all requirements in spite of q being strictly negative. Since this means that F is positive and concave this is only possible on bounded or semi-bounded intervals.

However, even on $\mathbb{R}$ one can have q negative on a large set. For example, with $F(x) = (1 + x^2)^{1/3}$ on $\mathbb{R}$ we have $1/F \in L^2(\mathbb{R})$, but note that

$$q = F''/F = \frac{1}{9}(6 - 2x^2)(1 + x^2)^{-2} .$$

Thus q is negative except on the compact interval $[-\sqrt{3}, \sqrt{3}]$.

Consider, finally, $F(x) = 1 + |x|^\alpha$. Then $1/F \in L^2(\mathbb{R})$ if $\alpha > 1/2$ and for $x \neq 0$ $F'(x) = \alpha |x|^{\alpha-2} x$ which is in L^2_{loc} if $\alpha > 1/2$. Differentiating pointwise for $x \neq 0$ we obtain $F''(x) = \alpha(\alpha - 1) |x|^{\alpha-2}$, which is negative if $0 < \alpha < 1$. Thus, in a neighborhood of any $x \neq 0$ we have $q(x) = F''(x)/F(x) = \alpha(\alpha - 1) |x|^{\alpha-2}/(1 + |x|^\alpha) < 0$ if $1/2 < \alpha < 1$, but in no neighborhood of zero is $\alpha(\alpha - 1) |x|^{\alpha-2}$ integrable if $1/2 < \alpha < 1$, so the formula does not represent the distribution F'' there.

---

[6] The transformation $h \mapsto q$ is often called the *Miura transformation*, see [168].

Since F' is locally square integrable but not of locally bounded variation (it has an infinite jump at 0) $F'' \in H_{\text{loc}}^{-1}(\mathbb{R})$ but is not a measure. Thus, for $x \neq 0$, $q(x) = \alpha(\alpha - 1) |x|^{\alpha-2} / (1 + |x|^{\alpha}) < 0$ but in a neighborhood of 0 we have $q \in H_{\text{loc}}^{-1}$ but not a measure, so it is meaningless to speak of the sign of q in a neighborhood of 0; cf. Theorem C.8.

Lemma 5.5.4. *Any measurable function u with finite semi-norm $\|u\|$ is in $H_{\text{loc}}^1(a, b)$ and $(u/F)' \in L^1(a, b)$ so that u/F is bounded and the limits of u/F at a and b exist finitely.*

Proof. Clearly $(u/F)' \in L_{\text{loc}}^2$ and by Cauchy–Schwarz we have

$$\int\limits_a^b |(u/F)'| = \int\limits_a^b |(u/F)'F| \, F^{-1} \leq \left(\int\limits_a^b F^{-2} \right)^{1/2} \|u\| = \|u\| , \qquad (5.5.2)$$

which proves the lemma. $\square$

Proposition 5.5.5. *$u \in H_{\text{loc}}^1$ is a solution of $-u'' + qu = 0$ precisely if the Wronskian $\mathcal{W}(u, F) = u'F - uF'$ is constant, and the set of all solutions equals the set of linear combinations of F and $F(x) \int_a^x F^{-2}$, and is therefore a two-dimensional linear space.*

Proof. By Theorem C.17, if $u \in H_{\text{loc}}^1(a, b)$ the Wronskian $\mathcal{W}(u, F)$ is well defined and may be differentiated using the product rule. Differentiating we obtain $u''F - uF'' = F(u'' - qu)$, so u is a solution if and only if $\mathcal{W}(u, F)$ is constant.

Now, u/F is a primitive of $\mathcal{W}(u, F)/F^2$, so for any solution u we have $u(x) = F(x)(A + B \int_a^x 1/F^2)$ for some constants A, B. $\square$

We shall need to introduce two distinguished solutions.

Definition 5.5.6. Set $F_-(x) = F(x) \int_a^x F^{-2}$ and $F_+(x) = F(x) \int_x^b F^{-2}$.

Proposition 5.5.7. *The solution F_- is 'subordinate' at a in the sense that $F_-/u \to 0$ at a for any linearly independent solution u. Similarly F_+ is subordinate at b, $F = F_- + F_+$ and $\mathcal{W}(F_-, F_+) = \mathcal{W}(F_-, F) = \mathcal{W}(F, F_+) = 1$.*

F_-, F_+ is a basis for the solutions of $-u'' + qu = 0$, $F_-(x) = F_+(x) \int_a^x F_+^{-2}$, $F_+(x) = F_-(x) \int_x^b F_-^{-2}$ and $\int_a^x F_-^{-2} \int_x^b F_+^{-2} = 1$ for all x.

Proof. Clearly $F = F_+ + F_-$ and any solution may be written

$$u(x) = F(x)(A + B \int\limits_a^x F^{-2}) = AF_+(x) + (A + B)F_-(x) ,$$

so F_+, F_- is a basis for the solutions. Now $\lim_a F_-/F_+ = \lim_b F_+/F_- = 0$, which proves the subordinacy claim. The formulas for the Wronskians are elementary calculations.

F_-/F_+ is a primitive of $\mathcal{W}(F_-, F_+)/F_+^2$ and since F_- is subordinate at a we obtain the formula for F_- and similarly the formula for F_+. Multiplying these formulas together yields the last formula. □

Lemma 5.5.8. *A function $u \in H^1_{\text{loc}}(a, b)$ is orthogonal to all functions in $C_0^\infty(a, b)$ if and only if it is a solution of $-u'' + qu = 0$.*

Proof. Suppose $\varphi \in C_0^\infty(a, b)$. We have

$$\int_a^b (u' - hu)\overline{(\varphi' - h\varphi)} = \int_a^b ((u' - hu)\overline{\varphi'} - h(u' - hu)\overline{\varphi}).$$

Thus $-(u' - hu)' - h(u' - hu) = -u'' + (h' + h^2)u = 0$ in a distributional sense if and only if u is orthogonal to $C_0^\infty(a, b)$. □

Definition 5.5.9. Let $\mathcal{H}_D$ be the set of all $u \in H^1_{\text{loc}}(a, b)$ for which $\|u\|$ is finite and $u/F \to 0$ at a and b.

Proposition 5.5.10. $\mathcal{H}_D$ *is a Hilbert space in which $C_0^\infty(a, b)$ is dense.*

Proof. Suppose $\{u_j\}$ is a Cauchy sequence in $\mathcal{H}_D$. Then $(u_j/F)'$ converges in the weighted L^2-space with weight F^2 to a function U. By (5.5.2) it is clear that $(u_j/F)'$ also converges in $L^1(a, b)$ to U, and from $u_j(x)/F(x) = \int_a^x (u_j/F)'$ it follows that u_j/F converges uniformly on (a, b) to $\int_a^x U$. In particular, the limit function, which we may write as u/F, has limits 0 at a and b. It also follows that $U = (u/F)'$, and since $\int |UF|^2 < \infty$ it follows that $u \in \mathcal{H}_D$.

By Theorem 5.5.8 the orthogonal complement of $C_0^\infty(a, b)$ in $\mathcal{H}_D$ are those solutions of the homogeneous equation which are in $\mathcal{H}_D$. However, a solution is of the form $u = AF_+ + BF_-$ so u/F tends to A at a and to B at b. Thus no nontrivial solution is in $\mathcal{H}_D$ and $C_0^\infty(a, b)$ is dense. □

Now define $g_0(x, y) = F_+(\max(x, y))F_-(\min(x, y))$, which is an important kernel for the sequel. Note in particular that by Proposition 5.5.7

$$g_0(x, x) = F_+(x)F_-(x) = F_+^2(x) \int_a^x F_+^{-2} = F_-^2(x) \int_x^b F_-^{-2} .$$

Lemma 5.5.11. *The function $g_0(x, \cdot) \in \mathcal{H}_D$ for every $x \in (a, b)$.*

Proof. For $y > x$ we have $g_0(x, y)/F(y) = F_-(x)F_+(y)/F(y) = F_-(x)\int_y^b F^{-2} \to 0$ as $y \to b$. Similarly as $y \to a$.

Moreover, $(g_0(x, \cdot)/F)'$ equals $F_-(x)\mathcal{W}(F_+, F)/F^2 = -F_-(x)/F^2$ in (x, b) and $F_+(x)/F^2$ in (a, x) so that $(g_0(x, \cdot)/F)'F \in L^2(a, b)$. Thus $g_0(x, \cdot) \in \mathcal{H}_D$. □

Theorem 5.5.12. *We have* $g_0(x, y) = \langle g_0(x, \cdot), g_0(y, \cdot) \rangle$ *and point evaluations are bounded linear forms on* $\mathcal{H}_D$*, given by* $u(x) = \langle u, g_0(x, \cdot) \rangle$ *and with norm* $\sqrt{g_0(x, x)}$.

Proof. Consider

$$\langle u, g_0(x, \cdot) \rangle = \int_a^b (u/F)'(g_0(x, \cdot)/F)' F^2 = \int_a^b (u/F)' \mathcal{W}(g_0(x, \cdot), F).$$

In (x, b) we have $\mathcal{W}(g_0(x, \cdot), F) = F_-(x)\mathcal{W}(F_+, F) = -F_-(x)$ and in (a, x) we have $\mathcal{W}(g_0(x, \cdot), F) = F_+(x)$ so that

$$\langle u, g_0(x, \cdot) \rangle = F_+(x) \int_a^x (u/F)' - F_-(x) \int_x^b (u/F)' = u(x).$$

In particular, we obtain $g_0(x, y) = \langle g_0(x, \cdot), g_0(y, \cdot) \rangle$ so that $\|g_0(x, \cdot)\| = \sqrt{g_0(x, x)}$. We also have $|u(x)| = |\langle u, g_0(x, \cdot) \rangle| \le \|g_0(x, \cdot)\| \|u\|$ with equality for $u = g_0(x, \cdot)$ so that $\|g_0(x, \cdot)\| = \sqrt{g_0(x, x)}$ is the norm of the evaluation form at x. $\qquad\square$

5.6 Spectral theory

Extending $\mathcal{H}_D$. Our definition of $\mathcal{H}_D$ builds a kind of Dirichlet boundary conditions into the space; cf. Section 1.4. From $\int_a^b F^{-2} = 1$ follows

$$\|AF_+ + BF_-\| = |A - B| \,,$$

so all solutions of $-u'' + qu = 0$ have a finite semi-norm. It follows now from Lemma 5.5.7 that our semi-norm is finite precisely on the orthogonal sum $\mathcal{H}_1 = \mathrm{Sp}(F_+, F_-) \oplus \mathcal{H}_D$. An element u of this space also gives a finite value for $\lim u/F$ at the interval endpoints a and b.

A scalar product on $\mathcal{H}_1$ which reduces to the scalar product (5.5.1) on elements of $\mathcal{H}_D$ is of the form

$$\langle u, v \rangle_Q = \langle u, v \rangle + Q(u, v) \,,$$

where Q is a Hermitian form in the limits of u/F and v/F at the endpoints. It will be convenient to use the following notation.

Definition 5.6.1. For any $u \in \mathcal{H}_1$ define $\tilde{u} = u/F$.

Using this notation we must therefore have

$$Q(u, v) = \alpha \tilde{u}(a)\overline{\tilde{v}(a)} + \beta \tilde{u}(b)\overline{\tilde{v}(b)} + \gamma \tilde{u}(a)\overline{\tilde{v}(b)} + \tilde{u}(b)\overline{\gamma \tilde{v}(a)}$$

with real α, β and complex γ. Note that the orthogonal complement of $\mathcal{H}_D$ will still be the solutions of the homogeneous equation. The form $\langle u, v \rangle_Q$ is positive definite on $\mathcal{H}_1$ only under certain conditions on α, β and γ, but all such choices give equivalent norms, since they differ only on a finite-dimensional space. For the same reason $\mathcal{H}_1$ is a Hilbert space if provided with such a norm.

We obtain a norm on $\mathcal{H}_1$ precisely if $\|u\|_Q^2 = \langle u, u \rangle_Q > 0$ for nontrivial solutions $u = AF_+ + BF_-$ of the homogeneous equation. A calculation shows that for such u

$$\|u\|_Q^2 = (\alpha + 1)|A|^2 + 2\,\mathrm{Re}((\gamma - 1)A\overline{B}) + (\beta + 1)|B|^2 ,$$

so the condition for positivity becomes $\alpha + 1 > 0$ and $(\alpha + 1)(\beta + 1) > |\gamma - 1|^2$. Again, we cannot use our original norm with $\alpha = \beta = \gamma = 0$. Any choice of $\alpha > -1$ and $\beta > -1$ such that $(\alpha + 1)(\beta + 1) > |\gamma - 1|^2$ will do.

Remark 5.6.2. As in Section 1.4 the boundary conditions that will give self-adjoint realizations of an equation $-u'' + qu = wf$ in $\mathcal{H}_1$ depend on the choice of Q, as we will see.

If we look for a *reproducing kernel*, i.e., a function $g_Q(x, y)$ such that $u(x) = \langle u, g_Q(x, \cdot) \rangle_Q$ for every $u \in \mathcal{H}_1$, it must be of the form

$$g_Q(x, y) = g_0(x, y)$$
$$+ d_{11}F_+(x)F_+(y) + d_{12}F_+(x)F_-(y) + d_{21}F_-(x)F_+(y) + d_{22}F_-(x)F_-(y) ,$$

with a Hermitian matrix (d_{jk}). The reason is that it has to reproduce $\mathcal{H}_D$, which means that for fixed x we can only add a solution of the homogeneous equation to the first term. Secondly, from the formula $g_Q(x, y) = \langle g_Q(x, \cdot), g_Q(y, \cdot) \rangle_Q$ it follows that $g_Q(x, y) = \overline{g_Q(y, x)}$. Now, the kernel g_Q will reproduce all elements of $\mathcal{H}_D$ for any choice of the constants d_{jk} . However, we must also require that the kernel reproduces all solutions of the equation $-u'' + qu = 0$. A simple calculation shows that this will be the case precisely if

$$d_{11} = \frac{\beta + 1}{(\alpha + 1)(\beta + 1) - |\gamma - 1|^2} ,$$

$$d_{12} = \overline{d_{21}} = \frac{1 - \gamma}{(\alpha + 1)(\beta + 1) - |\gamma - 1|^2} ,$$

$$d_{22} = \frac{\alpha + 1}{(\alpha + 1)(\beta + 1) - |\gamma - 1|^2} .$$

Another simple calculation shows that if $u \in \mathcal{H}_1$, then

$$\tilde{u}(a) = \langle u, d_{11}F_+ + d_{12}F_- \rangle_Q \quad \text{and} \quad \tilde{u}(b) = \langle u, d_{21}F_+ + d_{22}F_- \rangle_Q ,$$

so these values are bounded linear forms on $\mathcal{H}_1$.

Endpoints. We shall call an endpoint of the interval (a, b) *norm-regular* if F' is in L^2 near the endpoint. If $u(x) = F(x)(A + B\int_a^x F^{-2})$ we have $u'(x) = B/F(x) + F'(x)(A + B\int_a^x F^{-2})$, so since the last factor is bounded all solutions of $-u'' + qu = 0$ have square integrable derivatives near a norm-regular endpoint.

Since $1/F \in L^2(a, b)$ it follows that F'/F is integrable near a norm-regular endpoint, and therefore that $\log F$ has a finite limit there. In other words, F has a strictly positive, finite limit at the endpoint. Since $1/F \in L^2(a, b)$ this immediately implies that a norm-regular endpoint has to be finite. Thus all solutions of $-u'' + qu = 0$ have integrable derivatives near, and therefore finite limits at, a norm-regular endpoint.

If $u \in \mathcal{H}_1$ it therefore has a limit at a norm-regular endpoint, and if $u \in \mathcal{H}_D$ the limit is zero. It also follows that if a is norm-regular, then $\lim_a F_- = 0$ and $\lim_a F_+ > 0$, and if b is norm-regular, then $\lim_b F_+ = 0$ and $\lim_b F_- > 0$.

A symmetric relation. Any of the valid choices of α, β and γ will give a Hilbert space suitable for spectral theory for $-u'' + qu = \lambda wu$, with a suitable choice of w. We note that all the elements of $\mathcal{H}_1$ are functions with distributional derivative in $L^2_{\mathrm{loc}}(a, b)$; this means that $H^1_c(a, b) \subset \mathcal{H}_D \subset \mathcal{H}_1 \subset H^1_{\mathrm{loc}}(a, b)$, where $H^n_{\mathrm{loc}}(a, b)$, $n \in \mathbb{Z}$, is defined just before Theorem C.17 and we denote by $H^n_c(a, b)$ the compactly supported distributions in $H^n_{\mathrm{loc}}(a, b)$.

Lemma 5.6.3. *If $v \in H^{-1}_c(a, b)$ the linear form $\mathcal{H}_1 \ni u \mapsto v(u)$ is bounded.*

Proof. Since $F \in H^1_{\mathrm{loc}}(a, b)$ the distribution Fv has compact support and is in $H^{-1}_{\mathrm{loc}}(a, b)$ by Theorem C.17. Let V_1 be a primitive of Fv, so that $V_1 \in L^2_{\mathrm{loc}}(a, b)$ and constant in intervals not intersecting supp v.

Setting $V = V_1/F$ we then have $V \in L^2(a, b)$ since it equals a multiple of $1/F$ near each endpoint, and $1/F$ is bounded and continuous in any compact interval $\subset (a, b)$. For $\varphi \in C^\infty_0(a, b)$ we obtain

$$v(\varphi) = \tfrac{1}{F}V_1'(\varphi) = V'(\varphi) + hV(\varphi) = -V(\varphi' - h\varphi)$$

so that $|v(\varphi)| \leq \|V\|_{L^2}\|\varphi\|_Q$. Taking closure shows that this is true for any $\varphi \in \mathcal{H}_D$.

The orthogonal complement of $\mathcal{H}_D$ is the set of solutions of $-u'' + qu = 0$, a subspace of $H^1_{\mathrm{loc}}(a, b)$ of dimension two. If $\psi \in C^\infty_0(a, b)$ equals 1 on supp v we may define $v(u) = v(\psi u) = uv(\psi)$, the value of which is independent of the choice of ψ. On a space of finite dimension any linear form is bounded so this completes the proof. $\square$

If $v \in H^{-1}_c(a, b)$ the lemma shows that we may apply Riesz' representation theorem to find $v^* \in \mathcal{H}_1$ such that $v(u) = \langle u, \overline{v^*}\rangle_Q$. Clearly v^* depends linearly on v, so we have an operator $G_0 : H^{-1}_c(a, b) \to \mathcal{H}_1$ such that

$$v(\overline{u}) = \langle G_0 v, u\rangle_Q$$

for every $u \in \mathcal{H}_1$ and every $v \in H^{-1}_c(a, b)$. The restriction of G_0 to $H^1_c(a, b)$ is symmetric, since

$$\langle G_0 u, v\rangle_Q = \int u\overline{v} = \overline{\langle G_0 v, u\rangle_Q} = \langle u, G_0 v\rangle_Q \,.$$

Theorem C.17 shows that in order for the product wu to be a distribution for every $u \in \mathcal{H}_1$ we must assume that $w \in H^{-1}_{\mathrm{loc}}(a, b)$, and then $wu \in H^{-1}_{\mathrm{loc}}(a, b)$. Now define a linear relation in $\mathcal{H}_1$ by

$$T_c = \{(G_0(uw), u) : u \in \mathcal{H}_1 \text{ with compact support}\} \,.$$

Assumption 5.6.4 *The distribution* $w \in H^{-1}_{\mathrm{loc}}(a, b)$ *is real-valued, that is* $\overline{w(\varphi)} = w(\overline{\varphi})$ *for every test function* $\varphi \in C^{\infty}_0(a, b)$, *and* supp w *contains at least 2 points.*

The assumption on supp w is to avoid the degenerate case when the spectral theory takes place in a space of dimension at most 1. With these assumptions the relation T_c is symmetric, since if u and $v \in H^1_c(a, b)$ we have

$$\langle G_0(uw), v \rangle_Q = w(u\overline{v}) = \overline{w(v\overline{u})} = \overline{\langle G_0(vw), u \rangle_Q} = \langle u, G_0(vw) \rangle_Q \,.$$

The *minimal relation* T_0 is by definition the closure of T_c, and its adjoint T_1 is the *maximal relation*. This is a closed relation, so the symmetry of T_c implies that $T_0 \subset T_1$.

Theorem 5.6.5. *The relation* T_1 *is the maximal differential relation associated with the equation* $-u'' + qu = wf$ *in* $\mathcal{H}_1$. *This means that* T_1 *consists of all elements* $(u, f) \in \mathcal{H}_1 \oplus \mathcal{H}_1$ *satisfying the equation in a distributional sense.*

Proof. If $(u, f) \in T_1$ and $\varphi \in \mathcal{H}_c \supset C^{\infty}_0(a, b)$ we have, by definition,

$$\langle u, \varphi \rangle_Q = \langle f, G_0(\varphi w) \rangle_Q = w(f\overline{\varphi}) \,.$$

The left-hand side is $\int_a^b (u' - hu)(\overline{\varphi'} - h\overline{\varphi})$ so that as a distribution u satisfies

$$-(u' - hu)' = h(u' - hu) + wf \,,$$

which simplifies to $-u'' + qu = wf$ since $h' + h^2 = q$.

Conversely, if u and $f \in \mathcal{H}_1$ and satisfy the equation in the sense of distributions the same calculations in reverse show that $(u, f) \in T_1$. □

A normal form. Suppose $u \in \mathcal{H}_1$ and introduce the *Liouville transform*[7] $u \mapsto v$ where $v(y) = u(x)/F(x)$ and $y = \int_a^x F^{-2}$. Then a change of variable shows that

$$\|u\|^2_Q = \int\limits_0^1 |v'|^2 + \alpha\,|v(0)|^2 + \beta\,|v(1)|^2 + 2\operatorname{Re}(\gamma v(0)\overline{v(1)}) \,.$$

Thus, after a Liouville transform we might as well consider the Hilbert space $H^1(0, 1)$ of distributions in $\mathcal{D}'(0, 1)$ with derivative in $L^2(0, 1)$, and if we want to consider an operator associated with an equation $-u'' + qu = \lambda wu$ in $\mathcal{H}_1$ we may equivalently consider an operator associated with $-v'' = \lambda \breve{w} v$ in $H^1(0, 1)$ with the norm above,

---

[7] For details about such transformations, see Section 7.2.

where $\breve{w}(y) = F^4(x)w(x)$. We can therefore always reduce our study to this equation, called the *indefinite Kreĭn string* on $(0, 1)$.

However, important aspects of the original equation, like asymptotic properties of the coefficients, may be obscured by this transformation, so we will not do this. Nevertheless, it is worthwhile to note that *any* space of the type we have discussed can be transformed unitarily by a Liouville transform to *any other* such space. To do this, consider the functions F_- and F associated with $\mathcal{H}_D$ and similarly $\tilde{F}_-$ and $\tilde{F}$ associated with $\tilde{\mathcal{H}}_D$, and define a function t by the equation

$$\tilde{F}_-(t(x))/\tilde{F}(t(x)) = F_-(x)/F(x) \ .$$

Since F_-/F and $\tilde{F}_-/\tilde{F}$ are both strictly increasing and continuous with range $(0, 1)$ this defines a strictly increasing, continuous function t mapping the intervals associated with the two spaces bijectively one onto the other. Differentiating we obtain $t'(x) = (\tilde{F}(t(x))/F(x))^2$ so that t' is continuous. It then follows that t' is locally absolutely continuous and > 0. Setting $s = 1/\sqrt{t'}$ we obtain a Liouville transform unitarily transforming an element $\tilde{u} \in \tilde{\mathcal{H}}_D$ to an element $u \in \mathcal{H}_D$ via

$$u(x) = s(x)\tilde{u}(t(x)) \ .$$

5.7 Defect indices and boundary conditions

We begin with a definition.

Definition 5.7.1. Let W_F be a primitive of $F^2 w$.

Since $w \in H_{\mathrm{loc}}^{-1}(a, b)$ and F, and thus F^2, are in H_{loc}^1, it follows by Theorem C.17 that $F^2 w \in H_{\mathrm{loc}}^{-1}(a, b)$ so that $W_F \in L_{\mathrm{loc}}^2(a, b)$. We have the following existence and uniqueness theorem.

Theorem 5.7.2. *If $f \in H_{\mathrm{loc}}^1(a, b)$ the equation $-u'' + qu = \lambda w u + w f$ has a solution in $H_{\mathrm{loc}}^1(a, b)$, and if $f = 0$ the set of solutions is a linear space of dimension two.*

Proof. The equation $-u'' + qu = \lambda w u + w f$ is easily seen to be equivalent to $-(F^2 \tilde{u}')' = w F^2(\lambda \tilde{u} + \tilde{f})$ and then to the system

$$U' + \begin{pmatrix} \lambda W_F/F^2 & -1/F^2 \\ (\lambda W_F/F)^2 & -\lambda W_F/F^2 \end{pmatrix} U = \begin{pmatrix} -\tilde{f} W_F/F^2 \\ (f' - hf)W_F/F - \lambda(W_F/F)^2 \tilde{f} \end{pmatrix}$$

on setting $U = \begin{pmatrix} \tilde{u} \\ u_1 \end{pmatrix}$ where

$$u_1 = F^2 \tilde{u}' + W_F(\lambda \tilde{u} + \tilde{f}) = F(u' - hu) + (\lambda u + f)W_F/F \ . \tag{5.7.1}$$

The components of the right-hand side are locally integrable as are the elements of the coefficient matrix, so Theorem D.2 applies, a unique solution being determined by the values of $\tilde{u}$ and u_1 at any initial point in (a, b).   □

To determine the defect indices of T_1 first note that $n_\pm \leq 2$ by Theorem 5.7.2 and $n_+ = n_-$ since q and w are real, so $u \in D_\lambda$ if and only if $\overline{u} \in D_{\overline{\lambda}}$. Also note that any function u for which $u' - hu \in L^2(a, b)$ gives finite values to $\tilde{u}$ at both endpoints, as does any element of $\mathcal{H}_1$. Thus the defect indices are 0, 1 or 2 and independent of the choice of the form $Q(\cdot, \cdot)$. Also, both defect indices equal half the rank of the unmodified boundary form

$$\langle u, g \rangle - \langle f, v \rangle = (u_1 \overline{\tilde{g}} - \tilde{f} \overline{v_1})\big|_a^b, \qquad (5.7.2)$$

where (u, f) and $(v, g) \in T_1$ and v_1 is defined similarly to u_1. This formula follows on differentiating $u_1 \overline{\tilde{g}} - \tilde{f} \overline{v_1}$. The form is the difference of the two forms obtained by evaluating $u_1 \overline{\tilde{g}} - \tilde{f} \overline{v_1}$ at b and a, and we next prove that these forms are independent.

Theorem 5.7.3. *Any $(u, f) \in T_1$ may be written as $(u, f) = (u_a, f_a) + (u_b, f_b)$, where $(u_a, f_a), (u_b, f_b) \in T_1$, $(u_a, f_a) = 0$ near a, $(u_a, f_a) = (u, f)$ near b, $(u_b, f_b) = (u, f)$ near a and $(u_b, f_b) = 0$ near b.*

Proof. Let $[c, d] \subset (a, b)$ be such that supp w has at least two points in (c, d). Let ψ be the primitive vanishing on (a, c) of a function in $C_0^\infty(c, d)$ with integral 1 and let $f_0 = \psi f$. Then f_0 equals f in (d, b), vanishes in (a, c), and is in $\mathcal{H}_1$. By Theorem 5.7.2 we may find a solution v of $-v'' + qv = wf_0$ in (a, b).

In (a, c) the function v solves $-v'' + qv = 0$ so subtracting a solution of this equation we may assume $v = 0$ in (a, c). In (d, b) the function $u - v$ solves the same equation, so we may assume it equals $AF_+ - BF_-$ there. Let $\varphi \in C_0^\infty(c, d)$ and let Φ_- and Φ_+ be primitives of $\varphi F_- w$ respectively $\varphi F_+ w$ which vanish in (a, c). $\Phi_\pm$ are constant in (d, b) and we claim that φ may be chosen so that this constant is A for Φ_- and B for Φ_+.

Were no such φ to exist the distributions $F_\pm w$ would be linearly dependent in (c, d), so that we could find $(C, D) \neq (0, 0)$ such that $(CF_+ + DF_-)w$ is zero in (c, d). Since $CF_+ + DF_-$ has at most one zero this would contradict the assumption on supp w. We may therefore find such a φ, and setting $f_a = f_0 + \varphi$ and $u_a = v + F_+ \Phi_- - F_- \Phi_+$ it is easily verified that $-u_a'' + qu_a = wf_a$, $(u_a, f_a) \in T_1$, $(u_a, f_a) = (u, f)$ in (d, b) and $(u_a, f_a) = 0$ in (a, c). Setting $(u_b, f_b) = (u, f) - (u_a, f_a)$ finishes the proof. $\square$

The proof is essentially the same as that for Theorems 4.2.4 and 5.1.6, carried out with some care because of the low coefficient regularity. See also Exercise 5.7.1.

Corollary 5.7.4. *If W_F/F is square integrable near one endpoint the corresponding boundary form $u_1 \overline{\tilde{g}} - \tilde{f} \overline{v_1}$ has rank 2.*

Proof. Theorem 5.7.2 with $\lambda = 0$ implies that at any endpoint in a neighborhood of which W_F/F is square integrable the corresponding form has rank 2. For if a is such an endpoint and $a < c < d < b$ this condition is also satisfied in c and d, and the full boundary forms for (a, c), (c, d) and (a, d) all have rank 4 by the theorem. Thus the forms at a, c and d all have rank 2. Similarly if $W_F/F \in L^2$ near b. $\square$

Since the full boundary form has rank 0, 2 or 4 it follows that the form at any endpoint has rank 0 or 2. As in earlier cases we make the following definition.

Definition 5.7.5. An endpoint is said to be *limit-circle* if the corresponding form has rank 2 and *limit-point* if the rank is 0.

The following lemma is essentially a converse of Corollary 5.7.4.

Lemma 5.7.6. *Suppose* $-u'' + qu = wf$ *and that* $u' - hu$ *and* $f' - hf$ *are in* L^2 *near an endpoint where* $\tilde{f}$ *has a non-zero limit. Then* W_F/F *is square integrable near the endpoint in question.*

Proof. The equation $-u'' + qu = wf$ may also be written $-(u' - hu)' - h(u' - hu) = wf$. Multiplying by $F/\tilde{f}$ and rearranging gives

$$(F(u' - hu)/\tilde{f})' + (1/\tilde{f})^2(u' - hu)(f' - hf) + \lambda wF^2 = 0 \,.$$

The middle term is integrable near an endpoint near which $u' - hu$ and $f' - hf$ are in L^2 if $\tilde{f}$ has a non-zero limit there. Integrating we obtain

$$(u' - hu)/\tilde{f} + \frac{1}{F} \int (1/\tilde{f})^2(u' - hu)(f' - hf) + W_F/F = 0 \,,$$

the integral sign denoting a primitive. Here the two first terms are in L^2 near the endpoint, so the lemma follows.   □

We can now calculate the defect indices.

Theorem 5.7.7. *Defect indices for our equation in* $\mathcal{H}_1$ *are always equal, and have the following values.*

(1) *If* $W_F/F \in L^2(a, b)$ *then* $n_+ = n_- = 2$.

(2) *If* W_F/F *is in* L^2 *near* a *or* b *but not in* $L^2(a, b)$, *then* $n_+ = n_- = 1$.

(3) *If* W_F/F *is not in* L^2 *near either of the endpoints, then* $n_+ = n_- = 0$.

Proof. By Corollary 5.7.4 it is clear that the deficiency indices have at least the values claimed. Now suppose $c \in (a, b)$ with supp $w \cap (c, b) \neq \emptyset$. Let u be the solution of $-u'' + qu = \lambda u$ with $\lambda \neq 0$ for which $\tilde{u}(c) = 0$ and $u_1(c) = 1$. Then (5.7.2) with $u = v$ gives

$$\operatorname{Im} \lambda \int_c^b |u' - hu|^2 = \lim_b \operatorname{Im}(\lambda \tilde{u}\overline{u_1}) = \lim_b \operatorname{Im}(\lambda \tilde{u} F(\overline{u' - hu})) \,.$$

If $\operatorname{Im} \lambda > 0$ the right-hand side is therefore > 0 (possibly $+\infty$) so that the function $1/(\tilde{u}F(\overline{u' - hu}))$ is bounded near b. If the integral is finite it follows that the function $|u' - hu|^2 /(\tilde{u}F(\overline{u' - hu})) = \tilde{u}'/\tilde{u}$ is integrable near b so that $\tilde{u}$ has a non-zero limit at b. But Lemma 5.7.6 implies that then W_F/F is square integrable near b. Thus the integral cannot be finite and the full boundary form has rank < 4 so that the form at b has rank zero. It follows that the endpoint form has rank 0 if and only if W_F/F is not in L^2 near the endpoint. The theorem follows.   □

Boundary conditions. We next find explicit descriptions of all possible self-adjoint restrictions T_1 in terms of boundary conditions.

Suppose (u, f) and $(v, g) \in T_1$. By (5.7.2) we then have

$$\langle u, g \rangle_Q - \langle f, v \rangle_Q = \left(u_1 \overline{\tilde{g}} - \tilde{f} \, \overline{v_1} \right) \Big|_a^b + Q(u, g) - Q(f, v) = \left(u_2 \overline{\tilde{g}} - \tilde{f} \, \overline{v_2} \right) \Big|_a^b,$$

where $u_2 = u_1 - \alpha \tilde{u}(a) - \overline{\gamma} \tilde{u}(b)$ near a and $u_2 = u_1 + \beta \tilde{u}(b) + \gamma \tilde{u}(a)$ near b. Define v_2 similarly. Consider first the case when both endpoints are limit-circle, so that u_2 and v_2 are well defined at a and b. To obtain a self-adjoint restriction of T_1 we then have to impose two linear, homogeneous conditions on T_1 in such a way that the right-hand side vanishes. Since the right-hand side has a similar structure as in the standard left-definite case we already know the relevant conditions. They are either of the form

$$\tilde{f}(b) \cos \delta_b + u_2(b) \sin \delta_b = 0, \tag{5.7.3}$$

$$\tilde{f}(a) \cos \delta_a + u_2(a) \sin \delta_a = 0, \tag{5.7.4}$$

where δ_b and δ_a are fixed numbers in $[0, \pi)$, or else of the form

$$\begin{pmatrix} \tilde{f}(b) \\ u_2(b) \end{pmatrix} = S \begin{pmatrix} \tilde{f}(a) \\ u_2(a) \end{pmatrix},$$

where S is a symplectic matrix. However, one should note that if $\gamma \neq 0$ the first type of conditions cannot be called separated, since the value of $\tilde{u}(b)$ is then part of $u_2(a)$ and that of $\tilde{u}(a)$ is part of $u_2(b)$.

If a is limit-circle and b limit-point the form $u_1 \overline{\tilde{g}} - \tilde{f} \, \overline{v_1}$ vanishes at b, and since $\tilde{f}(b) = \tilde{g}(b) = 0$ by Lemma 5.7.6 the full boundary form is $\tilde{f}(a)\overline{v_2(a)} - u_2(a)\overline{\tilde{g}(a)}$. Conditions which restrict T_1 to a self-adjoint relation are then of the form (5.7.4). If instead a is limit-point and b limit-circle the conditions are of the form (5.7.3). Finally, if a and b are both limit-point the full boundary form vanishes, no boundary conditions are required, T_0 is essentially self-adjoint and T_1 self-adjoint.

Remark 5.7.8. If an endpoint is norm-regular, then the condition $W_F/F \in L^2$ for limit-circle may be simplified to $W \in L^2$, where W is a primitive of w, and one may prove that in this case $u' - hu + Wf$ has a limit at the endpoint. See Exercise 5.7.3. So, if $W \in L^2$ near an endpoint which is norm-regular the endpoint is limit-circle and the boundary conditions may be expressed in terms of the limits of f and $u' - hu + Wf$ at the endpoint.

It may therefore seem reasonable to call an endpoint *regular* if it is norm-regular and in addition $W \in L^2$ near the endpoint. It should be noted, however, that if q and w are locally integrable, the usual definition of a regular endpoint is that it is finite and q and w integrable in a neighborhood of the endpoint. These are stronger conditions than what is implied by our definition. Even if we assume q and w are locally integrable, our condition on w is weaker.

Exercises

Exercise 5.7.1 Referring to the last paragraph of the proof of Theorem 5.7.3, verify that $F_0 = F_+\Phi_- - F_-\Phi_+$ satisfies $-F_0'' + qF_0 = w\varphi$.

Exercise 5.7.2 Considering an equation $-(pu')' + qu = wf$ which for $f = 0$ has a positive solution, show that the left-definite theory presented earlier extends to this equation. In particular, find the theorem corresponding to Theorem 5.7.7 for such an equation.

Exercise 5.7.3 Prove the claims of Remark 5.7.8.

Hint: Find the first order system satisfied by $U = \left(\begin{smallmatrix} u \\ u'-hu+Wf \end{smallmatrix}\right)$.

5.8 Eigenfunction expansions

If the spectrum is discrete then any element of $\mathcal{H}$ has a series expansion in eigenfunctions converging locally uniformly, so we start by giving a criterion for compact resolvent analogous to Theorem 5.2.1.

Theorem 5.8.1. *Suppose that $W_F/F \in L^2(a,b)$. Then all self-adjoint extensions of T_0 have compact resolvent.*

By Theorem 5.7.7 the assumption $W_F/F \in L^2(a,b)$ is equivalent to both endpoints of (a,b) being in the limit-circle case, so limit-circle in both endpoints gives a compact resolvent also in the present situation.

Proof. We have

$$w(u\bar{v}) = -\int_a^b W_F(\tilde{u}'\bar{\tilde{v}} + \tilde{u}\bar{\tilde{v}}') .$$

As in the proof of Theorem 5.2.1 this together with our assumption may be used to prove that the form $\langle u, G_0(wv)\rangle$, defined if $u, v \in \mathcal{H}_1$ with supp v compact, extends to a bounded Hermitian form $B(u,v)$ on $\mathcal{H}_1$, yielding a bounded symmetric operator R_0 on $\mathcal{H}_1$ such that $B(u,v) = \langle u, R_0v\rangle$. Using that the linear forms $\mathcal{H}_1 \ni u \mapsto \tilde{u}(x)$ are uniformly bounded for $x \in (a,b)$ shows that R_0 is compact and the resolvent at 0 of a self-adjoint extension of T_0, the proof being very similar to that of Theorem 5.2.1; see Exercise 5.8.1.   □

We have the following result on pointwise convergence when the resolvent is compact.

Theorem 5.8.2. *Suppose T is a self-adjoint extension of T_0 which has an orthonormal basis of eigenfunctions.*

If $u \in \mathcal{H}$ and s_Nu are the partial sums of the generalized Fourier series, then $(s_Nu)/F$ converges uniformly in (a,b) to $\tilde{u}$. Near a norm-regular endpoint s_Nu converges uniformly.

The proof is very similar to that of Theorem 5.2.2; see Exercise 5.8.2. Moving on to cases where the spectrum of T may not be discrete, we need a detailed description of the resolvent R_λ. As in Section 5.2 we have $R_\lambda u(x) = \langle R_\lambda u, g_Q(x, \cdot)\rangle_Q = \langle u, R_{\overline{\lambda}} g_Q(x, \cdot)\rangle_Q$, where we may view $G(x, \cdot, \lambda) = \overline{R_{\overline{\lambda}} g_Q(x, \cdot)}$ as Green's function for our relation. As before it is convenient to introduce the kernel $g(x, y, \lambda) = G(x, y, \lambda) + g_Q(x, y)/\lambda$, so that we obtain

$$R_\lambda u(x) = \langle u, \overline{g(x, \cdot, \lambda)}\rangle_Q - u(x)/\lambda \,. \tag{5.8.1}$$

Note that $G(x, \cdot, \lambda) \in \mathcal{H}$ but this is not true of $g(x, \cdot, \lambda)$ if $x \notin \operatorname{supp} w$.

Proposition 5.8.3. *For all $x, y \in (a, b)$ we have both $\overline{G(x, y, \lambda)} = G(y, x, \overline{\lambda})$ and $g(x, y, \lambda) = g(y, x, \overline{\lambda})$.*

Proof. We have $g_Q(x, y) = g_Q(y, x)$ so the second claim follows from the first, and $\overline{G(x, y, \lambda)} = \overline{R_{\overline{\lambda}} g_Q(x, \cdot)(y)} = \overline{\langle R_{\overline{\lambda}} g_Q(x, \cdot), g_Q(y, \cdot)\rangle_Q} = \langle g_Q(x, \cdot), R_\lambda g_Q(y, \cdot)\rangle_Q = R_\lambda g_Q(y, \cdot)(x) = G(y, x, \overline{\lambda})$. $\quad\square$

Now let $c \in (a, b)$ and $\Phi(x, \lambda) = (\varphi(x, \lambda), \theta(x, \lambda))$ where

$$\begin{cases} \varphi(c, \lambda) = 0 \,, \\ (\varphi' - (h - \lambda W)\varphi)(c, \lambda) = 1 \,, \end{cases} \qquad \begin{cases} \lambda\theta(c, \lambda) = 1 \,, \\ (\theta' - (h - \lambda W)\theta)(c, \lambda) = 0 \,, \end{cases}$$

so that $\varphi(\cdot, \lambda), \theta(\cdot, \lambda)$ is a basis of the solutions for $-u'' + qu = \lambda wu$ for each $\lambda \neq 0$. We may view $\varphi' - (h - \lambda W)\varphi$ as a quasi-derivative of φ, and similarly for θ. We have the following description of Green's function.

Theorem 5.8.4. *Suppose $g(x, y, \lambda)$ is Green's function for a self-adjoint extension of T_0. Then there exists a unique 2×2 matrix-valued function $M(\lambda)$ defined for $0 \neq \lambda \in \rho(T)$ such that $M(\overline{\lambda}) = M^*(\lambda)$ and*

$$g(x, y, \lambda) = \Phi(y, \lambda)(M(\lambda) \pm \tfrac{1}{2}J)\Phi^*(x, \overline{\lambda}) \,,$$

where $J = \left(\begin{smallmatrix} 0 & -1 \\ 1 & 0 \end{smallmatrix}\right)$ and the sign in front of $\tfrac{1}{2}J$ agrees with that of $x - y$.

Proof. Suppose $\psi \in C_0^\infty(a, b)$. Since $(G_0(w\psi), \psi) \in T_0$ it follows that $G_0(w\psi) = R_\lambda(\psi - \lambda G_0(w\psi))$, which shows that $G_0(w\psi)(x) = \langle \psi - \lambda G_0(w\psi), \overline{g(x, \cdot, \lambda)}\rangle_Q - (\psi(x) - \lambda G_0(w\psi)(x))/\lambda$ or

$$\psi(x)/\lambda = \langle \psi, \overline{g(x, \cdot, \lambda)}\rangle - \lambda w(\psi g(x, \cdot, \lambda)) \,.$$

Thus $g(x, \cdot, \lambda)$ satisfies $-u'' + qu = \lambda wu + \delta_x/\lambda$, where δ_x is the Dirac measure at x. For $x < y$ it follows that $g(x, y, \lambda) = \Phi(y, \lambda)A(x, \lambda)$ for some 2×1 matrix-valued function A, and since $g(x, y, \lambda) = g(y, x, \overline{\lambda})$ that $A^*(y, \overline{\lambda}) = \Phi(y, \lambda)K(\lambda)$, where K is a 2×2 matrix depending only on λ. Thus

$$g(x, y, \lambda) = \Phi(y, \lambda)K^*(\overline{\lambda})\Phi^*(x, \overline{\lambda}) \,, \qquad x < y \,.$$

Again using $g(x, y, \lambda) = \overline{g(y, x, \overline{\lambda})}$ we obtain

$$g(x, y, \lambda) = \Phi(y, \lambda) K(\lambda) \Phi^*(x, \overline{\lambda}), \qquad x > y.$$

Setting $J = K(\lambda) - K^*(\overline{\lambda})$ the equation $g(x, \cdot, \lambda)$ satisfies implies that as $y \to x$

$$\Phi(x, \lambda) J \Phi^*(x, \overline{\lambda}) = 0,$$

$$(\Phi' + (\lambda W - h)\Phi)(x, \lambda) J \Phi^*(x, \overline{\lambda}) = -1/\lambda.$$

Differentiating these equations, using the differential equation for Φ in the second case, shows that

$$\Phi(x, \lambda) J (\Phi' + (\overline{\lambda} W - h)\Phi)^*(x, \overline{\lambda}) = 1/\lambda,$$

$$(\Phi' + (\lambda W - h)\Phi)(x, \lambda) J (\Phi' + (\overline{\lambda} W - h)\Phi)^*(x, \overline{\lambda}) = 0.$$

Evaluating the last four equations at $x = c$ shows that $J = \begin{pmatrix} 0 & -1 \\ 1 & 0 \end{pmatrix}$. Now define $M(\lambda) = \frac{1}{2}(K(\lambda) + K^*(\overline{\lambda}))$. Then clearly $M^*(\overline{\lambda}) = M(\lambda)$ and we obtain the desired formula for $g(x, y, \lambda)$. $\qquad\square$

Theorem 5.8.5. *The function M is analytic in $\rho(T) \setminus \{0\}$ and is a 2×2 matrix-valued Nevanlinna function.*

Proof. Since $g(x, y, \lambda) = G(x, y, \lambda) + g_Q(x, y)/\lambda$ and

$$G(x, y, \lambda) = \langle R_{\overline{\lambda}} g_Q(x, \cdot), g_Q(y, \cdot) \rangle_Q = \langle R_\lambda \overline{g_Q(y, \cdot)}, g_Q(x, \cdot) \rangle_Q$$

it is clear that $\lambda \mapsto g(x, y, \lambda)$ is analytic if $0 \ne \lambda \in \rho(T)$ for each x and y. Now, the components of $\Phi(x, \lambda)$ are linearly independent functions of x for all λ, and locally uniformly in x entire functions of λ, so it follows that the function K of the previous theorem, and thus M, is analytic wherever $g(x, y, \lambda)$ is.

The definitions of $G(x, y, \lambda)$ and $g(x, y, \lambda)$ and the resolvent relation $R_\lambda - R_\mu = (\lambda - \mu) R_\mu R_\lambda$ for $\mu = \overline{\lambda}$ easily show that

$$\Phi(y, \lambda)(M(\lambda) - M^*(\lambda)) \Phi^*(x, \overline{\lambda}) = (\lambda - \overline{\lambda}) \Big(\langle G(x, \cdot, \lambda), G(y, \cdot, \lambda) \rangle_Q + \frac{g_Q(x, y)}{|\lambda|^2} \Big)$$

for all x and y. By the linear independence of the components of $\Phi(\cdot, \lambda)$ it then follows easily that $(M(\lambda) - M^*(\lambda))/(\lambda - \overline{\lambda})$ is the matrix of a positive quadratic form, so that M is a matrix-valued Nevanlinna function. $\qquad\square$

Note that if a is a norm-regular limit-circle endpoint we may replace c by a in the proof above; similarly if b is norm-regular and limit-circle. If a is just limit-circle we can still replace c by a if we replace the initial data for φ, θ by

$$\begin{cases} \lambda \tilde{\varphi}(a, \lambda) = A, \\ (F^2 \tilde{\varphi}' + \lambda W_F \tilde{\varphi})(a, \lambda) = B, \end{cases} \qquad \begin{cases} \lambda \tilde{\theta}(a, \lambda) = C, \\ (F^2 \tilde{\theta}' + \lambda W_F \tilde{\theta})(a, \lambda) = D, \end{cases}$$

where (A, B) and (C, D) are linearly independent real vectors. Similarly if b is limit-circle. Finally, if the coefficient $\gamma = 0$ in the form Q used to define $\langle \cdot, \cdot \rangle_Q$ and a is limit-circle, and we use the boundary condition (5.7.4) at a, with the boundary condition (5.7.3) at b if b is limit-circle, we may choose $A = -\sin \delta_a$, $B = \cos \delta_a - \alpha \sin \delta_a / \lambda$, $C = \cos \delta_a$ and $D = \alpha \sin \delta_a / \lambda$. It is then easily seen that $M(\lambda) = \begin{pmatrix} m(\lambda) & 1/2 \\ 1/2 & 0 \end{pmatrix}$ where m is a scalar Nevanlinna function. Setting $\psi(\cdot, \lambda) = \theta(\cdot, \lambda) + m(\lambda)\varphi(\cdot, \lambda)$ the formula for Green's function simplifies to

$$g(x, y, \lambda) = \varphi(\min(x, y), \lambda)\psi(\max(x, y), \lambda) .$$

Definition 5.8.6. Let $\mathcal{H}_0 = \mathcal{H}_1 \ominus D_0$ where D_0 is the eigenspace of T associated with $\lambda = 0$, and let $\mathcal{H} = \mathcal{H}_0 \ominus \mathcal{H}_\infty$.

Furthermore, let $\mathcal{H}_\Xi$ be the elements of L_Ξ^2 which vanish at 0.

If $u \in \mathcal{H}_0$ has compact support we define the Fourier transform $\hat{u}(t) = \mathcal{F}u(t) = \langle u, \overline{\Phi(\cdot, t)} \rangle_Q$ with components $\hat{u}_\varphi(t) = \langle u, \varphi(\cdot, t) \rangle_Q$ and $\hat{u}_\theta(t) = \langle u, \overline{\theta(\cdot, t)} \rangle_Q$. We then obtain the following expansion theorem.

Theorem 5.8.7. *Suppose T is a self-adjoint extension of T_0 and that $M(\lambda)$ is the corresponding 2×2 m-function, with 2×2 spectral measure $d\Xi$. Then:*

(1) *The map $u \mapsto \mathcal{F}u$ defined for compactly supported elements of $\mathcal{H}_1$ extends to a map $\mathcal{F} : \mathcal{H}_0 \to \mathcal{H}_\Xi$ called the generalized Fourier transform for T.*

(2) *The restriction of $\mathcal{F}$ to $\mathcal{H}$ is unitary and the restriction to $\mathcal{H}_\infty$ vanishes, so that the Parseval formula $\langle u, v \rangle_Q = \langle \hat{u}, \hat{v} \rangle_\Xi$ is valid if at least one of u, v is in $\mathcal{H}$.*

(3) *If $(u, f) \in T$, then $\mathcal{F}(f)(t) = t\hat{u}(t)$. Conversely, if $\hat{u}$ and $t\hat{u}(t)$ are in L_Ξ^2, then $\hat{u} = \mathcal{F}u$ for some $u \in \mathcal{D}(T)$.*

(4) *If $\hat{u} \in \mathcal{H}_H$ the integral $\int_K \Phi^*(x, \cdot) d\Xi \hat{u}$ converges in $\mathcal{H}$ as $K \to \mathbb{R}$ through compact intervals. If $\hat{u} = \mathcal{F}(u)$ with $u \in \mathcal{H}$ the limit is u, so this defines the inverse, or adjoint, of $\mathcal{F} : \mathcal{H} \to \mathcal{H}_\Xi$.*

(5) *Let E_Δ denote the spectral projector of T for the interval (or Borel set) Δ. For $u \in \mathcal{H}$ then $E_\Delta u(x) = \int_\Delta \Phi^*(x, \cdot) d\Xi \hat{u}$.*

The proof is very similar to that of Theorem 5.3.3 and is left to the reader (Exercise 5.8.4).

Exercises

Exercise 5.8.1 Carry out the details of the proof of Theorem 5.8.1.

Exercise 5.8.2 Prove Theorem 5.8.2.

Exercise 5.8.3 Prove all statements made just after the proof of Theorem 5.8.4.

Exercise 5.8.4 Prove Theorem 5.8.7.

5.9 Notes and remarks

Since little can be found on left-definite differential equations in book form we shall try to give a snapshot of how the theory evolved. In 1906 Hilbert [102] created a theory of *polar* integral equations and applied it to left-definite (regular) Sturm–Liouville equations. The scalar product used is not a Dirichlet integral but in a sense a scalar product 'dual' to this. The first to use a Dirichlet integral as scalar product seems to have been Mason [163], who in 1907 proved an expansion theorem for a regular left-definite Sturm–Liouville equation. Singular left-definite equations were briefly treated by Weyl [223] in a second note following his fundamental paper Weyl [222] in 1910. Weyl used the same scalar product as Hilbert.

After these early efforts little seems to have happened until, in 1939, Kamke [118] gave an expansion theorem, using a Dirichlet integral as scalar product, for higher-order regular left-definite equations of the form $Su = \lambda Tu$, where S and T are formally symmetric differential expressions with the order of S higher than that of T. Although some left-definite elliptic partial differential equations were treated by Pleijel in the 1940s, mainly with a view of obtaining asymptotics of the spectral projectors, not many results for ordinary differential equations seem to have appeared until in the mid 1960s Schäfke and Schneider [197], [198], [199] in three papers developed a very general spectral theory for regular first order systems, incorporating both left- and right-definite equations.

Singular left-definite problems of a slightly more restricted type were treated by Schneider and Niessen [200], [201] in the mid 1970s. In a series of papers from 1969 onwards Pleijel [181, 182] treated the spectral theory of left- and right-definite, mostly singular, scalar equations of a form similar to the equations treated by Kamke. Expansion theorems for these equations are given in Bennewitz [15], and in [49] Daho and Langer gave an expansion theorem for a Sturm–Liouville equation which is not quite left-definite, owing to the boundary conditions imposed. After that one finds scattered results until the discovery of the Camassa–Holm equation created new interest in left-definite Sturm–Liouville equations, see for example Bennewitz [18]. Later developments centered primarily on inverse spectral and scattering theory. We defer a discussion of this to Chapter 7.

A different approach to left-definite problems was taken by Gesztesy and Weikard in [89]. Their approach relies on rewriting the eigenvalue problem $Su = \lambda Tu$ in the form $S^{-1/2}TS^{-1/2}v = \lambda^{-1}v$ with $v = S^{1/2}u$ when S is a positive operator.

Chapter 6
Oscillation, spectral asymptotics and special functions

The work of Sturm and Liouville on expansion in eigenfunctions is heavily dependent on the study of the zeros of solutions of the equation, nowadays called oscillation theory. Although we have chosen a different approach, the results of Sturm are still of considerable interest and in Section 6.1 we will prove modern versions of his results and then use them to study the *counting function* $N(\lambda)$, which is the number of eigenvalues less than λ. If the spectrum is not bounded from below one may instead let $N(\lambda)$ denote the number of eigenvalues between 0 and λ.

The expression *spectral asymptotics* refers to the behavior of any quantity depending on the spectral parameter as this becomes large. The primary example may be the asymptotic distribution of eigenvalues, given by the behavior of the counting function at infinity. Other relevant quantities are the spectral function, Green's function, the solutions $\varphi(x, \lambda)$ and $\psi(x, \lambda)$ for large λ, the kernels of the spectral projectors viewed as integral operators and, last but not least, the Weyl–Titchmarsh m-function.

In this chapter we shall deal exclusively with one-term asymptotics, but there are many results in the literature involving several terms or even an asymptotic series. However, in applications often the first term in such an expansion is the crucial one. We shall touch upon all the quantities mentioned, but in Sections 6.2–6.5 all our results are derived from the asymptotic behavior of the m-function for large $|\lambda|$, which will be studied in Sections 6.2 and 6.3.

In dealing with asymptotics we shall often use the following standard notation. If f and g are complex-valued function of a real variable defined in a (punctured) neighborhood of a, $-\infty \le a \le +\infty$, we write $f \sim g$ as $x \to a$ if $f(x)/g(x) \to 1$ as $x \to a$. One-sided limits are also possible. If f is positive we write $g(x) = O(f(x))$ as $x \to a$ if g/f is bounded in a neighborhood of a. Similarly we write $g(x) = o(f(x))$ as $x \to a$ if $g(x)/f(x) \to 0$ as $x \to a$.

The last section of the chapter is of a different character than the rest. In applications of Sturm–Liouville theory one deals with specific equations, and only a small number of different equations appear in the majority of such applications. Appropriately normalized solutions of these equations are nowadays called *special functions*. Naturally they are mostly the concern of engineers, physicists and applied

C. Bennewitz et al., *Spectral and Scattering Theory for Ordinary Differential Equations*, Universitext, https://doi.org/10.1007/978-3-030-59088-8_6

mathematicians, but we must say something about them. Since there is no hope of a comprehensive treatment we have chosen to focus on one of the most famous applications, namely the analysis of the hydrogen atom given by Erwin Schrödinger [202] in 1926.

6.1 The zeros of solutions

To study the zeros of solutions we will use a technique pioneered by Heinz Prüfer in 1926 [184]. Prüfer introduced polar coordinates in the phase plane and the change of coordinates is often called a *Prüfer transform*. Many variants exist, and we will use a few, adapted to the problems at hand. We consider only real-valued solutions, and begin with the equation

$$(pu')' + gu = 0$$

for real-valued, locally integrable $1/p$ and g. The basic Prüfer transform consists in introducing two new functions $r \geq 0$ and ϕ by setting

$$\begin{cases} u = r \sin \phi \,, \\ pu' = r \cos \phi \,. \end{cases} \tag{6.1.1}$$

Since u and pu' can only vanish simultaneously if u is identically 0, $r > 0$ in all other cases. Thus a non-trivial u has a zero only where ϕ is an integer multiple of π and pu' only where ϕ is an odd multiple of $\pi/2$. Differentiating the equations, use of the original equation gives the system

$$\begin{cases} r' \sin \phi + r\phi' \cos \phi = \frac{1}{p} r \cos \phi \,, \\ r' \cos \phi - r\phi' \sin \phi = -gr \sin \phi \,. \end{cases}$$

Viewing this as a linear system of equations in r' and ϕ' we obtain the equivalent system

$$\begin{cases} \phi' = \frac{1}{p} \cos^2 \phi + g \sin^2 \phi \,, \\ r' = r(1/p - g) \sin \phi \cos \phi \,. \end{cases} \tag{6.1.2}$$

The special features of this system are that the first equation of (6.1.2) concerns only the unknown angular variable ϕ, whereas the second equation is linear in the radial variable r. Thus

$$r(x) = r(a) \exp\left(\int_a^x (1/p - g) \sin \phi \cos \phi \right),$$

which can be useful when some knowledge of ϕ has been obtained from the first equation, and sometimes even without this since at least $|\sin \phi \cos \phi| \leq 1/2$ is always true.

However, the most important conclusions can be drawn from the first equation. Note that with $F(x, y) = \frac{1}{p(x)} \cos^2 y + g(x) \sin^2 y$ we have $|F(x, y_1) - F(x, y_2)| \leq |g(x) - 1/p(x)| |y_1 - y_2|$ so by Theorem D.2 the initial value problem for the angular equation has a unique, locally absolutely continuous solution in any interval where $1/p$ and g are locally integrable. We shall prove the following important theorem.

Theorem 6.1.1 (Sturm comparison theorem). *Suppose $1/p_j$ and g_j, $j = 1, 2$, are integrable in (a, b), that $(p_j u_j')' + g_j u_j = 0$, that $1/p_1 \geq 1/p_2 > 0$ and $g_1 \geq g_2$ a.e. Also assume that $u_2(b) = 0$ and either $u_2(a) = 0$ or that $p_1 u_1'(a)/u_1(a) \leq p_2 u_2'(a)/u_2(a)$.*

Then the function u_1 has a zero in (a, b) unless $p_1 = p_2$ a.e., $g_1 = g_2$ a.e. and $u_1 = u_2$.

For spectral theory our next theorem is particularly important.

Theorem 6.1.2 (Sturm oscillation theorem). *Suppose p and w are a.e. non-negative, that $1/p$, q and w are in $L^1(a, b)$ and that w is not a.e. zero. Then the equation $-(pu')' + qu = \lambda w u$ provided with separated, self-adjoint boundary conditions has a least eigenvalue and the eigenvalues are isolated.*

Numbering the eigenvalues $\lambda_0 < \lambda_1 < \lambda_2 < \ldots$ and assuming the eigenfunction to λ_0 has k zeros in (a, b), the eigenfunction to λ_n has precisely $k + n$ zeros in (a, b). If $\operatorname{supp} w = [a, b]$, then $k = 0$.

For coupled boundary conditions the result is somewhat less precise. Note that eigenfunctions associated with boundary conditions defined by a non-real symplectic matrix are not real-valued and never have zeros. The reason is that the imaginary part of a non-real symplectic matrix is invertible, so the eigenfunctions cannot be real-valued, but if an eigenfunction has a zero its real and imaginary parts have a common zero and both solve the same equation. They are therefore proportional, which implies that there is a real-valued eigenfunction.

Corollary 6.1.3. *Under the assumptions of Theorem 6.1.2 but with coupled boundary conditions given by a real symplectic matrix, and assuming $\operatorname{supp} w = [a, b]$, the real-valued eigenfunctions associated with λ_n have at most $n + 1$ and at least $n - 1$ zeros in (a, b).*

If $\lambda_n = \lambda_{n+1}$ a real-valued eigenfunction has n or $n + 1$ zeros in (a, b), and there are eigenfunctions with n zeros as well as those with $n + 1$ zeros.

We leave the proof to the reader in Exercise 6.1.2. In order to prove the theorems we need two elementary lemmas.

Lemma 6.1.4. *Suppose $\phi_j'(x) = F_j(x, \phi_j(x))$, $j = 1, 2$, where $x \mapsto F_j(x, y)$ is integrable in (a, b) for each y and $|F_1(x, y_1) - F_1(x, y_2)| \leq K(x) |y_1 - y_2|$ with K integrable in (a, b). Also assume $F_1(x, \phi_2(x)) \geq F_2(x, \phi_2(x))$ for $x \in (a, b)$.*

If $x_0 \in (a, b)$ and $\phi_1(x_0) \geq \phi_2(x_0)$, then $\phi_1 \geq \phi_2$ in (x_0, b). Also, if $\phi_1(x) = \phi_2(x)$ for some $x \in (x_0, b)$, then $\phi_1(x_0) = \phi_2(x_0)$ and $F_2(t, \phi_2(t)) = F_1(t, \phi_2(t))$ for almost all $t \in (x_0, x)$. In this case $\phi_1 = \phi_2$ in (x_0, x).

Similarly, if instead $\phi_1(x_0) \leq \phi_2(x_0)$, then $\phi_1 \leq \phi_2$ in (a, x_0). Also, if $\phi_1(x) = \phi_2(x)$ for some $x \in (a, x_0)$ then $\phi_1(x_0) = \phi_2(x_0)$ and $F_2(t, \phi_2(t)) = F_1(t, \phi_2(t))$ for almost all $t \in (x, x_0)$. In this case $\phi_1 = \phi_2$ in (x, x_0).

Proof. Put $f = \phi_1 - \phi_2$. Then $f' = gf + h$ where

$$g(x) = (F_1(x, \phi_1(x)) - F_1(x, \phi_2(x)))/(\phi_1(x) - \phi_2(x))$$

if $\phi_1(x) \neq \phi_2(x)$ and $g(x) = 0$ otherwise, and $h(x) = F_1(x, \phi_2(x)) - F_2(x, \phi_2(x)) \geq 0$ and integrable by Lemma D.1. Since g is measurable and $|g| \leq K$ it follows that g is integrable. All claims now follow from the fact that

$$f(x) = f(x_0) \exp\left(\int_{x_0}^{x} g\right) + \int_{x_0}^{x} h(t) \exp\left(\int_{t}^{x} g\right) dt .$$

$\square$

Lemma 6.1.5. *Suppose $\phi'(x) = F(x, \phi(x))$ and $x \mapsto F(x, \phi(x_0))$ is integrable and non-negative in a neighborhood of x_0. Suppose $|F(x, y_1) - F(x, y_2)| \leq K(x) |y_1 - y_2|$ for y_1, y_2 near $\phi(x_0)$, with K integrable near x_0. Then ϕ is increasing in x_0, and strictly increasing if there is a neighborhood of x_0 where $F(x, \phi(x_0))$ is positive almost everywhere.*

Proof. Setting $F_1 = F$, $F_2 \equiv 0$, $\phi_1 = \phi$ and $\phi_2 \equiv \phi(x_0)$ this is an immediate consequence of Lemma 6.1.4. $\square$

Corollary 6.1.6. *Suppose u is a non-trivial real solution of $(pu')' + gu = 0$ in (a, b) where g is locally integrable while $1/p$ and $g_+ = \max(0, g)$ are integrable and $p \geq 0$. Then there are only finitely many zeros of u in (a, b).*

Proof. Using the Prüfer transform (6.1.1) the Prüfer angle satisfies $\phi' = \frac{1}{p} \cos^2 \phi + g \sin^2 \phi$ and $\sin \phi(x_0) = 0$ if $u(x_0) = 0$. Thus Lemma 6.1.5 shows that ϕ is strictly increasing at x_0. There is therefore no zero $x > x_0$ of u until $\phi(x) - \phi(x_0) = \pi$. Since

$$\phi(x) - \phi(x_0) \leq \int_{x_0}^{x} \max(1/p, g_+)$$

the number of zeros in (a, b) does not exceed $1 + \frac{1}{\pi} \int_a^b \max(1/p, g_+)$. $\square$

The following result is an immediate consequence of Corollary 6.1.6 and the Sturm comparison theorem.

Corollary 6.1.7 (Sturm separation theorem). *Suppose $1/p$ and g are real-valued and locally integrable and $p \geq 0$, and consider the equation $(pu')' + gu = 0$. Then the zeros of linearly independent real-valued solutions interlace, i.e., the zeros are isolated and between consecutive zeros of one solution there is precisely one zero of the other.*

Proof (Theorem 6.1.1). Using the Prüfer transform (6.1.1) and the angular equation from (6.1.2) we set $F_j(x, y) = \frac{1}{p_j(x)} \cos^2 y + g_j(x) \sin^2 y$. It is no restriction to assume that $u_j(a) \geq 0$, $j = 1, 2$, since changing the sign of a solution does not change the location of its zeros. We may thus choose $\phi_j(a) \in [0, \pi)$. By Corollary 6.1.6 we may also assume that b is the first zero $> a$ of u_2. As in the proof of this corollary Lemma 6.1.5 shows that ϕ_2 can only reach an integer multiple of π from below, so that $\phi_2(b) = \pi$.

The assumption about the initial values of u_1 and u_2 shows that $\phi_2(a) \leq \phi_1(a)$. Now $F_1(x, \phi) - F_2(x, \phi) \geq 0$ so that Lemma 6.1.4 shows that $\phi_2(b) < \phi_1(b)$ unless $p_1 = p_2$ and $g_1 = g_2$ a.e. in (a, b) and $\phi_1(a) = \phi_2(a)$. Except in this case $\phi_1(x) = \pi$ for some $x \in (a, b)$, so that $u_1(x) = 0$. This completes the proof. □

Proof (Theorem 6.1.2). Let $u(x, \lambda)$ solve $-(pu')' + qu = \lambda wu$ with fixed initial data at a satisfying the relevant boundary condition, and let $\phi(x, \lambda)$ be the corresponding Prüfer angle satisfying $\phi' = \frac{1}{p} \cos^2 \phi + (\lambda w - q) \sin^2 \phi$ with $\phi(a, \lambda) = \alpha$ fixed in $[0, \pi)$ and determined by the boundary condition at a. Similarly, there is a constant $\beta \in (0, \pi]$ determined by the boundary condition at b such that λ is an eigenvalue precisely if $\phi(b, \lambda) - \beta$ is an integer multiple of π.

$\mathbb{R} \ni \lambda \mapsto \phi(x, \lambda)$ is continuous for fixed $x \in (a, b)$ by Theorem D.2 and increasing by Theorem 6.1.1, strictly increasing unless $w = 0$ a.e. in (a, x). Lemma 6.1.5 implies, as in the proof of Corollary 6.1.6, that $x \mapsto \phi(x, \lambda)$ can only reach an integer multiple of π from below. It follows that if $\lambda < \mu$ are consecutive eigenvalues, the eigenfunction associated with μ has precisely one more zero in (a, b) than the eigenfunction associated with λ. This also shows that there is[1] a least eigenvalue λ_0 (see also Exercise 4.5.1). If the eigenfunction associated with λ_0 has k zeros in (a, b) it follows by induction that the eigenfunction associated with λ_n has precisely $k + n$ zeros in (a, b), so it only remains to prove that $k = 0$ if supp $w = [a, b]$.

Since $\phi(x, \lambda) \geq 0$ is continuous and increases with λ there is a pointwise limit $\phi_0(x) \geq 0$ as $\lambda \to -\infty$ which is upper semi-continuous. By dominated convergence $\int_a^b w \sin^2 \phi(\cdot, \lambda) \to \int_a^b w \sin^2 \phi_0$ as $\lambda \to -\infty$. But if $\lambda \leq \lambda_0$, then $\phi(b, \lambda) - \phi(a, \lambda) \leq \beta + k\pi - \alpha$ so integrating the equation for ϕ we obtain

$$\left| \int_a^b w \sin^2(\cdot, \lambda)\phi \right| \leq \frac{1}{|\lambda|} \left((k+1)\pi + \int_a^b (|q| + 1/p) \right) \to 0$$

as $\lambda \to -\infty$. It follows that $\int_a^b w \sin^2 \phi_0 = 0$, so if supp $w = [a, b]$ we have $\sin \phi_0 = 0$ in a dense subset S of (a, b). For x in this set $\phi_0(x)$ is an integer multiple of π, and the multiple is non-decreasing as a function of x since $\phi(\cdot, \lambda)$ increases at integer multiples of π. Since $\phi(a, \lambda) \in [0, \pi)$ we have $\phi_0(x) \in [0, \pi)$ for x near a and hence $\phi_0(x) = 0$ if $x \in S$ and near a. On the other hand, if $\lambda < 0$ and $a < x_1 < x_2$ we have

---

[1] Moving towards smaller eigenvalues we lose one zero at each eigenvalue.

$$\phi(x_2, \lambda) - \phi(x_1, \lambda) \leq \int_{x_1}^{x_2} (1/p + |q|),$$

so also $\phi_0(x_2) - \phi_0(x_1)$ is no more than the right-hand side. Using this and that S is dense we find that ϕ_0 cannot increase in (a, b). Thus $\phi(b, \lambda) < \beta \leq \pi$ for λ close to $-\infty$, so that $\phi(b, \lambda_0) = \beta$ and the eigenfunction to λ_0 has no zeros in (a, b). □

For left-definite equations we have the following theorem.

Theorem 6.1.8. *Suppose q and w are in $L^1(a, b)$. Also assume q is non-negative and not a.e. zero, while w takes both signs on sets of positive measure. Consider the left-definite equation $-u'' + qu = \lambda wu$ provided with separated, self-adjoint boundary conditions and assume its strictly positive eigenvalues are $\lambda_0^+ < \lambda_1^+ < \lambda_2^+, \ldots$ while $\lambda_0^- > \lambda_1^- > \lambda_2^- > \ldots$ are the strictly negative[2] eigenvalues. Then the eigenfunctions associated with $\lambda_n^{\pm}$ have n zeros in (a, b).*

Proof. The form of separated boundary conditions in the left-definite case suggests that if $\lambda \neq 0$ a useful variant of the Prüfer transform for this case is

$$\begin{cases} \lambda u = r \sin \phi, \\ u' = r \cos \phi. \end{cases}$$

This gives the equation $\phi' = \lambda \cos^2 \phi + (w - q/\lambda) \sin^2 \phi$ for the Prüfer angle. The proof now closely mimics the proof of Theorem 6.1.2, the device of letting $\lambda \to -\infty$ being replaced by letting $\lambda \to 0$ from above when dealing with the positive eigenvalues and from below when dealing with the negative eigenvalues.

We leave the details to the reader in Exercise 6.1.3. □

One would perhaps expect something similar to be true for a right-definite equation with a p that takes both signs, but except in quite special cases we are not aware of any such results; it appears that the ability to count the zeros of eigenfunctions is closely connected with a positive p. See also Exercise 6.1.4.

In some cases similar results are valid also for singular Sturm–Liouville problems, in particular if the equation yields a semi-bounded operator. We shall only consider the following simple situation.

Theorem 6.1.9. *Suppose $q \in L^1(0, c)$ for every $c > 0$ and that $\underline{\lim} q \geq Q$ at ∞, where Q is finite or $+\infty$. Then the equation $-u'' + qu = \lambda u$ is limit-point at infinity, and provided with an appropriate boundary condition at 0 its spectrum is bounded below and discrete below Q.*

If the eigenvalues below Q are $\lambda_0 < \lambda_1 < \lambda_2 < \ldots$ the eigenfunction associated with λ_n has n zeros in $(0, \infty)$. The same is true for the equation on $\mathbb{R}$ if $\underline{\lim} q \geq Q$ at $\pm \infty$.

---

[2] If one or both boundary conditions are Dirichlet ($\alpha = 0$ in (5.1.5)) then $\lambda = 0$ is an eigenvalue, the corresponding eigenfunctions having one or no zero in (a, b).

Proof. Consider first the interval $(0, \infty)$. If Q is finite the essential spectrum is contained in $[Q, \infty)$, otherwise the spectrum is discrete by Theorem 4.5.8. In either case the spectrum is bounded from below, since if u is a solution of $-u'' + qu = \lambda u$ for $\lambda < Q$ and a is so large that $q(x) > \lambda$ for $x \geq a$, then the sign of u'' is the same as that of u in (a, ∞), so in this interval $|u|$ is convex. Thus $|u| \to \infty$ at infinity if there is a zero greater than a. If $u \in L^2(a, b)$ there are therefore *no* zeros in (a, ∞).

We may now employ the same arguments as in the proof of Theorem 6.1.2 to conclude that if there are eigenvalues below Q, then there is a smallest such eigenvalue λ_0 with the corresponding eigenfunction having no zeros. If there are more eigenvalues below Q, then numbering the eigenvalues in order of size it follows as before that the eigenfunction to λ_n has precisely n zeros.

Much the same arguments may be used when the base interval is $(-\infty, \infty)$. We shall leave the details to the reader. □

Under the assumptions of the theorem the equation is limit-point at infinite endpoints and there may be no, or only finitely many, eigenvalues below a finite Q, for example if $-q$ is the characteristic function of a compact interval. The situation is clarified by the following theorem.

Theorem 6.1.10. *Consider the equation $-u'' + qu = \lambda u$ on $(0, \infty)$, where 0 is a regular endpoint, and let $q_- = \max(0, -q)$ be the negative part of q. Given a self-adjoint, separated boundary condition at 0 and assuming $\lim_{x \to \infty} x \int_x^\infty q_- < 1/4$, there are only finitely many negative eigenvalues.*

Proof. If $\sup_{t \geq A} t \int_t^\infty q_- \leq 1/4$ and u is an absolutely continuous function with supp u a compact subset of (A, ∞), then

$$\int_A^\infty |u|^2 q_- = \int_A^\infty 2 \operatorname{Re}(u'(t)\overline{u(t)}) \int_t^\infty q_- \, dt$$

$$\leq \frac{1}{2} \int_A^\infty t^{-1} |u'(t)u(t)| \, dt \leq \frac{1}{2} \left(\int_A^\infty t^{-2} |u(t)|^2 \, dt \int_A^\infty |u'|^2 \right)^{1/2},$$

with strict inequality if $\sup_{t \geq A} t \int_t^\infty q_- < 1/4$. For $q_-(t) = 1/(4t^2)$ this gives

$$\int_A^\infty t^{-2} |u(t)|^2 \, dt \leq 2 \left(\int_A^\infty t^{-2} |u(t)|^2 \, dt \int_A^\infty |u'|^2 \right)^{1/2},$$

so that

$$\int_A^\infty t^{-2} |u(t)|^2 \, dt \leq 4 \int_A^\infty |u'|^2 .$$

Combining the first and last estimates we obtain

$$\int_A^\infty (|u'|^2 + q\,|u|^2) \ge \int_A^\infty (|u'|^2 - q_-\,|u|^2) > 0 .$$

By Proposition 5.5.1 it follows that solutions to $-u'' + qu = 0$ have at most one zero in (A, ∞) so there are only finitely many zeros in $(0, \infty)$ and thus a finite number of negative eigenvalues.   □

Distribution of eigenvalues. Use of variants of the Prüfer transform in many cases allow asymptotic estimates of the counting function $N(\lambda)$. We shall give a simple result of this type here. In general, particularly in the case of p that changes sign, we need other methods of proof; see Theorem 6.5.10.

Theorem 6.1.11. *Consider a self-adjoint realization of $-u'' + qu = \lambda u$ in a bounded interval (a, b) where $q \in L^1(a, b)$. Then $N(\lambda) = \frac{b-a}{\pi}\sqrt{\lambda} + O(1)$ as $\lambda \to \infty$. Equivalently, $\lambda_n = (\pi/(b-a))^2 n^2 + O(n)$ as $n \to \infty$.*

Proof. We use the Prüfer transform

$$\begin{cases} u = r \sin \phi , \\ u' = r\sqrt{\lambda} \cos \phi \end{cases}$$

for $\lambda > 0$. In this case the Prüfer angle satisfies $\phi' = \sqrt{\lambda} - \frac{1}{\sqrt{\lambda}} q \sin^2 \phi$. $N(\lambda_n) = n$ so $N(\lambda)$ deviates from the number of zeros of a solution by at most one. Thus $N(\lambda) = (\phi(b) - \phi(a))/\pi + O(1)$. This gives $N(\lambda) = \frac{b-a}{\pi}\sqrt{\lambda} + O(1)$ since $\left| \int_a^b q \sin^2 \phi \right| \le \int_a^b |q|$. Taking $\lambda = \lambda_n$ we can solve for λ_n to obtain the other formula.   □

Exercising more care in the calculations one may substantially improve these formulas; see Exercise 6.1.5.

We shall also use a Prüfer transform to prove similar results for the equation $-u'' + qu = \lambda u$ on $\mathbb{R}$ or $(0, \infty)$ with a regular endpoint at 0, in the case when q has a finite limit or tends to $+\infty$ at an infinite endpoint. Any corresponding self-adjoint operator is therefore bounded from below by Theorem 6.1.9 so the equation may be treated as right- or left-definite. The assumption on q implies that q is bounded below near any infinite endpoint which therefore is limit-point according to Theorems 4.2.7 and 5.1.9.

If the base interval is $[0, \infty)$ and $q \to Q \in (-\infty, \infty]$ at ∞ the spectrum is discrete if $Q = \infty$ by Theorem 4.5.8, but if Q is finite the essential spectrum is $[Q, \infty)$ by Corollary 4.5.6. In this case $N(\lambda)$ is used to count the eigenvalues below Q.

Theorem 6.1.12. *Let $-\infty < Q \le +\infty$ and assume:*

(1) *If the base interval is $[0, \infty)$ assume that $q \in L^1(0, c)$ for every $c \in (0, \infty)$ and that q is continuous and increasing near infinity with limit Q.*

(2) *If the base interval is $\mathbb{R}$ assume that $q \to Q$ at both endpoints or else at one of them, while $\underline{\lim} q > Q$ at the other endpoint. Also assume $q \in L^1_{\text{loc}}(\mathbb{R})$ and continuous and increasing near $+\infty$ if $\lim q = Q$ there, and continuous and decreasing near $-\infty$ if $\lim q = Q$ there.*

(3) *As x tends to any infinite interval endpoint where q has limit Q we have $x^{-2} = o(q(x) - q(kx))$ for any $k \in (0, 1)$.*

Then, as λ increases to Q, we have

$$N(\lambda) \sim \frac{1}{\pi} \int\limits_{q<\lambda} \sqrt{\lambda - q(t)} \, dt \ .$$

Proof. We shall prove the theorem for the interval $(0, \infty)$ and leave the modifications necessary when the interval is $(-\infty, \infty)$ to the reader in Exercise 6.1.6.

If Q is finite the integral of $\sqrt{\lambda - q}$ over a compact set is bounded as $\lambda \to Q$ while if $Q = +\infty$ it grows like $\sqrt{\lambda}$. As we shall see, the full integral always grows faster than that, so we may as well assume that q is continuous and increasing in $[0, \infty)$ with $q(0) = 0$. This means that q' is a diffuse, positive measure, see page 286.

Given λ let a satisfy $q(a) = \lambda$, which is possible as long as $\lambda < Q$, and $a \to \infty$ as $\lambda \to Q$ since assumption (3) excludes q being eventually constant. In the interval (a, ∞) a solution of $-u'' + qu = \lambda u$ has at most one zero since u'' has the same sign as u there. Thus by Theorem 6.1.9 $N(\lambda)$ differs by at most two from the number of zeros in $(0, a)$ of a solution of $-u'' + qu = \lambda u$. To exploit this we shall use a modified Prüfer transform in $(0, a)$, setting

$$\begin{cases} u = r \sin \phi \ , \\ u' = r\sqrt{\lambda - q} \cos \phi \ . \end{cases}$$

Differentiating $\tan \phi$ then leads to

$$\phi' = \sqrt{\lambda - q} + \frac{q'}{4(q - \lambda)} \sin(2\phi) \ .$$

Using $\lambda = q(a)$ and integrating we obtain, if $0 < x < a$,

$$\phi(x) = \phi(0) + \int\limits_0^x \sqrt{q(a) - q} + \frac{1}{4} \int\limits_0^x \sin(2\phi) \, df \ , \tag{6.1.3}$$

where $f = \log(q(a) - q)$. Now $N(\lambda) = O(1) + (\phi(a) - \phi(0))/\pi$ so one would like to estimate the last term in (6.1.3) for $x = a$. However, the integrator f in the last term has a singularity at a, so to avoid this difficulty we proceed as follows: Use of (6.1.3) shows that

$$N(\lambda) - \frac{1}{\pi} \int\limits_{q<\lambda} \sqrt{\lambda - q(t)} \, dt$$

$$= O(1) + \frac{\phi(a) - \phi(x)}{\pi} - \frac{1}{\pi} \int\limits_x^a \sqrt{q(a) - q} + \frac{1}{4\pi} \int\limits_0^x \sin(2\phi) \, df \ . \tag{6.1.4}$$

The theorem follows on choosing $x \in (0, a)$ as a function of a such that all terms to the right are $o\left(\int_0^a \sqrt{q(a) - q}\right)$ as $a \to \infty$. Now $q(a) - q(t) \le q(a) - q(x)$ for $t \in [x, a]$, so by the Sturm comparison theorem the number of zeros of u in $[x, a]$ is less than the number of zeros when q is replaced by its infimum $q(x)$ on (x, a). Thus

$$- \pi < \phi(a) - \phi(x) \le (a - x)\sqrt{q(a) - q(x)} + \pi . \tag{6.1.5}$$

We also have

$$0 \le \int_x^a \sqrt{q(a) - q} \le (a - x)\sqrt{q(a) - q(x)} \tag{6.1.6}$$

and

$$\int_0^a \sqrt{q(a) - q} \ge \int_0^x \sqrt{q(a) - q} \ge x\sqrt{q(a) - q(x)} . \tag{6.1.7}$$

If $Q = \infty$ we obtain $\sqrt{q(a) - q(x)} \le \frac{1}{x} \int_0^a \sqrt{q(a) - q}$ and for fixed x the left member is asymptotic to $\sqrt{\lambda} = \sqrt{q(a)}$ as $\lambda \to \infty$ so $\sqrt{\lambda} \le \frac{2}{x} \int_0^a \sqrt{q(a) - q}$ for large λ. As x can be arbitrarily large we obtain $\sqrt{\lambda} = o\left(\int_0^a \sqrt{q(a) - q}\right)$, as promised. On the other hand, setting $x = a/2$ assumption (3) shows that the left member of (6.1.7) is always unbounded, thus taking care of the first term to the right in (6.1.4).

Now, if we choose x in (6.1.4) so that $a - x = o(a)$, i.e., $x/a \to 1$ as $a \to \infty$, the inequalities (6.1.5)–(6.1.7) show that this will take care of the next two terms. The function f is decreasing so we estimate the last integral in (6.1.4) as

$$\left| \int_0^x \sin(2\phi) \, df \right| \le - \int_0^x df = \log(q(a) - q(0)) - \log(q(a) - q(x)) .$$

The first term to the right is clearly $o(\sqrt{q(a) - q(1)})$ if it is unbounded, so can be ignored since setting $x = 1$ in (6.1.7) shows that $\sqrt{q(a) - q(1)} \le \int_0^a \sqrt{q(a) - q}$ if $a > 1$. We now wish to choose x so that $x/a \to 1$ and

$$\log(q(a) - q(x)) = o\left(\int_0^a \sqrt{q(a) - q}\right) . \tag{6.1.8}$$

To do this, choose x so that $q(x) = q(a) - a^{-2}$. Then if $0 < k < 1$ assumption (3) shows that $q(a) - q(x) = a^{-2} = o(q(a) - q(ka))$ so eventually, since q increases, $ka < x < a$ given any $k < 1$. Thus $x/a \to 1$ or $a - x = o(a)$. Next, assumption (3) for $x = 2t$ and $k = 1/2$ shows that $t^{-1} = o(\sqrt{q(2t) - q(t)})$, and integrating this we obtain

$$\int_{\sqrt{a}}^{a/2} \frac{dt}{t} = o\left(\int_{\sqrt{a}}^{a/2} \sqrt{q(2t) - q(t)} \, dt\right) = o\left(\int_0^{a/2} \sqrt{q(a) - q}\right) .$$

The left-hand side is $\log(a/2) - \log \sqrt{a} = \frac{1}{2} \log a - \log 2$, so in view of the definition of x this means that (6.1.8) is satisfied and the proof is finished. $\square$

In Titchmarsh [212] it is assumed that $q \in C^1$ increases to infinity and $x^3 q'(x) \to \infty$ as $x \to \infty$. Clearly a convex C^1 function tending to infinity always satisfies this condition, since q' is then positive and non-decreasing. Hartman [99] also mentions this condition on q', but shows that it suffices that q increases to infinity and

$$(q(v) - q(u))/\int_u^v t^{-3} \, dt \to \infty \tag{6.1.9}$$

as u and $v \to \infty$. This condition is clearly implied by the derivative condition. On the other hand, Hartman's condition (6.1.9) implies assumption (3) since $\int_{kx}^x t^{-3} \, dt = \frac{1-k^2}{2k^2} x^{-2}$, but our condition (3) is slightly weaker than (6.1.9).

Example 6.1.13.

1. For positive integers j let $q(x) = 2j$ if $2j \leq x < 2j + 1$ and $q(x) = 2(x - j - 1)$ if $2j + 1 \leq x < 2j + 2$, for which $q(x) - q(kx) \geq 1$ if $k \in (0, 1)$ and $x \geq 2/(1 - k)$. Thus assumption (3) is satisfied, but for $u = 2j$, $v = 2j + 1$ Hartman's condition (6.1.9) is violated.

2. Consider the case $Q = \infty$. Clearly the condition $x^3 q'(x) \to \infty$, which implies assumption (3), is satisfied for any increasing product of iterated exponentials, powers or iterated logarithms, since derivatives of these functions are eventually monotone and may tend to zero, but never as fast as $x^{-\alpha}$ for $\alpha > 1$. This would imply integrability of q' and so prevent $q \to \infty$.

3. If q has a finite limit at infinity we may as well assume the limit to be 0, which only involves translating the spectrum. Consider first $q(x) = -(1 + x)^{-\alpha}$ for some $\alpha > 0$. Then $x^2(q(x) - q(kx)) \sim x^{2-\alpha}(k^{-\alpha} - 1)$, which tends to $+\infty$ as $x \to \infty$ for $0 < \alpha < 2$ if $0 < k < 1$. If we consider an arbitrary negative power of an iterated logarithm the derivative behaves like $1/x$ multiplied by a very slowly decreasing function, so such functions always satisfy the condition. But of course functions that tend to zero faster than x^{-2} may have very small derivatives and not satisfy any of the mentioned conditions.

This is, however, not a defect of Theorem 6.1.12 in view of Theorem 6.1.10, which shows that rapidly decaying potentials do not produce infinitely many eigenvalues below zero. In particular, if $q(x) = -C(1 + x)^{-2}$ we have $\overline{\lim}_\infty x \int_x^\infty q_- = C$, so if $C < 1/4$ there are only finitely many (in fact at most one) negative eigenvalues. By Exercise 6.1.7 this is true also for $C = 1/4$, but if $C > 1/4$ there are infinitely many negative eigenvalues, so in this respect Theorem 6.1.12 is nearly optimal.

4. We have $\int_0^a \sqrt{q(a) - q(x)} \, dx = a\sqrt{|q(a)|} \int_0^1 \sqrt{|1 - q(at)/q(a)|} \, dt$. If $q(x) \sim Cx^\alpha$ for large x or, more generally, q has limit-order $\alpha > -2$ (see the introduction to Section 6.3) the integral converges to $\int_0^1 \sqrt{|1 - t^\alpha|} \, dt$ as $a \to \infty$, which is non-zero and finite if $-2 < \alpha \neq 0$, so in such cases the asymptotic formula becomes $N(\lambda) \sim C_\alpha q^{-1}(\lambda)\sqrt{|\lambda|}$, where[3] $C_\alpha = \frac{1}{\pi} \int_0^1 \sqrt{|1 - t^\alpha|}\, dt$. See also Exercise 6.1.8.

5. If $q(x) = x^\alpha$ for some $\alpha > 0$, this shows that $n = N(\lambda_n) \sim C_\alpha \lambda_n^{(\alpha+2)/(2\alpha)}$ or equivalently $\lambda_n \sim (n/C_\alpha)^{2\alpha/(\alpha+2)}$. Since the resolvent is a Hilbert–Schmidt operator precisely if $\sum |\lambda_n + i|^{-2} < \infty$, the resolvent is Hilbert–Schmidt precisely if $4\alpha/(\alpha + 2) > 1$ or $\alpha > 2/3$.

Exercises

Exercise 6.1.1 A different approach to the Sturm theorems is to use the *Picone identity*

$$\left(\frac{u_1}{u_2}(p_1 u_1' u_2 - u_1 p_2 u_2')\right)' = (p_1 - p_2)(u_1')^2 + (g_2 - g_1)(u_1)^2 + p_2 u_2^2 ((u_1/u_2)')^2,$$

where $(p_j u_j')' + g_j u_j = 0$; see Picone [180]. Show this, and then use it to prove versions of the three Sturm theorems.

Exercise 6.1.2 Prove Corollary 6.1.3.

Hint: A real-valued eigenfunction is also an eigenfunction for some separated boundary conditions. Consider the dimension of the range of a spectral projector.

Exercise 6.1.3 Finish the proof of Theorem 6.1.8.

Hint: For $\lambda = 0$ there are solutions without zeros. Show that for $\lambda \neq 0$ but close to 0 there are solutions satisfying the boundary condition at a without zeros, and that the corresponding Prüfer angle at b can take any value in $(0, \pi]$ if $\lambda > 0$ and in $[0, \pi)$ if $\lambda < 0$.

Exercise 6.1.4 Consider the equation $-(pu')' = \lambda w u$ in $[a, b]$, with $w > 0$ a.e., p taking both signs on sets of positive measure and $1/p$ and w integrable in (a, b). Show that if u is an eigenfunction the zero-count of pu' satisfies conditions similar to those of the eigenfunctions in Theorem 6.1.8.

Exercise 6.1.5 To improve Theorem 6.1.11, show that the boundary condition at a requires $\phi(a) = \pi/2 + \lambda_n^{-1/2}\cot\alpha + O(\lambda_n^{-3/2})$ except if the boundary condition at a is Dirichlet ($\alpha = 0$), when $\phi(a) = 0$. Similarly $\phi(b) = (2n+1)\pi/2 + \lambda_n^{-1/2}\cot\beta + O(\lambda_n^{-3/2})$ except if the condition at b is Dirichlet, when $\phi(b) = (n+1)\pi$. This gives a more accurate value for $\phi(b) - \phi(a)$ when $\lambda = \lambda_n$.

Next consider $\int_a^b q \sin^2\phi = \frac{1}{2}\int_a^b q - \frac{1}{2}\int_a^b q \cos(2\phi)$ and use that $\phi(x) = \phi(a) + \frac{x-a}{\pi}\lambda_n^{1/2} + O(\lambda_n^{-1/2})$ and the Riemann–Lebesgue Lemma 4.5.1 to show that the

---

[3] C_α is easily expressed in terms of the Euler gamma function.

second integral tends to zero as $n \to \infty$ by estimating the difference $\cos(2\phi(x)) - \cos(2\phi(a) + 2(x-a)\lambda_n^{1/2}/\pi)$, leading to the formula

$$n = N(\lambda_n) = \frac{b-a}{\pi} \lambda_n^{1/2} + \frac{1}{\pi}\left(\cot\alpha - \cot\beta + \frac{1}{2}\int_a^b q\right)\lambda_n^{-1/2} + o(\lambda_n^{-1/2})$$

when neither boundary condition is Dirichlet and slightly modified formulas in other cases. By solving for λ_n show that

$$\lambda_n = \left(\frac{\pi}{b-a}\right)^2 n^2 - \frac{1}{b-a}\left(2(\cot\alpha - \cot\beta) + \int_a^b q\right) + o(1)$$

if none of the boundary conditions are Dirichlet and slightly modified formulas otherwise.

Exercise 6.1.6 Carry out the proof of Theorem 6.1.12 for the case of the interval $(-\infty, \infty)$.

Exercise 6.1.7 Show that $-u'' - C(1+x)^{-2}u = 0$ has solutions on $[0, \infty)$ of the form $u(x) = (1+x)^\alpha$ where α is real if $C \leq 1/4$ and non-real if $C > 1/4$. Use this to show that there are infinitely many negative eigenvalues if $C > 1/4$ but not otherwise.

Exercise 6.1.8 If $q(x) = \alpha \log(1 + x)$ then $q(at)/q(a) \to 1$ as $a \to \infty$ so Example 6.1.13 **4** only gives $N(\lambda) = o(\alpha\sqrt{q(\alpha)})$. Show that if $\exp(q(at))/\exp(q(a)) \to t^\alpha$ as $a \to \infty$, then $q(a) - q(at) \to -\alpha \log t$ as $a \to \infty$ so that

$$N(\lambda) \sim q^{-1}(\lambda)\sqrt{\alpha}\int_0^1 \sqrt{-\log t}.$$

Also show that $\int_0^1 \sqrt{-\log t} = \sqrt{\pi}/2$.

This may be generalized as follows. Let $q_0 = q$ and $q_k(x) = \exp(q_{k-1}(x))$, $k = 1, 2, \dots$. Assuming $k > 1$ and $q_k(at)/q_k(a) \to t^\alpha$ as $a \to \infty$, show that

$$N(\lambda) \sim \left(\frac{\alpha\pi}{2}\right)^{1/2}\frac{q^{-1}(\lambda)}{\prod_{j=1}^{k-1}\sqrt{q_j(q^{-1}(\lambda))}} \quad \text{as } \lambda \to \infty.$$

Hint: If $k = 2$ we have

$$q(a) - q(at) = -\log(q_1(at)/q_1(a))$$
$$= -\log(1 + \log(q_2(at)/q_2(a))/q_1(a)) \sim -\alpha \log t/q_1(a).$$

6.2 Order of magnitude estimates for $m(\lambda)$

We shall deal with the classical Sturm–Liouville equation (4.1.1) on an interval (a, b) with the standard assumptions that $1/p$, q and w are real-valued and integrable over any interval (a, c) with $c \in (a, b)$, so that a is a regular endpoint.

Definition 6.2.1. For $a \leq x \leq y < b$, let $P(x, y) = \int_x^y 1/p$, $W(x, y) = \int_x^y w$, $Q(x, y) = \int_x^y q$ and $\tilde{Q}(x, y) = \int_x^y |q|$. If $a \leq x \leq y < b$ we also define

$$\tilde{P}(x, y) = \sup_{x \leq s \leq t \leq y} \left| \int_s^t 1/p \right|, \qquad \tilde{W}(x, y) = \sup_{x \leq s \leq t \leq y} \left| \int_s^t w \right|. \tag{6.2.1}$$

For fixed x clearly $P(x, \cdot)$, $W(x, \cdot)$, $Q(x, \cdot)$ and $\tilde{Q}(x, \cdot)$ are locally absolutely continuous and vanish at x. To exclude trivial cases we always assume that w does not vanish a.e. on (a, b) so that $\tilde{P}(a, \cdot)$ and $\tilde{W}(a, \cdot)$ are increasing, but not necessarily strictly so. Clearly, if $p > 0$ a.e. then $\tilde{P} = P$, and if $w \geq 0$ a.e. then $\tilde{W} = W$. In this section the first variable in all these functions will always stay equal to a, so to simplify the notation we will here always suppress the first variable, writing $\tilde{P}(x)$ for $\tilde{P}(a, x)$, etc.

Lemma 6.2.2. $\tilde{P}$ and $\tilde{W}$ are absolutely continuous on any interval (a, c), $c \in (a, b)$.

Proof. If $a \leq s \leq t \leq y$ and $x \leq y$, then if $s \leq x \leq t$

$$\left| \int_s^t 1/p \right| \leq \left| \int_s^x 1/p \right| + \left| \int_x^t 1/p \right| \leq \tilde{P}(x) + \int_x^y 1/|p| .$$

If instead $t < x$, the left-hand side is less than $\tilde{P}(x)$, while if $a \leq x < s$, it is less than $\int_x^y 1/|p|$. Since $\tilde{P}$ is non-decreasing, taking the least upper bound over s, t gives $0 \leq \tilde{P}(y) - \tilde{P}(x) \leq \int_x^y 1/|p|$. Thus $\int_a^x 1/|p| - \tilde{P}(x)$ is non-decreasing, so that $\tilde{P}$ generates a positive measure smaller than that with density $1/|p|$. By the comment before Corollary B.8.5 $\tilde{P}$ is thus locally absolutely continuous. Similarly for $\tilde{W}$. □

The function $x \mapsto \tilde{P}(x)\tilde{W}(x)$ is well defined and vanishes at a. It is non-decreasing in (a, b), not identically zero and not a positive constant on an interval unless both $\tilde{P}$ and $\tilde{W}$ are. Thus the following definition makes sense.

Definition 6.2.3. For $r > \lim_{x \to b} (\tilde{P}(x)\tilde{W}(x))^{-1}$ define the functions f and g by $f(r) = \tilde{P}(x)$ and $g(r) = \tilde{W}(x)$ when $r\tilde{P}(x)\tilde{W}(x) = 1$.

It is clear that f and g are decreasing, but perhaps not strictly so. It follows that $r \mapsto rf(r) = 1/g(r)$ and $r \mapsto rg(r) = 1/f(r)$ are non-decreasing. If $a \in \mathrm{supp}\, w$, so that $\tilde{W}(x) > 0$ for $x > a$, then since also $\tilde{P}(x) > 0$ for $x > a$ it follows that $f(r) \to 0$ and $g(r) \to 0$ while $rf(r) \to \infty$ and $rg(r) \to \infty$ as $r \to \infty$.

Remark 6.2.4. There is a concept of *generalized inverse* of a monotone but perhaps not strictly monotone function F, defined on an interval I, as follows. The generalized inverse is defined on the convex hull of $F(I)$ as $F^{-1}(x) = \inf\{y \mid F(y) \geq x\}$ if F increases and by $F^{-1} = -(-F)^{-1}$ if F decreases.

This makes the generalized inverse left-continuous, and one may prove that the functions f and g are the generalized inverses of $x \mapsto 1/(x\tilde{W}(\tilde{P}^{-1}(x)))$ and $x \mapsto 1/(x\tilde{P}(\tilde{W}^{-1}(x)))$ respectively, $\tilde{P}^{-1}$ being the generalized inverse of $\tilde{P}$ and $\tilde{W}^{-1}$ of $\tilde{W}$ (Exercise 6.2.1). We will not explicitly use any of this, however.

We first consider the right-definite case and thus assume $w \geq 0$ so that $\tilde{W} = W$, and study m-functions belonging to the Neumann boundary condition at a. This means that $m(\lambda)$ satisfies

$$\int_a^b |\psi(\cdot, \lambda)|^2 \, w = \frac{\operatorname{Im} m(\lambda)}{\operatorname{Im} \lambda} \tag{6.2.2}$$

where $\psi(\cdot, \lambda) = \theta(\cdot, \lambda) + m(\lambda)\varphi(\cdot, \lambda)$ and $\theta(\cdot, \lambda)$ and $\varphi(\cdot, \lambda)$ solve the equation $-(pv')' + qv = \lambda wv$ with initial data

$$\begin{cases} \varphi(a, \lambda) = -1, \\ p\varphi'(a, \lambda) = 0, \end{cases} \qquad \begin{cases} \theta(a, \lambda) = 0, \\ p\theta'(a, \lambda) = 1. \end{cases} \tag{6.2.3}$$

If $\inf \operatorname{supp} w = c > a$ we denote by $m(c, \cdot)$ the m-function for (c, b) corresponding to the same boundary condition at b and with the Neumann boundary condition at c. Since the Weyl solution for an interval $[c, b)$ with, if needed, a given boundary condition at b is determined up to multiples independent of c and the boundary condition imposed there, it follows that $m(c, \lambda) = -\psi(c, \lambda)/p\psi'(c.\lambda)$. We obtain the following proposition.

Proposition 6.2.5. *Suppose* $w(x) = 0$ *for* $a \leq x \leq c$. *Then in this interval the solutions defined in* (6.2.3) *are both independent of* λ, *real-valued and*

$$m(\lambda) = -\frac{p\theta'(c)m(c, \lambda) + \theta(c)}{p\varphi'(c)m(c, \lambda) + \varphi(c)}.$$

One must solve the equation $-(pu')' + qu = 0$ to find the explicit connection between m and $m(c, \cdot)$, but then any asymptotic information on $m(c, \cdot)$ will give asymptotic information on m. In what follows we will therefore assume that $a \in \operatorname{supp} w$. It follows that the function f of Definition 6.2.3 satisfies $f(r) \downarrow 0$ and $rf(r) \uparrow \infty$ as $r \to \infty$.

Consider the interval $[a, b)$, where a is regular and b either limit-point, or we consider a fixed, separated boundary condition at b. Using the boundary condition (4.2.2) at a we denote the corresponding Weyl solution by ψ_α and the m-function by m_α. Since there is only a one-dimensional set of solutions of $-(pu')' + qu = \lambda wu$ satisfying the boundary condition at b, the Weyl solutions for all α are proportional, so we have $m_{\pi/2}(\lambda) = -\psi_\alpha(a, \lambda)/p\psi'_\alpha(a, \lambda)$ for any α. But $\psi_\alpha(a, \lambda) = \cos \alpha -$

$m_\alpha(\lambda) \sin \alpha$ and $p\psi'_\alpha(a, \lambda) = \sin \alpha + m_\alpha(\lambda) \cos \alpha$. Thus we may express m_α in terms of $m_{\pi/2}$, and a brief calculation shows that

$$m_\alpha(\lambda) = \frac{\cos \alpha + m_{\pi/2}(\lambda) \sin \alpha}{\sin \alpha - m_{\pi/2}(\lambda) \cos \alpha}. \tag{6.2.4}$$

If we instead consider an interval $(a, b]$, where the regular endpoint b is to the right of the possibly singular endpoint a, then according to Remark 4.3.6 the calculations above need to be slightly modified. We leave the details to Exercise 6.2.2

Returning to the case of a being the regular endpoint we shall show that if $a \in \operatorname{supp} w$ any Neumann m-function (there are many if b is a limit-circle endpoint) tends to zero as λ tends to infinity in any double sector $|\operatorname{Re} \lambda| \le C |\operatorname{Im} \lambda|$. Referring to (6.2.4) it follows that any asymptotic information on $m_{\pi/2}$ may be translated into asymptotic information on m_α, so it is enough to study the case $m = m_{\pi/2}$. Note in particular that $m_0 = -1/m_{\pi/2}$ so $m_0(\lambda) \to \infty$ and if $\alpha \in (0, \pi)$, then $m_\alpha(\lambda) - \cot \alpha \sim m_{\pi/2}(\lambda)/\sin^2 \alpha$ as $\lambda \to \infty$ in such a sector.

Theorem 6.2.6. *For a right-definite Neumann m-function associated with an interval* $[a, b)$ *we have:*

(1) *Given a sufficiently large* $|\lambda|$*, let* $c \in (a, b)$ *satisfy*

$$\tilde{P}(c)(|\lambda| W(c) + \tilde{Q}(c)) = 1/5. \tag{6.2.5}$$

Then as soon as $W(c) > 0$ *it follows that*

$$|m(\lambda)| \le 22/(9 |\operatorname{Im} \lambda| W(c)).$$

(2) *If* $a \in \operatorname{supp} w$*, then for all sufficiently large* $|\lambda|$

$$|m(\lambda)| \le 13 f(|\lambda|)/|\sin(\arg \lambda)|, \tag{6.2.6}$$

where f *is given by Definition 6.2.3.*

(3) *Suppose* $a \in \operatorname{supp} w$ *and that there is a constant* L *such that*

$$\int_a^x \tilde{P}^2 w \le L^2 \int_a^x P^2 w \tag{6.2.7}$$

for all $x > a$ *sufficiently close to* a*. Then we have*

$$|m(\lambda)| \ge |\sin(\arg \lambda)| (2(L+1))^{-4} f(|\lambda|) \tag{6.2.8}$$

for all sufficiently large $|\lambda|$*.*

Note that (6.2.7) holds trivially with $L = 1$ when p is of one sign. It is also possible for (6.2.7) to hold for wildly oscillating p. For example, if $a = 0$, $w \equiv 1$ and $\int_0^x 1/p = x^\alpha \sin(1/x)$ for some $\alpha > 1$, then $L = 3$ would suffice in (6.2.7)

(Exercise 6.2.3). To prove Theorem 6.2.6 we shall need the following estimate close to the initial point a of solutions of $-(pu')' + qu = \lambda wu$.

Lemma 6.2.7. *Suppose $-(pu')' + qu = \lambda wu$ in (a, b) and that $w \geq 0$. If we have $\int_a^x \tilde{P}(|\lambda| w + |q|) \leq k < 1$ it follows that*

$$|u(x) - u(a) - P(x)pu'(a)|$$
$$\leq (1 - k)^{-1}\tilde{P}(x)\big((|\lambda| W(x) + \tilde{Q}(x)) |u(a)| + k |pu'(a)|\big) .$$

Proof. Integrating the equation for u gives $(pu')(x) = (pu')(a) + \int_a^x (q - \lambda w)u$. Dividing by p and integrating again we see after an integration by parts that

$$u(x) = u(a) + P(x)pu'(a) + \int\limits_a^x \left(\int\limits_t^x \frac{1}{p}\right)(q(t) - \lambda w(t))u(t) \, dt .$$

If we set $F(x) = \sup_{a \leq t \leq x} |u(t) - u(a) - P(t)pu'(a)|/\tilde{P}(t)$ we obtain from this, provided $\int_a^x \tilde{P}(|\lambda| w + |q|) \leq k$,

$$F(x) \leq F(x) \int\limits_a^x \tilde{P}(|\lambda| w + |q|) + \int\limits_a^x |u(a) + Ppu'(a)| \, (|\lambda| w + |q|)$$
$$\leq kF(x) + (|\lambda| W(x) + \tilde{Q}(x)) |u(a)| + k |pu'(a)| ,$$

from which the lemma follows. $\qquad\square$

Proof (Theorem 6.2.6). From (6.2.2) it follows that for any x, $a < x < b$,

$$\int\limits_a^x |\theta + m\varphi|^2 w \leq \text{Im } m(\lambda)/\text{Im }\lambda .$$

It follows from this that

$$|m|^2 \int\limits_a^x |\varphi|^2 w - |m| \left\{2\Big|\int\limits_a^x \varphi\bar{\theta} w\Big| + 1/|\text{Im }\lambda|\right\} + \int\limits_a^x |\theta|^2 w \leq 0 ,$$

or, using the Cauchy–Schwarz inequality and setting $\|u\|_x^2 = \int_a^x |u|^2 w$,

$$|m|^2 \|\varphi\|_x^2 - |m| \, (2\|\varphi\|_x\|\theta\|_x + 1/|\text{Im }\lambda|) + \|\theta\|_x^2 \leq 0 .$$

Dropping the last, respectively first, term we obtain

$$|m(\lambda)| \leq 2\|\theta\|_x/\|\varphi\|_x + (|\text{Im }\lambda| \, \|\varphi\|_x^2)^{-1} \qquad (6.2.9)$$

and

$$|m(\lambda)| \geq \left\{ 2\|\varphi\|_x / \|\theta\|_x + (|\mathrm{Im}\,\lambda|\, \|\theta\|_x^2)^{-1} \right\}^{-1} . \tag{6.2.10}$$

To get bounds for m from this we use the estimate of Lemma 6.2.7. If we have $\int_a^x \tilde{P}(|\lambda|\, w + |q|) \leq k < 1$ this gives

$$|\varphi(x) + 1| \leq \tilde{P}(x)(|\lambda|\, \tilde{W}(x) + \tilde{Q}(x))/(1 - k) . \tag{6.2.11}$$

Further, we also deduce

$$|\theta(x) - P(x)| \leq k\, \tilde{P}(x)/(1 - k) \tag{6.2.12}$$

so that

$$|\theta(x)| \leq |\theta(x) - P(x)| + |P(x)| \leq \tilde{P}(x)/(1 - k) . \tag{6.2.13}$$

We obtain an upper bound for m from (6.2.9) by choosing $x = c$ where c satisfies $\tilde{P}(c)(|\lambda|\, W(c) + \tilde{Q}(c)) = 1/5$. Thus $\int_a^c \tilde{P}(|\lambda|\, w + |q|) \leq 1/5$, (6.2.11) implies $3/4 \leq |\varphi(x)| \leq 5/4$ and from (6.2.13) we get $|\theta(x)| \leq 5\,\tilde{P}(x)/4$ for $a \leq x \leq c$. Thus $\frac{3}{4}\sqrt{W(c)} \leq \|\varphi\|_c \leq \frac{5}{4}\sqrt{W(c)}$ and $\|\theta\|_c \leq \frac{5}{4}\tilde{P}(c)\sqrt{W(c)}$. This gives

$$|m(\lambda)| \leq \frac{10\tilde{P}(c)}{3} + \frac{16}{9\,|\mathrm{Im}\,\lambda|\, W(c)} .$$

Now $\tilde{P}(c) \leq 1/(5\,|\lambda|\, W(c))$ so this gives (6.2.5). If $a \in \mathrm{supp}\, w$ we have $c \to 0$ as $\lambda \to \infty$, so setting $\tilde{k} = |\lambda|\, \tilde{P}(c)W(c)$ we have $1/5 - \tilde{k} = \tilde{P}(c)\tilde{Q}(c) \to 0$ as $|\lambda| \to \infty$ and $\tilde{P}(c) = f(|\lambda|\,/\tilde{k}) \leq f(|\lambda|)$ for large $|\lambda|$. Furthermore $1/(|\lambda|\, W(c)) = f(|\lambda|\,/\tilde{k})/\tilde{k} \leq f(|\lambda|)/\tilde{k}$. This leads to the inequality

$$|m(\lambda)| \leq \left(\frac{10}{3} + \frac{16}{9\tilde{k}} \right) \frac{f(|\lambda|)}{|\sin \arg \lambda|} ,$$

which gives (6.2.6) if $\tilde{k}$ is sufficiently close to $1/5$, i.e., if $|\lambda|$ is sufficiently large.

Next we show that a similar lower bound for m is obtained by assuming (6.2.7) in addition to $a \in \mathrm{supp}\, w$. Fix k and $\tilde{k}$ such that $0 < \tilde{k} < k < 1$, and choose $x = d$ in (6.2.10) where $d \leq b$ satisfies

$$|\lambda| \int_a^d \tilde{P} w = \tilde{k} , \tag{6.2.14}$$

which is possible for sufficiently large $|\lambda|$ since $\tilde{P}(x) > 0$ for $x > a$ and $a \in \mathrm{supp}\, w$. Thus $\int_a^d \tilde{P}(|\lambda|\, w + |q|) \leq k$ if also $\tilde{P}(d)\tilde{Q}(d) \leq k - \tilde{k}$, which will certainly be true for sufficiently large $|\lambda|$. We then have (6.2.11) and (6.2.12) for $a \leq x \leq d$ and large $|\lambda|$. It follows from (6.2.13) and (6.2.7) that

$$\|\theta\|_d \geq \|P\|_d - \|\theta - P\|_d \geq M\|\tilde{P}\|_d .$$

Here we assume k chosen so that $M = 1/L - k/(1-k) > 0$. By (6.2.11) we have $\|\varphi\|_d \leq \sqrt{W(d)} + (|\lambda| W(d) + \tilde{Q}(d))\|\tilde{P}\|_d/(1-k)$. From (6.2.14) it follows that $\|\tilde{P}\|_d \sqrt{W(d)} \geq \int_a^d \tilde{P} w = \tilde{k}/|\lambda|$ and so $\|\theta\|_d^{-2} \leq (M\|\tilde{P}\|_d)^{-2} \leq (|\lambda|/(\tilde{k}M))^2 W(d)$. Similarly we may estimate

$$\|\varphi\|_d/\|\theta\|_d \leq \frac{|\lambda| W(d)}{\tilde{k}M} + \frac{|\lambda| W(d) + \tilde{Q}(d)}{(1-k)M} .$$

Now $\tilde{k} \leq |\lambda| \tilde{P}(d)W(d)$ so $\tilde{Q}(d)/(|\lambda| W(d)) \leq (k - \tilde{k})/\tilde{k}$. We thus obtain

$$\|\varphi\|_d/\|\theta\|_d \leq \frac{|\lambda| W(d)}{\tilde{k}M(1-k)} ,$$

so that our final estimate is

$$|m(\lambda)| \geq \frac{(\tilde{k}M)^2(1-k)}{2\tilde{k}M + 1 - k} \frac{|\sin \arg \lambda|}{|\lambda| W(d)} .$$

This is a lower bound of the same order of magnitude as $f(|\lambda|)$, which is clear if $1/(|\lambda| W(d)) \geq f(|\lambda|)$. If $1/(|\lambda| W(d)) < f(|\lambda|)$ then $d \geq x$, where $|\lambda| \tilde{P}(x)W(x) = 1$. Thus

$$W(d) = W(x) + \int_x^d w \leq W(x) + \int_a^d \tilde{P} w/\tilde{P}(x) = (1 + \tilde{k})W(x) .$$

In all cases it follows that $1/(|\lambda| W(d)) \geq f(|\lambda|)/(1 + \tilde{k})$. We now obtain $|m(\lambda)| \geq K f(|\lambda|)$, where

$$K = \frac{(\tilde{k}M)^2(1-k)}{(2\tilde{k}M + 1 - k)(1 + \tilde{k})} .$$

It is easy to see that choosing $k = 1/(2L + 1)$ and $\tilde{k} = 1/(2L + 2)$ we obtain $M = 1/(2L)$ and $K > (2(L+1))^{-4}$. In any double sector $|\text{Re } \lambda| \leq C |\text{Im } \lambda|$ the upper and lower bounds (6.2.6) and (6.2.8) are of the same order of magnitude. The proof is now complete. $\qquad\square$

Now turn to the left-definite case and so assume $p \equiv 1$ and $q \geq 0$, but make no positivity assumption about w. Thus $\tilde{W}$, $\tilde{P}(x) = P(x) = x - a$ and $\tilde{Q} = Q$ are all increasing and locally absolutely continuous. Consider an m-function satisfying

$$\int_a^b (|\theta' + m\varphi'|^2 + q |\theta + m\varphi|^2) = \text{Im } m(\lambda)/\text{Im } \lambda \qquad (6.2.15)$$

where $\varphi(\cdot, \lambda)$ and $\theta(\cdot, \lambda)$ are solutions of $-v'' + qv = \lambda w v$ with initial data

$$\begin{cases} \varphi(a, \lambda) = 0, \\ \varphi'(a, \lambda) = 1, \end{cases} \qquad \begin{cases} \lambda\theta(a, \lambda) = 1, \\ \theta'(a, \lambda) = 0, \end{cases}$$

so that we now consider an m-function belonging to the left-definite Dirichlet boundary condition at a. We shall prove that if $a \in \operatorname{supp} w$ such an m-function always tends to zero as $\lambda \to \infty$ in a double sector $|\operatorname{Re} \lambda| \leq C |\operatorname{Im} \lambda|$. Also in the left-definite case there are simple formulas connecting m-functions for different boundary conditions at a (Exercise 6.2.4), so it will be enough to consider the case of a Dirichlet condition. Recall the function g of Definition 6.2.3. Thus, if $r(x - a)\tilde{W}(x) = 1$ we have $g(r) = \tilde{W}(x) = 1/(r(x - a)) = 1/(rf(r))$.

Theorem 6.2.8. *For a left-definite Dirichlet m-function on $[a, b]$ we have:*

(1) *Given a sufficiently large $|\lambda|$, let $c \in (a, b)$ satisfy*

$$P(c)(|\lambda| \tilde{W}(c) + Q(c)) = 1/5 \,.$$

Then we have

$$|m(\lambda)| \leq 31/(9 |\operatorname{Im} \lambda| P(c)) \,.$$

(2) *For all sufficiently large $|\lambda|$*

$$|m(\lambda)| \leq 13 \, g(|\lambda|)/|\sin(\arg \lambda)|$$

where g is as above.

(3) *Assume that there is a constant L such that*

$$\int_a^x \tilde{W}^2 \leq L^2 \int_a^x W^2$$

holds for all $x > a$ sufficiently close to a. Then we have

$$|m(\lambda)| \geq |\sin(\arg \lambda)| \, (2(L + 1))^{-4} g(|\lambda|)$$

for sufficiently large $|\lambda|$.

Turning now to a brief indication of the proofs in the left-definite case one first obtains

$$u'(x) = u'(a) + (Q(x) - \lambda W(x))u(a) + \int_a^x \left(\int_t^x (\lambda w - q) \right) u'(t) \, dt$$

by integrating $-u'' + qu = \lambda wu$ and then integrating by parts. By a process very similar to that in the right-definite case this gives

$$|\varphi'(x) - 1| \leq (|\lambda| \tilde{W}(x) + Q(x))P(x)/(1-k) ,$$
$$|\theta'(x) + W(x)| \leq (k\tilde{W}(x) + Q(x)/|\lambda|)/(1-k) ,$$
$$|\theta'(x)| \leq (\tilde{W}(x) + Q(x)/|\lambda|)/(1-k) ,$$

provided $\int_a^x (|\lambda| \tilde{W} + Q) \leq k$. One obtains estimates for φ and θ by use of these estimates in $u(x) = u(a) + \int_a^x u'$. This gives $|\varphi(x) - (x-a)| \leq k(x-a)/(1-k)$ and $|\theta(x) - 1/\lambda| \leq k/(|\lambda|(1-k))$. We may now proceed to get upper and lower bounds in close analogy with the right-definite case, using (6.2.15) instead of (6.2.2).

Exercises

Exercise 6.2.1 Prove that f is the generalized inverse of $x \mapsto 1/(x\tilde{W} \circ \tilde{P}^{-1}(x))$ and g of $x \mapsto 1/(x\tilde{P} \circ \tilde{W}^{-1}(x))$, where $\tilde{P}^{-1}$ and $\tilde{W}^{-1}$ are generalized inverses of $\tilde{P}$ and $\tilde{W}$ respectively, using the definition in Remark 6.2.4.

Exercise 6.2.2 Redo the calculations leading to (6.2.4) in the case of an interval $(a, b]$ where the regular endpoint b is used to define φ and θ and the possibly singular endpoint a is to the left of b, using the definition of the m-function in Remark 4.3.6.

Exercise 6.2.3 Show that if $a = 0$, $\int_0^x 1/p = x^\alpha \sin(1/x)$ for some $\alpha > 1$, and $w \equiv 1$, then $L = 3$ would suffice in (6.2.7).

Exercise 6.2.4 Prove a formula analogous to (6.2.4) for the left-definite case.

6.3 Asymptotic estimates for $m(\lambda)$

The results of the previous section indicate that in many cases one would expect a right-definite Neumann m-function to satisfy a formula $m(r\lambda) \sim \tilde{m}(\lambda)f(r)$ as $r \to \infty$, where $\tilde{m}$ is some Nevanlinna function. If such a formula were true, then $f(rs)/f(r) \to \tilde{m}(s\lambda)/\tilde{m}(\lambda)$ as $r \to \infty$ for positive s. The limit is independent of λ so $\tilde{m}$ must be homogeneous. Thus $f(rs)/f(r)$ tends to some homogeneous function s^α. This property of f is called *regular variation* at infinity, see Bingham, Goldie and Teugels [24].

We shall then say that f has *limit-order* α at ∞, and it is not very hard to prove that this is the case if $\tilde{P} \circ W^{-1}$ has limit-order $\alpha/(1-\alpha)$ at 0. Indeed, in Bennewitz [16] it is proved that assuming $\tilde{P} \sim P$ and $\tilde{P} \circ W^{-1}$ is of regular variation at 0 one has an asymptotic formula for the m-function of the kind mentioned. However, we shall be content to prove a simpler result, which is a slight extension of a result given by Everitt [68].

We first consider the right-definite equation $-(pu')' + qu = \lambda wu$ on (a, b), where a is a regular endpoint. We will then give a similar result (Theorem 6.3.3) for the left-definite case.

Recall that two functions F and G of a complex variable λ are said to be asymptotic as $\lambda \to \infty$ in an unbounded domain $\Omega \subset \mathbb{C}$ if $F(\lambda)/G(\lambda) \to 1$ as $\lambda \to \infty$ in Ω. We write this as $F \sim G$ at infinity.

Theorem 6.3.1. *Suppose a is a Lebesgue point (see page 301) for both $1/p$ and w, and that $1/p(a)$ and $w(a)$ are both non-zero. Then a Neumann m-function for the right-definite case of $-(pu')' + qu = \lambda wu$ is asymptotic to $1/(p(a)\sqrt{-\lambda w(a)/p(a)})$ as $\lambda \to \infty$ in any double sector $|\mathrm{Re}\,\lambda| \le C\,|\mathrm{Im}\,\lambda|$. The square root is the square root with a positive real part.*

Recall that, as explained on page 186, an asymptotic formula for Neumann m-functions implies asymptotic formulas for all other boundary conditions at the initial point. For the proof we need the following simple lemma.

Lemma 6.3.2. *The equation $-(pu')' = \lambda wu$, where $p \in \mathbb{R}$ and $w > 0$ are non-zero constants, is limit-point at infinity, and its Neumann m-function on $(0, \infty)$ is $1/(p\sqrt{-\lambda w/p})$. The square root is the root with a positive real part.*

Proof. Solutions in $L^2(0, \infty)$ for $\mathrm{Im}\,\lambda \ne 0$ are all multiples of $u(x) = e^{-x\sqrt{-\lambda w/p}}$, where the square root is the root with a positive real part, so the equation is limit-point at infinity. As mentioned on page 186, the Neumann m-function on $(0, \infty)$ is $-u(0)/pu'(0) = 1/(p\sqrt{-\lambda w/p})$. $\square$

Proof (Theorem 6.3.1). In addition to $-(pu')' + qu = \lambda wu$ on (a, b) we consider the equation

$$- (p_r u')' + q_r u = \mu w_r u \text{ on } (0, (b-a)\sqrt{r}) \,, \tag{6.3.1}$$

where $r > 0$, and define

$$\begin{cases} p_r(x) = p(a + x/\sqrt{r}) \,, \\ w_r(x) = w(a + x/\sqrt{r}) \,, \\ q_r(x) = q(a + x/\sqrt{r})/r \,, \end{cases} \quad \text{and} \quad \begin{cases} \varphi_r(x, \mu) = \varphi(a + x/\sqrt{r}, r\mu) \,, \\ \theta_r(x, \mu) = \sqrt{r}\,\theta(a + x/\sqrt{r}, r\mu) \,, \\ m_r(\mu) = \sqrt{r}\,m(r\mu) \,. \end{cases}$$

Then φ_r and θ_r solve (6.3.1) with the same initial data at 0 as φ and θ at a (recall that we assume a Neumann condition at a). Furthermore, a change of variable in (4.3.5) shows that

$$\int_0^c |\theta_r(\cdot, \mu) + m_r(\mu)\varphi_r(\cdot, \mu)|^2 \, w_r \le \frac{\mathrm{Im}\, m_r(\mu)}{\mathrm{Im}\,\mu} \tag{6.3.2}$$

if r is so large that $c_r = a + c/\sqrt{r} \le b$. Thus m_r is in the Weyl circle for (6.3.1) on the interval $(0, c)$ for any given $c > 0$ if r is large enough. As $r \to \infty$ the coefficients $1/p_r$, w_r and q_r converge in $L^1(0, c)$ to $1/p(a)$, $w(a)$ and 0 respectively for every $c > 0$. For, we have

$$\int\limits_{0}^{c} |w_r(x) - w(a)| \ dx = \sqrt{r} \int\limits_{a}^{c_r} |w(y) - w(a)| \ dy \to 0 \text{ as } r \to \infty$$

since we have assumed a is a Lebesgue point for w, with a similar calculation for p_r and a simpler one for q_r. By Theorem D.2 it follows that φ_r and θ_r converge uniformly in $(0, c)$ and locally uniformly in μ for every $c > 0$ as $r \to \infty$ to solutions of $-(p(a)u')' = \mu w(a)u$ with unchanged initial data. By Lemma 6.3.2 this equation is limit-point at ∞ with Neumann m-function $1/(p(a)\sqrt{-\mu w(a)/p(a)})$ on $(0, \infty)$. If μ stays in a compact set not intersecting $\mathbb{R}$, then by (6.3.2) m_r is asymptotically in the *Weyl disk* for this equation for the interval $(0, c)$ for any $c > 0$, so $m_r(\mu) \to 1/(p(a)\sqrt{-\mu w(a)/p(a)})$ as $r \to \infty$. But this is the claim of the theorem.   □

A similar result holds for the left-definite case with a very similar proof, so we shall be content to state the theorem without proof (Exercise 6.3.1).

Theorem 6.3.3. *Suppose a is a Lebesgue point for w, and that $w(a) \neq 0$. Then a Dirichlet m-function for the left-definite case of $-u'' + qu = \lambda w u$ on (a, b) is asymptotic to $w(a)/\sqrt{-\lambda w(a)}$ as $\lambda \to \infty$ in any double sector $|\text{Re } \lambda| \leq C |\text{Im } \lambda|$. The square root is the root with a positive real part.*

Exercise

Exercise 6.3.1 Prove Theorem 6.3.3.

6.4 Asymptotics of solutions

The principal aim of this section is to prove estimates for solutions of $-(pu')' + qu = \lambda w u$ as λ becomes large. An estimate for the Weyl solution ψ, as defined for the right- and left-definite cases, is given in Theorem 6.4.1. An easy consequence of this is an estimate for Green's function of a boundary value problem associated with our equation. The results are stated in Theorem 6.4.4. We will also deduce an estimate for the radius of the *Weyl disk*, Theorem 6.4.3. The basic result is the following.

Theorem 6.4.1. *In both the right- and left-definite cases and for any x, $a < x < b$, we have*

$$\psi(x, \lambda) = \exp\left\{ -\int\limits_{a}^{x} \sqrt{-\lambda w/p} + o(\sqrt{|\lambda|}) \right\}$$

as $\lambda \to \infty$ in a fixed double sector $|\text{Re } \lambda| \leq C |\text{Im } \lambda|$. Here ψ is the Weyl solution for any separated boundary conditions at a and b appropriate to the right- and *left-definite cases respectively. The error in the exponent is locally uniform in x and the root is that with a positive real part. The same estimate holds for $p\psi'(x, \lambda)$.*

The theorem shows that $\psi(x, \lambda)$ and $p\psi'(x, \lambda)$ tend exponentially to 0 as $\lambda \to \infty$ in any double sector $|\text{Re } \lambda| \le C |\text{Im } \lambda|$ provided supp w intersects (a, x). Before turning to the proof we will state and prove some simple consequences.

Corollary 6.4.2. *Let u be a solution of $-(pu')' + qu = \lambda wu$ with fixed non-trivial initial data in a. If A, B are constants, not both equal to 0, we then have*

$$Au(x, \lambda) + Bpu'(x, \lambda) = \exp\left\{ \int_a^x \sqrt{-\lambda w/p} + o(\sqrt{|\lambda|}) \right\}.$$

The estimate is valid as $\lambda \to \infty$ in any double sector $|\text{Re } \lambda| \le C |\text{Im } \lambda|$ with the error in the exponent locally uniform in x.

Proof. $\psi = \theta + m\varphi$ where θ and φ are the solutions with initial data (4.3.3) for $\alpha = \pi/2$. Thus $\mathcal{W}_p(\varphi, \psi) = 1$ so that in the right-definite case setting

$$\begin{cases} m(x, \lambda) = -\psi(x, \lambda)/p\psi'(x, \lambda) , \\ \tilde{m}(x, \lambda) = \varphi(x, \lambda)/p\varphi'(x, \lambda) \end{cases} \tag{6.4.1}$$

we obtain

$$\begin{aligned} 1 = \mathcal{W}_p(\varphi, \psi) &= p\varphi'(x, \lambda)\psi(x, \lambda) - \varphi(x, \lambda)p\psi'(x, \lambda) \\ &= -p\psi'(x, \lambda)p\varphi'(x, \lambda)(m(x, \lambda) + \tilde{m}(x, \lambda)) . \end{aligned}$$

Here $m(x, \lambda)$ is a Neumann m-function for the interval $[x, b)$ and $\tilde{m}(x, \lambda)$ a Neumann m-function for the interval $(a, x]$ (with initial point x).

Thus $m(x, \lambda) + \tilde{m}(x, \lambda)$ is a Nevanlinna function for every $x \in (a, b)$ and therefore its logarithm is $O(\log |\lambda|)$ by Theorem E.1.6. It follows from Theorem 6.4.1 that $p\varphi'(x, \lambda)$ has the claimed growth. Since $\varphi(x, \lambda) = \tilde{m}(x, \lambda)p\varphi'(x, \lambda)$, where $\tilde{m}(x, \lambda) = \exp(O(\log |\lambda|)$, the estimate of growth for $\varphi(x, \lambda)$ is the same.

For a general solution u with fixed initial data it is easily seen that $u(x, \lambda) = pu'(a)\psi(x, \lambda) - (u(a) + m(\lambda)pu'(a))\varphi(x, \lambda)$ with a similar formula for $pu'(x, \lambda)$. Since $u(a) + m(\lambda)pu'(a) = \exp(O(\log |\lambda|))$ and $\psi(x, \lambda) = o(|\varphi(x, \lambda)|)$ with a similar estimate for the quasi-derivative we have

$$\begin{aligned} Au(x, \lambda) &+ Bpu'(x, \lambda) \\ &= \exp(O(\log |\lambda|))(A\varphi(x, \lambda)(1 + o(1)) + Bp\varphi'(x, \lambda)(1 + o(1))) \\ &= \exp(O(\log |\lambda|))(A\tilde{m}(x, \lambda)(1 + o(1)) + B(1 + o(1)))p\varphi'(x, \lambda) \\ &= \exp(O(\log |\lambda|))p\varphi'(x, \lambda) \end{aligned}$$

from which the claim follows in the right-definite case. In the left-definite case a factor λ must be added in the denominators of (6.4.1), but does not materially affect the calculations. □

In the next theorem the Weyl disk at x refers to the disk in the m-plane defined by $\|\theta(\cdot, \lambda) + m\varphi(\cdot, \lambda)\|_x^2 \leq \operatorname{Im} m / \operatorname{Im} \lambda$, where $\|\cdot\|_x$ is as in Section 6.2 in the right-definite case and $\|u\|_x^2 = \int_a^x (|u'|^2 + q|u|)$ in the left-definite case.

Theorem 6.4.3. *In both right- and left-definite cases, and for any symmetric boundary condition at a, the radius $R(x, \lambda)$ of the Weyl disk at x satisfies*

$$R(x, \lambda) = \exp\left\{-2 \int_a^x \operatorname{Re} \sqrt{-\lambda w/p} + o(\sqrt{|\lambda|})\right\}$$

as $\lambda \to \infty$ in any fixed double sector $|\operatorname{Re} \lambda| \leq C |\operatorname{Im} \lambda|$. The error in the exponent is locally uniform in x.

Proof. The radius of the Weyl disk at x is easily seen to be $(2 |\operatorname{Im} \lambda| \|\varphi\|_x^2)^{-1}$ in both the right- and left-definite cases, where φ is a solution of $-(p\varphi')' + q\varphi = \lambda w\varphi$ satisfying the appropriate initial conditions at a; see Exercise 6.4.2. In the right-definite case we have

$$\operatorname{Im} \lambda \|\varphi\|_x^2 = \operatorname{Im}(p\varphi'(x, \lambda)\overline{\varphi(x, \lambda)}) = |\varphi(x, \lambda)|^2 \operatorname{Im}(p\varphi'(x, \lambda)/\varphi(x, \lambda)) .$$

Thus $p\varphi'/\varphi$ is a non-trivial Nevanlinna function, so the last factor is $\exp(O(|\lambda|))$ by Theorem E.1.6 whereas the first factor has the desired form by Corollary 6.4.2. Very similar calculations prove the theorem in the left-definite case. □

Theorem 6.4.4. *Denote by $g(x, y, \lambda)$ Green's function for a right-definite (separated) boundary value problem for $-(pu')' + qu = \lambda wu$. Then, if $x < y$ and with locally uniform error in the exponent as $\lambda \to \infty$ in any double sector $|\operatorname{Re} \lambda| \leq C |\operatorname{Im} \lambda|$,*

$$g(x, y, \lambda) = \exp\left\{-\int_x^y \sqrt{-\lambda w/p} + o(|\sqrt{\lambda}|)\right\} = g(y, x, \lambda) .$$

If x is a Lebesgue point for $1/p$ and w, thus a.e. in supp w,

$$g(x, x, \lambda) = \frac{1}{2p(x)\sqrt{-\lambda w(x)/p(x)}} + o(|\lambda|^{-1/2}) .$$

We could easily give a result analogous to Theorem 6.4.4 but for the kernel of $u \mapsto p(R_\lambda u)'$ or the similar kernels for the left-definite case; this will be obvious from the proof; see Exercise 6.4.1. Some more information about the behavior of Green's function on the diagonal $y = x$ will be given in the course of the proof.

Proof. In the right-definite case we have

$$g(x, y, \lambda) = \psi_-(x, \lambda)\psi_+(y, \lambda)/(m_+ + m_-)$$

when $x < y$ by Theorem 4.4.3. Referring to the notation used in this theorem the 'anchor point' c is arbitrary in (a, b) so choose $c = x$. Applying Theorem 6.4.1 to $\psi_+(y, \lambda)$ and noting that $g(x, y, \lambda)/\psi_+(y, \lambda)$ equals $m_-(\lambda)/(m_+(\lambda) + m_-(\lambda))$, the logarithm of the absolute value of which is $O(\log|\lambda|)$ by Theorem E.1.6, we immediately get the first part of the theorem.

Similarly, for $y = x$ we have $g(x, x, \lambda) = m_+ m_-/(m_+ + m_-)$. It is therefore clear that any asymptotic information on m_+ and m_- may be translated into information about Green's function on the diagonal. In particular the results of Sections 6.2 and 6.3 are available, and the statement of the theorem is a direct consequence of Theorem 6.3.1.

The proof in the left-definite case is similar.   □

For the proof of Theorem 6.4.1 we need the notation of Definition 6.2.1 and the following simple lemma.

Lemma 6.4.5. *The functions of Definition 6.2.1 are continuous functions of (x, y) for $x, y \in (a, b)$.*

Proof. This is clear for P, W, Q and $\tilde{Q}$. It is also clear that if $a \le x \le y < b$ then $\tilde{P}(x, y) \le \int_x^y 1/|p|$ and $\tilde{W}(x, y) \le \int_x^y |w|$. Since $\tilde{W}(x, y) = \tilde{W}(y, x)$ we need only prove this for $x \le y$. If $a \le x \le y \le b$ and $a \le u \le v < b$ we put $c = \min(x, u)$ and $d = \max(y, v)$ so that if $c \le d$, then

$$\tilde{W}(x, y) \le \tilde{W}(c, d) \le \tilde{W}(c, u) + \tilde{W}(u, v) + \tilde{W}(v, y) \le \left|\int_x^u |w|\right| + \tilde{W}(u, v) + \left|\int_v^y |w|\right|.$$

Interchanging (x, y) and (u, v) we obtain $\left|\tilde{W}(x, y) - \tilde{W}(u, v)\right| \le \left|\int_x^u |w|\right| + \left|\int_v^y |w|\right|$, thereby proving the lemma.   □

Proof (Theorem 6.4.1). Let $m(t, \lambda)$ be the function defined in (6.4.1). If ψ is a Neumann Weyl solution and ψ_α a Weyl solution for the boundary condition (4.2.2) with the same boundary condition at b, it follows easily from (6.2.4) that $\psi/\psi_\alpha = \sin\alpha - m_{\pi/2}\cos\alpha$ so that $\psi(\cdot, \lambda)/\psi_\alpha(\cdot, \lambda) = \exp(O(\log|\lambda|))$ by Theorem E.1.6. It follows that it is enough to prove the theorem when ψ is a Neumann Weyl solution, which we now assume.

Multiplying (6.4.1) by $\lambda w - q$ and using the differential equation we obtain $(p\psi')'(\cdot, \lambda) = (\lambda w - q)m(\cdot, \lambda)p\psi'(\cdot, \lambda)$ so since $p\psi'(a, \lambda) = 1$ we have

$$p\psi'(x, \lambda) = \exp\left\{\int_a^x (\lambda w - q)m(\cdot, \lambda)\right\}.$$

The theorem will follow, in the right-definite case, by calculating the asymptotic form of the exponent. The left-definite case follows in the same way from the formula $\psi(x, \lambda) = \exp\{\int_a^x \lambda m(\cdot, \lambda)\}$, where now $m(t, \lambda) = \psi'(t, \lambda)/(\lambda\psi(t, \lambda))$ and ψ is a left-definite Weyl solution for the Dirichlet condition at a. We will leave the details of the proof of this case to the reader in Exercise 6.4.3.

If $x < \sup(\operatorname{supp} w)$ we can for any t, $a \leq t \leq x$, find $c = c(t, |\lambda|)$, $t < c < b$ so $\tilde{P}(c, t)(|\lambda| W(c, t) + \tilde{Q}(c, t)) = 1/5$ if $|\lambda|$ is sufficiently large. In fact, it is clear that if we determine $B > 0$ so that we can find c for $|\lambda| \geq B$ when $t = x$, then a c exists for such λ and all $t \in [a, x]$, and $c(t, |\lambda|) \leq c(x, |\lambda|) \leq c(x, B)$. We write $d = c(x, B) < b$.

By Theorem 6.2.6 (1) $|m(t, \lambda)| \leq D/(|\lambda| W(c, t))$ for some constant D if λ is in the double sector with $|\lambda| \geq B$ if only $W(c, t) > 0$. To ensure this we make the assumption, which will be removed later, that there is no subinterval (t, s) of (a, d) with $t \leq x$ such that $w = 0$ a.e. in (t, s) while $\tilde{P}(c, t)\tilde{Q}(c, t) \geq 1/10$. This means a restriction on the size of the gaps in $\operatorname{supp} w \cap (a, d)$, i.e., on the size of subintervals of (a, d) on which $w = 0$ a.e.

Now consider the set

$$K = \left\{ (t, s) : a \leq t \leq x, \ t \leq s \leq d \text{ and } \tilde{P}(s, t)\tilde{Q}(s, t) \geq 1/10 \right\}.$$

It follows that K is a compact subset of $\mathbb{R}^2$. Thus $W(s, t)$ has a minimum M on K which is strictly positive since otherwise our restriction on the size of gaps in $\operatorname{supp} w$ would be violated.

Now fix $t \in [a, x]$ and (assuming $|\lambda| \geq B$) determine c so that $\tilde{P}(c, t)(|\lambda| W(c, t) + \tilde{Q}(c, t)) = 1/5$. If $(t, c) \in K$ we have $W(c, t) \geq M$, and if $(t, c) \notin K$ we obtain $\tilde{P}(c, t)\tilde{Q}(c, t) < 1/10$, so that $|\lambda| \tilde{P}(c, t)W(c, t) \geq 1/10$. It follows that we always have $1/(|\lambda| W(c, t)) \leq 10\tilde{P}(c, t) + 1/(|\lambda| M)$, so by Theorem 6.2.6 (1)

$$|m(t, \lambda)| \leq \tilde{D}(\tilde{P}(c, t) + 1/|\lambda|) \leq \tilde{D}(\tilde{P}(c(x, B), a) + 1/B)$$

for some constant $\tilde{D}$ if $|\lambda| \geq B$ and λ is in the double sector.

This ensures that $m(\cdot, \lambda)q$ is bounded by a constant multiple of $|q|$ as soon as $|\lambda| \geq B$ and λ is in the double sector so that $\int_a^x m(\cdot, \lambda)q$ is bounded for such λ.

We also obtain $\sqrt{|\lambda|}\, |m(t, \lambda)| w \leq \tilde{D}\sqrt{|\lambda|}\tilde{P}(c, t)w + w\tilde{D}/\sqrt{B}$, where the last term is integrable over (a, x) and independent of λ. Now $|\lambda| \tilde{P}(c, t)W(c, t) \leq 1/5 < 1$ so that

$$\sqrt{|\lambda|}\tilde{P}(c, t) \leq \sqrt{\tilde{P}(c, t)/W(c, t)} \leq \sqrt{P_w(t)},$$

where P_w is the (one-sided) *maximal function*

$$P_w(t) = \sup_{t \leq s \leq d} \frac{\int_t^s 1/|p|}{\int_t^s w}.$$

According to the Hardy maximal theorem (Lemma B.8.9) we have

$$\int_{M_s} w \leq \frac{3}{s} \int_a^d 1/|p|, \tag{6.4.2}$$

where $M_s = \{t \in (a, d) : P_w(t) > s > 0\}$. Now

$$\int_a^d \sqrt{P_w}\, w = \int_a^d \left(\int_0^{\sqrt{P_w}} ds \right) w = \int_0^\infty \int_{M_{s^2}} w\, ds = \left(\int_0^1 + \int_1^\infty \right) \int_{M_{s^2}} w\, ds$$

$$\leq \int_a^d w + \int_a^d \frac{1}{|p|} w \int_1^\infty \frac{3}{s^2}\, ds = \int_a^d w + 3 \int_a^d \frac{1}{|p|} < \infty .$$

It follows that $\sqrt{|\lambda|}\, |m(t, \lambda)|$ is bounded on (a, x) by a function integrable with weight w. The theorem now follows, under the restriction of the size of gaps of supp w assumed, by dominated convergence from the pointwise result of Theorem 6.3.1.

To complete the proof we must remove this restriction. We have just proved that the theorem is true for x up to and including the left endpoint d in the first gap violating the restriction. For x in this gap we have

$$\psi(x, \lambda) = \psi(d, \lambda)u(x) + p\psi'(d, \lambda)v(x) ,$$

where u and v are solutions of $-(pu')' + qu = 0$ with appropriate λ-independent initial data in d. Since $m(d, \cdot)$ is a non-trivial Nevanlinna function it follows that $p\psi'(x, \lambda)/p\psi'(d, \lambda) = \exp\{O(\log |\lambda|)\}$ so the theorem holds up to and including the right endpoint of the gap and therefore up to and including the first point in the next large gap (if any). Since the support of w can only have a finite number of such large gaps in any compact interval this completes the proof. $\square$

Exercises

Exercise 6.4.1 Give a result analogous to Theorem 6.4.4 but for the kernel of $u \mapsto p(R_\lambda u)'$. Also prove corresponding results for the left-definite case.

Exercise 6.4.2 Prove that the radius of the circle defined by $\|\theta(\cdot, \lambda) + m\varphi(\cdot, \lambda)\|_x^2 \leq \operatorname{Im} m / \operatorname{Im} \lambda$ in both the right- and left-definite cases is given by $(2 |\operatorname{Im} \lambda| \|\varphi\|_x^2)^{-1}$.

Hint: The inequality $\|\psi\|_x^2 \leq \operatorname{Im} m / \operatorname{Im} \lambda$ may be rewritten as $|m - c|^2 \leq r^2$, where $r^2 = A(2 \operatorname{Im} \lambda \|\varphi\|_x^2)^{-2}$ and

$$A = |2\mathrm{i} \operatorname{Im} \lambda \langle \theta, \varphi \rangle_x + 1|^2 + 2\mathrm{i} \operatorname{Im} \lambda \|\theta\|_x^2\, 2\mathrm{i} \operatorname{Im} \lambda \|\varphi\|_x^2 .$$

Use the differential equation and integration by parts to write $\lambda \langle \theta, \varphi \rangle_x$ as an integrated term plus $\overline{\lambda} \langle \theta, \varphi \rangle_x$. Similarly for $\lambda \|\theta\|_x^2$ and $\lambda \|\varphi\|_x^2$. Now rearrange the resulting terms to show that $A = |\mathcal{W}_p(\varphi, \theta)|^2 = 1$.

Exercise 6.4.3 Prove Theorem 6.4.1 in the left-definite case.

6.5 Other spectral quantities

In this section we will use the results of Sections 6.3 and 6.4 to obtain asymptotic estimates of some other important spectral quantities. These results are all obtained by use of the following simple theorem of Tauberian type.

Theorem 6.5.1. *Suppose m and $\tilde{m}$ are Nevanlinna functions with spectral functions η and $\tilde{\eta}$ respectively. Let f be a positive function defined for large r and assume that*

$$m(r\lambda) \sim f(r)\tilde{m}(\lambda) \tag{6.5.1}$$

as $r \to \infty$, locally uniformly for non-real λ.

Then $\eta(\pm t) = (\tilde{\eta}(\pm 1) + o(1)) \, t \, f(t)$ as $t \to \infty$ except if $\tilde{\eta}$ jumps at 0. In this case only $\eta(t) - \eta(-t) \sim (\tilde{\eta}(1) - \tilde{\eta}(-1)) \, t \, f(t)$ may be concluded.

Remark 6.5.2. Arguing as in the beginning of Section 6.3 the function f must have a limit-order at infinity and $\tilde{m}$ is homogeneous of that order. It is easy to see that the theorem can be phrased as a (bilateral) Tauberian theorem of a more conventional kind; it then becomes a bilateral version of a well-known Tauberian theorem of Karamata [119]; see also Bingham, Goldie and Teugels [24, Theorem 1.7.4].

Proof. Since $\tilde{m}$ is homogeneous so is its spectral function $\tilde{\eta}$, and of degree one more, as is seen by the Stieltjes inversion formula Theorem E.1.5. Now $\tilde{m}$ is a Nevanlinna function so the degree of homogeneity is in $[-1, 1]$. It follows that $\tilde{\eta}$ is continuous unless the degree[4] is -1, in which case $d\tilde{\eta}$ is a positive multiple of the Dirac measure. Setting $\eta_r(t) = \eta(rt)/(r \, f(r))$ the theorem follows if we can prove that $\eta_r \to \tilde{\eta}$ pointwise at every point of continuity of $\tilde{\eta}$ as $r \to \infty$.

We shall prove that any sequence tending to infinity has a subsequence $r_1, r_2, \ldots$ such that η_{r_j} converges pointwise, and that the limit always equals $\tilde{\eta}$ at points of continuity of $\tilde{\eta}$. Thus $d\eta_r \to d\tilde{\eta}$ weakly[5] as $r \to \infty$ along the subsequence. Thus $r \mapsto d\eta_r$ would have a single point of accumulation as $r \to \infty$ so that $d\eta_r \to d\tilde{\eta}$ with no need of taking subsequences.

To prove this we first need to estimate η_r. Assuming η and $\tilde{\eta}$ are normalized by $\eta(0) = \tilde{\eta}(0) = 0$ so that also $\eta_r(0) = 0$ we have, for $s > 0$,

$$\frac{\eta_r(s)}{s^2 + 1} \leq \int\limits_0^s \frac{d\eta_r(t)}{t^2 + 1} \leq \int\limits_{-\infty}^\infty \frac{d\eta_r(t)}{t^2 + 1} = \frac{1}{f(r)} \int\limits_{-\infty}^\infty \frac{r \, d\eta(t)}{t^2 + r^2} \leq \frac{1}{f(r)} \, \mathrm{Im} \, m(ir) \to \mathrm{Im} \, \tilde{m}(i)$$

as $r \to \infty$. A similar calculation may be made for $s < 0$, so for large r we have a bound $|\eta_r(s)| \leq C(s^2 + 1)$ where $C > \mathrm{Im} \, \tilde{m}(i)$ is independent of (r, s).

Now η_r is increasing for fixed r so by a simple variant of Helly's theorem (Lemma E.1.4) (see Exercise 6.5.1) every sequence $\to \infty$ has a subsequence $r_1, r_2, \ldots$ such that η_{r_j} tends pointwise to an increasing function η_∞. Helly's theorem

---

[4] If the degree is 1, then $\tilde{m}(\lambda) = B\lambda$, $d\tilde{\eta} = 0$ and we only get $\eta(\pm t) = o(t f(t))$.

[5] That is, $\int_{-\infty}^\infty \varphi \, d\eta_r \to \int_{-\infty}^\infty \varphi \, d\tilde{\eta}$ for every $\varphi \in C_0(\mathbb{R})$.

also shows the weak convergence of $d\eta_{r_j}$ to $d\eta_\infty$ and, if I is a compact interval, that

$\operatorname{Im} \tilde{m}(i) \geq \int_I \frac{d\eta_{r_j}(t)}{t^2+1} \to \int_I \frac{d\eta_\infty(t)}{t^2+1}$. Thus $\int_{-\infty}^\infty \frac{d\eta_\infty(t)}{t^2+1} \leq \operatorname{Im} \tilde{m}(i) < \infty$.

From $\operatorname{Im} m(r\lambda)/f(r) \to \operatorname{Im} \tilde{m}(\lambda)$ it follows, if $\lambda = x + iy$, that

$$\int\limits_{-\infty}^\infty \frac{y \, d\eta_\infty(t)}{(t-x)^2+y^2} = \lim_{j\to\infty} \int\limits_{-\infty}^\infty \frac{y \, d\eta_{r_j}(t)}{(t-x)^2+y^2} = \int\limits_{-\infty}^\infty \frac{y \, d\tilde{\eta}(t)}{(t-x)^2+y^2}.$$

By the Stieltjes inversion formula (Theorem E.1.5) it follows that $d\eta_\infty = d\tilde{\eta}$, so if 0 is a point of continuity for $\tilde{\eta}$, integrating from 0 gives $\eta_\infty(t) = \tilde{\eta}(t)$. If $\tilde{\eta}$ jumps at 0, integrating from $-t$ to t gives just $\eta_\infty(t) - \eta_\infty(-t) = \tilde{\eta}(t) - \tilde{\eta}(-t)$.   □

Our first result concerns the spectral function.

Theorem 6.5.3. *Let η_α be any spectral function associated with a right-definite equation (4.1.1) and the boundary condition $u(a)\cos\alpha + pu'(a)\sin\alpha = 0$, $0 \leq \alpha < \pi$. Under the assumptions of Theorem 6.3.1 we then have*

$$\begin{cases} \eta_0(t) \sim \dfrac{2}{3\pi}\sqrt{t^3 w(a)p(a)} \\[2mm] \eta_0(-t) = o(|t|^{3/2}) \end{cases} \qquad \text{as } tp(a) \to +\infty.$$

If $0 < \alpha < \pi$, then

$$\begin{cases} \eta_\alpha(t) \sim \dfrac{2}{\pi\sin^2\alpha}\sqrt{t/(w(a)p(a))} \\[2mm] \eta_\alpha(-t) = o(\sqrt{|t|}) \end{cases} \qquad \text{as } tp(a) \to +\infty.$$

Theorem 6.5.4. *Let η_α be any spectral function for a left-definite case of (4.1.1) and the boundary condition $\lambda u(a)\sin\alpha + pu'(a)\cos\alpha = 0$, $0 \leq \alpha < \pi$. Under the assumptions of Theorem 6.3.3 the claims of Theorem 6.5.3 then hold, but for $w(a)t \to +\infty$.*

Lemma 6.5.5. *The spectral functions of $(-\lambda)^{\pm 1/2}$ are $\frac{2}{\pi}t^{1/2}$ respectively $\frac{2}{3\pi}t^{3/2}$ for $t > 0$ and 0 for $t \leq 0$, while the spectral functions of $\lambda^{\pm 1/2}$ are 0 for $t > 0$ and $-\frac{2}{\pi}(-t)^{1/2}$ respectively $-\frac{2}{3\pi}(-t)^{3/2}$ for $t \leq 0$.*

Proof. By the Stieltjes inversion formula (Theorem E.1.5) the spectral function for $(-\mu)^{-1/2}$ is $1/\pi$ times the limit as $\nu \to 0+$ of

$$\int\limits_0^t \operatorname{Im}(-\mu - i\nu)^{-1/2} \, d\mu = \operatorname{Im}(-2(-\mu - i\nu)^{1/2}\big|_0^t \to -2\operatorname{Im}(-t)^{1/2},$$

which is $2t^{1/2}$ if $t > 0$ and 0 otherwise. Similar calculations prove the other claims.   □

It is then obvious that Theorems 6.5.3 and 6.5.4 are direct consequences of this, of Theorems 6.3.1 and 6.3.3 respectively, and of the Tauberian theorem.

Remark 6.5.6. If $p \geq 0$ the estimates on the negative half-axis in Theorem 6.5.3 can be considerably improved; in fact the spectral functions are all bounded from below in this case. See the remark after Corollary 3 in Atkinson [9]. An even stronger statement can be made in the left-definite case if $w \geq 0$. In this case it is easy to see that the spectrum is bounded from below so that every spectral function is constant near minus infinity.

We next consider a self-adjoint realization of the right-definite equation $-(pu')' + qu = \lambda w u$. To allow for interval endpoints of arbitrary character and also coupled boundary conditions we base our considerations on Theorem 4.4.4. If Δ is a bounded interval we have $\mathcal{F}(E_\Delta u) = \chi_\Delta \hat{u}$, where χ_Δ is the characteristic function of Δ. Thus

$$E_\Delta u(x) = \int_\Delta \Phi^*(x, \cdot) \, d\Xi \, \hat{u} = \langle u, \mathcal{F}^{-1}(\chi_\Delta \Phi(x, \cdot)) \rangle$$

so with $e(x, y, \Delta) = \mathcal{F}^{-1}(\chi_\Delta \Phi(x, \cdot))(y)$ we have $E_\Delta u(x) = \langle u, e(x, \cdot, \Delta) \rangle$. Thus E_Δ may be viewed as an integral operator with kernel $e(x, y, \Delta)$. We shall consider the kernel $e(x, y, t)$ which equals $e(x, y, (0, t])$ if $t > 0$ and $e(x, y, (t, 0])$ otherwise. Thus

$$e(x, y, t) = \int_{(0,t]} \Phi^*(x, \cdot) \, d\Xi \, \Phi(y, \cdot) \tag{6.5.2}$$

if $t > 0$, replacing $(0, t]$ by $(t, 0]$ if $t \leq 0$.

The asymptotic behavior as $t \to \pm\infty$ of this kernel is of some interest. It may for example be used to investigate summability of the generalized Fourier expansion, although we shall not give such results here. To prove an asymptotic formula for $e(x, y, t)$ we need the following lemma.

Lemma 6.5.7. $t \mapsto e(x, x, t)$ *is non-decreasing and* $\lambda \mapsto g(x, x, \lambda)$ *is a Nevanlinna function with spectral function* $e(x, x, t)$ *for every* $x \in (a, b)$.

Proof. The first claim follows from (6.5.2), and Theorem 4.3.13 then implies that $\langle u, \overline{g(x, \cdot, \lambda)} \rangle = R_\lambda u(x) = \mathcal{F}^{-1}(\hat{u}(t)/(t - \lambda))(x)$.

Thus $g(x, y, \lambda) = \mathcal{F}^{-1}(\Phi(x, t)/(t - \lambda))(y)$ since $\Phi(x, t)$ is real for real t. For $\lambda = i$ this shows that $\int_{-\infty}^{\infty} \frac{\Phi^*(x,t) \, d\Xi(t) \Phi(x,t)}{t^2 + 1} < \infty$ so that $\int_{-\infty}^{\infty} \frac{\Phi^*(x,t) \, d\Xi(t) \Phi(y,t)}{(t - \lambda)(t - \mu)}$ is absolutely convergent for non-real λ and μ.

This integral equals $\langle g(x, \cdot, \lambda), g(y, \cdot, \overline{\mu}) \rangle$ and $\frac{1}{t - \lambda} - \frac{1}{t - \mu} = \frac{\lambda - \mu}{(t - \lambda)(t - \mu)}$ so we obtain

$$g(x, y, \lambda) - g(x, y, \mu) = (\lambda - \mu) \langle g(x, \cdot, \lambda), g(y, \cdot, \overline{\mu}) \rangle,$$

a formula corresponding to $R_\lambda - R_\mu = (\lambda - \mu) R_\lambda R_\mu$.

For $x = y$ and $\mu = \overline{\lambda}$ this gives

$$\text{Im}\, g(x, x, \lambda) = \text{Im}\, \lambda \|g(x, \cdot, \lambda\|^2 = \int_{-\infty}^{\infty} \frac{\phi^*(x,t)\, d\Xi(t)\Phi(x,t)}{|t-\lambda|^2},$$

together with (6.5.2) showing that $g(x,x,\lambda)$ is a Nevanlinna function with spectral function $e(x,x,t)$.   □

Theorem 6.5.8. *Let $e(x,y,t)$ be the kernel defined above. Then for almost all x and y with respect to $w(x)\, dx$ the following hold*

$$e(x,x;t) = \frac{\sqrt{(w/p)_+(x)}}{\pi w(x)}\sqrt{t} + o(\sqrt{t}) \ \text{as}\ t \to \infty,$$

$$e(x,x;-t) = -\frac{\sqrt{(w/p)_-(x)}}{\pi w(x)}\sqrt{t} + o(\sqrt{t}) \ \text{as}\ t \to \infty,$$

$$e(x,y;\pm t) = o(\sqrt{t}) \ \text{as}\ t \to \infty \ \text{if}\ x \neq y,$$

where $(w/p)_+ = \max(w/p, 0)$ and $(w/p)_- = \max(-w/p, 0)$.

Proof. The main term in the formula for $g(x,x,\lambda)$ in Theorem 6.4.4 may be written $(-\lambda)^{-1/2}\frac{1}{w}\sqrt{(w/p)_+} - \lambda^{1/2}\frac{1}{w}\sqrt{(w/p)_-}$, so the first claims follow directly from this, Lemmas 6.5.5 and 6.5.7 and Theorem 6.5.1. By (6.5.2) and Cauchy–Schwarz we have, for a bounded interval Δ,

$$2\,|e(x,y,\Delta)| \leq 2(e(x,x,\Delta)e(y,y,\Delta))^{1/2} \leq e(x,x,\Delta) + e(y,y,\Delta)$$

so that $f(t) = e(x,x,t) - 2\,\text{Re}(Ce(x,y,t)) + e(y,y,t)$ is increasing if $|C| = 1$. By Theorem 6.4.4, $g(x,x,\lambda - 2\,\text{Re}(Cg(x,y,\lambda)) + g(y,y,\lambda)$ is asymptotically equal to $g(x,x,\lambda) + g(y,y,\lambda)$, so by use of Theorem 6.5.1 $f(t)$ is asymptotically equal to $e(x,x,t) + e(y,y,t)$. Choosing $C = 1$ and $C = i$ now proves the third claim.   □

We shall finally deal with the asymptotic distribution of the eigenvalues of a regular problem. The result is very well known in simple cases, e.g. if $p = w \equiv 1$, when it can be proved by elementary means as we did in Theorem 6.1.11. It seems to be less well known in the case when p changes sign arbitrarily, but for a related result see Atkinson and Mingarelli [11]. We shall need a lemma.

Lemma 6.5.9. *Suppose that $\lambda \mapsto F(\lambda, r)$ and G are analytic in some domain $\Omega \subset \mathbb{C}$ for large r, and that $F(\cdot, r) \to G$ as $r \to \infty$, locally uniformly in Ω. Then $\partial^k F(\lambda, r)/(\partial\lambda)^k \to G^{(k)}(\lambda)$ locally uniformly in Ω, $k = 1, 2, \ldots$ as $r \to \infty$.*

Proof. Let γ be the positively oriented boundary of a closed disk in Ω. If K is a compact subset of the interior of the disk and $\lambda \in K$ we have

$$\frac{\partial^k F(\lambda, r)}{\partial\lambda^k} = \frac{k!}{2\pi i}\oint_\gamma \frac{F(\zeta, r)}{(\zeta - \lambda)^{k+1}}\, d\zeta$$

and a similar formula for $G^{(k)}$. Thus, if d is the distance from K to γ and R is the radius of γ we obtain

$$\sup_K \left| \frac{\partial^k F(\lambda, r)}{\partial \lambda^k} - G^{(k)}(\lambda) \right| \le k! d^{-k-1} R \sup_\gamma |F(\cdot, r) - G| \to 0$$

as $r \to \infty$ proving the claim. □

Theorem 6.5.10. *Let $N(t)$ be the number of eigenvalues between 0 and t of a right- or left-definite regular, self-adjoint boundary value problem of the form (4.1.1). Let $(w/p)_\pm$ denote the positive and negative parts of the function w/p as in Theorem 6.5.8. Then*

$$N(t) = \frac{\sqrt{t}}{\pi} \int_a^b \sqrt{(w/p)_+} + o(\sqrt{t}) \ \text{as} \ t \to \infty \,,$$

$$N(t) = \frac{\sqrt{-t}}{\pi} \int_a^b \sqrt{(w/p)_-} + o(\sqrt{-t}) \ \text{as} \ t \to -\infty \,.$$

Proof. We give the details for the right-definite case; the left-definite case is very similar. In the case of separated boundary conditions, if $\varphi(\cdot, \lambda)$ is the solution of (4.1.1) satisfying the initial conditions (4.3.3), the eigenvalues are the zeros of a certain fixed linear combination $z(\lambda)$ of $\varphi(b, \lambda)$ and $p\varphi'(b, \lambda)$. In the case of coupled boundary conditions we introduce linear independent solutions φ and θ given by fixed initial data at a. An eigenfunction is then a non-trivial linear combination of these solutions, the coefficients of which satisfies a homogeneous linear system of equations with coefficients made up from the values of φ and θ and their quasi-derivatives at b. Thus the eigenvalues are the zeros of the determinant $z(\lambda)$ of this system.

The function z has only real zeros and by Lemma D.6 it is an entire function of order at most $1/2$. Hence, by Theorem E.2.8 it has a product expansion

$$z(\lambda) = C\lambda^k \prod_j \left(1 - \frac{\lambda}{\lambda_j}\right)$$

where C is constant, k is the multiplicity of 0 as a zero of z and we order the other zeros λ_j by size, positive zeros by positive and negative zeros by negative indices. If we introduce $N(t)$ as the number of zeros strictly between 0 and t we obtain

$$\log z(\lambda) = \log C + k \log \lambda + \int_{-\infty}^\infty \log\left(1 - \frac{\lambda}{t}\right) dN(t) \,.$$

In the case of separated boundary conditions we may apply Corollary 6.4.2 to estimate $\log z$. A simple extension gives the same estimate in the case of coupled boundary conditions; see Exercise 6.5.2. For $\lambda = r\mu$ and differentiating according to

Lemma 6.5.9 we obtain

$$\sqrt{r} \int\limits_{-\infty}^{\infty} \frac{dN(t)}{t - r\mu} \to \frac{1}{2\sqrt{\mu}} \int\limits_{a}^{b} \sqrt{(w/p)_-} - \frac{1}{2\sqrt{-\mu}} \int\limits_{a}^{b} \sqrt{(w/p)_+} \;.$$

The theorem now follows from Theorem 6.5.1.   □

Exercises

Exercise 6.5.1 Prove the variant of Helly's theorem used in the proof of Theorem 6.5.1.

Exercise 6.5.2 Prove the extension of Corollary 6.4.2 which is needed to prove Theorem 6.5.10 in the case of coupled boundary conditions.

Also carry out the details of the proof of Theorem 6.5.10 for the left-definite case.

6.6 Special functions

In applications many specific Sturm–Liouville equations turn up repeatedly, and the solutions of these equations were intensively studied in the 19th century. These functions are known as *special functions*. In order to illustrate how special functions turn up in applications we shall discuss the solution of the Schrödinger equation for the hydrogen atom, as given[6] by Erwin Schrödinger [202] in 1926. This was the first spectacular success of the new wave mechanics based on the equation invented by Schrödinger, establishing it as a theory to be taken seriously.

The hydrogen atom. The motion of an electron in the electrical field of a proton (which itself does not move) is described by Schrödinger's equation

$$i\hbar\Psi_t = H\Psi \;,$$

where the operator H, the Hamiltonian, is given by

$$H = -\frac{\hbar^2}{2m}\Delta - \frac{e^2}{4\pi\varepsilon_0 r} \;.$$

Here Ψ is a function, normalized in $L^2(\mathbb{R}^3)$, of the space variable $\vec{r} = (x, y, z) \in \mathbb{R}^3$, the location of the electron, and the time variable $t \in [0, \infty)$. By Remark 3.1.8 its physical significance is that $\int_M |\Psi(\cdot, t)|^2$ gives the probability of finding the electron in the set $M \subset \mathbb{R}^3$ at time t. The physical constants m, e, $2\pi\hbar$, and ε_0 are, respectively,

---

[6] Our development is substantially different.

the mass of the electron, the charges (of opposite signs) of the electron and proton, Planck's constant, and the vacuum permittivity. Δ is the Laplace operator

$$\Delta u = u_{xx} + u_{yy} + u_{zz} ,$$

and $r = \sqrt{x^2 + y^2 + z^2}$ is the distance of a point $\vec{r} = (x, y, z)$ to the origin, which is the location of the proton. We seek solutions $\Psi(\cdot, t) \in L^2(\mathbb{R}^3)$ for which $\Psi(\cdot, 0)$ equals a given function ψ. Normalizing Ψ in $L^2(\mathbb{R}^3)$ the quantity $\hbar^2/(2m) \int_{\mathbb{R}^3} |\nabla \Psi(\cdot, t)|^2$ is the kinetic energy of the electron while $-e^2/(4\pi\varepsilon_0) \int_{\mathbb{R}^3} r^{-1} |\Psi(r, t)|^2 \, dr$ is the potential energy of the electron. Thus we also require that $|\Psi(r, t)|^2 / r$ and $|\nabla \Psi(\cdot, t)|^2$ are in $L^1(\mathbb{R}^3)$ for all $t \geq 0$.

The formal operator H with domain $C_0^\infty(\mathbb{R}^3)$ is essentially self-adjoint in $L^2(\mathbb{R}^3)$ and by Remark 3.1.8 $\Psi(\vec{r}, t) = \exp(-itH/\hbar)\Psi(\vec{r}, 0)$, so to understand the dynamics of the system we need to study the spectral properties of H, in a similar way to our example of heat conduction in Section 1.1. After choosing the units of measurement so that $\hbar^2/(2m) = e^2/(4\pi\epsilon_0) = 1$ we arrive at the time-independent Schrödinger equation

$$(-\Delta - \frac{1}{r})\psi = E\psi , \tag{6.6.1}$$

where ψ is a function of the space variables and E is a separation constant, describing the total energy of the system. The equation is *elliptic* in $\mathbb{R}^3 \setminus \{0\}$, which means any solution is smooth outside the origin; see e.g. Folland [73] or John [114].

According to Remark 3.1.8 we are especially interested in stationary states, i.e., eigenfunctions of (6.6.1). Since the potential is radially symmetric we separate variables further and assume that $\psi(\vec{r}) = r^{-1} R(r) P(\xi) \Phi(\varphi)$ where $x = r\sqrt{1 - \xi^2} \cos\varphi$, $y = r\sqrt{1 - \xi^2} \sin\varphi$, and $z = r\xi$ with $r \in [0, \infty)$, $\xi \in [-1, 1]$ and $\varphi \in [0, 2\pi)$. It is then easy to see (Exercise 6.6.1) that $\psi \in L^2(\mathbb{R}^3)$ precisely if $R \in L^2(0, \infty)$, $P \in L^2(-1, 1)$ and $\Phi \in L^2(0, 2\pi)$.

Remark 6.6.1. To understand why it is so important to determine the eigenvalues of Schrödinger's model of the atom, note that for any atom few things are possible to observe except its emission and absorption spectra. These are caused by changes from one stationary energy level to another. Every such change causes the atom to emit or absorb a photon of energy, i.e., color, equal to the change in energy. But the stationary energy levels are the eigenvalues of the Schrödinger equation, so the model is evaluated by comparing calculated eigenvalue differences with observed light-spectra.

Expressed in the coordinates (r, ξ, φ) the Laplace operator reads

$$\Delta u = \frac{1}{r^2}(r^2 u_r)_r + \frac{1}{r^2}((1 - \xi^2)u_\xi)_\xi + \frac{1}{r^2(1 - \xi^2)}u_{\varphi\varphi} ,$$

and the Schrödinger equation becomes

$$-r^2(\frac{R''}{R} + \frac{1}{r} + E) = \frac{1}{P}((1 - \xi^2)P')' + \frac{1}{\Phi(1 - \xi^2)}\Phi'' .$$

The terms on the left do not depend on ξ or φ while the terms on the right do not depend on r. Therefore both left- and right-hand sides of this equation equal a constant, which we denote by $-\lambda$ and we arrive at the equations

$$- R'' + (\frac{\lambda}{r^2} - \frac{1}{r})R = ER \qquad (6.6.2)$$

and

$$\frac{1-\xi^2}{P}((1-\xi^2)P')' + \lambda(1-\xi^2) = -\frac{1}{\Phi}\Phi'' .$$

Again we get the existence of a constant, say μ, such that

$$\Phi'' + \mu\Phi = 0 \qquad (6.6.3)$$

and

$$- ((1-\xi^2)P')' + \frac{\mu}{1-\xi^2}P = \lambda P . \qquad (6.6.4)$$

We see that each of the equations (6.6.2) – (6.6.4) is a Sturm–Liouville equation. The simplest equation is (6.6.3), which is defined on $(0, 2\pi)$. In order for ψ to be smooth we must require solutions of (6.6.3) and their derivatives to be periodic with period 2π, so we consider (6.6.3) in $L^2(0, 2\pi)$ with periodic boundary conditions. This is a regular self-adjoint problem with eigenvalues $\mu_m = m^2$ where $m = 0, 1, 2, \ldots$. If $m = 0$ the eigenvalue is simple, otherwise double.

Equation (6.6.4) is the *(associated) Legendre equation*. Smoothness of ψ requires that the values of $P(\xi)\Phi(\varphi)$ for $\xi = \pm 1$ are independent of φ. For $\mu = 0$ this requires at least continuity of P at ± 1, and for $\mu = m^2$ with $m = 1, 2, 3, \ldots$ we must have $P \to 0$ at ± 1. As we will see below the corresponding restriction of the maximal operator in $L^2(-1, 1)$ is self-adjoint with eigenvalues $\lambda_\ell = (\ell+m)(\ell+m+1)$, where ℓ is a non-negative integer.

Finally, consider the radial equation (6.6.2) for $\lambda \geq 0$. The equation is limit-point at infinity for all λ by Theorem 4.2.7, and we are only looking for solutions in $L^2(0, \infty)$. We shall show that unless $\ell = m = 0$, so that $\lambda = 0$, the equation is limit-point at 0, but if $\lambda = 0$ the equation is limit-circle at 0. However, we shall see that the requirement of finite potential energy forces a boundary condition at 0 which will determine a self-adjoint realization of the equation in $L^2(0, \infty)$.

The essential spectrum is $[0, \infty)$ for all $\lambda \geq 0$, but there are negative eigenvalues $E_n = -(2n)^{-2}$, where $n - \ell - m \geq 1$ is an integer.

Although we will leave the proof to the reader in Exercise 6.6.2, we finally obtain the following theorem.

Theorem 6.6.2. *The essential spectrum of the Schrödinger operator for the hydrogen atom is $[0, \infty)$ and the discrete spectrum is*

$$\sigma_d = \{-1/(4n^2) : n = 1, 2, 3, \ldots\} .$$

The eigenvalue $-1/(4n^2)$ has geometric multiplicity n^2.

The high multiplicity of the eigenvalues for large n is due to the high degree of symmetry of the system.

The Legendre equation. Given a number $m \geq 0$ let $\mathcal{P}$ be the formal operator

$$\mathcal{P}u(x) = -((1 - x^2)u'(x))' + \frac{m^2}{1 - x^2}u(x)$$

and let T_1 with domain $\mathcal{D}_1$ be the corresponding maximal operator in $L^2(-1, 1)$. If $m = 0$, the equation $\mathcal{P}u = \lambda u$ is called Legendre's equation, otherwise the associated Legendre equation. Since we are only interested in solutions of $\mathcal{P}u = \lambda u$ in $L^2(-1, 1)$ we are looking for solutions in $\mathcal{D}_1$. However, we also want our solutions to be continuous at the endpoints of $(-1, 1)$ and if $m > 0$ even vanish there.

Lemma 6.6.3. T_1 *is self-adjoint if $m \geq 1$ and then all solutions of $\mathcal{P}u = \lambda u$ in $\mathcal{D}_1$ are continuous at the endpoints of $(-1, 1)$ and vanish there.*

On the other hand, if $0 \leq m < 1$ there is a unique self-adjoint restriction of T_1 such that all eigenfunctions are continuous at the endpoints of $(-1, 1)$ but no other solutions of $\mathcal{P}u = \lambda u$ are. If $0 < m < 1$ all eigenfunctions vanish at the endpoints.

Proof. The equation $\mathcal{P}u = \lambda u$ may be written

$$(x + 1)^2 u'' + 2x(x - 1)^{-1}(x + 1)u' + (\lambda(1 - x^2) - m^2)(x - 1)^{-2}u = 0,$$

so according to Theorem D.10 its indicial equation at $x = -1$ is $\alpha^2 = (m/2)^2$ with roots $\alpha = \pm m/2$. At $x = 1$ we obtain the same indicial equation. It follows by Corollary 4.2.11 that T_1 is self-adjoint precisely if $m \geq 1$, but limit-circle at both endpoints if $0 \leq m < 1$. If $m \geq 1$ all eigenfunctions vanish at the endpoints.

By the Frobenius method of page 328 there is a solution $\psi_-(x, \lambda) \sim (x + 1)^{m/2}$ at $x = -1$, with $(1 - x^2)\psi'_-(x, \lambda) \sim m(x + 1)^{m/2}$ if $m > 0$. If $m = 0$, the quasi-derivative tends to 0 at -1. If $m > 0$ all solutions u linearly independent of ψ_- satisfy $u(x) \sim A(x + 1)^{-m/2}$ and $(1 - x^2)u'(x) \sim -Am(x + 1)^{-m/2}$ at $x = -1$, with $A \neq 0$. If $m = 0$ we instead have $u(x) \sim A\log(x + 1)$ and $(1 - x^2)u'(x) \to 2A$ at $x = -1$. Corresponding to ψ_- there is a solution $\psi_+(x, \lambda) \sim (1 - x)^{m/2}$ at $x = 1$.

Consider $0 < m < 1$ and let $u_- = \psi_-(\cdot, 0)$ and $u_+ = \psi_+(\cdot, 0)$. For $p = 1 - x^2$ we then have $\mathcal{W}_p(\psi_-, u_-) \to 0$ as $x \to -1$ while for the generic solution u we obtain $\mathcal{W}_p(u, u_-) \to -2Am \neq 0$. Similarly, the condition $\mathcal{W}_p(u, u_+) \to 0$ at 1 selects those solutions u which are multiples of ψ_+, i.e., tend to 0 at 1. However, according to Theorem 4.4.2 the set

$$\mathcal{D} = \{u \in \mathcal{D}_1 : \mathcal{W}_p(u, u_-) \to 0 \text{ at } -1 \text{ and } \mathcal{W}_p(u, u_+) \to 0 \text{ at } 1\}$$

is the domain of a self-adjoint realization of $\mathcal{P}$ in $L^2(-1, 1)$.

It remains to consider the case $m = 0$. Then $u_0 \equiv 1$ is a solution for $\lambda = 0$ so $\mathcal{W}_p(\psi_-, u_0) \to 0$ while $\mathcal{W}_p(u, u_0) \to 2A \neq 0$ as $x \to -1$, with a similar situation at $x = 1$, so the conditions $\mathcal{W}_p(u, u_0) \to 0$ at ± 1 selects precisely the solutions which are continuous at ± 1. Now, according to Theorem 4.4.2, the set

$$\mathcal{D} = \{u \in \mathcal{D}_1 : \mathcal{W}_p(u, u_0) \to 0 \text{ as } x \to \pm 1\}$$

is for $m = 0$ the domain of a self-adjoint realization of $\mathcal{P}$ in $L^2(-1, 1)$ and we are done. $\square$

We must determine the spectrum of these self-adjoint realizations of $\mathcal{P}$ for all $m \geq 0$. Applying the Frobenius method to $\mathcal{P}$ will yield a three-term recursion formula for the coefficients of the expansion which is difficult to analyze. We can simplify this by setting $g(x) = (1 - x^2)^{-m/2}u(x)$. If $\mathcal{P}u = \lambda u$, then g satisfies

$$(1 - x^2)g'' - 2(m + 1)xg' + (\lambda - m(m + 1))g = 0.$$

The points ± 1 are regular singular points, both with indicial equation $\alpha(\alpha + m) = 0$ with roots $\alpha = 0$ and $\alpha = -m \leq 0$. According to Theorem D.10 there exists a solution of the form $g(x) = \sum_{n=0}^{\infty} g_n(1 + x)^{n+\alpha}$ with $g_0 \neq 0$ and $\alpha = 0$. The general theory tells us there is another solution of the indicated form for $\alpha = -m$ unless m is an integer ≥ 0, in which case the second solution may contain a logarithm but still blows up at -1 like $(1 + x)^{-m}$ or, if $m = 0$, like $\log(1 + x)$.

This means that $u(x) = (1 - x^2)^{m/2}g(x)$ is not continuous at -1 unless g is a multiple of the solution for $\alpha = 0$. By symmetry the situation at 1 is precisely the same, so u is an eigenfunction of our equation if and only if g is bounded.

It will be convenient to write $g_n = a_n 2^{-n}$ so that the series for g becomes $g(x) = \sum_0^{\infty} a_n((1 + x)/2)^n$. Inserting this into the differential equation gives a recurrence equation $a_{n+1} = b_n a_n$, $n = 0, 1, 2, \ldots$, for the coefficients, where

$$b_n = \frac{(n + m)(n + m + 1) - \lambda}{(n + 1)(n + m + 1)}.$$

All $b_n \neq 0$ unless $\lambda = (\ell + m)(\ell + m + 1)$ for some non-negative integer ℓ, when $b_\ell = 0$ but all other $b_n \neq 0$. In this case $a_n = 0$ for $n > \ell$, and g is a polynomial of degree ℓ. Since polynomials are bounded on $[-1, 1]$ these polynomials multiplied by $(1 - x^2)^{m/2}$ are all eigenfunctions, the polynomial of degree ℓ being associated with the eigenvalue $\lambda_\ell = (\ell + m)(\ell + m + 1)$. Given $m \geq 0$ we denote the set of these eigenvalues by σ_m.

If λ is real but not in σ_m and $a_0 \neq 0$ and real, then all a_n are real and $\neq 0$. Clearly $b_n > 0$ for all large n, so by possibly changing the sign of a_0 we may, and will, assume $a_n > 0$ for all large n when λ is real and not in σ_m.

Lemma 6.6.4. *If $m \geq 0$ and $\lambda \in \mathbb{R} \setminus \sigma_m$ there is a constant $C > 0$ such that $g(x) \geq h(x) - C\log(1 - x)$ where h is a polynomial.*

Proof. We have

$$\frac{(n + 1)b_n}{n} = 1 + \frac{m(n + m + 1) - \lambda}{n(n + m + 1)},$$

which is ≥ 1 for large n if $m > 0$ or if $m = 0$ and $\lambda \leq 0$.

If $m = 0$ and $\lambda > 0$, then $(n + 1)b_n/n = 1 - \lambda/(n(n + 1))$, which is > 0 for large n. By Proposition E.2.3 the product $\prod_{k=N}^{\infty}(1 - \frac{\lambda}{k(k+1)})$ converges absolutely to a number $D > 0$ if N is large. The partial products decrease, so they are all $\geq D$. It follows that

$$n\,a_n = \prod_{k=N}^{n-1}\left(1 - \frac{\lambda}{k(k+1)}\right) N\,a_N \geq DN\,a_N$$

for $n > N$. Setting $C = N\,a_N$ if $m > 0$ or $m = 0$ and $\lambda \leq 0$, and $C = DN\,a_N$ if $m = 0$ and $\lambda > 0$, and choosing N so large that $a_n > 0$ for $n \geq N$, we obtain $n\,a_n \geq C > 0$ for all $n \geq N$. It follows that

$$\sum_{n=N}^{\infty} a_n\left(\frac{1+x}{2}\right)^n \geq -C\log(1 - (1+x)/2) - s(x) = -C\log(1-x) + C\log 2 - s(x),$$

for $-1 \leq x < 1$, where s is the sum of the first $N - 1$ terms in the expansion of $-C\log((1-x)/2)$ around $x = -1$. $\square$

Theorem 6.6.5. *The spectra of the operators of Lemma 6.6.3 are purely discrete. For a given $m \geq 0$ the eigenvalues are all simple and are the numbers in the set*

$$\sigma_m = \{(\ell + m)(\ell + m + 1) : \ell = 0, 1, 2, \dots\}.$$

Proof. For $0 \leq m < 1$ both interval endpoints are limit-circle so the spectrum is discrete by Corollary 4.3.2, and Lemma 6.6.4 shows that we have found all eigenvalues. If $m \geq 1$ we have still found all eigenvalues, but we must show that there is no essential spectrum.

We have shown that there are solutions $\psi_\pm$ of $\mathcal{P}u = \lambda u$ with $\psi_-(x,\lambda) \sim (1+x)^{m/2}$ at $x = -1$ and $\psi_+(x,\lambda) \sim (1-x)^{m/2}$ at $x = 1$, so ψ_- satisfies the boundary condition at -1 and ψ_+ the boundary condition at 1. The functions $\psi_\pm(\cdot,\lambda)$ are linearly independent if $\lambda \notin \sigma_m$, so for such λ Green's function as given in Theorem 4.4.3 is well defined. There is therefore no essential spectrum. $\square$

The radial equation. Define the formal operator $\mathcal{R}$ for $\lambda \geq 0$ by

$$\mathcal{R}u(r) = -u''(r) + (\lambda/r^2 - 1/r)u(r),$$

and let T_1 be the associated maximal operator in $L^2(0, \infty)$. The condition that an eigenfunction of Schrödinger's equation should have finite potential energy requires that we only consider solutions of $\mathcal{R}u = Eu$ for which

$$\int_0^{\infty} r^{-1}|u(r)|^2\,dr < \infty. \tag{6.6.5}$$

Lemma 6.6.6. *All self-adjoint realizations of $\mathcal{R}$ in $L^2(0, \infty)$ have essential spectrum $[0, \infty)$. If $\lambda \geq 3/4$, then T_1 is self-adjoint and the condition (6.6.5) is satisfied for all eigenfunctions u.*

If $0 \leq \lambda < 3/4$ there is a unique self-adjoint restriction of T_1 such that (6.6.5) is satisfied for all eigenfunctions but for no other solutions of $\mathcal{R}u = Eu$ in $\mathcal{D}_1$.

Proof. By Theorem 4.2.7 T_1 is limit-point at infinity. Consider now the equation $\mathcal{R}u = Eu$ at the regular singular endpoint 0. The indicial equation for $\lambda \geq 0$ is $\lambda = \alpha(\alpha - 1)$ with roots $\alpha \geq 1$ and $\alpha' = 1 - \alpha \leq 0$. It follows that T_1 is limit-point also at 0, and thus self-adjoint, for $\alpha' \leq -1/2$, giving $\lambda \geq 3/4$, and limit-circle for $-1/2 < \alpha' \leq 0$, giving $0 \leq \lambda < 3/4$.

The Frobenius method shows there is a solution ψ such that $\psi(x, E) \sim x^\alpha$ at 0, with derivative $\psi'(x, E) \sim \alpha x^{\alpha-1}$. All solutions u linearly independent of $\psi(\cdot, E)$ are asymptotic to $Ax^{1-\alpha}$ with derivative asymptotic to $A(1 - \alpha)x^{-\alpha}$ at 0 and $A \neq 0$ constant. Let $u_0 = \psi(\cdot, 0)$. Then $\mathcal{W}(u, u_0) \to A(1 - 2\alpha) \neq 0$ at 0 for solutions u linearly independent to $\psi(\cdot, E)$, but $\mathcal{W}(\psi, u_0) \to 0$.

To satisfy the condition (6.6.5) we must exclude those solutions of $\mathcal{R}u = Eu$ in $\mathcal{D}_1$ which are not asymptotic to a multiple of x^α at 0. If $\lambda \geq 3/4$ this is automatic, but if $0 \leq \lambda < 3/4$ we need to restrict $\mathcal{D}_1$. Setting

$$\mathcal{D} = \{u \in \mathcal{D}_1 : \mathcal{W}(u, u_0) \to 0 \text{ at } 0\}$$

accomplishes precisely what is required, and $\mathcal{D}$ is the domain of a self-adjoint restriction of T_1 by Theorem 4.4.2.

If $0 \leq \lambda < 3/4$ the equation $\mathcal{R}u = Eu$ is limit-circle at 0. Thus considered in $L^2(0, 1)$ both endpoints are limit-circle so the essential spectrum is empty. By the comparison Theorem 6.1.1 all solutions of $\mathcal{R}u = Eu$ have finitely many zeros even for $\lambda \geq 3/4$. Hence there can be no essential spectrum below any E, and the essential spectrum has to be empty. Since $\lambda/r^2 - 1/r \to 0$ at infinity Corollary 4.5.6 shows that any self-adjoint realization of $\mathcal{R}$ in $L^2(1, \infty)$ has essential spectrum $[0, \infty)$, so by Proposition 4.5.4 this is also the essential spectrum for any self-adjoint restriction of T_1.   □

We must find the discrete spectrum of the restriction with domain $\mathcal{D}$. Using the Frobenius method to determine a solution of $\mathcal{R}u = Eu$ leads to a three-term recursion formula. As in the case of the Legendre equation we shall therefore transform the equation into one for which we obtain a two-term recursion. Let $v(x) = e^{\beta x}u(x)$ for $\beta = \sqrt{-E} > 0$. Since $\lambda = \alpha(\alpha - 1)$ the equation becomes

$$x^2 v'' - 2\beta x^2 v' + (x - \alpha(\alpha - 1))v = 0 \ . \tag{6.6.6}$$

Theorem 6.6.7. *The eigenvalues in the discrete spectrum of the operators given in Lemma 6.6.6 are all simple and equal* $-(2(n + \alpha))^{-2}$, $n = 0, 1, 2, \ldots$. *In particular, if* $\lambda = (\ell + m + 1)(\ell + m)$ *as in Theorem 6.6.5 the eigenvalues are the numbers* $-(2(n + \ell + m + 1))^{-2}$, $n = 0, 1, 2, \ldots$.

Proof. We are looking for a solution $v(x) = \sum_{k=0}^{\infty} a_k x^{k+\alpha}$ with $a_0 \neq 0$ of (6.6.6). This gives a recursion formula

$$(k + 1)(k + 2\alpha)a_{k+1} = (2\beta(k + \alpha) - 1)a_k \ , \qquad k = 0, 1, 2, \ldots \ .$$

If $2\beta(n+\alpha) = 1$ for some integer $n \geq 0$, then $a_k = 0$ for $k > n$ and the solution is x^α times a polynomial q_n of degree n, so that the corresponding $u_n(x) = q_n(x)x^\alpha e^{-\beta x}$ is in $L^2(0, \infty)$. Such solutions are therefore eigenfunctions, and since $\beta = \sqrt{-E}$ the corresponding eigenvalue is $E_n = -(2(n + \alpha))^{-2}$.

We must show that there are no more eigenvalues < 0. We then note that the recursion formula shows that $(k + 1)! h^{-k-1} a_{k+1} = b_k k! h^{-k} a_k$, where h is a parameter to be chosen below, and

$$b_k = \frac{2\beta(k + \alpha) - 1}{h(k + 2\alpha)} \, .$$

Now $b_k \to 2\beta/h$ as $k \to \infty$, so for large k we have $b_k \geq 1$ if $0 < h < 2\beta$. Thus, if N is large enough we have $k! h^{-k} a_k \geq N! h^{-N} a_N = C > 0$ if $k \geq N$, so if $x > 0$

$$\sum_{k=0}^\infty a_k x^{k+\alpha} \geq Cx^\alpha \sum_{k=N}^\infty \frac{(hx)^k}{k!} + x^\alpha \sum_{k=0}^{N-1} a_k x^k = Cx^\alpha e^{hx} - x^\alpha s(x) \, ,$$

where $s(x)$ is a polynomial. We obtain $u(x) \geq x^\alpha e^{-\beta x}(Ce^{hx} - s(x))$ which is certainly not in $L^2(0, \infty)$ if $\beta < h < 2\beta$. We have therefore found all eigenvalues < 0. $\qquad\qquad\square$

Note that the Schrödinger equation has eigenfunctions $r^{-1}R(r)P(\xi)\Phi(\varphi)$ with eigenvalue $1/(4n^2)$ if $n = k + \ell + m + 1$ where Φ is an eigenfunction of $-\Phi'' + m^2\Phi$ with eigenvalue m^2, P is an eigenfunction of the Legendre equation for this value of m with eigenvalue $\lambda = (\ell+m)(\ell+m+1)$ and R is an eigenfunction of the radial equation for this value of λ with eigenvalue k. We have n possible choices of $j = \ell + m$ and for each such j we have either $m = 0$, which is a simple eigenvalue, or j choices of m with a double eigenvalue, altogether $\sum_{j=0}^{n-1}(2j + 1) = n^2$ different ways to achieve $0 \leq \ell + m \leq n - 1$. Thus the multiplicity of the eigenvalues are as stated in Theorem 6.6.2.

Other equations. The Legendre equation and the radial equation are typical of two classes of equations determining special functions. The first class, to which the Legendre equation belongs, has precisely three regular singular points and no other singularities. This is the smallest number of singularities, all regular singular, a second order equation with meromorphic coefficients can have and still not be solvable by elementary means; see Exercise 6.6.3.

Applying a Möbius transform to the independent variable one may place the three singularities at arbitrary points in the extended plane, conventionally at 0, 1 and infinity. Multiplying the dependent variable by powers of z and $z - 1$ allows arbitrary translations of the indicial roots for 0 and 1; requiring 0 as an indicial root at both points one arrives at

$$z(1 - z)w''(z) + (c - (a + b + 1)z)w'(z) - ab\, w(z) = 0,$$

where a, b and c are arbitrary constants. This equation is called the *hypergeometric equation* and was first given by Euler, and investigated in detail by many mathematicians, beginning with Gauss. Another form of the same equation, with the finite singularities at ± 1, is the *Jacobi equation* $(1-z^2)u'' + (\tilde{b} - \tilde{a} - (\tilde{a} + \tilde{b} + 2)z)u' + \lambda u = 0$, where $\tilde{a}$, $\tilde{b}$ and λ are arbitrary constants. Various special or slightly transformed cases have their own names attached to them, like Chebychev, Gegenbauer and others.

The second class, to which our radial equation belongs, is a limiting case of the hypergeometric equation, when one of the finite regular singularities moves to infinity, creating a more serious singularity there. Placing the remaining regular singularity at 0 one obtains the *confluent hypergeometric equation*, also called *Kummer's equation*,

$$zu'' + (b - z)u' - au = 0 ,$$

where a and b are arbitrary constants. Also here various special cases or different forms of the equation have their own names attached to them, like Airy, Bessel, Hermite and Laguerre. A modern text on special functions is Beals and Wong [13]. The best source for information on the enormous amount of known facts on special functions is the NIST Digital Library of Mathematical Functions [55].

Exercises

Exercise 6.6.1 Show that using spherical coordinates in $\mathbb{R}^3$ as in the beginning of the section the function $r^{-1}R(r)P(\xi)\Phi(\varphi) \in L^2(\mathbb{R}^3)$ if and only if $R \in L^2(0, \infty)$, $P \in L^2(-1, 1)$ and $\Phi \in L^2(0, 2\pi)$.

Exercise 6.6.2 Prove Theorem 6.6.2.

Hint: Find a 'nice' class of functions dense in $L^2(\mathbb{R}^3)$ such that applying the expansion theorems for our three types of self-adjoint Sturm–Liouville equations successively results in an expansion theorem for the original Schrödinger equation. Use this to find an explicit resolution of the identity for this equation and to conclude that no eigenvalues have been missed.

Exercise 6.6.3 Find all second order equations with one regular singularity (on the whole Riemann sphere) and no other singularities. Similarly for two regular singular points. Consider also the 'confluent' case where two regular singular points move together and there are no other singularities.

Exercise 6.6.4 Work out what the conditions of finite kinetic and potential energy for the Schrödinger equation of the hydrogen atom mean for the solutions Φ, P and R and check that our solutions satisfy these conditions. Also show that, after a possible translation of the spectrum, the corresponding equations may be considered left-definite. Finally, check whether these equations are limit-point or limit-circle at the endpoints as left-definite equations. Conclusions?

6.7 Notes and remarks

All the results of Section 6.1 are now classical; even Theorem 6.1.8 may be found, for example, in Ince [112, p. 228]. Leighton [142] generalized the Sturm comparison theorem by replacing the pointwise conditions $1/p_1 \leq 1/p_2$ and $g_1 \leq g_2$ by the integral condition

$$\int \left((p_2 - p_1)u_1'^2 - (g_2 - g_1)u_1^2\right) < 0 ,$$

where $(p_1 u_1')' + g_1 u_1 = 0$, to conclude that any solution of $(p_2 u')' + g_2 u = 0$ has a zero between any consecutive zeros of u_1. Ghatasheh and Weikard [93] extended Leighton's approach to cover differential equations of the form

$$-(p(u' + su))' + rp(u' + su) + qu = 0$$

with real and integrable $1/p, r, s$, and q. More details on comparison and oscillation theory can be found, for instance, in the textbook [210] by Swanson.

Results like Theorem 6.1.12 first appeared in 1929 in Milne [167]. See also Atkinson [7]. De Wet and Mandl [220] used min-max arguments, which also allowed them to deal with higher-dimensional Schrödinger equations. Such arguments are also used by Glazman [96, p. 92]. The best result of this kind was obtained by Hartman [99], and our version is a mild sharpening of this. All these papers deal with the case $Q = \infty$, while the first result for finite Q seems to be by Rosenfeld [194], who used assumptions similar to those of Milne and was apparently unaware of the results of Hartman.

Sections 6.2–6.5 are taken from Bennewitz [16] although the results corresponding to Section 6.3 are much stronger there. A precursor is Kac [116], [117] and some of these results are already in Kasahara [121]. The order of magnitude estimates for the m-function given in Section 6.2 are generalizations of the estimates given for the Schrödinger equation in Hille [104, Theorem 10.2.1].

More accurate asymptotic results for the solutions of Sturm–Liouville equations than those of Section 6.4 may be obtained if the equation has smoother coefficients than we assumed. This may be achieved by use of the *Liouville–Green* or *WKB transformation*, a special case of the Liouville transforms considered in Section 7.2. Independently of each other, Liouville [153] and Green [97] introduced the transformation $u = f^{-1/4} v \circ t$, where $t = \int_a^x f^{1/2}$, to study the differential equation $u'' = fu$ for a positive and twice differentiable function f in the vicinity of a point a. Note that this transformation is a special case of the Liouville transformation discussed in Section 7.2. It yields the differential equation $v'' = (1 + h)v$, where $h \circ t = (4ff'' - 5f'^2)/(16f^3)$. Note that, if f is multiplied by the parameter λ, then h is multiplied by $1/\lambda$. Hence, as λ tends to infinity, h becomes negligible and the Liouville–Green transformation is useful in establishing the asymptotic behavior of solutions when λ tends to infinity. In particular, to first order the solutions are linear combinations of $f^{-1/4}e^{\pm t(x)}$ and show exponential behavior. This approach may be generalized in various directions. For instance, the equation $u'' = (\lambda f + g)u$ turns into $v'' = (1 + \lambda^{-1}h_1)v$, where $h_1 \circ t = h \circ t + g/f$ and h is as before. Another generalization concerns the case when f is negative, in which one establishes oscillatory behavior instead of exponential.

In 1924 Jeffreys [113] investigated the impact of turning points, i.e., points where f has a simple zero, and established so-called connection formulas linking the oscillatory behavior on one side of the turning point to the exponential behavior on the other. The simplest case is, of course, the equation $u'' = xu$, whose solutions are Airy functions. Consequently, even in more general situations the connection formulas involve Airy functions. A little later, in 1926, the physicists Wentzel [219], Kramers [131], and Brillouin [35] (hence WKB) used these connection formulas in the context of quantum mechanics. Here, of course, a major role is played by *Schrödinger's equation* (see also Section 6.6) for a single particle of mass m

$$\frac{-\hbar}{2m} u'' + Vu = Eu \,,$$

where $\hbar = h/(2\pi)$ and h is Planck's constant. Thinking of Planck's constant as a small parameter yields what physicists call the semi-classical approximation to quantum mechanics. More details about the WKB method may be found, for instance, in Olver [175].

Our arguments in Section 6.6 are more ad hoc than those of Schrödinger [202], who used various much stronger results for special functions available in standard references.

Chapter 7
Uniqueness of the inverse problem

Problems in Sturm–Liouville spectral theory fall into two broad categories. So-called direct problems address properties of the spectral data (spectrum, spectral measures *etc.*) of the equation, given its coefficients and boundary conditions, while inverse problems start with certain known properties of the eigenvalues and eigenfunctions or other spectral data of the equation and use this information to deduce properties of its coefficients and boundary data. Up to this point we have dealt exclusively with direct problems.

For inverse problems there are three main questions that are investigated. The first is the uniqueness of the coefficients given certain spectral data, and the second the characterization of those data that are in fact spectral data for some equation. A third problem is the *reconstruction* of the equation from the spectral data, i.e., some method which, at least in principle, allows the determination of the equation.

Historically such investigations seem to have started with Ambarzumian [4] who showed that if the eigenvalue problem $-y'' + qy = \lambda y$, $y'(0) = y'(\pi) = 0$ has eigenvalues $\lambda_n = n^2$, $n \in \mathbb{Z}_+$, then $q = 0$. Much more general results were obtained in Göran Borg's thesis [32]. This dealt with the determination of the potential q for a regular problem $-y'' + qy = \lambda y$ given the spectrum. Borg showed that in general the spectrum is not enough to determine q unless q is symmetric around the interval midpoint with a similar symmetry in the (separated) boundary conditions, but that knowing *two* spectra for the same equation with the same boundary condition at one endpoint while at the other endpoint one condition is Dirichlet and the other different determines the potential. Borg also showed a 'local' characterization, in other words, if a sequence of numbers is sufficiently close to a spectrum, it will also be a spectrum.

Levinson [147] gave another proof of Borg's 'two-spectra' theorem (dropping the condition that at the endpoint with different boundary conditions one of them must be Dirichlet) by first showing that one spectrum may be used to determine the so-called normalization constants (see page 217) associated with the other spectrum. He then showed that eigenvalues together with normalization constants determine the potential, using a simplified version of a method he had earlier [148] used to solve a problem in inverse scattering (see Chapter 8). There are many later results which

C. Bennewitz et al., *Spectral and Scattering Theory for Ordinary Differential Equations*, Universitext, https://doi.org/10.1007/978-3-030-59088-8_7

show uniqueness given just one spectrum, or even part of a spectrum, provided one is given extra information on the potential. Hochstadt and Lieberman [108] obtain such a result if q is known in one half of the base interval. Simon and Gesztesy [85] gave many improvements on that result and the reader may consult their paper for further references.

For an equation with discrete spectrum, knowing the spectrum and the normalization constants is equivalent to knowing the spectral measure (Proposition 7.1.1), and this points the way to inverse theory for cases with continuous spectra. In [33] Borg takes his starting point from the Weyl–Titchmarsh m-function while Marčenko [161] takes his from the spectral measure. Since the m-function is essentially the Stieltjes transform of the spectral measure, these different approaches are closely related and it is usual to credit this early uniqueness result as the Borg–Marčenko uniqueness theorem.

Somewhat later, in 1951, Gel'fand and Levitan [82] (English translation [83]) showed the uniqueness of the potential, a fairly complete characterization of possible spectral measures, and that reconstruction is possible using an integral equation. Both Marčenko and Gel'fand–Levitan made decisive use of the so-called *transformation operators* introduced by Delsarte [53, 54] and first employed in spectral theory by Povzner [183].

All the results mentioned deal with the simplest Sturm–Liouville equation $-y'' + qy = \lambda y$ on an interval with (at least) one regular endpoint. For more general Sturm–Liouville equations we shall give some uniqueness results in Sections 7.3, 7.4 and 7.5.

7.1 The Borg–Marčenko theorem

In this section we consider the equation $-u'' + qu = \lambda u$, often called the one-dimensional Schrödinger equation, on an interval with one regular endpoint. We shall give some basic uniqueness results on the determination of the potential q and the boundary conditions from the spectral measure.

Our object of study is the eigenvalue problem

$$- u'' + qu = \lambda u \text{ on } [0, b) , \tag{7.1.1}$$

$$u(0) \cos \alpha + u'(0) \sin \alpha = 0 . \tag{7.1.2}$$

Here α is a fixed number in $[0, \pi)$, so that the boundary condition is an arbitrary separated boundary condition. We assume q is integrable on any interval $[0, c]$ with $c \in (0, b)$, so that 0 is a regular endpoint for the equation. The other endpoint b may be infinite or finite, in the latter case singular or regular.

If the defect indices for the equation in $L^2(0, b)$ equal 1 the operator corresponding to (7.1.1), (7.1.2) is self-adjoint; if they equal 2 a boundary condition at b is required to obtain a self-adjoint operator. In that case we assume a choice of boundary condition at b is made, so that we are dealing with a fixed self-adjoint operator which we shall call T.

If the defect indices are 2 we know by Corollary 4.3.2 that the spectrum is discrete, but when the defect indices are 1 the spectrum can be of any type. As in Chapter 4, let φ and θ be solutions of (7.1.1) satisfying initial conditions

$$
\begin{cases} \varphi(0, \lambda) = -\sin \alpha\,, \\ \varphi'(0, \lambda) = \cos \alpha\,, \end{cases}
\qquad
\begin{cases} \theta(0, \lambda) = \cos \alpha\,, \\ \theta'(0, \lambda) = \sin \alpha\,. \end{cases}
$$

Now let m be the Weyl–Titchmarsh function and $d\eta$ the spectral measure for our equation.

In the case of a discrete spectrum the spectral function η is a step function, with a jump at each eigenvalue. Suppose the eigenvalues are $\lambda_1, \lambda_2, \ldots$ and that the size of the jump at λ_j is $c_j = \lim_{\varepsilon \downarrow 0}(\eta(\lambda_j + \varepsilon) - \eta(\lambda_j - \varepsilon))$. The inverse generalized Fourier transform then takes the form

$$
u(x) = \sum_{j=1}^{\infty} \hat{u}(\lambda_j)\varphi(x, \lambda_j)c_j\,,
$$

where $\hat{u}(\lambda_j) = \langle u, \varphi(\cdot, \lambda_j)\rangle$. For $u = \varphi(\cdot, \lambda_j)$ the expansion becomes $\varphi(x, \lambda_j) = \|\varphi(\cdot, \lambda_j)\|^2 \varphi(x, \lambda_j)c_j$. It follows that $c_j = \|\varphi(\cdot, \lambda_j)\|^{-2}$. Note that $\varphi(\cdot, \lambda_j)$ is an eigenfunction associated with λ_j, so the jump c_j of η at λ_j is the so-called *normalization constant* for the eigenfunction. The name comes from the fact that a normalized eigenfunction is given by $e_j = \sqrt{c_j}\,\varphi(\cdot, \lambda_j)$. We have shown the following proposition.

Proposition 7.1.1. *In the case of a discrete spectrum knowledge of the spectral measure $d\eta$ is equivalent to knowing the eigenvalues and the corresponding normalization constants.*

Given q, b, and the boundary conditions, one may in principle determine m and thus $d\eta$. We shall take as our basic inverse problem to determine q (and possibly b and the boundary conditions) when $d\eta$ is given. We shall not discuss here the Gel'fand–Levitan theory of [82] mentioned above, which is fully treated in Levitan [149] (English translation [150]). Instead we shall confine ourselves to the problem of *uniqueness*, i.e., to show that two different operators cannot have the same spectral measure. This is the Borg–Marčenko theorem mentioned in the introduction.

To state the theorem we introduce, in addition to the operator T, another operator $\tilde{T}$ of the same kind, corresponding to a boundary condition of the form (7.1.2), but with a parameter $\tilde{\alpha} \in [0, \pi)$, an interval $[0, \tilde{b})$, a potential $\tilde{q}$ and, if needed, a boundary condition at $\tilde{b}$. Let the corresponding spectral measure be $d\tilde{\eta}$.

Theorem 7.1.2 (Borg–Marčenko). *If $d\tilde{\eta} = d\eta$, then $\tilde{T} = T$, i.e., $\tilde{\alpha} = \alpha$, $\tilde{b} = b$, $\tilde{q} = q$ and any boundary condition at b required is the same for $\tilde{T}$ and T.*

In 1999 Barry Simon [204] (see also Gesztesy and Simon [86]) proved a 'local' version of this uniqueness theorem. This was a product of a new strategy devised by Simon for obtaining the results of Gel'fand and Levitan. We shall give the proof from [17], which is quite simple and does not use the machinery of Simon. To prove Theorem 7.1.2 we shall use the same idea, which may be said to be a development of the ideas in Levinson [147] and is similar to the method used by Borg in [33].

In order to state Simon's theorem, one should first note that knowing m is essentially equivalent to knowing $d\eta$, at least if the boundary condition (7.1.2) is known. In fact, $d\eta$ is determined from m via the Stieltjes inversion formula (Theorem E.1.5), and knowing $d\eta$ one may calculate the integral in the representation (4.3.6) of m. By Theorem 6.3.1 we always have $B = 0$, and A may be determined (if $\alpha \neq 0$) since we also have $m(iv) \to \cot\alpha$ as $v \to \pm\infty$. We denote the m-functions associated with T and $\tilde{T}$ by m and $\tilde{m}$ respectively, and by a *non-real ray* we always mean a half-line from the origin not part of $\mathbb{R}$. Then Simon's theorem is the following.

Theorem 7.1.3 (B. Simon). *Suppose that* $0 < a \leq \min(b, \tilde{b})$. *Then* $\alpha = \tilde{\alpha}$ *and* $q = \tilde{q}$ *a.e. on* $(0, a)$ *if* $(m(\lambda) - \tilde{m}(\lambda))e^{2(a-\varepsilon)\,\mathrm{Re}\,\sqrt{-\lambda}} \to 0$ *for every* $\varepsilon > 0$ *as* $\lambda \to \infty$ *along some non-real ray. Conversely, if* $\alpha = \tilde{\alpha}$ *and* $q = \tilde{q}$ *on* $(0, a)$, *then* $(m(\lambda) - \tilde{m}(\lambda))e^{2(a-\varepsilon)\,\mathrm{Re}\,\sqrt{-\lambda}} \to 0$ *for every* $\varepsilon > 0$ *as* $\lambda \to \infty$ *along any non-real ray.*

The crucial point in the proofs of Theorems 7.1.2 and 7.1.3 is the following proposition.

Proposition 7.1.4. *For any fixed* $x \in (0, b)$ $\varphi(x, \lambda)\psi(x, \lambda) \to 0$ *as* $\lambda \to \infty$ *along a non-real ray.*

Note that $\varphi(x, \lambda)\psi(x, \lambda)$ is Green's function on the diagonal $x = y$, so this is an immediate consequence of Theorem 6.4.4. We now need good estimates of the solutions of the initial value problem for large λ.

Lemma 7.1.5. *If* u *solves* $-u'' + qu = \lambda u$ *with* λ-*independent initial data at* 0, *then*

$$u(x, \lambda) = \left(u(0)\cosh(x\sqrt{-\lambda}) + \frac{u'(0)}{\sqrt{-\lambda}}\sinh(x\sqrt{-\lambda})\right)(1 + o(1)),$$

locally uniformly in x *as* $\lambda \to \infty$.

Proof. Solving the equation $u'' + \lambda u = f$ and then replacing f by qu gives

$$u(x) = \cosh(kx)u(0) + \frac{\sinh(kx)}{k}u'(0) + \int_0^x \frac{\sinh(k(x-t))}{k}q(t)u(t)\,dt,$$

where we have written k for $\sqrt{-\lambda}$. Setting

$$g(x) = \left|u(x) - \cosh(kx)u(0) - \frac{\sinh(kx)}{k}u'(0)\right|e^{-x\,\mathrm{Re}\,k}$$

easy estimates give

$$g(x) \le \frac{1}{|k|} \int\limits_0^x |q|(g + c(\lambda)) \,,$$

where $c(\lambda) = |u(0)| + |u'(0)|/|k|$. Applying the Gronwall inequality (Lemma D.3) to $g(x) + c(\lambda)$ we obtain

$$g(x) \le c(\lambda)\left(\exp\left(\int\limits_0^x |q|/|k|\right) - 1\right) = c(\lambda)o(1)$$

locally uniformly in x as $k \to \infty$. The desired estimate follows. □

We have the following important corollary to Proposition 7.1.4.

Corollary 7.1.6. *If $\alpha = \tilde{\alpha} = 0$ or $\alpha \ne 0 \ne \tilde{\alpha}$ both $\tilde{\varphi}(x, \lambda)\psi(x, \lambda)$ and $\varphi(x, \lambda)\tilde{\psi}(x, \lambda)$ tend to 0 locally uniformly in x as $\lambda \to \infty$ along a non-real ray.*

Proof. For fixed x and $\tilde{\alpha} \ne 0$ we have, by Lemma 7.1.5,

$$\varphi(x, \lambda)/\tilde{\varphi}(x, \lambda) \to \sin\alpha/\sin\tilde{\alpha} \text{ as } \lambda \to \infty$$

along a non-real ray. If $\alpha = \tilde{\alpha} = 0$ we instead obtain the limit 1, so the corollary follows from Proposition 7.1.4. □

Proof (Theorem 7.1.2). According to the Nevanlinna representation formula (E.1.1) and Theorem 4.3.17 the difference $m - \tilde{m}$ is a real constant A. In particular, since Dirichlet m-functions are unbounded near ∞ on a non-real ray but no other m-functions are by (6.2.4) and Theorem 6.3.1, we must have either $\alpha = \tilde{\alpha}$ or $\alpha \ne 0 \ne \tilde{\alpha}$ if $d\eta = d\tilde{\eta}$. Thus, according to Corollary 7.1.6, the functions $\tilde{\varphi}(x, \lambda)\psi(x, \lambda)$ and $\varphi(x, \lambda)\tilde{\psi}(x, \lambda)$ tend to 0 as $\lambda \to \infty$ along a non-real ray. Their difference is

$$\tilde{\varphi}(x, \lambda)\theta(x, \lambda) - \varphi(x, \lambda)\tilde{\theta}(x, \lambda) + A\varphi(x, \lambda)\tilde{\varphi}(x, \lambda) \,, \tag{7.1.3}$$

which is an entire function of λ tending to 0 along non-real rays, and it may be bounded by a multiple of $e^{C|\lambda|^{1/2}}$ for some constant C by Lemma 7.1.5 (or Theorem 6.5.9). By Theorem E.4.3 such a function is bounded in any sector determined by two non-real rays and thus in the entire plane. It is therefore constant by Liouville's Theorem E.2.13, so since the limit is zero along the rays the function is identically zero. It follows that

$$\theta(x, \lambda)/\varphi(x, \lambda) = \tilde{\theta}(x, \lambda)/\tilde{\varphi}(x, \lambda) - A$$

for all x, λ. Differentiating with respect to x and using the fact that $\mathcal{W}(\varphi, \theta) = \mathcal{W}(\tilde{\varphi}, \tilde{\theta}) = 1$, we obtain $\varphi^2(x, \lambda) = \tilde{\varphi}^2(x, \lambda)$. Taking the logarithmic derivative of this we obtain $\varphi'(x, \lambda)/\varphi(x, \lambda) = \tilde{\varphi}'(x, \lambda)/\tilde{\varphi}(x, \lambda)$.

Letting $x \to 0$ this gives $\alpha = \tilde{\alpha}$. Differentiating once more we obtain $\varphi''/\varphi = \tilde{\varphi}''/\tilde{\varphi}$, which means that $q = \tilde{q}$ on $(0, \min(b, \tilde{b}))$. From this follows that $\varphi = \tilde{\varphi}$ and $\theta = \tilde{\theta}$, so (7.1.3) being identically zero shows that $A = 0$ and thus $m = \tilde{m}$. It follows that $\psi = \tilde{\psi}$ on $\min(b, \tilde{b})$. This implies that $b = \tilde{b}$, since otherwise ψ (or $\tilde{\psi}$) would satisfy self-adjoint boundary conditions both at b and $\tilde{b}$, so for $\operatorname{Im}\lambda \neq 0$ ψ would be an eigenfunction to a self-adjoint operator with a non-real eigenvalue. Since $\psi = \tilde{\psi}$ the boundary conditions at $b = \tilde{b}$ (if any) are the same. It follows that $T = \tilde{T}$. $\square$

Proof (Theorem 7.1.3). If $\alpha = \tilde{\alpha} = 0$ or $\alpha \neq 0 \neq \tilde{\alpha}$ then $\tilde{\varphi}(x, \lambda)\psi(x, \lambda)$ and $\varphi(x, \lambda)\tilde{\psi}(x, \lambda)$ both tend to 0 as $\lambda \to \infty$ along a non-real ray by Corollary 7.1.6. Their difference is

$$\tilde{\varphi}(x, \lambda)\theta(x, \lambda) - \varphi(x, \lambda)\tilde{\theta}(x, \lambda) + (m(\lambda) - \tilde{m}(\lambda))\varphi(x, \lambda)\tilde{\varphi}(x, \lambda) . \qquad (7.1.4)$$

Suppose first that $\alpha = \tilde{\alpha}$ and $q = \tilde{q}$ on $(0, a)$. Then the first two terms cancel on $(0, a)$, so that $(m(\lambda) - \tilde{m}(\lambda))\varphi(x, \lambda)\tilde{\varphi}(x, \lambda) \to 0$ as $\lambda \to \infty$ along non-real rays if $x \in (0, a)$. By Lemma 7.1.5 this implies that $(m(\lambda) - \tilde{m}(\lambda))e^{2(a-\varepsilon)\operatorname{Re}\sqrt{-\lambda}} \to 0$ as $\lambda \to \infty$ along any non-real ray.

Conversely, the estimate for $m - \tilde{m}$ implies that $m - \tilde{m} \to 0$ along a non-real ray, so that $\alpha = \tilde{\alpha} = 0$ or $\alpha \neq 0 \neq \tilde{\alpha}$ and then that for $0 < x < a$ the last term of (7.1.4) tends to 0 according to assumption and Lemma 7.1.5. Thus the entire function $\tilde{\varphi}(x, \lambda)\theta(x, \lambda) - \varphi(x, \lambda)\tilde{\theta}(x, \lambda)$ of λ also tends to 0 along a non-real ray, and by symmetry also along its conjugate. However, as in the proof of Theorem 7.1.2 this entire function is bounded by $e^{C|\lambda|^{1/2}}$ for some constant C, so by the Phragmén–Lindelöf and Liouville theorems it vanishes for all $x \in (0, a)$. It follows that $\alpha = \tilde{\alpha}$ and $q = \tilde{q}$ in $(0, a)$ exactly as in the proof of Theorem 7.1.2. $\square$

7.2 Liouville transforms

In attempting to give a generalization of the Borg–Marčenko theorem to the more general equation (4.1.1) one is faced by the fact that the coefficients p, q and w cannot all be determined by the spectral measure. This seems intuitively reasonable, but a more solid reason is that one can always transform (4.1.1) to a similar equation with different coefficients but the same spectral measure via a *Liouville transform*. In this section we will discuss such transforms.

Suppose u satisfies

$$-(pu')' + qu = wf \text{ on } (a, b) \qquad (7.2.1)$$

and introduce a new independent variable $t = t(x)$ with an increasing[1] function t defined on (a, b), and new dependent variables $\tilde{u}$ and $\tilde{f}$ by setting

$$u(x) = s(x)\tilde{u}(t(x)) , \qquad f(x) = s(x)\tilde{f}(t(x)) ,$$

where s is a fixed, real-valued function. Formal differentiation gives

---

[1] One could also use a decreasing t.

$$pu' = ps'\tilde{u} \circ t + pst'\tilde{u}' \circ t \,,$$
$$(pu')' = (ps')'\tilde{u} \circ t + 2ps't'\tilde{u}' \circ t + s(pt'\tilde{u}' \circ t)' \,,$$
$$= (ps')'\tilde{u} \circ t + (ps^2t'\tilde{u}' \circ t)'/s \,.$$

Setting $ps^2t' = \tilde{p} \circ t$ we obtain $(pu')' = (ps')'\tilde{u} \circ t + t'(\tilde{p}\tilde{u}')' \circ t/s$. Thus we get

$$0 = -(pu')' + qu - wf = \frac{t'}{s}\left(-(\tilde{p}\tilde{u}')' \circ t + \frac{s}{t'}((-(ps')' + qs)\tilde{u} \circ t - ws\tilde{f} \circ t)\right) \,,$$

so if $t'/s \neq 0$ the equation connecting $\tilde{u}$ and $\tilde{f}$ becomes

$$-(\tilde{p}\tilde{u}')' + \tilde{q}\tilde{u} = \tilde{w}\tilde{f} \text{ on } (\tilde{a}, \tilde{b}) \,, \tag{7.2.2}$$

where $\tilde{a} = t(a)$, $\tilde{b} = t(b)$ and

$$\tilde{p} \circ t = ps^2t' \,,$$
$$\tilde{q} \circ t = \frac{s}{t'}(-(ps')' + qs) \,, \tag{7.2.3}$$
$$\tilde{w} \circ t = ws^2/t' \,.$$

If we assume the standard conditions that $1/p$, q and w are locally integrable, we therefore need to require that s and ps' are locally absolutely continuous, that s does not vanish on the base interval and also that t is locally absolutely continuous with derivative > 0 a.e. in order for $\tilde{p}$, $\tilde{q}$ and $\tilde{w}$ as functions of t to satisfy the same conditions. Assuming this it is easily verified that the formal calculations leading to (7.2.3) are valid. A transform of this type is called a *Liouville transform*. It is readily seen that such a transform is linear and has an inverse of the same kind; see also Exercise B.8.3.

Since t maps (a, b) onto $(\tilde{a}, \tilde{b})$ a change of variable gives

$$\int_a^b |u|^2 w = \int_{\tilde{a}}^{\tilde{b}} |\tilde{u}|^2 \tilde{w}$$

so the forms associated with the right-hand side of the equations are transformed into each other. Thus in the right-definite case any Liouville transform is *unitary* between the relevant spaces. Similarly, after integrating by parts, we obtain

$$\int_a^b (p|u'|^2 + q|u|^2) = \int_{\tilde{a}}^{\tilde{b}} (\tilde{p}|\tilde{u}'|^2 + \tilde{q}|\tilde{u}|^2) + |u|^2 ps'/s\Big|_a^b \,.$$

The forms associated with the left-hand side of the equations are thus also transformed into each other provided u has compact support in (a, b). If we consider a left-definite equation we may assume $p \equiv 1$, so if a is finite and q integrable in (a, c) for every $c \in (a, b)$, then by (5.1.2) $u(a)$ is finite for any $u \in \mathcal{H}_1$. Thus if $s(a) \neq 0$

while $s'(a) = 0$ the forms agree assuming only that u vanishes near b, i.e., that u is a *finite function* in the terminology of Section 5.2.

Similarly, if both endpoints are finite and s does not vanish at either endpoint, while $s'(a) = s'(b) = 0$ we need only assume u to be in the Hilbert space $\mathcal{H}_1$ for the forms to agree. If the equations are left-definite the Liouville transform is then unitary if restricted to the closure of compactly supported elements of the space, respectively to the space $\mathcal{H}_0$ of Definition 5.2.7 if a is limit-circle, $s(a) \neq 0$ and $s'(a) = 0$. If also b is limit-circle it is also required that $s(b) \neq 0$, $s'(b) = 0$.

It is also easy to see that if $\tilde{u}_j$ is transformed to u_j, then $pu_1'u_2 - u_1 pu_2' = \tilde{p}\tilde{u}_1'\tilde{u}_2 - \tilde{u}_1\tilde{p}\tilde{u}_2'$ if evaluated at corresponding points, so the Wronskians are invariant under the transform.

As soon as we have a unitary transform $\mathcal{L}$ of one Hilbert space onto another an operator T which is self-adjoint in one space will be transformed into a self-adjoint operator $\tilde{T}$ in the other space. Since the two operators are *similar* in the sense that $\mathcal{L}\tilde{T} = T\mathcal{L}$ the two operators have the same spectra and spectral multiplicities. In particular, if the unitary Liouville transform $\mathcal{L}$ takes the equation (7.2.1) into (7.2.2), then any self-adjoint realization T of (7.2.1) is transformed into a self-adjoint realization $\tilde{T}$ of (7.2.2).

For equations with one regular endpoint and separated boundary conditions we have defined a unique spectral measure by introducing the solutions φ and θ of (4.3.3) in the right- and of (5.2.5) in the left-definite case. One may ask what is required of the Liouville transform to preserve the spectral measure, or even the m-function.

Let $\varphi(\cdot, \lambda) = \mathcal{L}\tilde{\varphi}(\cdot, \lambda)$, so that $-(\tilde{p}\tilde{\varphi}')' + \tilde{q}\tilde{\varphi} = \lambda\tilde{w}\tilde{\varphi}$. Since $\mathcal{L}$ is unitary we have $\hat{u}(\lambda) = \langle u, \varphi(\cdot, \overline{\lambda})\rangle = \langle \tilde{u}, \tilde{\varphi}(\cdot, \overline{\lambda})\rangle_\sim$ if $u = \mathcal{L}\tilde{u}$, at least if u is finite. We also have $u(x) = \langle \hat{u}, \varphi(x, \cdot)\rangle_\eta = s(x)\langle \hat{u}, \tilde{\varphi}(t(x), \cdot)\rangle_\eta$ so $\tilde{u}(t) = \langle \hat{u}, \tilde{\varphi}(t, \cdot)\rangle_\eta$. We therefore have an expansion theorem with the same transform space for (7.2.2) as for (7.2.1). However, there is no guarantee that the initial data of $\tilde{\varphi}$ is of the form (4.3.3) respectively (5.2.5); the measure $d\eta$ may not be a spectral measure of (7.2.2) as we have defined it. The Liouville transform is

$$\begin{pmatrix} u \\ pu' \end{pmatrix} = \begin{pmatrix} s & 0 \\ ps' & 1/s \end{pmatrix}\begin{pmatrix} \tilde{u} \circ t \\ \tilde{p}\tilde{u}' \circ t \end{pmatrix}, \tag{7.2.4}$$

so if $\tilde{\alpha}$ is the boundary condition parameter for (7.2.2) we require

$$\begin{pmatrix} -\sin\alpha \\ \cos\alpha \end{pmatrix} = \begin{pmatrix} s(a) & 0 \\ ps'(a) & 1/s(a) \end{pmatrix}\begin{pmatrix} -\sin\tilde{\alpha} \\ \cos\tilde{\alpha} \end{pmatrix}$$

in the right-definite case. In the left-definite case the sines should be divided by λ, but since in this case $ps'(a) = 0$ this makes no difference. If $\tilde{\alpha} = 0$ this means also $\alpha = 0$ and $s(a) = 1$ while $ps'(a)$ is arbitrary in the right-definite and equals 0 in the left-definite case. If $\tilde{\alpha} \neq 0$ we also have $\alpha \neq 0$ since we must have $s(a) \neq 0$. We obtain $s(a) = \sin\alpha/\sin\tilde{\alpha}$ and $ps'(a) = \cos\tilde{\alpha}/\sin\alpha - \cos\alpha/\sin\tilde{\alpha}$. In the left-definite case we must also have $ps'(a) = 0$ so this requires $\sin(2\alpha) = \sin(2\tilde{\alpha})$. More explicitly, we have either $\alpha = \tilde{\alpha}$, $0 < \alpha = \pi/2 - \tilde{\alpha}$ or $\pi/2 < \alpha = 3\pi/2 - \tilde{\alpha}$.

If we wish the Liouville transform to preserve not just the spectral measure, but the m-function, there is an additional requirement, for even though Im m is determined by the spectral measure, Re m is only determined up to addition of a real number.

Denote the m-function for (7.2.1) by m and that for (7.2.2) by $\tilde{m}$. If the equations have the same spectral measure and $\theta + m\varphi$, respectively $\tilde{\theta} + \tilde{m}\tilde{\varphi}$, are the Weyl solutions for the two equations, we have seen that $\mathcal{L}\tilde{\varphi} = \varphi$, and since $\mathcal{L}$ is unitary and $B = 0$ in (5.2.6) by Theorems 4.3.17 respectively 5.2.17 we have

$$\frac{\operatorname{Im} m(\lambda)}{\operatorname{Im} \lambda} = \frac{\operatorname{Im} \tilde{m}(\lambda)}{\operatorname{Im} \lambda} = \|\tilde{\theta} + \tilde{m}\tilde{\varphi}\|_\sim^2 = \|\mathcal{L}\tilde{\theta} + \tilde{m}\varphi\|^2 .$$

Now $\mathcal{W}_p(\varphi, \mathcal{L}\tilde{\theta}) = \mathcal{W}_{\tilde{p}}(\tilde{\varphi}, \tilde{\theta}) = 1 = \mathcal{W}_p(\varphi, \theta)$ so $\mathcal{W}_p(\varphi, \mathcal{L}\tilde{\theta} - \theta) = 0$. Thus we have $\mathcal{L}\tilde{\theta} = \theta + A\varphi$ for some constant A, so that $m = A + \tilde{m}$. Here $A = \mathcal{W}_p(\theta, \mathcal{L}\tilde{\theta}) = \cot \tilde{\alpha} - \cot \alpha$ if α and $\tilde{\alpha}$ are both $\neq 0$, otherwise $A = 0$. Summarizing, we have proved the following theorem.

Theorem 7.2.1. *Consider right- or left-definite equations* (7.2.1) *and* (7.2.2) *with a regular endpoint at a and $\tilde{a}$ respectively. Assume $1/p$, q and w are integrable in (a, c) for every $c \in (a, b)$ and $1/\tilde{p}$, $\tilde{q}$ and $\tilde{w}$ integrable in $(\tilde{a}, \tilde{c})$ for every $\tilde{c} \in (\tilde{a}, \tilde{b})$. Suppose the equations are connected via a unitary Liouville transform $u(x) = s(x)\tilde{u}(t(x))$. Then s, ps' and t are locally absolutely continuous with $t' > 0$ a.e. and $s \neq 0$ in (a, b), and the coefficients are connected via* (7.2.3).

Let T and $\tilde{T}$ be self-adjoint realizations of (7.2.1) *respectively* (7.2.2) *given by separated conditions with boundary condition parameters α at a respectively $\tilde{\alpha}$ at $\tilde{a}$. Then T and $\tilde{T}$ have the same spectral measure only if one of the following two alternatives hold:*

- *$\alpha = \tilde{\alpha}$ and $s(a) = 1$ and in the left-definite case $ps'(a) = 0$, whereas $ps'(a)$ is arbitrary in the right-definite case.*

- *Both α and $\tilde{\alpha} \neq 0$ and*

$$s(a) = \sin \alpha / \sin \tilde{\alpha} , \quad ps'(a) = \cos \tilde{\alpha} / \sin \alpha - \cos \alpha / \sin \tilde{\alpha} .$$

In the left-definite case it is also required that $ps'(a) = 0$ so that $\alpha = \tilde{\alpha}$, $0 < \alpha = \pi/2 - \tilde{\alpha}$ or $\pi/2 < \alpha = 3\pi/2 - \tilde{\alpha}$.

If the equations are limit-point at the right endpoint this is also a sufficient condition for equal spectral measures. If one or both equations are limit-circle at the right endpoint the right-hand boundary conditions need also be connected via (7.2.4).

If the spectral measures coincide, then $m = \tilde{m}$ if $\alpha = \tilde{\alpha} = 0$, while if α and $\tilde{\alpha}$ are both $\neq 0$, then $m = \cot \tilde{\alpha} - \cot \alpha + \tilde{m}$.

It is clear from the theorem that if one wants to prove that a given spectral measure can only come from a unique spectral problem, one needs to find a 'normal form' for the equation, such that every equation can be transformed into precisely one of normal form by a Liouville transform. In general this is not feasible, as can be seen from the following examples.

1. $\tilde{p} \equiv 1$ requires $t' = (ps^2)^{-1}$; in particular we must require $p > 0$. If we only require $|\tilde{p}| \equiv 1$ we may take $t' = (|p|s^2)^{-1}$.

2. $\tilde{w} \equiv 1$ requires $t' = s^2 w$; in particular we must require $w > 0$. Again, if $w \neq 0$ a.e. and we only ask $|\tilde{w}| \equiv 1$ we may take $t' = s^2 |w|$.

3. $\tilde{p} \equiv \tilde{w}$ requires $t' = \sqrt{w/p}$; in particular that $w/p > 0$, but if we only require $|\tilde{p}| = |\tilde{w}|$ we may take $t' = \sqrt{|w/p|}$ if $w \neq 0$ a.e.

4. $\tilde{q} \equiv \mu \tilde{w}$ requires that s solves the equation $-(ps')' + qs = \mu ws$. By translating the spectrum and making an appropriate Liouville transformation we may therefore eliminate the potential q, provided we can find such a solution s which *does not vanish* on the base interval. If $p > 0$ and we are dealing with a regular problem, this can always be done by choosing μ sufficiently small. Also in a singular case this is true provided the spectrum of the original equation is bounded from below.

5. Combining **1.** and **2.** or **3.**, so that $\tilde{p} \equiv \tilde{w} \equiv 1$ and the transformed equation is a Schrödinger equation, requires that $t' = \sqrt{w/p}$ and $s = (pw)^{-1/4}$. One must therefore require $w/p > 0$, and to obtain the required smoothness of s that pw and $(pw)'/w$ are locally absolutely continuous. Smoothness is required even if we only ask $|\tilde{p}| \equiv |\tilde{w}| \equiv 1$

6. Combining **1.** and **4.**, i.e., $\tilde{p} \equiv 1$ and $\tilde{q} \equiv 0$, the equation becomes a so-called *Kreĭn string* $-\ddot{u} = \tilde{w}\tilde{f}$. This requires $p > 0$ and that there is a real-valued solution of $-(ps')' + qs = \mu ws$ without zeros.

7. Combining **2.** and **4.**, i.e., $\tilde{w} \equiv 1$ and $\tilde{q} \equiv 0$, the equation becomes $-(\tilde{p}\tilde{u}')' = \tilde{f}$. This requires $w > 0$ and that there is a real-valued solution of $-(ps')' + qs = \mu ws$ without zeros.

We may think of Sturm–Liouville equations as equivalent if they can be transformed into each other by a Liouville transform. To find a normal form we would have to find a special class of Sturm–Liouville equations such that each equivalence class contains precisely one equation in the special class. The cases above where s and t' are uniquely determined are the last three, and they all require properties not every Sturm–Liouville equation possesses. One might think of a few other possible normal forms, for example by combining **3.** and **4.**, but will soon find that no generally applicable normal form can be found.

The best one can hope to prove is therefore that if the spectral measure is given, then the equivalence class of the boundary value problem is determined. The next section is devoted to proving such a theorem for the right-definite case, and Section 7.4 to a similar theorem for the left-definite case. Additional information may determine a specific boundary value problem, and we shall also give such results.

7.3 Right-definite Sturm–Liouville equations

Here we shall deal with the following question: *To what extent is a self-adjoint, right-definite Sturm–Liouville operator T with separated boundary conditions and one regular endpoint determined by its spectral measure?* That is, to what extent can we determine the base interval[2] $[0, b)$, the coefficients p, q and w, and the boundary condition parameter α given the spectral measure $d\eta$? To answer this question the Liouville transforms introduced in the previous section will be central.

We consider boundary value problems of the kind

$$M[u] := -(pu')' + qu = \lambda u \text{ on } [0, b) , \qquad (7.3.1)$$

$$u(0) \cos \alpha + pu'(0) \sin \alpha = 0 , \qquad (7.3.2)$$

where $1/p$ and q are real-valued and integrable in $(0, a)$ for every $a \in (0, b)$, and $\alpha \in [0, \pi)$ is fixed, and we will freely use notation and results from Chapter 4.

We could similarly treat the equation $-(pu')' + qu = \lambda w u$ with $w \geq 0$, but the methods of this section then require that $w > 0$ a.e., and in this case the change of variable $t = \int_0^x w$ turns the equation into (7.3.1) with unchanged spectral function and m-function, as follows from Theorem 7.2.1 with $s \equiv 1$. The case when w is allowed to vanish on a set of positive measure requires slightly more sophisticated methods, similar to those of Chapter 8.

If (7.3.1) is limit-point at b then together with (7.3.2) it generates a self-adjoint operator T. If not, we impose a symmetric boundary condition at b to obtain a self-adjoint operator T. The possibility that the problem is regular is not excluded.

In addition to the operator T we consider a similar operator $\tilde{T}$, acting on $L^2(0, \tilde{b})$ and generated by a differential expression similar to (7.3.1) but with p and q replaced by $\tilde{p}$ and $\tilde{q}$ respectively. Furthermore, a boundary condition of the form (7.3.2), but with α replaced by $\tilde{\alpha}$, is imposed at 0, and, if needed, a boundary condition at $\tilde{b}$ making $\tilde{T}$ self-adjoint. Our main result is then the following.

Theorem 7.3.1. *Suppose the operators T and $\tilde{T}$ have the same spectral measure. Then there is a unitary Liouville transform $\mathcal{L}$ mapping $\tilde{T}$ to T, i.e., $\mathcal{L}\tilde{T} = T\mathcal{L}$.*

The new variable $t(x)$ is the composition of $x \mapsto \int_0^x |p|^{-1/2}$ followed by the inverse of $x \mapsto \int_0^x |\tilde{p}|^{-1/2}$ and the multiplier $s(x) = \sqrt{t'(x)} = |\tilde{p}(t(x))/p(x)|^{1/4}$ is strictly positive. The functions s and ps' are locally absolutely continuous in $[0, b)$.

Theorem 7.2.1 gives further information on the properties of s and t and possible combinations of boundary conditions, and (7.2.3) gives formulas connecting the coefficients of the two equations.

It is noteworthy that the map $\mathcal{L}$ is independent of any sign changes in p and $\tilde{p}$. In other words, all the information about sign changes in p is implicit in the spectral measure. The following corollary shows that given the absolute value of p the operator T is uniquely determined by the spectral measure.

---

[2] We assume the left interval endpoint to be 0 since clearly translating the interval and the coefficients by an arbitrary amount will not change the spectral measure.

Corollary 7.3.2. *Suppose T and $\tilde{T}$ have the same spectral measure and that $|p| = |\tilde{p}|$ in $[0, \min(b, \tilde{b}))$. Then $\tilde{T} = T$, i.e., $\tilde{\alpha} = \alpha$, $\tilde{b} = b$, $\tilde{p} = p$, $\tilde{q} = q$, and any boundary condition at $\tilde{b} = b$ is the same for T and $\tilde{T}$.*

The corollary is an immediately consequence of Theorem 7.3.1, since the Liouville transform $\mathcal{L}$ is the identity if $|p| = |\tilde{p}|$. It even follows that if $|p| = |\tilde{p}|$ in $[0, a]$, where $a \in (0, \min(b, \tilde{b}))$, then T and $\tilde{T}$ have the same boundary condition at 0, and both $p = \tilde{p}$ and $q = \tilde{q}$ in $[0, a]$. In particular, if p and $\tilde{p}$ are both constant equal to 1 it follows from the corollary that $T = \tilde{T}$, so that we also retrieve the Borg–Marčenko Theorem 7.1.2.

Our main tool in proving Theorem 7.3.1 is a theorem of Paley–Wiener type, which we now turn to. The original Paley–Wiener theorems [178] deal with the classical Fourier transform and are today perhaps best known in a variant valid for the (multi-dimensional) Fourier transform of compactly supported distributions. The Fourier transform of a compactly supported function or distribution is an entire function of order 1, and the theorem characterizes the convex hull of the support of a distribution in terms of the growth properties of its Fourier transform; see Hörmander [111, Theorem 7.3.1].

At the heart of our method for obtaining uniqueness theorems for Sturm–Liouville inverse spectral theory lies a generalization of the one-dimensional version of this to the generalized Fourier transforms associated with such equations. Before we state this theorem it will be convenient to introduce the following special class of entire functions.

Definition 7.3.3. Suppose h is measurable and $\sqrt{|h|} \in L^1(0, a)$ for every $a \in [0, b)$. Let $\mathcal{A}(h)$ be the set of entire functions $\hat{u}$ of order[3] $\leq 1/2$ which satisfy

$$\varlimsup_{t \to +\infty} t^{-1} \log |\hat{u}(t^2 \lambda)| \leq \int_0^a \operatorname{Re} \sqrt{-\lambda h} \qquad (7.3.3)$$

for some $a \in (0, b)$ and all non-real λ, where $\sqrt{}$ is the root with a positive real part.

Our generalization of the Paley–Wiener theorem is as follows.

Theorem 7.3.4. *Let $\hat{u}$ be the generalized Fourier transform of $u \in L^2(0, b)$. Then $\hat{u}$ has at most one entire continuation in $\mathcal{A}(1/p)$. If $\sup(\operatorname{supp} u) = a < b$ then $\hat{u}(\lambda) = \int_0^a u(x) \varphi(x, \lambda) \, dx$ is such a continuation.*

Conversely, if $\hat{u}$ has an entire continuation of order $\leq 1/2$ satisfying (7.3.3) for $h = 1/p$ and λ on two non-real rays, then $\sup(\operatorname{supp} u) \leq a$.

Note that the uniqueness of an entire continuation of $\hat{u}$ is by no means obvious. The difference of two continuations will have to vanish in $\operatorname{supp} d\eta$, so if the support of the spectral measure has a finite point of accumulation, there can be at most one such continuation. On the other hand, if the spectrum of T is discrete, so that $\operatorname{supp} d\eta$

---

[3] See section E.2.

consists of isolated points, there are always non-trivial entire functions that vanish on supp $d\eta$, and then an entire continuation is never unique. However, the theorem shows that uniqueness is restored if we only consider entire continuations with a sufficiently restricted growth at infinity, *i.e.*, continuations in $\mathcal{A}(1/p)$.

The main steps in the proof of Theorem 7.3.4 are in the following two lemmas. We shall always assume that $a \in (0, b)$.

Lemma 7.3.5. *Suppose* $u \in L^2(0, b)$ *and* $\sup(\mathrm{supp}\, u) \le a$. *Then*

$$\hat{u}(\lambda) = \int\limits_0^a u(x)\varphi(x, \lambda)\, dx$$

is an entire function of order $\le 1/2$ *for which* $\hat{u}(\lambda) = o(|\varphi(a, \lambda)|)$ *as* $\lambda \to \infty$ *along any non-real ray.*

Proof. The function $\varphi(x, \lambda)/\varphi(a, \lambda)$ satisfies the boundary condition (7.3.2) and has the value 1 for $x = a$. Hence it equals the Weyl solution for the operator generated by M in $L^2(0, a)$, considering a as the initial point and provided with Dirichlet's boundary condition there and the boundary condition (7.3.2) at 0. We denote this function by $\psi_{(0,a]}$.

If $\sup(\mathrm{supp}\, u) \le a$, then the integral defining $\hat{u}(\lambda)$ is absolutely convergent for any $\lambda \in \mathbb{C}$ so it defines an entire function of order $\le 1/2$, since φ is such a function, locally uniformly in x. Furthermore, for non-real λ we have

$$\hat{u}(\lambda) = \int\limits_0^a u(x)\varphi(x, \lambda)\, dx = \varphi(a, \lambda) \int\limits_0^a u(x)\psi_{(0,a]}(x, \lambda)\, dx\,,$$

and by Theorem 4.3.17 the last integral is $o(1)$ as $\lambda \to \infty$ along any non-real ray. $\square$

Lemma 7.3.6. *Suppose* $u \in L^2(0, b)$, *that* $\hat{u}$ *has a continuation to an entire function of order* $\le 1/2$, *and that* $\hat{u}(\lambda) = \mathcal{O}(1/|\psi(a, \lambda)|)$ *as* $\lambda \to \infty$ *along two different non-real rays. Then* $\sup(\mathrm{supp}\, u) \le a$ *and* $\hat{u}(\lambda) = \int_0^a u(x)\varphi(x, \lambda)\, dx$.

Proof. Consider the function $A(\lambda) = \langle R_\lambda u, v \rangle - \hat{u}(\lambda)\langle \psi(\cdot, \lambda), v \rangle$ for $\lambda \notin \mathbb{R}$ and v in the domain of T. We shall show that this function has a continuation to an entire function of order $\le 1/2$, and that it tends to 0 as $\lambda \to \infty$ along the given non-real rays if $\mathrm{supp}\, v \subset (a, b)$. Applying the Phragmén–Lindelöf principle (Theorem E.4.3) to the two sectors defined by the rays it follows that A is bounded in the entire plane, and so by Liouville's Theorem E.2.13 it is constant.

The constant is 0 since this is the limit along the rays. It follows that $R_\lambda u - \hat{u}(\lambda)\psi(\cdot, \lambda)$ is orthogonal to every v in the domain of T which has support in (a, b). But the set of such functions is dense in $L^2(a, b)$, for example because it contains the domain of the minimal operator generated by the differential expression M in $L^2(a, b)$. Hence $R_\lambda u(x) - \hat{u}(\lambda)\psi(x, \lambda) = 0$ for $x \in (a, b)$, and applying $M - \lambda$ to this we find that $u = 0$ in (a, b). Hence, for $x \ge a$ we also have

$$0 = R_\lambda u(x) - \hat{u}(\lambda)\psi(x, \lambda) = \psi(x, \lambda)\left\{ \int_0^a u(y)\varphi(y, \lambda)\,dy - \hat{u}(\lambda) \right\},$$

so that $\hat{u}(\lambda) = \int_0^a u\varphi(\cdot, \lambda)$.

To show that A is entire of order $\leq 1/2$ we apply Parseval's formula, using Lemma 4.3.13 and Theorem 4.3.17, to obtain

$$A(\lambda) = \int_{-\infty}^{\infty} \frac{\hat{u}(t) - \hat{u}(\lambda)}{t - \lambda}\, \overline{\hat{v}(t)}\, d\eta(t)\,.$$

The integrand is an entire function of λ for each t, since it is clearly analytic with a removable singularity at t. We need to estimate the integrand. Now $\hat{u}(t) - \hat{u}(\lambda) = \int_\lambda^t \hat{u}'$, integrating over the line segment connecting λ and t. For $|t - \lambda| \leq 1$ this shows that the integrand may be estimated by $|\hat{v}(t)| \sup_{|z| \leq 1} |\hat{u}'(\lambda + z)|$.

On the other hand, for $|t - \lambda| > 1$ the integrand may be estimated by $|\hat{u}(t)|\,|\hat{v}(t)| + |\hat{u}(\lambda)|\,|\hat{v}(t)|$, so that the integrand is less than

$$|\hat{u}(t)|\,|\hat{v}(t)| + (|\hat{u}(\lambda)| + \sup_{|z| \leq 1} |\hat{u}'(\lambda + z)|)\,|\hat{v}(t)|\,,$$

for all $t \in \mathbb{R}$. Thus dominated convergence shows that A is an entire function for which we have the estimate

$$|A(\lambda)| \leq \|u\|\|v\| + (|\hat{u}(\lambda)| + \sup_{|z| \leq 1} |\hat{u}'(\lambda + z)|) \int_{-\infty}^{\infty} |\hat{v}(t)|\, d\eta(t)\,.$$

Note that since v is in the domain of T its Fourier transform is integrable with respect to $d\eta$, for if $f = (T - i)v$, then $\hat{f}(t) = (t - i)\hat{v}(t) \in L_\eta^2$ so that $\hat{v}(t) = \hat{f}(t)/(t - i)$, where both factors are in L_η^2. Since $\hat{u}$, and by Theorem E.2.7 also $\hat{u}'$, have order $\leq 1/2$ it follows that so has A.

We finally need to show that A tends to 0 along the given rays if $\operatorname{supp} v \subset (a, b)$. But if $\operatorname{supp} v \subset (a, b)$ we have

$$\langle \psi(\cdot, \lambda), v \rangle = \psi(a, \lambda) \int_a^b \psi_{[a,b)}(x, \lambda)\overline{v(x)}\, dx\,,$$

where $\psi_{[a,b)}$ denotes the Weyl solution for the operator in $L^2(a, b)$ generated by (7.3.1), Dirichlet's boundary condition at the initial point a, and the original boundary condition at b. By Theorem 4.3.17 we thus have $\langle \psi(\cdot, \lambda), v \rangle = o(|\psi(a, \lambda)|)$, so that the assumed bound on $\hat{u}$ and Theorem 2.3.7 (2) imply that

$$|A(\lambda)| \leq \|u\|\|v\|/|\operatorname{Im}\lambda| + o(1)$$

as $\lambda \to \infty$ along the given rays, and the lemma is finally proved. $\qquad\square$

Combining these lemmas with the asymptotics of Corollary 6.4.2 it is now easy to prove Theorem 7.3.4.

Proof (Theorem 7.3.4). It is clear that from Lemma 7.3.5 and Corollary 6.4.2 it follows that if supp $u \subset [0, a]$, then $\int_0^a u(x)\varphi(x, \lambda)\, dx$ is an entire continuation of $\hat{u}$ of order $\le 1/2$ such that

$$\varlimsup_{t\to\infty} t^{-1}\log\left|\hat{u}(t^2\lambda)\right| \le \lim t^{-1}\log\left|\varphi(a, t^2\lambda)\right| = \int_0^a \operatorname{Re}\sqrt{-\lambda/p}$$

for all non-real λ. On the other hand, suppose there is an entire continuation of order $\le 1/2$ such that

$$\varlimsup_{t\to\infty} t^{-1}\log\left|\hat{u}(t^2\lambda)\right| \le \int_0^a \operatorname{Re}\sqrt{-\lambda/p}$$

for λ on two different rays. If $\varepsilon > 0$, then according to Theorem 6.4.1, $\hat{u}(\lambda) = \mathcal{O}(|\psi(a + \varepsilon, \lambda)|^{-1})$ for large λ on two non-real rays if $0 < \varepsilon < b - a$, so that according to Lemma 7.3.6 supp $u \subset [0, a + \varepsilon]$ for small $\varepsilon > 0$ and thus also for $\varepsilon = 0$. Lemma 7.3.6 also shows the uniqueness of the continuation. $\qquad\square$

For the proof of Theorem 7.3.1 we require the following lemma.

Lemma 7.3.7. *Suppose F is a unitary map on $L^2(0, b)$ with the property that* $\sup(\operatorname{supp} Fu) = \sup(\operatorname{supp} u)$ *for every compactly supported $u \in L^2(0, b)$. Then F is multiplication by a measurable function of absolute value 1.*

Proof. First note that F^{-1} has the same support property as F. Thus F maps $L^2(0, a)$ onto itself for every $a \in (0, b)$. Hence, if $u = 0$ in $(0, a)$ then so is its image, since it is orthogonal to $L^2(0, a)$. It follows that F actually preserves the convex hull of the support of u (in fact, one sees similarly that the support itself is preserved).

Since F preserves convex hulls of supports, it is clear by the linearity of F that the values of Fu in an open subinterval ω of $(0, b)$ only depend on the values of u in ω, so that $\int_\omega |Fu|^2 = \int_\omega |u|^2$. If $u = 1$ in an open set, it follows by the Lebesgue differentiation theorem, Theorem B.8.8, that $h(x) = Fu(x)$ has absolute value 1 almost everywhere in this set. Clearly this defines h uniquely as an element of $L^\infty(0, b)$ with absolute value 1. If u is a step function, it follows by the linearity of F that $Fu(x) = h(x)u(x)$ a.e., and since step functions are dense in $L^2(0, b)$, this holds in general. $\qquad\square$

We finally turn to the proof of Theorem 7.3.1.

Proof (Theorem 7.3.1). If T and $\tilde{T}$ have the same spectral measure, then the generalized Fourier transforms for T and $\tilde{T}$ have the same target space, so composing the Fourier transform for T with the inverse transform for $\tilde{T}$ we obtain a unitary map

$\mathcal{U} : L^2(0, b) \to L^2(0, \tilde{b})$ which takes T into $\tilde{T}$. We need to show that $\mathcal{U}$ is a Liouville transform. We shall prove this by constructing a unitary Liouville transform which is the inverse of $\mathcal{U}$.

Applying Theorem 7.3.4 for the rays generated by $\pm i$ it is clear that if $\tilde{a} \in (0, \tilde{b})$ and $u \in L^2(0, b)$, then $\sup(\operatorname{supp} u) = a$ if and only if $\sup(\operatorname{supp} \mathcal{U}u) = \tilde{a}$, where $\int_0^a |p|^{-1/2} = \int_0^{\tilde{a}} |\tilde{p}|^{-1/2}$, provided[4] there is such an $a \in (0, b)$. This will certainly be the case if $\tilde{a}$ is sufficiently close to 0.

Suppose for some $\tilde{a} \in (0, \tilde{b})$ we have $\int_0^b |p|^{-1/2} \leq \int_0^{\tilde{a}} |\tilde{p}|^{-1/2}$. Then, since functions of compact support are dense in $L^2(0, b)$, the range of $\mathcal{U}$ would be orthogonal to all elements of $L^2(0, \tilde{b})$ with supports in $(\tilde{a}, \tilde{b})$, contradicting the fact that $\mathcal{U}$ is unitary. A similar argument applied to $\mathcal{U}^{-1}$ shows that the mapping $t : [0, b) \ni a \mapsto \tilde{a} \in [0, \tilde{b})$ is strictly increasing and bijective.

The function t is the composition of the map $x \mapsto \int_0^x |p|^{-1/2}$ with the inverse of $x \mapsto \int_0^x |\tilde{p}|^{-1/2}$, and since an absolutely continuous, increasing function has an absolutely continuous inverse precisely if its derivative is a.e. > 0, as follows from Theorem B.8.3 (see Exercise B.8.3), it is clear that t and t^{-1} are both locally absolutely continuous.

Let $\mathcal{L}$ be a unitary Liouville transform $L^2(0, \tilde{b}) \to L^2(0, b)$ generated by using t for the change of independent variable. The composition $\mathcal{L}\mathcal{U}$ is then a unitary map on $L^2(0, b)$ such that $\sup(\operatorname{supp} \mathcal{L}\mathcal{U}u) = \sup(\operatorname{supp} u)$. It follows from Lemma 7.3.7 that $\mathcal{L}\mathcal{U}$ is multiplication by a function $h \in L^\infty(0, b)$ of absolute value 1. If we replace the multiplier s in the definition of $\mathcal{L}$ by $\bar{h}s$, it follows that $\mathcal{L}$ will be the inverse of $\mathcal{U}$. We may therefore choose s so that $\mathcal{L} = \mathcal{U}^{-1}$.

The other claimed properties of s and t now follow from Theorem 7.2.1, and (7.2.3) gives formulas connecting the coefficients of the two equations.   □

Exercise

Exercise 7.3.1 Extend Theorem 7.3.1 to the case when 0 is not a regular endpoint but limit-circle.

7.4 Left-definite Sturm–Liouville equations

We shall here deal with the same question as in Section 7.3, but now for the left-definite equation. We will freely use notation and results from the first two sections of Chapter 5. Liouville transforms and a Paley–Wiener type theorem will again be central to the answer, but there are some technical differences in the proofs.

We consider the left-definite boundary value problem

---

[4] Note that $\operatorname{Re} \sqrt{\pm i / p} = 1/\sqrt{2 |p|}$.

$$-u'' + qu = \lambda wu , \qquad (7.4.1)$$
$$\lambda u(0) \cos \alpha + u'(0) \sin \alpha = 0 , \qquad (7.4.2)$$

where $q \geq 0$ and w is real-valued and integrable over $(0, a)$ for all $a \in (0, b)$, $q \not\equiv 0$ and $\alpha \in [0, \pi)$ fixed. If b is limit-point for this problem we obtain a self-adjoint relation in the Hilbert space $\mathcal{H}_0$ of Definition 5.2.7 with norm-square $\|u\|^2 = \int_0^b (|u'|^2 + q |u|^2)$. If b is not limit-point we fix a self-adjoint boundary condition at b so that we always have a self-adjoint realization T of the equation in $\mathcal{H}_0$.

Now consider another relation $\tilde{T}$ of the same type as T, with interval $[0, \tilde{b})$, boundary condition parameter $\tilde{\alpha}$, coefficients $\tilde{q}$ and $\tilde{w}$, and Hilbert space $\tilde{\mathcal{H}}_0$. Note that finite functions are dense in $\mathcal{H}_0$ unless b is limit-circle. Similarly for $\tilde{\mathcal{H}}_0$. We shall make the following additional assumption, which simplifies the argument but is not essential.

Assumption 7.4.1 *The coefficients w and $\tilde{w}$ are a.e. $\neq 0$ in $(0, b)$ and $(0, \tilde{b})$ respectively.*

The assumption implies that $\mathcal{H}_\infty = \{0\}$ so T, and similarly $\tilde{T}$, is an operator. Our main theorem is the following.

Theorem 7.4.2. *Suppose that T and $\tilde{T}$ have the same spectral measure $d\eta$. Then there is a unitary Liouville transform $\mathcal{L}$ mapping $\tilde{T}$ to T, i.e., such that $\mathcal{L}\tilde{T} = T\mathcal{L}$.*

The new variable $t(x)$ is the composition of $x \mapsto \int_0^x |w|^{1/2}$ followed by the inverse of $x \mapsto \int_0^x |\tilde{w}|^{1/2}$ and the multiplier $s(x) = 1/\sqrt{t'(x)} = |w(x)/\tilde{w}(t(x))|^{1/4}$ is strictly positive. Furthermore, $s \in C^1(0, b)$, s' is locally absolutely continuous in $[0, b)$ and $s'(0) = 0$.

Theorem 7.2.1 gives further information on the possible combinations of boundary conditions and formulas connecting the coefficients of the two equations. The following corollaries show that given some additional information the spectral measure determines the operator T uniquely.

Corollary 7.4.3. *Suppose T and $\tilde{T}$ have the same spectral measure and that $|w| = |\tilde{w}|$ in $[0, \min(b, \tilde{b}))$. Then $T = \tilde{T}$, i.e., $b = \tilde{b}$, $\alpha = \tilde{\alpha}$, $q = \tilde{q}$, $w = \tilde{w}$ and any boundary condition at b is the same for T and $\tilde{T}$.*

Note that only the absolute value of w need be known, so that all information about sign changes in w is encoded in the spectral measure. Also note that if $|w| = |\tilde{w}|$ only in $[0, a)$, where $0 < a < \min(b, \tilde{b})$, we still have $\alpha = \tilde{\alpha}$ and $q = \tilde{q}$, $w = \tilde{w}$ in $[0, a)$.

Corollary 7.4.4. *Suppose T and $\tilde{T}$ have the same spectral measure, that $q = \tilde{q}$ on $[0, \min(b, \tilde{b}))$, and that either $b = \tilde{b}$ or $\alpha = \tilde{\alpha}$. Then $T = \tilde{T}$, i.e., $b = \tilde{b}$, $\alpha = \tilde{\alpha}$, $q = \tilde{q}$, $w = \tilde{w}$ and any boundary condition at b is the same for T and $\tilde{T}$.*

We will prove the corollaries after the proof of Theorem 7.4.2, which is based on the following Paley–Wiener type theorem.

Theorem 7.4.5. *Let $\hat{u}$ be the generalized Fourier transform of $u \in \mathcal{H}$. Then $\hat{u}$ has at most one entire continuation in the class $\mathcal{A}(w)$ of Definition 7.3.3, and if $\sup(\operatorname{supp} u) = a < b$, such a continuation is given by*

$$\hat{u}(\lambda) = \int_0^a (u'\varphi'(\cdot, \lambda) + qu\varphi(\cdot, \lambda)) \, .$$

Conversely, if $\hat{u}$ has an entire continuation of order $\leq 1/2$ satisfying (7.3.3) for λ on two non-real rays, then $\sup(\operatorname{supp} u) \leq a$.

The following lemma implies the simple direction of Theorem 7.4.5.

Lemma 7.4.6. *Suppose $u \in \mathcal{H}_0$ and $\operatorname{supp} u \subset [0, a]$. Then*

$$\hat{u}(\lambda) = \int_0^a (u'\varphi'(\cdot, \lambda) + qu\varphi(\cdot, \lambda))$$

is an entire function of order $\leq 1/2$ for which $\hat{u}(\lambda) = o(|\lambda\varphi(a + \varepsilon, \lambda)|)$ as λ tends to infinity along any non-real ray, for any ε, $0 < \varepsilon < b - a$.

Proof. We have $\langle u, \varphi(\cdot, \bar{\lambda}) \rangle = -u(0) \cos \alpha + \int_0^b u\lambda\varphi(\cdot, \lambda)w$ for finite u. This is entire of order $\leq 1/2$ since $\lambda \mapsto \lambda\varphi(x, \lambda)$ is of order $\leq 1/2$, locally uniformly in x. Thus

$$\hat{u}(\lambda) = -u(0) \cos \alpha + \lambda\varphi(a + \varepsilon, \lambda) \int_0^a u\varphi(\cdot, \lambda)w/\varphi(a + \varepsilon, \lambda) \, .$$

The function $\varphi(x, \lambda)/\varphi(a + \varepsilon, \lambda)$ tends to zero uniformly for $x \in [0, a]$ and we have $\lambda\varphi(a + \varepsilon, \lambda) \to \infty$ according to Corollary 6.4.2 as $\lambda \to \infty$ along a non-real ray. The lemma follows   □

The hard direction of Theorem 7.4.5 follows from the next lemma.

Lemma 7.4.7. *Suppose $u \in \mathcal{H}_0$, that $\hat{u}$ has an entire continuation of order $\leq 1/2$, and that $\hat{u}(\lambda) = \mathcal{O}(1/|\psi(a, \lambda)|)$ as $\lambda \to \infty$ along two non-real rays.*
Then $\sup(\operatorname{supp} u) \leq a$ and $\hat{u}(\lambda) = \int_0^a (u'\varphi'(\cdot, \lambda) + qu\varphi(\cdot, \lambda))$.

Proof. Consider $A(\lambda) = \langle R_\lambda u, v \rangle - \hat{u}(\lambda)\langle \psi(\cdot, \lambda), v \rangle$ where $v = G_0(wf)$, G_0 is the operator defined in (5.1.3) and $f \in C_0^\infty(a, b)$. In particular $v \in D_T$. As in the proof of Lemma 7.3.6 (see Exercise 7.4.1) it follows that A has an entire continuation of order $\leq 1/2$ which tends to 0 along the given rays. Applying the Phragmén–Lindelöf Theorem E.4.3 to the two sectors bounded by the rays it follows that A is bounded in the whole plane and is therefore constant by Liouville's Theorem E.2.13, thus identically 0.

Now $A(\lambda) = \int_0^b (R_\lambda u - \hat{u}(\lambda)\psi(\cdot, \lambda))\bar{f}w$ so since this vanishes for all $f \in C_0^\infty(a, b)$ it follows that $(R_\lambda u - \hat{u}(\lambda)\psi(\cdot, \lambda))w = 0$ a.e. in (a, b). Since $\operatorname{supp} w = [0, b)$ the continuous function $R_\lambda u - \hat{u}(\lambda)\psi(\cdot, \lambda)$ has support in $[0, a]$. Applying the differential equation it follows that u also has support in $[0, a]$. For $x > a$ the formula (5.2.2) and Theorem 5.2.4 show that $R_\lambda u(x) = \psi(x, \lambda)\langle u, \varphi(\cdot, \lambda)\rangle$ so that $\psi(x, \lambda)(\hat{u}(\lambda) - \langle u, \varphi(\cdot, \lambda)\rangle) = 0$ and the proof is finished. $\quad\square$

Theorem 7.4.5 is a simple consequence of these lemmas.

Proof (Theorem 7.4.5). If $\operatorname{supp} u \subset [0, a]$, Lemma 7.4.6 and Corollary 6.4.2 imply that $\hat{u}(\lambda) = \langle u, \varphi(\cdot, \bar{\lambda})\rangle$ is an entire continuation of $\hat{u}$ of order $\leq 1/2$ such that

$$\overline{\lim_{t \to \infty}} \, t^{-1} \log \left|\hat{u}(t^2\lambda)\right| \leq \lim_{t \to \infty} t^{-1} \log \left|\varphi(a + \varepsilon, t^2\lambda)\right| = \int_0^{a+\varepsilon} \operatorname{Re} \sqrt{-\lambda w}$$

for non-real λ and small $\varepsilon > 0$. Thus $\hat{u} \in \mathcal{A}(w)$.

On the other hand, suppose there is an entire continuation of $\hat{u}$ of order $\leq 1/2$ and such that

$$\overline{\lim_{t \to \infty}} \, t^{-1} \log \hat{u}(t^2\lambda) \leq \int_0^a \operatorname{Re} \sqrt{-\lambda w}$$

for λ on two different non-real rays from the origin. If $0 < \varepsilon < b - a$, then by Corollary 6.4.1 this implies that $\hat{u}(\lambda) = \mathcal{O}(1/|\psi(a + \varepsilon, \lambda)|)$ as $\lambda \to \infty$ on these rays. Lemma 7.4.7 now shows that $\operatorname{supp} u \subset [0, a + \varepsilon]$ for small $\varepsilon > 0$ and thus for $\varepsilon = 0$. The uniqueness of the continuation also follows from Lemma 7.4.7. The proof is now complete. $\quad\square$

We now turn to the proof of Theorem 7.4.2. We first need a lemma.

Lemma 7.4.8. *Let* $t : [0, b) \to [0, \tilde{b})$ *be increasing and* $t(0) = 0$. *Suppose the operator* $\mathcal{U} : \mathcal{H}_0 \to \tilde{\mathcal{H}}_0$ *is linear with the properties that* $\mathcal{U}u(0) = 0$ *if* $u(0) = 0$, *that* $\operatorname{supp} \mathcal{U}u \subset [0, t(x)]$ *if* $\operatorname{supp} u \subset [0, x]$, *and that* $\operatorname{supp} \mathcal{U}u \subset [t(x), \tilde{b})$ *if* $\operatorname{supp} u \subset [x, b)$. *Then there exists a function* s *such that* $u(x) = s(x)\mathcal{U}u(t(x))$ *for all* $u \in \mathcal{H}_0$.

Proof. Fix $x \in [0, b)$. Suppose $u, v \in \mathcal{H}_0$ and that $u(x) = v(x)$. We will first show that $\mathcal{U}(u - v)(t(x)) = 0$. If $x = 0$, this is by assumption.

For $x > 0$ we define[5] $u_- = \chi_{[0,x]}(u - v)$ and $u_+ = \chi_{[x,b)}(u - v)$. These are elements of $\mathcal{H}_0$. Thus $\operatorname{supp} \mathcal{U}u_- \subset [0, t(x)]$ and $\operatorname{supp} \mathcal{U}u_+ \subset [t(x), b)$ so that the functions $\mathcal{U}u_\pm$ vanish at $t(x)$. Adding them gives $\mathcal{U}(u - v)(t(x)) = 0$, as desired.

It follows that the value of $\mathcal{U}u$ at $t(x)$ only depends on the value of u at x. Thus, for each fixed $x \in [0, b)$, the map $\mathcal{U}u(t(x)) \mapsto u(x)$ is well-defined and linear on $\mathbb{C}$, so we may find $s(x)$ so that $\mathcal{U}u(t(x)) = s(x)u(x)$. $\quad\square$

Proof (Theorem 7.4.2). Note first that by Theorem 5.2.15 we must have either $\alpha = \tilde{\alpha} = 0$ or else $\alpha \neq 0 \neq \tilde{\alpha}$.

Let $\mathcal{H}_0$ and $\tilde{\mathcal{H}}_0$ denote the Hilbert spaces and $\mathcal{F}$ and $\tilde{\mathcal{F}}$ the generalized Fourier transforms associated with the two equations, and put $\mathcal{U} = \tilde{\mathcal{F}}^{-1} \circ \mathcal{F} : \mathcal{H}_0 \to \tilde{\mathcal{H}}_0$,

---

[5] χ_M denotes the characteristic function of a set M.

which is unitary since the target space is L_η^2 for both $\mathcal{F}$ and $\tilde{\mathcal{F}}$. By Lemma 5.2.15 we have $u(0) = -\hat{u}(0) = \mathcal{U}u(0)$ if $\alpha = \tilde{\alpha} = 0$, and if $\alpha \neq 0 \neq \tilde{\alpha}$ and P and $\tilde{P}$ are the orthogonal projections of $\mathcal{H}_1$ onto $\mathcal{H}_0$ respectively $\tilde{\mathcal{H}}_1$ onto $\tilde{\mathcal{H}}_0$, we have $\mathcal{U}PF_+ = \tilde{P}\tilde{F}_+ \sin\alpha/\sin\tilde{\alpha}$ by Theorem 5.2.17. Since $\langle u, PF_+ \rangle = \langle u, F_+ \rangle = u(0)$ for finite u if $\alpha \neq 0$ and similarly in $\tilde{\mathcal{H}}_0$ it follows that

$$u(0) = \frac{\sin\alpha}{\sin\tilde{\alpha}} \mathcal{U}u(0) \,, \tag{7.4.3}$$

if $\alpha \neq 0 \neq \tilde{\alpha}$. Thus, in all cases $\mathcal{U}u(0) = 0$ if and only if $u(0) = 0$.

Now, applying Theorem 7.4.5 for the rays generated by $\pm i$, it is clear that if $\tilde{a} \in (0, \tilde{b})$ and $u \in \mathcal{H}$, then $\sup(\operatorname{supp} u) = a$ if $\sup(\operatorname{supp}\mathcal{U}u) = \tilde{a}$, where $\int_0^a |w|^{1/2} = \int_0^{\tilde{a}} |\tilde{w}|^{1/2}$, provided there is such an $a \in (0, b)$.[6] This will certainly be the case if $\tilde{a}$ is sufficiently close to 0. Suppose for some $\tilde{a} \in (0, \tilde{b})$ we have $\int_0^b |w|^{1/2} \leq \int_0^{\tilde{a}} |\tilde{w}|^{1/2}$. Then, since finite functions are dense in $\mathcal{H}_0$, the range of $\mathcal{U}$ would be orthogonal to all elements of $\tilde{\mathcal{H}}_0$ with supports in $(\tilde{a}, \tilde{b})$, contradicting the fact that $\mathcal{U}$ is unitary.

A similar argument applied to $\mathcal{U}^{-1}$ shows that the function

$$t : [0, b) \ni a \mapsto \tilde{a} \in [0, \tilde{b})$$

is bijective, and $\sup(\operatorname{supp}\mathcal{U}u) = \tilde{a} = t(a)$ if and only if $\sup(\operatorname{supp} u) = a$.

We also have $\inf \operatorname{supp} u = a$ if and only if $\inf \operatorname{supp} \mathcal{U}u = t(a)$. To see this, note that what we have already proved implies that if $\inf \operatorname{supp} u = a > 0$, then $\mathcal{U}u$ is orthogonal to all elements of $\tilde{\mathcal{H}}$ with support in $[0, t(a)]$. This means that in this interval $\mathcal{U}u$ is a multiple of $\tilde{F}_-$. However, since $u(0) = 0$ we also have $\mathcal{U}u(0) = 0$, so that the multiple is 0, and thus $\inf \operatorname{supp} \mathcal{U}u \geq t(a)$. A similar reasoning applied to $\mathcal{U}^{-1}$ proves the other direction.

It follows that $\mathcal{U}$ as well as $\mathcal{U}^{-1}$ have the properties required in Lemma 7.4.8. Thus there is a non-vanishing function s such that

$$u(x) = s(x)\mathcal{U}u(t(x)) \,.$$

The unitary map $\mathcal{U}$ therefore has an inverse which is a Liouville transform. Regularity properties of s and its connection with t, as well as formulas connecting the coefficients of the two equations follow from Theorem 7.2.1, so this completes the proof.   □

Finally we have to prove Corollaries 7.4.3 and 7.4.4.

Proof (Corollary 7.4.3). The assumptions together with Theorem 7.4.2 show that $t(x) = x$, so that $b = \tilde{b}$ and $s(x) = 1$. Since $\mathcal{F}\psi(\cdot, \lambda) = (t - \lambda)^{-1} = \tilde{\mathcal{F}}\tilde{\psi}(\cdot, \lambda)$, it follows that $\psi = \tilde{\psi}$. Thus the boundary conditions at b for T and $\tilde{T}$, if any, are the same, so that T and $\tilde{T}$ are identical.   □

Proof (Corollary 7.4.4). The function $\tilde{s} = -\varphi_0$ solves $-\tilde{s}'' + q\tilde{s} = 0$ with initial data $\tilde{s}(0) = 1$, $\tilde{s}'(0) = 0$. Since $q \geq 0$ this solution is strictly positive on $[0, b)$, so we

---

[6] Note that Re $\sqrt{\pm iw} = \sqrt{|w|}/2$.

may put $\tilde{t}(x) = \int_0^x 1/\tilde{s}^2$. The pair of functions $\tilde{s}$, $\tilde{t}$ defines a Liouville transform $\mathcal{L}_0$ mapping $[0, b)$ onto some interval $[0, c)$ and $[0, \tilde{b})$ onto $[0, \check{c})$, and transforming the equations into $-u_0'' = \lambda w_0 u_0$ and $-\tilde{u}_0'' = \lambda \tilde{w}_0 \tilde{u}_0$, respectively. Thus $\mathcal{L}_0 \mathcal{L} \mathcal{L}_0^{-1}$, where $\mathcal{L}$ is the Liouville transform of Theorem 7.4.2, transforms one of these equations into the other.

Being a composition of Liouville transforms this is itself a Liouville transform given, say, by $u_0(x) = s_1(x)\tilde{u}_0(t_1(x))$. By construction we obtain $s_1(0) = s(0)$, $s_1'(0) = 0$, and $s_1^2 t_1' \equiv 1$. Since both potentials are identically 0 it follows that $s_1'' = 0$. This means that $s_1 \equiv s(0)$ and $t_1(x) = x/(s(0))^2$.

If $\alpha = \tilde{\alpha}$, then by Theorem 7.2.1 $s(0) = 1$ so that $\mathcal{L}_0 \mathcal{L} \mathcal{L}_0^{-1}$ is the identity, implying that $\mathcal{L}$ is also the identity. Similarly, if $b = \tilde{b}$, then $c = \check{c}$ so that $s(0) = 1$, unless $c = \check{c} = \infty$. We will show that c is always finite, and then it again follows that $\mathcal{L}$ is the identity.

Now $c = \int_0^b 1/\tilde{s}^2$, so we need to show that this integral is finite. Put $H = \tilde{s}'\tilde{s}$ and differentiate. We obtain $H' = (\tilde{s}')^2 + \tilde{s}''\tilde{s} = (\tilde{s}')^2 + q\tilde{s}^2$. Thus H is non-decreasing and not constant so eventually > 0. It follows that $1/\tilde{s}^2 = (\tilde{s}')^2/H^2 \le H'/H^2$ so that $\int_d^b 1/\tilde{s}^2 \le 1/H(d) < \infty$ if d is sufficiently close to b. This completes the proof. See also Section 5.5 $\qquad\qquad$ □

Exercise

Exercise 7.4.1 In the proof of Lemma 7.4.7 some details are missing. Supply them.

Exercise 7.4.2 Extend Theorem 7.4.2 to the case when 0 is not a regular endpoint but limit-circle.

7.5 Two singular endpoints

Here we shall show that if we are dealing with a Sturm–Liouville equation on an interval (a, b) with two singular endpoints the theories of the previous sections can be used to get similar uniqueness theorems for this case. We assume that we have a right- or left-definite operator determined by separated boundary conditions; the most interesting case is, of course, when both a and b are limit-point. It is always assumed that the coefficient $w \neq 0$ a.e. in (a, b).

Referring to the spectral theories of Sections 4.4 and 5.3 we select an anchor point $c \in (a, b)$. We then have an m-matrix M of the form

$$M = (m_{jk}) = \frac{1}{m_+ + m_-} \begin{pmatrix} m_+ m_- & \frac{1}{2}(m_- - m_+) \\ \frac{1}{2}(m_- - m_+) & -1 \end{pmatrix}. \tag{7.5.1}$$

There is a unique representation

$$M(\lambda) = A + B\lambda + \int\limits_{-\infty}^{\infty} \left(\frac{1}{t-\lambda} - \frac{t}{t^2+1} \right) d\Xi(t) \qquad (7.5.2)$$

where A is Hermitian, $B \geq 0$ and $d\Xi$ is a positive, 2×2 matrix-valued measure. Since $M = (m_{jk})$ equals its transpose the uniqueness of the representation shows that this is also true of $A = (a_{jk})$, $B = (b_{jk})$ and $d\Xi = (d\xi_{jk})$, which are therefore real. We shall show that $B = 0$.

Theorem 7.5.1. *For $x = c$ the kernels $g(x, y, \lambda)$ and $g_1(x, y, \lambda) = \frac{d}{dx} g(x, y, \lambda)$ of Theorem 4.4.3, respectively Theorem 5.2.17, have Fourier transforms*

$$\mathcal{F}(g(c, \cdot, \lambda))(t) = \tfrac{1}{t-\lambda} \left(\begin{smallmatrix} 1 \\ 0 \end{smallmatrix} \right) \text{ and } \mathcal{F}(g_1(c, \cdot, \lambda))(t) = \tfrac{1}{t-\lambda} \left(\begin{smallmatrix} 0 \\ 1 \end{smallmatrix} \right)$$

and in (7.5.2) we have $B = 0$.

Proof. The formulas for the Fourier transforms are obtained in the same way as similar formulas in Theorems 4.3.17 and 5.2.17; see Exercise 7.5.1. Putting $h_1(x, \lambda) = g(c, x, \lambda)$ and $h_2(x, \lambda) = g_1(c, x, \lambda)$ direct calculation then shows that

$$\langle h_j, h_k \rangle = \frac{\operatorname{Im} m_{jk}(\lambda)}{\operatorname{Im}\lambda} = b_{jk} + \int\limits_{-\infty}^{\infty} \frac{d\xi_{jk}(t)}{|t-\lambda|^2} \,,$$

using the fact that h_1, h_2 satisfy the boundary conditions at a and b. However, Parseval gives

$$\langle h_j, h_k \rangle = \langle \hat{h}_j, \hat{h}_k \rangle_\Xi = \int\limits_{-\infty}^{\infty} \frac{d\xi_{jk}(t)}{|t-\lambda|^2} \,.$$

It follows that $B = 0$. □

Now suppose we have another equation with the same spectral measure $d\Xi$ and m-matrix $\tilde{M}$, given by (7.5.1) by replacing A, B and $m_\pm$ by $\tilde{A}$, 0 respectively $\tilde{m}_\pm$. By (7.5.2) and Theorem 7.5.1, $\tilde{M}(\lambda) = \breve{A} + M(\lambda)$ for some real symmetric matrix $\breve{A} = (\breve{a}_{jk})$.

The relation between the m-matrices may, after subtracting $\frac{1}{2} \left(\begin{smallmatrix} 0 & -1 \\ 1 & 0 \end{smallmatrix} \right)$ from both sides, be written

$$\frac{1}{\tilde{m}_+ + \tilde{m}_-} \begin{pmatrix} \tilde{m}_+\tilde{m}_- & \tilde{m}_- \\ -\tilde{m}_+ & -1 \end{pmatrix} = \frac{1}{m_+ + m_-} \begin{pmatrix} m_+m_- & m_- \\ -m_+ & -1 \end{pmatrix} + \begin{pmatrix} \breve{a}_{11} & \breve{a}_{12} \\ \breve{a}_{12} & \breve{a}_{22} \end{pmatrix} \,.$$

Multiplying from the left by $\left(1 \; \tilde{m}_- \right)$ and from the right by $\left(\begin{smallmatrix} -1 \\ m_+ \end{smallmatrix} \right)$ this gives

$$\breve{a}_{22}\tilde{m}_-m_+ + \breve{a}_{12}(m_+ - \tilde{m}_-) - \breve{a}_{11} = 0 \,. \qquad (7.5.3)$$

Consider first the right-definite case, when $m_\pm(\lambda)$, being Dirichlet m-functions, both tend to ∞ as $\lambda \to \infty$ along the imaginary axis by Theorem 6.2.6 (2). Thus all terms in (7.5.3) except the first is $o(\tilde{m}_+m_-)$ so that we must have $\breve{a}_{22} = 0$. It follows that

$\tilde{m}_+ + \tilde{m}_- = m_+ + m_-$. It also follows that *either* $\breve{a}_{12} = 0$, which implies $\breve{a}_{11} = 0$ so that $\breve{A} = 0$, *or* $\tilde{m}_- = m_+ - \breve{a}_{11}/\breve{a}_{12}$. In the first case $\tilde{M} = M$ so that $\tilde{m}_+ = m_+$ and $\tilde{m}_- = m_-$. Thus m_+ has the same spectral measure as $\tilde{m}_+$ and m_- the same spectral measure as $\tilde{m}_-$.

Consider now the second alternative, when $\tilde{m}_- = m_+ - \breve{a}_{11}/\breve{a}_{12}$. Making the change of variable $x \mapsto -x$ in (7.2.2) interchanges $\tilde{m}_+$ and $\tilde{m}_-$, and m_+ and $\tilde{m}_-$ are both Dirichlet m-functions. By Theorem 7.3.1 there is a Liouville transform taking one equation into the other, so by Theorem 7.2.1 we must have $\tilde{m}_- = m_+$ so that $\breve{a}_{11} = 0$. Since $\tilde{m}_+ + \tilde{m}_- = m_+ + m_-$ we then also have $\tilde{m}_+ = m_-$. From $\tilde{M} = \breve{A} + M$ therefore follows $m_- = m_+ + \breve{a}_{12}$. Again, $m_\pm$ are both Dirichlet m-functions so by Theorem 7.2.1 it follows that $\breve{a}_{12} = 0$ so also in this case $\breve{A} = 0$. Thus $\tilde{m}_+ = m_+$ and $\tilde{m}_- = m_-$ in all cases.

In the left-definite case we have a very similar situation, but now $m_\pm$ tend to zero along the imaginary axis, so we instead obtain $\breve{a}_{11} = 0$ but apart from this the arguments are much the same, and it follows that $-1/\tilde{m}_+ = -1/m_+$ and $-1/\tilde{m}_- = -1/m_-$.

Appealing now to Theorem 7.3.1 in the right-definite case and Theorem 7.4.2 in the left-definite case for the intervals (a, c) and (c, b) we finally obtain the following theorem.

Theorem 7.5.2. *Suppose given two right- or left-definite self-adjoint realizations of Sturm–Liouville equations (7.2.1) and (7.2.2) with two singular endpoints and separated boundary conditions. Assume the two equations have the same (matrix-valued) spectral measure and that $w \neq 0$ a.e. in (a, b) and $\tilde{w} \neq 0$ a.e. in $(\tilde{a}, \tilde{b})$.*

Then the corresponding m-matrices are equal and there is a unitary Liouville transform which takes one equation into the other, mapping the interval endpoints and any boundary conditions there as well as the anchor points of the two equations into each other.

There are also corollaries corresponding to Corollaries 7.3.2, 7.4.3 and 7.4.4; see Exercise 7.5.2.

Exercises

Exercise 7.5.1 Show the formulas for the Fourier transforms in Theorem 7.5.1.

Hint: Look at the proofs of Theorems 4.3.17 and 5.2.17.

Exercise 7.5.2 Carry out the details of the arguments leading up to Theorem 7.5.2 in the left-definite case. Also formulate and prove analogues of Corollaries 7.3.2, 7.4.3 and 7.4.4 for the case of two singular endpoints.

7.6 Notes and remarks

The approach to the Borg–Marčenko theorem of Section 7.1 owes much to Borg [33].
In [17] Bennewitz gave a particularly concise proof of Simon's local Borg–Marčenko
theorem. This idea was extended to cover complex potentials by Brown, Peacock,
and Weikard [41] and to Jacobi difference operators by Weikard [217].

In addition to the Gel'fand–Levitan theory mentioned in the introduction, a few
other approaches to inverse spectral theory for equations with one regular endpoint
exist. In the 1950s M. G. Kreĭn wrote a series of papers on the *string equation*
$-u'' = \lambda w u$, where w is a positive measure, including a spectral as well as inverse
spectral theory [138, 136, 134, 135, 137]. There is some connection between the
work of Kreĭn and the de Branges [34] theory of Hilbert spaces of entire functions,
which gives an inverse spectral theory for 2×2 first order canonical systems of
differential equations. This theory may be applied to Kreĭn strings, as was done by
Dym and McKean [57], as well as to right-definite Sturm–Liouville equations as
was done by Remling [186] and Eckhardt [59]. It may also be applied to left-definite
Sturm–Liouville equations, see Eckhardt [58]. A completely different approach to
the results of the Gel'fand–Levitan theory was presented in the papers Simon [204]
and Simon and Gesztesy [86] in 1999–2000.

A Gel'fand–Levitan theory for the Schrödinger equation on the full real line
was first considered by Bloh [29]. Rofe-Beketov [192] provided a complete inverse
spectral theory for this case.

Inverse theory for left-definite Sturm–Liouville problems has been investigated in
a series of papers of Eckhardt and collaborators. They have, among other work, used
de Branges' theory to establish uniqueness [58] of coefficients. They also established
connections between the left-definite Sturm–Liouville problem, the moment problem
and the Kreĭn string [62, 64, 61, 60]. In particular they show the Kreĭn string problem
with a quadratic spectral term [63] is related to the flow of the conservative Camassa–
Holm system, a generalization of the Camassa–Holm equation.

For coupled boundary conditions, on the other hand, the knowledge of two spectra
is not sufficient to establish uniqueness of the potential q. In fact, if q has period 1,
consider the potentials $q_a : [0, 1] \to \mathbb{R}$ defined by $q_a(x) = q(x+a)$ for $a \in [0, 1]$. Fix
$\theta \in [0, 2\pi)$ and pose the boundary conditions $y(1) = e^{i\theta} y(0)$ and $y'(1) = e^{i\theta} y'(0)$.
The corresponding eigenvalues will then be independent of a. To expand a little on
this claim, let $c_a(\cdot, \lambda)$ and $s_a(\cdot, \lambda)$ be the solutions of $-y'' + q_a y = \lambda y$ satisfying initial
conditions $c_a(0, \lambda) = s'_a(0, \lambda) = 1$ and $c'_a(0, \lambda) = s_a(0, \lambda) = 0$. The monodromy
matrix $\Phi_a(1, \lambda) = \begin{pmatrix} c_a(1,\lambda) & s_a(1,\lambda) \\ c'_a(1,\lambda) & s'_a(1,\lambda) \end{pmatrix}$ then has determinant 1 and its trace, called the
Floquet discriminant, is denoted by $D(\lambda)$. The eigenvalues of the boundary value
problem mentioned above are the zeros of the function $D - 2\cos(\theta)$ and, in particular,
the periodic eigenvalues are the zeros of $D - 2$. Being an entire function of growth
order $1/2$, D is determined by the periodic eigenvalues up to a multiplicative constant
(which may be found from its asymptotic behavior as λ tends to infinity along, say,
the imaginary axis). Since, clearly, the periodic eigenvalues do not depend on a,

this proves that D does not depend on a and neither does any of the spectra for the boundary conditions mentioned above.

To uniquely identify a periodic potential one may now use the Borg–Marčenko theorem and note that the Floquet discriminant almost determines the Neumann data of the Dirichlet eigenvalues if the latter are given. Indeed, if μ is a Dirichlet eigenvalue then $\Phi(1, \mu)$ is lower triangular, so that its diagonal elements are also its eigenvalues, implying that $s'(1, \mu)$ is one of the values $(D(\mu) \pm \sqrt{D(\mu)^2 - 4})/2$. Hence the potential is uniquely determined by the Floquet discriminant, the Dirichlet eigenvalues, and an assignation of a sign for each of the latter. Since the Dirichlet eigenvalues lie in the instability intervals, i.e., the intervals where $|D(\lambda)| \geq 2$, this implies that the set of periodic potentials with the same discriminant is a torus (in general, infinite-dimensional).

Another variation of this theme is tightly connected with the integration of the Korteweg–de Vries equation. Borg [32] had already shown that a periodic potential q of the Schrödinger equation without any spectral gaps must necessarily be constant when, in 1965, Hochstadt [107] proved that a periodic potential with precisely one spectral gap (and hence two spectral bands) must satisfy the differential equation

$$q'^2 - 2q^3 + Aq^2 + Bq + C = 0$$

for appropriate constants A, B, and C, and hence must be an elliptic function. McKean and van Moerbeke [164] explored the relationship of finite-band potentials with the hyperelliptic curve $p^2 + (\lambda - \lambda_0)...(\lambda - \lambda_{2n}) = 0$ where the λ_j are the simple periodic and semi-periodic eigenvalues associated with q. Later McKean and Trubowitz [165] investigated the situation where infinitely many bands are present. A different viewpoint of such matters was pursued by Gesztesy and Weikard, who showed that among elliptic functions the set of finite-band potentials is characterized by the property that, for each fixed λ, the differential equation $-y'' + qy = \lambda y$ has only meromorphic solutions. We refer, in particular, to [87] and the survey [88]. For even more background on all these matters we mention the monograph [84] by Gesztesy and Holden.

Chapter 8
Scattering

8.1 Introduction

Broadly speaking, scattering theory concerns the effect a perturbation (obstacle or 'scatterer') has on the time evolution of an evolving system, which we will assume is described by a linear evolution equation as in Exercise 3.1.8. By separation of the time and space variables one may instead consider a time independent equation $Tu = \lambda u$, and then the question is how the spectral properties of T are influenced by the scatterer. We shall only consider the case of one space variable and a perturbation with negligible effect at spatial infinity.

There are then primarily two situations of interest. One may deal with a radial equation on $(0, \infty)$ obtained by separation of variables of a spherically symmetric equation in higher dimensions, as in Section 6.6, or an equation defined on $(-\infty, \infty)$, perhaps modelling a wave guide. We shall only consider the latter case, which is crucial for dealing with many infinite-dimensional integrable systems, as described in Appendix F. For each solution u of the unperturbed equation we then have two solutions $u_\pm$ of the perturbed equation such that $u_- \sim u$ near $-\infty$ and $u_+ \sim u$ near $+\infty$. The aim of scattering theory is to describe the relation between u_- and u_+ for a given perturbation.

In many applications it is only this relation which is accessible to observation, and then the important problem is that of *inverse scattering theory*, where knowledge of the relation between u_- and u_+ is used to obtain information on the perturbation.

More concretely, suppose the unperturbed equation (the 'model equation') is $-u'' = \lambda u$ and the perturbed equation is the one-dimensional Schrödinger equation $-u'' + qu = \lambda u$ where the potential q represents the perturbation. If $|q|$ is small at $\pm\infty$ the essential spectrum is $[0, \infty)$ and there may be discrete spectrum in $(-\infty, 0)$.

Setting $k = \sqrt{\lambda}$ for $\lambda > 0$ the solutions of the model equation are linear combinations of $e^{\pm ikx}$, and the Schrödinger equation will have solutions asymptotic to $e^{\pm ikx}$ as $x \to -\infty$ as well as solutions with these asymptotics at $+\infty$. Scattering theory concerns the relations between these solutions.

© Springer Nature Switzerland AG 2020

C. Bennewitz et al., *Spectral and Scattering Theory for Ordinary Differential Equations*, Universitext, https://doi.org/10.1007/978-3-030-59088-8_8

A complete theory of inverse scattering for this equation on $(-\infty, \infty)$ was given by Faddeev [69, 70]. Further developments are due to Marčenko [156], Deift and Trubowitz [52], Melin [166] and others. These results are essential for the theory of the Korteweg–de Vries (KdV) equation, the first of many infinite-dimensional integrable systems discovered from the late 1960s onwards.

To each infinite-dimensional integrable system in the sense of Appendix F there is an associated spectral problem. In each case an inverse scattering theory is required, but at present there are still many spectral problems where this is missing, even though great strides were made by Beals and Coifman [12] in the 1980s. In this chapter we shall give some results in the scattering and inverse scattering theory for left-definite Sturm–Liouville equations which are applicable to the spectral problems associated with the Camassa–Holm equation, as is discussed in Appendix F.

8.2 Jost solutions

Consider two Sturm–Liouville equations

$$-u'' + qu = \lambda w u, \tag{8.2.1}$$

$$-u'' + q_0 u = \lambda u, \tag{8.2.2}$$

on $\mathbb{R}$, where q and w are locally integrable, real-valued functions and q_0 a real constant. Like in the latter part of Chapter 5 we could allow less regular q and w, for example assume that they are measures or at least in H^{-1}_{loc}. The results of this chapter would still be true, with minimal changes in statements and proofs, but in the interest of minimizing technicalities we will confine ourselves to locally integrable coefficients.

The solutions of (8.2.2) are linear combinations of $e^{\pm ikx}$, where $k = \sqrt{\lambda - q_0}$. The equation (8.2.2) will be viewed as a model equation for (8.2.1), assuming that $q - q_0$ and $w - 1$ are small at $\pm\infty$. We then expect that there are solutions of (8.2.1) asymptotic to $e^{\pm ikx}$ at $\pm\infty$. Defining the *first moment* of a function f locally integrable on $\mathbb{R}$ to be $\int_{-\infty}^{\infty} |t f(t)| \, dt$ we have the following theorem.

Theorem 8.2.1. *Using the notation above, assume that $q - q_0$ and $w - 1$ are in $L^1(\mathbb{R})$ and that $k = \sqrt{\lambda - q_0}$, using the square root with argument in $[0, \pi)$. Then if $\lambda \neq q_0$, so that $k \neq 0$, there exist unique solutions $f_\pm(\cdot, \lambda)$ of (8.2.1) such that*

$$f_+(x, \lambda) = e^{ikx}(1 + o(1)) \quad and \quad f'_+(x, \lambda) = e^{ikx}(ik + o(1)) \quad as \; x \to \infty,$$

$$f_-(x, \lambda) = e^{-ikx}(1 + o(1)) \quad and \quad f'_-(x, \lambda) = e^{-ikx}(-ik + o(1)) \quad as \; x \to -\infty.$$

If $q - q_0 w$ has a finite first moment this is valid also for $\lambda = q_0$, i.e., $k = 0$.

As functions of λ these functions are analytic in $\mathbb{C}$ outside the real interval $[q_0, \infty)$, with continuous limits as λ approaches (q_0, ∞) from above or below. If $q - q_0 w$ has a finite first moment this is true also as $\lambda \to q_0$. All this is true locally uniformly in

x. For $a \in \mathbb{R}$ and $x \geq a$ there is a constant C_a such that $|f_\pm(x,\lambda)| \leq \exp(C_a |\lambda|^{1/2})$ for large $|\lambda|$.

We shall call solutions of this type *Jost solutions*. For the proof we shall need the following lemma, the proof of which we leave to Exercise 8.2.1.

Lemma 8.2.2. *If* $\operatorname{Re} z \leq 0$, *then* $|e^z - 1| \leq \min(2, |z|)$.

Proof (Theorem 8.2.1). Writing $f_+(x,\lambda) = g_+(x,\lambda)e^{ikx}$ the function g_+ should satisfy $g_+''(x,\lambda) - 2ikg_+'(x,\lambda) = Q(x,\lambda)g_+(x,\lambda)$ where $Q(\cdot,\lambda) = q - q_0 - \lambda(w-1) = q - q_0 w + k^2(1 - w)$. Solving the differential equation, viewing the right-hand side as known, we arrive at the integral equation

$$g_+(x,\lambda) = 1 + \int_x^\infty \frac{e^{2ik(t-x)} - 1}{2ik} Q(t,\lambda)g_+(t,\lambda)\, dt$$

for g_+. Note that for $k \neq 0$ we have $|Q(\cdot,\lambda)/k| \leq |q - q_0 w|/|k| + |k|\,|w - 1| \in L^1(\mathbb{R})$.

We solve the integral equation by successive approximations, setting $g_0 \equiv 1$ and inductively

$$g_{j+1}(x,\lambda) = \int_x^\infty \frac{e^{2ik(t-x)} - 1}{2ik} Q(t,\lambda)g_j(t,\lambda)\, dt, \text{ for } j = 0, 1, 2 \ldots$$

By Lemma 8.2.2 $\left|e^{2ik(t-x)} - 1\right| \leq 2$ so that $|g_{j+1}(x,\lambda)| \leq \int_x^\infty |Q(\cdot,\lambda)/k|\,|g_j(\cdot,\lambda)|$. Here $Q(x,\lambda)/k$ is integrable for $k \neq 0$, so if $x \geq a$ induction shows that

$$|g_j(x,\lambda)| \leq \left(\int_x^\infty |Q(\cdot,\lambda)/k|\right)^j / j! \leq \left(\int_a^\infty |Q(\cdot,\lambda)/k|\right)^j / j!.$$

Thus all g_j have the analyticity and continuity properties required of f_+, and $g_+(x,\lambda) = \sum_{j=0}^\infty g_j(x,\lambda)$ is absolutely and uniformly convergent for $x \geq a$ and $|\lambda - q_0| \geq \varepsilon > 0$. Since

$$\sum_{j=0}^{n+1} g_j(x,\lambda) = 1 + \int_x^\infty \frac{e^{2ik(t-x)} - 1}{2ik} Q(t,\lambda) \sum_{j=0}^n g_j(t,\lambda)\, dt$$

this shows that g_+ is a solution, and that for large $|\lambda|$ and $x \geq a$ we have

$$|f_+(x,\lambda)| \leq \exp\left(-a \operatorname{Im} k + \int_a^\infty |Q(\cdot,\lambda)/k|\right) \leq \exp(C_a |\lambda|^{1/2})$$

for some constant C_a. The uniform convergence shows that $f_+(x,\lambda) = g_+(x,\lambda)e^{ikx}$ has all the required properties. Uniqueness follows by applying the Gronwall Lemma D.3 to the absolute value of the difference of two solutions.

Finally, Lemma 8.2.2 shows that $\int_x^\infty |\frac{e^{2ik(t-x)}-1}{2ik}(q-q_0w)|$ may be estimated by the integral of $(1+|t|)|q-q_0w|$, so if this is finite all the claims made are also correct in a neighborhood of $\lambda = q_0$. The proof of the existence of a unique f_- with the appropriate properties is completely analogous.   □

Note that our definition of $k = \sqrt{\lambda - q_0}$ shows that $\overline{f_\pm(x,\lambda)} = f_\pm(x,\bar\lambda)$ so that for $t > q_0$ we have $f_\pm(x, t + i0) = f_\pm(x, t)$ but $f_\pm(x, t - i0) = \overline{f_\pm(x, t)}$. On the other hand, if λ is real the conjugate of any solution of (8.2.1) is also a solution. If $\lambda < q_0$, then $ik < 0$ so that $g_\pm$ and thus also $f_\pm$ are real-valued. If $\lambda > q_0$, then ik is purely imaginary and $\overline{f_\pm}$ are solutions satisfying

$$\overline{f_+}(x,\lambda) \sim e^{-ikx} \quad \text{and} \quad \overline{f_+'}(x,\lambda) \sim -ike^{-ikx} \quad \text{as } x \to +\infty\,,$$

$$\overline{f_-}(x,\lambda) \sim e^{ikx} \quad \text{and} \quad \overline{f_-'}(x,\lambda) \sim ike^{ikx} \quad \text{as } x \to -\infty\,.$$

The Wronskian $\mathcal{W}(f_+, \overline{f_+})$ is independent of x, and since $\mathcal{W}(f_+, \overline{f_+}) \sim 2ik$ at infinity it follows that $\mathcal{W}(f_+, \overline{f_+}) = 2ik$. Similarly $\mathcal{W}(f_-, \overline{f_-}) = -2ik$.

Inserting this in the identity $\mathcal{W}(f_+, f_-)\mathcal{W}(\overline{f_+}, \overline{f_-}) = \mathcal{W}(f_+, \overline{f_+})\mathcal{W}(f_-, \overline{f_-}) - \mathcal{W}(f_+, \overline{f_-})\mathcal{W}(f_-, \overline{f_+})$ (see the footnote on page 112) shows that

$$|\mathcal{W}(f_+, f_-)|^2 = 4k^2 + \left|\mathcal{W}(f_+, \overline{f_-})\right|^2 > 0 \tag{8.2.3}$$

so that the pair $f_+(\cdot, \lambda)$, $f_-(\cdot, \lambda)$ is a basis for the solutions. For $\lambda \in (q_0, \infty)$ we may thus find coefficients $\mathfrak{T}_\pm$ and $\mathfrak{R}_\pm$ depending only on λ such that

$$\begin{cases} \mathfrak{T}_+ f_- = \mathfrak{R}_+ f_+ + \overline{f_+}\,, \\ \mathfrak{T}_- f_+ = \mathfrak{R}_- f_- + \overline{f_-}\,. \end{cases} \tag{8.2.4}$$

Taking the Wronskians of the first equation by f_+ and f_- we obtain $\mathfrak{T}_+\mathcal{W}(f_+, f_-) = \mathcal{W}(f_+, \overline{f_+}) = 2ik$ and $\mathfrak{R}_+\mathcal{W}(f_+, f_-) = \mathcal{W}(f_-, \overline{f_+})$. Doing the same to the second equation we arrive at the formulas

$$\begin{cases} \mathfrak{T}_+ = \mathfrak{T}_- = 2ik/\mathcal{W}(f_+, f_-)\,, \\ \mathfrak{R}_+ = \mathcal{W}(f_-, \overline{f_+})/\mathcal{W}(f_+, f_-)\,, \\ \mathfrak{R}_- = \mathcal{W}(\overline{f_-}, f_+)/\mathcal{W}(f_+, f_-)\,. \end{cases} \tag{8.2.5}$$

Since $\mathfrak{T}_+ = \mathfrak{T}_-$ we drop the signs and denote $\mathfrak{T}_\pm$ by $\mathfrak{T}$. From (8.2.5) it follows that $\mathfrak{R}_+\overline{\mathfrak{T}} + \mathfrak{T}\overline{\mathfrak{R}_-} = 0$, and (8.2.3) now reads $|\mathfrak{T}|^2 + |\mathfrak{R}_\pm|^2 = 1$ so the *scattering matrix*

$$\mathfrak{S} = \begin{pmatrix} -\mathfrak{R}_+ & \mathfrak{T} \\ \mathfrak{T} & -\mathfrak{R}_- \end{pmatrix}$$

is unitary. The coefficient $\mathfrak{T}$ is called the *transmission coefficient* and the coefficients $\mathfrak{R}_\pm$ are the *reflection coefficients* associated with our equation (8.2.1). Note that if $q \equiv q_0$ and $w \equiv 1$, then $\overline{f_\pm} = f_\mp$ so that $\mathfrak{T} \equiv 1$ and $\mathfrak{R}_\pm \equiv 0$. Thus (8.2.4) states that a wave $\overline{f_+}$ coming in from $+\infty$ is split into a wave $\mathfrak{R}_+ f_+$ reflected from the obstacle and a wave $\mathfrak{T}f_-$ transmitted through the obstacle. A similar interpretation may be made of the second equation. Note that $|\mathfrak{T}|^2 + |\mathfrak{R}_\pm|^2 = 1$ says that all of the wave[1] is either reflected or transmitted. Since $\mathfrak{T}$ and $\mathfrak{R}_\pm$ are complex-valued the reflected and transmitted waves also contain phase shifts.

Corollary 8.2.3. *The transmission coefficient $\mathfrak{T}$ extends uniquely to a function meromorphic outside $[q_0, \infty)$ and with continuous, non-zero limits on approaching (q_0, ∞) from above or below. The transmission coefficient has poles precisely where $f_\pm$ are linearly dependent.*

Proof. The existence of the extension of $\mathfrak{T}$ follows from the expression for $\mathfrak{T}$ in (8.2.5). The uniqueness is a standard fact, for the difference of two extensions vanishes on $[q_0, \infty)$. By the reflection principle[2] the difference extends to a function analytic in a neighborhood ω of any $x \in (q_0, \infty)$, and since it vanishes on the real part of ω it is identically zero.

For $\lambda \neq [q_0, \infty)$, the poles of $\mathfrak{T}$ are the zeros of the Wronskian $\mathcal{W}(f_+, f_-)$, i.e., those λ for which $f_+(\cdot, \lambda)$ and $f_-(\cdot, \lambda)$ are linearly dependent. $\qquad\square$

Only in exceptional cases is it possible to extend the domains of $\mathfrak{R}_\pm$ similarly.

Exercise

Exercise 8.2.1 Prove Lemma 8.2.2.

Hint: Show that $\left|e^{x+iy} - 1\right|^2 = (e^x - 1)^2 + 2e^x(1 - \cos y)$ and that if $x \le 0$, then the first term is less than x^2 while the second is less than y^2.

8.3 The Jost transform

Our approach to inverse scattering is to view it as an instance of inverse spectral theory. This is done by proving an expansion theorem in which spectral data are scattering data. To do this we shall derive an expansion theorem with respect to the Jost solutions. We restrict our analysis to left-definite equations, i.e., we assume $q \ge 0$ and not identically zero, so also $q_0 \ge 0$. However, there are analogous results for right-definite equations, with very similar proofs. We relegate this to a few exercises.

First note that a left-definite equation (8.2.1) on $\mathbb{R}$ is limit point at $\pm\infty$ by Theorem 5.1.8. We denote the corresponding self-adjoint relation by T. The following theorem describes the spectrum of T.

---

[1] $|\mathfrak{T}|$ and $|\mathfrak{R}_\pm|$ are the intensities of the waves, so their power contents are the squares of this.
[2] See Ahlfors [1, Chapter 4.6.5].

Theorem 8.3.1.

(1) *T has essential spectrum* $[q_0, \infty)$.

(2) *The eigenvalues of T are all simple, located in* $(-\infty, q_0]$, *and can only accumulate at* $-\infty$ *and* q_0.

(3) *Points in* $(-\infty, q_0)$ *are eigenvalues if and only if they are poles of* $\mathfrak{T}$.

(4) *If* $w \geq 0$ *all eigenvalues are strictly positive, but if* $w < 0$ *on a set of positive measure there are infinitely many negative eigenvalues accumulating at* $-\infty$.

Proof. If $\lambda > q_0$ the real and imaginary parts of $f_+(\cdot, \lambda)$ are linearly independent solutions of (8.2.1), and their asymptotics show that no non-trivial linear combination f makes $\int_0^\infty (|f'|^2 + q\,|f|^2)$ finite, so there are no eigenvalues in (q_0, ∞). On the other hand, if $\lambda < q_0$ and $f = f_+(\cdot, \lambda)$ the integral is finite, so the essential spectrum of the equation on $[0, \infty)$ is $[q_0, \infty)$. For similar reasons the essential spectrum for the equation on $(-\infty, 0]$ is the same. By an analogue of Proposition 4.5.4 (see Exercise 8.3.1) the essential spectrum for the whole line is also the same, which proves (1).

Since the equation is limit point at $\pm\infty$ all eigenvalues are simple and located in $(-\infty, q_0]$. The asymptotics of $f_\pm(\cdot, \lambda)$ shows that $\lambda < q_0$ is an eigenvalue precisely if $f_\pm(\cdot, \lambda)$ are linearly dependent, i.e., $\mathcal{W}(f_+(\cdot, \lambda), f_-(\cdot, \lambda)) = 0$. Since the Wronskian is analytic outside $[q_0, \infty)$ eigenvalues can only accumulate as stated and those in $(-\infty, q_0)$ are poles of $\mathfrak{T}$, proving (2) and (3).

Finally (4) is an immediate consequence of Theorem 5.4.1.   □

We also have the following theorem.

Theorem 8.3.2. *If* $q_0 > 0$ *it may be an eigenvalue, but* 0 *is never an eigenvalue of T, nor can eigenvalues accumulate at* 0.

If $q - q_0 w$ *has a finite first moment then* q_0 *is not an eigenvalue, nor can eigenvalues accumulate at* q_0.

Proof. The proof that $q_0 > 0$ may be an eigenvalue is left to Exercise 8.3.2. By Theorem 5.1.9 (4) dim $D_0 = 0$ so 0 is not an eigenvalue.

An eigenfunction to $\lambda \neq q_0$ is proportional to $f_\pm$ so the asymptotics for these functions show that it converges to 0 at $\pm\infty$ but is different from zero outside a compact interval. By Corollary 6.1.6 such an eigenfunction has a finite number of zeros. Writing the differential equation as $(\lambda^{-1} u')' + (w - q/\lambda)u = 0$ for $\lambda > 0$ and similarly for $\lambda < 0$ it follows from the Sturm comparison theorem on page 173 that eigenfunctions lose one zero as we move from one eigenvalue to the next one in the direction of 0. It follows that eigenvalues cannot accumulate at 0, even if $q_0 = 0$.

If we can find a solution for $\lambda = q_0$ with only finitely many zeros any eigenfunction for $0 < \lambda < q_0$ has strictly fewer zeros, so there would only be finitely many eigenvalues in $(0, q_0)$. But if $q - q_0 w$ has a finite first moment we have Jost solutions $f_\pm(\cdot, q_0)$ which at $\pm\infty$ are asymptotic to 1, so f_- has a first zero a and f_+ a last zero b, if any.

By Corollary 6.1.7 $f_-(\cdot, q_0)$ can have at most one zero $> b$, and in $[a, b]$ it has finitely many zeros by Corollary 6.1.6. Supposing $f_-(\cdot, q_0)$ has altogether n zeros it follows, again by the Sturm separation theorem, that all eigenfunctions for a positive eigenvalue have fewer zeros, so that there are precisely n eigenvalues in $(0, q_0)$. For missing details in the proof we refer to Exercise 8.3.3. □

We next need to express Green's function in terms of the Jost solutions.

Proposition 8.3.3. *With* $\mathcal{W}(\lambda) = \mathcal{W}(f_-(\cdot, \lambda), f_+(\cdot, \lambda))$ *the resolvent* R_λ *of* T *is given for* λ *outside the spectrum by* $R_\lambda u(x) = \langle u, g(x, \cdot, \lambda) \rangle - u(x)/\lambda$, *where*

$$g(x, y, \lambda) := \frac{f_-(\min(x, y), \lambda) f_+(\max(x, y), \lambda)}{\lambda \mathcal{W}(\lambda)}.$$

This is just (5.3.1) if we choose $\psi_\pm = f_\pm$. For compactly supported $u \in \mathcal{H}_1$ a short calculation shows that

$$R_\lambda u(x) = \frac{f_+(x, \lambda) \langle u, \overline{f_-(\cdot, \lambda)} \rangle}{\lambda \mathcal{W}(\lambda)} + \int_x^\infty u \varphi(x, \cdot, \lambda) w, \qquad (8.3.1)$$

where $\varphi(x, y, \lambda) = (f_-(x, \lambda) f_+(y, \lambda) - f_+(x, \lambda) f_-(y, \lambda))/\mathcal{W}(\lambda)$. Now $\varphi(x, x, \lambda) = 0$, $\frac{d}{dy}\varphi(x, y, \lambda)$ equals -1 for $y = x$ and $y \mapsto \varphi(x, y, \lambda)$ solves (8.2.1). Thus $\lambda \mapsto \varphi(x, y, \lambda)$ has an extension which is entire, locally uniformly in (x, y). Similarly it follows that $\frac{d}{dx}\varphi(x, \cdot, \lambda)$ has analogous properties. The last term of (8.3.1) and its derivative are therefore entire functions of λ, locally uniformly in x.

Definition 8.3.4. Suppose $u \in \mathcal{H}_1$ has compact support and define, for $\lambda \neq q_0$, $\hat{u}_\pm(\lambda) = \int_{-\infty}^\infty (u' f'_\pm(\cdot, \lambda) + qu f_\pm(\cdot, \lambda))$, which we write as $\langle u, \overline{f_\pm(\cdot, \lambda)} \rangle$.

If $\lambda \neq q_0$ integration by parts shows that $\hat{u}_\pm(\lambda) = \lambda \int_{-\infty}^\infty u f_\pm(\cdot, \lambda) w$. Thus, $\hat{u}_\pm$ are analytic at 0 and vanish there if $0 < q_0$.

Taking the scalar product of (8.3.1) with $v \in \mathcal{H}_1$ of compact support we obtain

$$\langle R_\lambda u, v \rangle = \frac{1}{\lambda \mathcal{W}(\lambda)} \hat{u}_-(\lambda) \overline{\hat{v}_+(\overline{\lambda})} + \Phi(u, v, \lambda),$$

where $\lambda \mapsto \Phi(u, v, \lambda)$ is entire. Thus, if γ is a closed contour in $\mathbb{C}$ we obtain

$$\oint_\gamma \langle R_\lambda u, v \rangle \, d\lambda = \oint_\gamma \frac{1}{\lambda \mathcal{W}(\lambda)} \hat{u}_-(\lambda) \overline{\hat{v}_+(\overline{\lambda})} \, d\lambda \qquad (8.3.2)$$

if (one of) the integrals exist. We shall need the following lemma.

Lemma 8.3.5. *Let* $\{E_t\}_{t \in \mathbb{R}}$ *be the resolution of the identity for* T, $u, v \in \mathcal{H}_1$ *and* γ *an axis-parallel rectangle intersecting* $\mathbb{R}$ *at the endpoints of a compact interval* Δ *such that* $t \mapsto \langle E_t u, v \rangle$ *is differentiable at these endpoints. Then*

$$\oint_{\gamma} \langle R_{\lambda} u, v \rangle \, d\lambda = -2\pi \mathrm{i} \langle E_{\Delta} u, v \rangle \, .$$

Proof. We have, as in Lemma 3.1.4 (3), $\langle R_{\lambda} u, v \rangle = \int_{-\infty}^{\infty} d\langle E_t u, v \rangle / (t - \lambda)$ and

$$\oint_{\gamma} \int_{-\infty}^{\infty} \frac{d\langle E_t u, v \rangle}{t - \lambda} \, d\lambda = \int_{-\infty}^{\infty} \oint_{\gamma} \frac{d\lambda}{t - \lambda} \, d\langle E_t u, v \rangle = -2\pi \mathrm{i} \langle E_{\Delta} u, v \rangle$$

by absolute convergence (using Lemma 4.3.9) and the fact that the inner integral is 0 for t outside and $-2\pi \mathrm{i}$ for t inside γ. If t is an endpoint of Δ the inner integral is not defined, but these points are nullsets for $d\langle E_t u, v \rangle$ so this is irrelevant. $\qquad \square$

If Δ does not intersect $[q_0, \infty)$ and its endpoints are not eigenvalues but γ contains precisely one eigenvalue λ_n with normalized eigenfunction e_n, then (8.3.2) and Lemma 8.3.5 show that

$$\oint_{\gamma} \frac{\hat{u}_{-}(\lambda) \overline{\hat{v}_{+}(\overline{\lambda})}}{\lambda \mathcal{W}(\lambda)} \, d\lambda = -2\pi \mathrm{i} \langle u, e_n \rangle \overline{\langle v, e_n \rangle} = -2\pi \mathrm{i} \frac{\hat{u}_{-}(\lambda_n) \overline{\hat{v}_{+}(\lambda_n)}}{\langle f_{-}(\cdot, \lambda_n), f_{+}(\cdot, \lambda_n) \rangle} \, , \qquad (8.3.3)$$

since $f_{\pm}(\cdot, \lambda_n)$ are real-valued and proportional to e_n. Thus $-\langle f_{-}(\cdot, \lambda_n), f_{+}(\cdot, \lambda_n) \rangle^{-1}$ is the residue of $(\lambda \mathcal{W}(\lambda))^{-1}$ at λ_n. Since $\mathfrak{T}(\lambda) = 2ik / \mathcal{W}(\lambda)$ it follows that $\mathfrak{T}$ has simple poles with residue $-2ik\lambda / \langle f_{-}, f_{+} \rangle$ at all eigenvalues different from q_0. We may also write the last quotient of (8.3.3) as $\hat{u}_{+}(\lambda_n) \overline{\hat{v}_{+}(\lambda_n)} / \| f_{+}(\cdot, \lambda_n) \|^2$ or $\hat{u}_{-}(\lambda_n) \overline{\hat{v}_{-}(\lambda_n)} / \| f_{-}(\cdot, \lambda_n) \|^2$, and if Δ contains several eigenvalues each contributes a corresponding term to the integral.

As before we shall call $\| f_{+}(\cdot, \lambda_n) \|^2$ the *normalization constant* for the eigenvalue λ_n. One could also use $\| f_{-}(\cdot, \lambda_n) \|^2$ as the normalization constant. We may also consider the *matching constant* α_n for which $f_{-}(\cdot, \lambda_n) = \alpha_n f_{+}(\cdot, \lambda_n)$. If we assume $\mathfrak{T}$ known, one of these constants determines the other two, since the residue of $\mathfrak{T}$ at λ_n determines $\langle f_{+}(\cdot, \lambda_n), f_{-}(\cdot, \lambda_n) \rangle$. We make the following definition.

Definition 8.3.6. The *scattering data* associated with T consist of the scattering matrix, the eigenvalues and for each eigenvalue the two normalization constants and the matching constant.

Since the scattering matrix is unitary and $\mathfrak{T}$ has a unique meromorphic extension outside q_0 with poles precisely at the eigenvalues, the scattering data are known if the transmission coefficient, one of the reflection coefficients and for each eigenvalue one of the normalization constants or the matching constant are known.

If Δ is increased to $(-\infty, q_0)$ through a sequence of intervals whose endpoints are not eigenvalues, we obtain

$$\langle E_{(-\infty, q_0)} u, v \rangle = \sum \frac{\hat{u}_{-}(\lambda_n) \overline{\hat{v}_{+}(\lambda_n)}}{\langle f_{-}(\cdot, \lambda_n), f_{+}(\cdot, \lambda_n) \rangle} = \sum \frac{\hat{u}_{+}(\lambda_n) \overline{\hat{v}_{+}(\lambda_n)}}{\| f_{+}(\cdot, \lambda_n) \|^2} \, ,$$

the sum being over all eigenvalues outside $[q_0, \infty)$. The sum is absolutely convergent by the Bessel inequality (page 12) if there are infinitely many eigenvalues.

If instead Δ is contained in (q_0, ∞) we may move the horizontal sides of the contour γ towards the real axis, the integrals being unchanged by analyticity. Since the coefficients in (8.2.1) are real-valued we have $f_\pm(x, \overline{\lambda}) = \overline{f_\pm(x, \lambda)}$. If $t > q_0$ we have defined $f_\pm$ so that $f_\pm(x, t+i0) = f_\pm(x, t)$, while $f_\pm(x, t-i0) = \overline{f_\pm(x, t)}$. It follows that $\mathcal{W}(t - i0) = \overline{\mathcal{W}(t)}$. By (8.2.4) this means that $\hat{u}_\pm(t + i0) = \langle u, f_\pm(\cdot, t) \rangle = \hat{u}_\pm(t)$ and $\hat{u}_\pm(t - i0) = \mathfrak{T}(t)\hat{u}_\mp(t) - \mathfrak{R}_\pm(t)\hat{u}_\pm(t)$. All this shows that

$$\langle E_\Delta u, v \rangle = \int_\Delta (\hat{u}_+\overline{\hat{v}_+} + \hat{u}_-\overline{\hat{v}_-}) \, d\xi \,, \tag{8.3.4}$$

where $d\xi(t) = |\mathfrak{T}(t)|^2 \, (4\pi t\sqrt{t - q_0})^{-1} \, dt$, so $d\xi$ is a locally absolutely continuous positive measure on (q_0, ∞).

If the left endpoint of Δ is not an eigenvalue and q_0 is interior to Δ we introduce two new vertical sides in γ with real parts $q_0 - \varepsilon$ and $q_0 + \varepsilon$ for small $\varepsilon > 0$ and such that $q_0 - \varepsilon$ is not an eigenvalue. This divides γ into three rectangles γ_1, γ_2, and γ_3, corresponding to intervals Δ_1, Δ_2, and Δ_3. Here Δ_1 only contains eigenvalues, $\Delta_2 = [q_0 - \varepsilon, q_0 + \varepsilon]$, and $\Delta_3 \subset (q_0, \infty)$. Since $d\xi$ has a continuous density the assumptions of Lemma 4.3.9 are satisfied for each γ_j. As $\varepsilon \to 0$ through a sequence such that $q_0 - \varepsilon$ is never an eigenvalue, we obtain $E_{\Delta_2} \to E_{\{q_0\}}$, so that

$$\langle E_\Delta u, v \rangle = \sum_{\lambda_n \in \Delta \cap (-\infty, q_0)} \frac{\hat{u}_+(\lambda_n)\overline{\hat{v}_+(\lambda_n)}}{\|f_+(\cdot, \lambda_n)\|^2} + \langle E_{\{q_0\}} u, v \rangle + \int_{\Delta \cap (q_0, \infty)} (\hat{u}_+\overline{\hat{v}_+} + \hat{u}_-\overline{\hat{v}_-}) \, d\xi.$$

Here $E_{\{q_0\}} = 0$ unless q_0 is an eigenvalue.

We now define a space $L^2_{\mathcal{J}}$ of equivalence classes of $\mathbb{C}^2$-valued functions, with scalar product $\langle \cdot, \cdot \rangle_{\mathcal{J}}$ and norm $\|\cdot\|_{\mathcal{J}}$, as a direct sum of three spaces corresponding to the terms above. This means that the elements of the space are functions $\hat{u} = \begin{pmatrix} \hat{u}_+ \\ \hat{u}_- \end{pmatrix}$ defined on $\mathbb{R}$ which give a finite value to

$$\|\hat{u}\|^2_{\mathcal{J}} = \sum \frac{|\hat{u}_+(\lambda_n)|^2}{\|f_+(\cdot, \lambda_n)\|^2} + |\hat{u}_+(q_0)|^2 + \int_{q_0}^\infty (|\hat{u}_+|^2 + |\hat{u}_-|^2) \, d\xi \,.$$

The sum is over all eigenvalues $\lambda_n < q_0$. The middle term should be dropped unless q_0 is an eigenvalue, in which case $\hat{u}_+(q_0) = \langle u, e_0 \rangle$, where e_0 is a normalized eigenfunction to q_0. The space $L^2_{\mathcal{J}}$ is the completion of $C_0(\mathbb{R}; \mathbb{C}^2)$ in this norm, and two functions $\hat{u}$ and $\hat{v}$ with finite norm represent the same element of $L^2_{\mathcal{J}}$ precisely if $\|\hat{u} - \hat{v}\| = 0$. This means that for $t \le q_0$ the second component of $\hat{u}$ is completely undetermined while the first component $\hat{u}_+$ is only determined at the eigenvalues. If $t > q_0$ both components of $\hat{u}$ are determined a.e.

Define the *Jost transform* of a compactly supported $u \in \mathcal{H}_1$ as $\mathcal{J}u = \hat{u} = \begin{pmatrix} \hat{u}_+ \\ \hat{u}_- \end{pmatrix}$ where, except at q_0,

$$\hat{u}_\pm(\lambda) = \langle u, f_\pm(\cdot, \overline{\lambda}) \rangle = \lambda \int_{-\infty}^{\infty} u f_\pm(\cdot, \lambda) w \, ,$$

where the second equality follows on integrating by parts. If q_0 is an eigenvalue, $\hat{u}_+(q_0)$ is defined as above. We have shown that $\hat{u} \in L^2_\mathcal{J}$ and $\langle E_\Delta u, v \rangle = \langle \chi_\Delta \hat{u}, \hat{v} \rangle_\mathcal{J}$ if $v \in \mathcal{H}_1$ also has compact support, where χ_Δ is the characteristic function of Δ. This leads to the following expansion theorem in terms of the Jost transform, where $F(\cdot, \lambda)$ denotes the 2×1 matrix-valued function with rows $f_+(\cdot, \lambda)$ and $f_-(\cdot, \lambda)$ for $\lambda \neq q_0$ and, if q_0 is an eigenvalue, $F(\cdot, q_0)$ has rows e_0 and 0 with e_0 as above. Thus, for compactly supported $u \in \mathcal{H}_1$ we may write the Jost transform $\hat{u}(\lambda) = \langle u, \overline{F(\cdot, \lambda)} \rangle = \lambda \int_{-\infty}^{\infty} u F(\cdot, \lambda) w$.

Theorem 8.3.7.

(1) *The map $u \mapsto \mathcal{J}u$, defined for compactly supported $u \in \mathcal{H}_1$, extends by continuity to a map $\mathcal{J} : \mathcal{H}_1 \to L^2_\mathcal{J}$ called the* Jost transform *of u and denoted by $\mathcal{J}u$ or $\hat{u}$. We write this as $\hat{u}(t) = \langle u, \overline{F(\cdot, t)} \rangle$ although the integral may not converge pointwise.*

(2) *The map $\mathcal{J} : \mathcal{H} \to L^2_\mathcal{J}$ is unitary and $\mathcal{J}\mathcal{H}_\infty = 0$. If χ_Δ is the characteristic function of the interval (or Borel set) Δ, then $\langle E_\Delta u, v \rangle = \langle \chi_\Delta \hat{u}, \hat{v} \rangle_\mathcal{J}$. In particular, the Parseval formula $\langle u, v \rangle = \langle \hat{u}, \hat{v} \rangle_\mathcal{J}$ is valid if $u, v \in \mathcal{H}_1$ and at least one of them is in $\mathcal{H}$.*

(3) *For fixed x the function $t \mapsto F(x, t)$ is in $L^2_\mathcal{J}$. Moreover, if $\hat{u} \in L^2_\mathcal{J}$, then $x \mapsto \langle \hat{u}, F(x, \cdot) \rangle_\mathcal{J}$ is an element $u \in \mathcal{H}$. The map $\hat{u} \mapsto u$ is the adjoint of $\mathcal{J} : \mathcal{H}_1 \to L^2_\mathcal{J}$ and thus the inverse of the restriction of $\mathcal{J}$ to $\mathcal{H}$. We shall call $\mathcal{J}^*$ the* inverse Jost transform.

(4) *If $(u, f) \in T$, then $\hat{f}(t) = t\hat{u}(t)$. Conversely, if $\hat{u}$ and $t\hat{u}(t)$ are in $L^2_\mathcal{J}$, then $\mathcal{J}^*\hat{u} \in \mathcal{D}(T)$.*

As the proof is very similar to the proofs of several expansion theorems we have proved earlier, we leave it to the reader as Exercise 8.3.4.

Remark 8.3.8. It is clear from Theorem 8.3.7 that the spectrum of T consists of an essential spectrum of multiplicity 2 in $[q_0, \infty)$ in addition to simple eigenvalues located in $(-\infty, q_0]$. The point q_0 may be an eigenvalue, but in (q_0, ∞) there are no eigenvalues and the spectrum is absolutely continuous, i.e., the spectral measure has a locally Lebesgue integrable, even continuous, density.

Exercises

Exercise 8.3.1 Prove an analogue of Proposition 4.5.4 valid for left-definite equations.

Exercise 8.3.2 Show that if $q_0 > 0$ it may be an eigenvalue.

Hint: To construct examples with arbitrary $q_0 > 0$, let $q = q_0$ and find w such that $1/(x^2 + 1)$ is an eigenfunction to $\lambda = q_0$.

Exercise 8.3.3 Fill in all missing details in the proof of Theorem 8.3.2.

Exercise 8.3.4 Prove Theorem 8.3.7.

Hint: Look at the proof of Theorem 5.3.3.

Exercise 8.3.5 State and prove a theorem analogous to Theorem 8.3.7 but for right-definite equations.

8.4 Inverse theory

In this section we shall add the following assumption.

Assumption 8.4.1 *In addition to the assumptions of Theorem 8.2.1 assume that the first moment of $q - q_0 w$ is finite.*

This assumption will be in force throughout the section. By Theorem 8.3.2 it implies that q_0 is not an eigenvalue, nor a point of accumulation of eigenvalues, and by Theorem 8.2.1 that $\lambda \mapsto f_\pm(x, \lambda)$ is continuous also at q_0, locally uniformly in x.

To prove that scattering data determine the equation up to certain Liouville transforms we follow the approach of Chapter 7 and assume that we have two equations with the same scattering data. In particular, they have the same Jost transform space $L^2_{\mathcal{J}}$. However, instead of (as in Chapter 7) relying on high energy asymptotics for the Jost solutions we shall use a lemma of de Branges. We will also use a little of the Nevanlinna theory of functions of *bounded type*. For these things we refer to Appendix E, sections E.3 and E.4.

We shall need the following lemma.

Lemma 8.4.2. *As functions of λ the Jost solutions $f_\pm(x, \lambda)$ are functions of* bounded type *in the upper and lower half-planes for every $x \in \mathbb{R}$.*

Proof. The Jost solutions are continuous in the closed upper and lower half-planes (with values $\overline{f_\pm(x, t)}$ at $t \geq q_0$ in the closed lower half-plane) and analytic outside $\mathbb{R}$. Furthermore, by Theorem 8.2.1 we have $\log_+ |f_\pm(x, \lambda)| \leq C_x |\lambda|^{1/2}$ for large $|\lambda|$. It immediately follows that

$$\int_{-\infty}^{\infty} \frac{\log_+ |f_\pm(x, t)|}{t^2 + 1} \, dt < \infty$$

so that Theorem E.3.7 implies that $\lambda \mapsto f_\pm(x, \lambda)$ are of bounded type in the two half-planes. $\qquad \square$

Assuming $u \in \mathcal{H}_1$ has a Jost transform $\hat{u}$ such that $\hat{u}_-$ has an extension outside $[q_0, \infty)$ which is analytic and such that $\hat{u}_-(t + i0) = \hat{u}_-(t)$ is continuous for $t \geq q_0$ we define an auxiliary function A_- by

$$A_-(u, x, \lambda) = R_\lambda u(x) - \frac{f_+(x, \lambda)}{\lambda \mathcal{W}(\lambda)} \hat{u}_-(\lambda). \tag{8.4.1}$$

Similarly, if $\hat{u}_+$ has such an extension we define

$$A_+(u, x, \lambda) = R_\lambda u(x) - \frac{f_-(x, \lambda)}{\lambda \mathcal{W}(\lambda)} \hat{u}_+(\lambda).$$

We need to establish that under appropriate assumptions the auxiliary functions, which are clearly analytic outside $\mathbb{R}$, have extensions which are entire.

Theorem 8.4.3. *Suppose $\hat{u} \in L_{\mathcal{J}}^2$ and $\hat{u}_-$ has an extension analytic outside $[q_0, \infty)$ such that $\hat{u}_-(\lambda)/\lambda$ is bounded near 0. For $t > q_0$ also assume $\hat{u}_-(t + i0) = \hat{u}(t)$ and $\hat{u}_-(t - i0) = \mathfrak{T}(t)\hat{u}_+(t) - \mathfrak{R}_-(t)\hat{u}_-(t)$.*

Then $A_-(u, x, \lambda)$ extends to an entire function. If we make analogous assumptions on $\hat{u}_+$ then $A_+(x, u, \lambda)$ extends to an entire function.

Proof. If γ is a rectangle cutting out an interval $\Delta \subset (q_0, \infty)$, calculations in reverse to those leading from (8.3.2) to (8.3.4) show that

$$R_\lambda u(x) = \langle \chi(t)\hat{u}(t)/(t - \lambda), F(x, t) \rangle_{\mathcal{J}} + \frac{i}{2\pi} \oint_\gamma \frac{1}{\mu - \lambda} \frac{f_+(x, \mu)}{\mu W(\mu)} \hat{u}_-(\mu) \, d\mu$$

if λ is non-real and outside γ, where χ is the characteristic function of $\mathbb{R} \setminus \Delta$. If γ expands to contain λ there is an additional term coming from the residue of the second integrand at λ, which is $f_+(x, \lambda)\hat{u}_-(\lambda)/(\lambda W(\lambda))$. For such λ we have

$$A_+(u, x, \lambda) = \langle \chi(t)\hat{u}(t)/(t - \lambda), F(x, t) \rangle_{\mathcal{J}} + \frac{i}{2\pi} \oint_\gamma \frac{1}{\mu - \lambda} \frac{f_+(x, \mu)}{\mu W(\mu)} \hat{u}_-(\mu) \, d\mu \, .$$

Both terms are clearly analytic functions of λ as long as λ is inside γ. Thus A_- has an analytic extension in a neighborhood of (q_0, ∞).

By assumption and using Assumption 8.4.1 and Theorem 8.3.2 the point q_0 is a removable singularity of A_-. In $(-\infty, q_0)$ both terms of (8.4.1) have poles at the eigenvalues, but the residues of the two terms cancel. We leave the proof of this to Exercise 8.4.1. Consequently $A_-(x, u, \lambda)$ extends to an entire function of λ.

Similar arguments apply to A_+.   □

To deduce further properties of $A_\pm$ we shall need a few lemmas. For their proofs it will be convenient to introduce the matrix-valued measure $d\Xi$ which in (q_0, ∞) is the 2×2 unit matrix times $d\xi$ and in $(-\infty, q_0]$ has appropriate multiples of the Dirac measures at the eigenvalues in the upper left corner, and 0 in other positions. We may then write $\langle \hat{u}, \hat{v} \rangle_{\mathcal{J}} = \int_{-\infty}^\infty \hat{v}^* d\Xi \hat{u}$, and the inversion formula becomes $u(x) = \int_{-\infty}^\infty F^*(x, \cdot) d\Xi \hat{u}$.

Lemma 8.4.4. *Suppose* $u \in \mathcal{H}_1$. $R_\lambda u(x)$ *is then, for every* $x \in \mathbb{R}$, *of bounded type in the upper and lower half-planes, and is* $O(1/|\mathrm{Im}\,\lambda|)$ *as* $\lambda \to \infty$.

Proof. $R_\lambda u(x) = \langle R_\lambda u, g_0(x, \cdot) \rangle$ so that $|R_\lambda u(x)| \leq \|u\| \|g_0(x, \cdot)\|/|\mathrm{Im}\,\lambda|$. $\qquad\square$

Lemma 8.4.5. $\lambda \mapsto f_+(x, \lambda) f_-(x, \lambda)/\mathcal{W}(\lambda)$ *is, for every* $x \in \mathrm{supp}\,w$, *of bounded type in the upper and lower half-planes and is* $o(|\lambda/\mathrm{Im}\,\lambda|)$ *as* $\lambda \to \infty$.

Proof. By Proposition 8.3.3 we are dealing with the function $\lambda g(x, x, \lambda)$. Now $R_{\overline{\lambda}} u \in \mathcal{H}$ and $\mathcal{J}(R_{\overline{\lambda}} u)(t) = \hat{u}(t)/(t - \overline{\lambda})$ so applying the inverse transform we obtain, with absolute convergence, $R_{\overline{\lambda}} u(x) = \langle \hat{u}(t), F(x, t)/(t - \lambda) \rangle_{\mathcal{J}}$. We also have $\overline{g(x, y, \overline{\lambda})} = g(y, x, \lambda)$ so that

$$R_{\overline{\lambda}} u(x) = \langle u, g(\cdot, x, \lambda) \rangle - u(x)/\overline{\lambda} = \langle \hat{u}, \mathcal{J}(g(\cdot, x, \lambda)) - F(x, \cdot)/\lambda \rangle_{\mathcal{J}},$$

using the inversion formula. Thus $\mathcal{J}(g(\cdot, x, \lambda))(t) - F(x, t)/\lambda = F(x, t)/(t - \lambda)$, so that $\mathcal{J}(\lambda g(\cdot, x, \lambda)) = t F(x, t)/(t - \lambda)$. The value of any $v \in \mathcal{H}_1$ coincides with the value of $E_{\mathbb{R}} v \in \mathcal{H}$ at every $x \in \mathrm{supp}\,w$ by Theorem 5.1.11, so using the inversion formula for such x we obtain

$$f_+(x, \lambda) f_-(x, \lambda)/\mathcal{W}(\lambda) = \int\limits_{-\infty}^{\infty} \frac{t F^*(x, t) d\Xi(t) F(x, t)}{t - \lambda}.$$

Thus the claim follows from Lemma E.3.4. $\qquad\square$

It will be convenient to make the following definition.

Definition 8.4.6. For $a \in \mathrm{supp}\,w$, let $\mathcal{H}(-\infty, a)$ be those $u \in \mathcal{H}$ for which $\mathrm{supp}\,u \subset (-\infty, a]$ and $\mathcal{H}(a, \infty)$ those $u \in \mathcal{H}$ for which $\mathrm{supp}\,u \subset [a, \infty)$.

Lemma 8.4.7. *Suppose* $a \in \mathrm{supp}\,w$ *and* $u \in \mathcal{H}(-\infty, a)$. *Then* $\hat{u}_-$ *has an analytic extension outside* $[q_0, \infty)$ *satisfying*

$$\hat{u}_-(t + i0) = \hat{u}_-(t) \quad and \quad \hat{u}_-(t - i0) = \mathfrak{T}(t)\hat{u}_+(t) - \mathfrak{R}_-(t)\hat{u}_-(t)$$

for $t > q_0$, *and such that* $\hat{u}_-(\lambda)/(\lambda f_-(a, \lambda))$ *is of bounded type in the upper and lower half-planes and is* $o(|\lambda/\mathrm{Im}\,\lambda|)$ *as* $\lambda \to \infty$ *in the open half-planes.*

Similarly, if $u \in \mathcal{H}(a, \infty)$, *then* $\hat{u}_+$ *has an analytic extension outside* $[q_0, \infty)$ *satisfying*

$$\hat{u}_+(t + i0) = \hat{u}_+(t) \quad and \quad \hat{u}_+(t - i0) = \mathfrak{T}(t)\hat{u}_-(t) - \mathfrak{R}_+(t)\hat{u}_+(t)$$

for $t > q_0$. *The function* $\lambda \mapsto \hat{u}_+(\lambda)/(\lambda f_+(a, \lambda))$ *is of bounded type in the upper and lower half-planes and is* $o(|\lambda/\mathrm{Im}\,\lambda|)$ *as* $\lambda \to \infty$ *in the open half-planes.*

Proof. The first part is an immediate consequence of the corresponding properties of f_- and the scattering matrix.

For $\operatorname{Im}\lambda \neq 0$ we have $f_-(x,\lambda) = \lambda f_-(a,\lambda)\frac{f_-(x,\lambda)}{\lambda f_-(a,\lambda)}$, where we denote the last factor by $\psi_{(-\infty,a]}(x,\lambda)$, since this is the Weyl solution for the left-definite Dirichlet problem (5.1.1) on $(-\infty,a]$ with[3] initial point a. By Theorem 5.2.17 we have

$$\langle u, \overline{\psi_{(-\infty,a]}(\cdot,\lambda)}\rangle = \int\limits_{\mathbb{R}} \frac{\tilde{u}(t)}{t-\lambda}\,d\sigma(t)$$

with absolute convergence, where $\tilde{u}$ is the generalized Fourier transform of u associated with the Dirichlet problem on $(-\infty,a]$ and $d\sigma$ the corresponding spectral measure. Thus the claims for $\hat{u}_-$ follow from Lemma E.3.4, and the proof of the claims for $\hat{u}_+$ is entirely analogous.   □

It will be convenient to make the following definition.

Definition 8.4.8. Let $\operatorname{Op}(q_0)$ denote the set of self-adjoint operators associated with a left-definite equation (8.2.1) on $\mathbb{R}$ such that $q-q_0$ and $w-1$ are in $L^1(\mathbb{R})$ and the first moment of $q-q_0 w$ is finite.

All operators in $\operatorname{Op}(q_0)$ are therefore associated with equations which have the model equation (8.2.2). The main result of this chapter is the following uniqueness theorem for inverse scattering.

Theorem 8.4.9. *Suppose T and $\tilde{T}$ in $\operatorname{Op}(q_0)$ are associated with the equations (8.2.1) respectively $-\tilde{u}'' + \tilde{q}\tilde{u} = \lambda\tilde{w}\tilde{u}$ and that T and $\tilde{T}$ have the same scattering matrix, the same eigenvalues and the same normalization constants.*

Then there is a Liouville transform given by $u = s\tilde{u} \circ t$ transforming T into $\tilde{T}$, so that

$$\tilde{q} \circ t = s^3(-s'' + qs), \qquad \tilde{w} \circ t = s^4 w\,,$$

where $t : \mathbb{R} \to \mathbb{R}$ is strictly increasing and continuously differentiable, and $s = (t')^{-1/2}$ as well as s' are locally absolutely continuous. Moreover, $t(x) - x \to 0$ and $s(x) \to 1$ as $x \to \pm\infty$.

Conversely, if the operators T and $\tilde{T}$ have coefficients related in this way, then they have the same scattering matrix, eigenvalues and norming constants.

It is to be expected that one cannot retrieve both coefficients q and w from the scattering data. The following corollaries show how the conclusion may be improved with some *a priori* information on the coefficients.

Corollary 8.4.10. *Suppose T and $\tilde{T}$ in $\operatorname{Op}(q_0)$ have the same scattering matrix, eigenvalues and norming constants, and that $|\tilde{w}| = |w|$. Then $\tilde{w} = w$ and $\tilde{q} = q$ on $\operatorname{supp} w$. In particular, if $\operatorname{supp} w = \mathbb{R}$ or if $\tilde{q} = q$ outside $\operatorname{supp} w$, then $T = \tilde{T}$.*

Corollary 8.4.11. *Suppose T and $\tilde{T}$ in $\operatorname{Op}(q_0)$ have the same scattering matrix, eigenvalues and norming constants, and that $\tilde{q} = q$. Then $\tilde{T} = T$, i.e., $\tilde{w} = w$.*

---

[3] If $q \equiv 0$ in $(-\infty, a]$ one has to use the norm introduced in Exercise 5.1.3 for this problem.

The results above are of interest for the spectral problem $-u'' + \frac{1}{4}u = \lambda w u$ on $\mathbb{R}$ which is the spectral problem associated with the Camassa–Holm equation (see Appendix F) in the case $\kappa = 1$. Here $q = q_0 = 1/4$, so we have absolutely continuous spectrum $[1/4, \infty)$ and discrete spectrum below $1/4$. The condition on w in this case becomes $(1 + |x|)(1 - w(x)) \in L^1(\mathbb{R})$. There may be a finite number of positive eigenvalues, but unless $w \geq 0$ a.e. there are also infinitely many negative eigenvalues accumulating only at $-\infty$.

Turning now to the proofs, we will by $f_\pm$ denote the Jost solutions associated with T and by $\tilde{f}_\pm$ those associated with $\tilde{T}$. A similar convention is adopted for all other quantities associated with the two operators. We shall first prove the easy direction of Theorem 8.4.9.

Proof (Converse part of Theorem 8.4.9). Let $f(x, \lambda) = s(x)\tilde{f}_+(t(x), \lambda)$. It follows that $e^{-ikx} f(x, \lambda)$ is asymptotic to $s(x)e^{ik(t(x)-x)}$ which by assumption is asymptotic to 1 as $x \to \infty$. Furthermore, by Section 7.2 f satisfies $-f'' + qf = \lambda w f$ so that $f = f_+$. Similarly one shows that $s(x)\tilde{f}_-(t(x), \lambda) = f_-(x, \lambda)$.

It follows that the two equations have the same scattering matrix and, since the eigenvalues are the poles of $\mathfrak{T}$, also the same eigenvalues. If λ_n is such an eigenvalue and $\tilde{f}_-(\cdot, \lambda_n) = \alpha_n \tilde{f}_+(\cdot, \lambda_n)$ it follows that $f_-(\cdot, \lambda_n) = \alpha_n f_+(\cdot, \lambda_n)$. Thus the two equations have the same scattering data.   □

To prove the direct part of Theorem 8.4.9, note that the assumptions imply that T and $\tilde{T}$ have the same Jost transform space $L^2_{\mathcal{J}}$. We may therefore define a unitary operator $\mathcal{U} : \mathcal{H} \to \tilde{\mathcal{H}}$ by $\mathcal{U} = \tilde{\mathcal{J}}^* \mathcal{J}$. Note that it follows that if $\tilde{u} = \mathcal{U}u$, $\tilde{f} = \mathcal{U}f$ and $Tu = f$, then $\tilde{T}\tilde{u} = \tilde{f}$. We need to prove that $\mathcal{U}$ is a Liouville transform, which will require a number of lemmas.

Lemma 8.4.12. *Suppose $\lambda \mapsto A_-(z, v, \lambda)$ is entire of exponential type for every $z \geq x$ and that $A_-(x, v, \cdot) = 0$. Then $A_-(z, v, \cdot) = 0$ for $z \geq x$.*

Similarly, if $\lambda \mapsto A_+(z, v, \lambda)$ is entire of exponential type for every $z \leq x$ and $A_+(x, v, \cdot) = 0$, then $A_+(z, v, \cdot) = 0$ for $z \leq x$.

Proof. If $A_-(x, v, \cdot) = 0$ we find that

$$A_-(z, v, \lambda) = R_\lambda u(z) - \psi_{[x,\infty)}(z, \lambda)\lambda R_\lambda u(x) ,$$

where $\psi_{[x,\infty)}(z, \lambda) = f_+(z, \lambda)/(\lambda f_+(x, \lambda))$ is the Weyl solution for the Dirichlet problem[4] on $[x, \infty)$. This function tends to zero locally uniformly in z as $\lambda \to \infty$ in a double sector $|\mathrm{Re}\,\lambda| \leq C\,|\mathrm{Im}\,\lambda|$ by Theorem 5.2.17. By Lemma 8.4.4 $R_\lambda u(z)$ tends to zero while $\lambda R_\lambda u(x)$ remains bounded in such a sector.

It follows that $A_-(z, v, \lambda) \to 0$ as $\lambda \to \infty$ in the double sector, so by the Phragmén–Lindelöf principle (Theorem E.4.3) it is bounded and thus by Liouville's Theorem E.2.13 constant. The constant is zero since this is the limit in the double sector.

The claims about A_+ are proved similarly.   □

---

[4] If $q \equiv 0$ in $[x, \infty)$ one has to use the norm introduced in Exercise 5.1.3 for this problem.

We expect to be able to find a strictly increasing function t such that $u \in \mathcal{H}(x, \infty)$ precisely if $\tilde{u} \in \tilde{\mathcal{H}}(t(x), \infty)$. Therefore, if $\tilde{v} \in \tilde{\mathcal{H}}(y, \infty)$ we should have either $y \leq t(x)$ or $y \geq t(x)$, i.e., either $v \in \mathcal{H}(x, \infty)$ or $\tilde{u} \in \tilde{\mathcal{H}}(y, \infty)$. We shall find the variable change $x \mapsto t(x)$ by first showing that this is true for arbitrary $x \in \operatorname{supp} w$ and $y \in \operatorname{supp} \tilde{w}$.

Lemma 8.4.13. *Suppose $x \in \operatorname{supp} w$ and $y \in \operatorname{supp} \tilde{w}$. Also suppose $u \in \mathcal{H}(x, \infty)$ and $\tilde{v} \in \tilde{\mathcal{H}}(y, \infty)$ and let $\tilde{u} = \mathcal{U}u$, $v = \mathcal{U}^{-1}\tilde{v}$. Then either $\tilde{u} \in \tilde{\mathcal{H}}(y, \infty)$ or $v \in \mathcal{H}(x, \infty)$.*

Also, if $u \in \mathcal{H}(-\infty, x)$ and $\tilde{v} \in \tilde{\mathcal{H}}(-\infty, y)$, then $\tilde{u} \in \tilde{\mathcal{H}}(-\infty, y)$ or $v \in \mathcal{H}(-\infty, x)$.

Proof. $A_-(x, v, \cdot)$ is entire since $\hat{v} = \tilde{\mathcal{J}}\tilde{v}$ and $\hat{v}_-$ by Lemma 8.4.7 has the properties required in Theorem 8.4.3. Now

$$A_-(x, v, \lambda) = R_\lambda v(x) - \frac{f_+(x, \lambda) f_-(x, \lambda)}{\mathcal{W}(\lambda)} \frac{\tilde{f}_-(y, \lambda)}{f_-(x, \lambda)} \frac{\hat{v}_-(\lambda)}{\lambda \tilde{f}_-(y, \lambda)},$$

so using lemmas 8.4.4, 8.4.5, and 8.4.2 it follows that $A_-(x, v, \cdot)$ is of bounded type in the upper and lower half-planes. By Theorem E.3.8 it is therefore of exponential type. Similar statements are valid for $\tilde{A}_-(y, \tilde{u}, \cdot)$. We obtain

$$A_-(x, v, \lambda) = \left(1 + \left|\frac{\tilde{f}_-(y, \lambda)}{f_-(x, \lambda)}\right|\right) o(|\lambda/\operatorname{Im}\lambda|^2) \text{ as } \lambda \to \infty,$$

$$\tilde{A}_-(y, \tilde{u}, \lambda) = \left(1 + \left|\frac{f_-(x, \lambda)}{\tilde{f}_-(y, \lambda)}\right|\right) o(|\lambda/\operatorname{Im}\lambda|^2) \text{ as } \lambda \to \infty .$$

Now, f_- and $\tilde{f}_-$ are both of bounded type and have the symmetry property $\overline{f(\bar{\lambda})} = f(\lambda)$, so the quotient $\tilde{f}_-(y, \lambda)/f_-(x, \lambda)$ has the same mean type in the upper and lower half-planes. If this is τ the mean type of the reciprocal is $-\tau$, and if $\tau < 0$ it follows by the Phragmén–Lindelöf principle (Theorem E.4.3) that $A_-(x, v, \cdot)$ is bounded, and therefore constant by Liouville's Theorem E.2.13. The limit along $i\mathbb{R}$ being zero, it follows that $A_-(x, v, \cdot) = 0$. If $\tau > 0$ we instead obtain $\tilde{A}_-(y, \tilde{u}, \cdot) = 0$, and if $\tau = 0$ it follows from Theorem E.3.8 that both functions are of exponential type 0. By the estimates above

$$\min(|A_-(v, x, \lambda)|, |\tilde{A}_-(\tilde{u}, y, \lambda)|) = o(|\lambda/\operatorname{Im}\lambda|^2) \text{ as } \lambda \to \infty ,$$

so by Lemma E.4.1 it follows that also in this case one of the functions vanishes identically in λ. If $A_-(x, v, \cdot) = 0$ Lemma 8.4.12 shows that $A_-(z, v, \cdot) = 0$ for $z \geq x$, and applying $-\frac{d^2}{dz^2} + q - \lambda w$ to this it follows that $wv = 0$ in $[x, \infty)$. Since $x \in \operatorname{supp} w$, Theorem 5.1.11 implies that $v \equiv 0$ in $[x, \infty)$, i.e., $v \in \mathcal{H}(-\infty, x)$. If instead $\tilde{A}_-(y, \tilde{u}, \cdot) = 0$ we find that $\tilde{u} \in \tilde{\mathcal{H}}(-\infty, y)$.

The arguments for A_+ and $\tilde{A}_+$ are completely analogous. $\qquad \square$

We next show how supports of elements of $\mathcal{H}$ are related to the supports of their images under $\mathcal{U}$.

Lemma 8.4.14. *If T and $\tilde{T}$ have the same scattering data, then there are strictly increasing, bijective maps $t_\pm : \operatorname{supp} w \to \operatorname{supp} \tilde{w}$ such that $\tilde{\mathcal{H}}(t_+(x), \infty) = \mathcal{U}\mathcal{H}(x, \infty)$ and $\tilde{\mathcal{H}}(-\infty, t_-(x)) = \mathcal{U}\mathcal{H}(-\infty, x)$ for all $x \in \operatorname{supp} w$.*

Proof. If $x \in \operatorname{supp} w$ there exists non-zero $u \in \mathcal{H}(x, \infty)$. $\mathcal{U}$ is unitary so $\tilde{u} \notin \tilde{\mathcal{H}}(y, \infty)$ for some $y \in \operatorname{supp} \tilde{w}$. By Lemma 8.4.13 this means that $v \in \mathcal{H}(x, \infty)$ for every $\tilde{v} \in \tilde{\mathcal{H}}(y, \infty)$. Now let $t_+(x)$ be the infimum of all $y \in \operatorname{supp} \tilde{w}$ for which this is true.

Since $\tilde{w} - 1$ is integrable, $\operatorname{supp} \tilde{w}$ is unbounded in both directions, so the projection onto $\tilde{\mathcal{H}}$ of a compactly supported element of $\tilde{\mathcal{H}}_1$ has compact support. Since compactly supported elements of $\tilde{\mathcal{H}}_1$ are dense, their projections onto $\tilde{\mathcal{H}}$ are dense in $\tilde{\mathcal{H}}$. Consequently, if $t_+(x) = -\infty$, then $\tilde{\mathcal{H}} \subset \mathcal{U}\mathcal{H}(x, \infty)$. However, $\operatorname{supp} w$ is also unbounded in both directions, so $\mathcal{H}(x, \infty) \neq \mathcal{H}$, contradicting the fact that $\mathcal{U}$ is unitary. Thus $t_+(x)$ is finite, so $t_+(x) \in \operatorname{supp} \tilde{w}$.

Note that if $t_+(x)$ is the left endpoint of a gap in $\operatorname{supp} \tilde{w}$, then the infimum defining $t_+(x)$ is attained. If it is not, there are points of $\operatorname{supp} \tilde{w}$ to the right of and arbitrarily close to $t_+(x)$. But then we may approximate elements in $\tilde{\mathcal{H}}(t_+(x), \infty)$ arbitrarily well by elements of some $\tilde{\mathcal{H}}(y, \infty)$ with $y > t_+(x)$ (Exercise 8.4.2). It follows that $\tilde{\mathcal{H}}(t_+(x), \infty) \subset \mathcal{U}\mathcal{H}(x, \infty)$.

On the other hand, if $y = -\infty$ or $\operatorname{supp} \tilde{w} \ni y < t_+(x)$ there exists a $\tilde{v} \in \tilde{\mathcal{H}}(y, \infty)$ such that $\mathcal{U}^{-1}\tilde{v} \notin \mathcal{H}(x, \infty)$ and thus, by Lemma 8.4.13, $\mathcal{U}\mathcal{H}(x, \infty) \subset \tilde{\mathcal{H}}(y, \infty)$. Since this is true for all $y \in \operatorname{supp} \tilde{w}$ with $y < t_+(x)$ we have in fact $\mathcal{U}\mathcal{H}(x, \infty) \subset \tilde{\mathcal{H}}(t_+(x), \infty)$ unless $t_+(x)$ is the right endpoint of a gap in $\operatorname{supp} \tilde{w}$. In the latter case we may choose $y \geq -\infty$ so that $(y, t_+(x))$ is a gap in $\operatorname{supp} \tilde{w}$.

In this case $\tilde{\mathcal{H}}(y, \infty)$ is a one-dimensional extension of $\tilde{\mathcal{H}}(t_+(x), \infty)$, so if there exists a $u \in \mathcal{H}(x, \infty)$ with $\operatorname{supp} \mathcal{U}u$ intersecting $(y, t_+(x))$, then $\mathcal{U}^{-1}\tilde{\mathcal{H}}(y, \infty) \subset \mathcal{H}(x, \infty)$. But this would mean that $t_+(x) \leq y$. It follows that $\mathcal{U}\mathcal{H}(x, \infty) = \tilde{\mathcal{H}}(t_+(x), \infty)$ for all $x \in \operatorname{supp} w$.

The function t_+ has range $\operatorname{supp} \tilde{w}$, since if not let y be in this set but not in the range of t_+. An argument analogous to that defining t_+ determines $x \in \operatorname{supp} w$ such that $\tilde{\mathcal{H}}(y, \infty) = \mathcal{U}\mathcal{H}(x, \infty)$. But then $y = t_+(x)$, contradicting the choice of y.

Analogous reasoning proves the existence of the function t_-. $\qquad\square$

We can now show that $\mathcal{U}$ is given by a Liouville transform except in gaps of $\operatorname{supp} w$.

Lemma 8.4.15. *There exist real-valued maps s, t defined in $\operatorname{supp} w$ such that s has no zeros and $t : \operatorname{supp} w \to \operatorname{supp} \tilde{w}$ is increasing, bijective, and such that $u = s \mathcal{U}u \circ t$ on $\operatorname{supp} w$ for any $u \in \mathcal{H}$.*

Proof. If $x \in \operatorname{supp} w$ and $v \in \mathcal{H}$ with $v(x) = 1$ we may, given any $u \in \mathcal{H}$, write $u = u_- + u_+ + u(x)v$, where $u_- \in \mathcal{H}(-\infty, x)$ and $u_+ \in \mathcal{H}(x, \infty)$. Applying $\mathcal{U}$ we obtain from Lemma 8.4.14 that $\tilde{u} = \tilde{u}_- + \tilde{u}_+ + u(x)\tilde{v}$, where $\tilde{u}_- \in \tilde{\mathcal{H}}(-\infty, t_-(x))$ and $\tilde{u}_+ \in \tilde{\mathcal{H}}(t_+(x), \infty)$.

If $t_\pm$ are both defined at x we cannot have $t_-(x) < t_+(x)$ since then the restrictions of elements of $\tilde{\mathcal{H}}$ to $(t_-(x), t_+(x))$ would be spanned by multiples of the restrictions of $\tilde{v}$ and thus one-dimensional. This would imply that $(t_-(x), t_+(x))$ is an unbounded gap of $\operatorname{supp} \tilde{w}$, contradicting the fact that $t_\pm(x)$ are in $\operatorname{supp} \tilde{w}$.

A similar reasoning but starting from $\tilde{u} \in \tilde{\mathcal{H}}$ and using the inverses of $t_{\pm}$ shows that we cannot have $t_-(x) > t_+(x)$ either, so that we define $t = t_+ = t_-$ in supp w. It now follows that $\tilde{u}(t(x)) = \tilde{v}(t(x))u(x)$, and $\tilde{v}(t(x)) \neq 0$ since not all elements of $\tilde{\mathcal{H}}$ vanish at $t(x)$. We may now set $s(x) = 1/\tilde{v}(t(x))$ and the proof is finished. □

It remains to define the Liouville transform in each gap.

Lemma 8.4.16. *Suppose (a, b) is a gap in supp w. Then $(t(a), t(b))$ is a gap in supp $\tilde{w}$ and there is a Liouville transform mapping restrictions to $[t(a), t(b)]$ of elements $\tilde{u} \in \tilde{\mathcal{H}}$ to the restrictions to $[a, b]$ of the pre-images $u = \mathcal{U}^{-1}\tilde{u}$.*

Proof. Since t is strictly increasing we have $t(a) < t(b)$, and t takes no values in $(t(a), t(b))$, while $t(a)$ and $t(b) \in$ supp $\tilde{w}$. Thus $(t(a), t(b))$ is a gap in supp $\tilde{w}$.

Let φ and θ be solutions of $-f'' + qf = 0$ with $\varphi(a) = 1$, $\varphi(b) = 0$ and $\theta(a) = 0$, $\theta(b) = 1$. Such solutions exist since a solution is convex where it is positive, so a solution u with $u(a) = 0$, $u'(a) = 1$ is positive in (a, b). We may therefore put $\theta = u/u(b)$. Similarly we may find φ, so we have a basis of positive solutions with no zeros of their derivatives. Thus $\varphi' < 0$ and $\theta' > 0$ throughout $[a, b]$ so that $\mathcal{W}(\theta, \varphi) > 0$.

Now let $\tilde{\varphi}$ and $\tilde{\theta}$ be the analogous solutions of $-f'' + \tilde{q}f = 0$ in $[t(a), t(b)]$. By Theorem 5.1.11 it is clear that $\tilde{\varphi}/s(a)$ and $\tilde{\theta}/s(b)$ are images under $\mathcal{U}$ of φ and θ respectively, in the following sense: Any element of $\mathcal{H}$ whose restriction to $[a, b]$ is φ is mapped to an element of $\tilde{\mathcal{H}}$ whose restriction to $[t(a), t(b)]$ is $\tilde{\varphi}/s(a)$, and similarly for θ.

If we extend φ by 0 in $[b, \infty)$ and θ by 0 in $(-\infty, a]$, the images will have analogous properties. The scalar product of these extensions is $\int_a^b (\varphi'\overline{\theta'} + q\varphi\overline{\theta}) = \varphi'(b) < 0$. Since $\mathcal{U}$ is unitary this is equal to $(s(a)\overline{s(b)})^{-1} \int_{t(a)}^{t(b)} (\tilde{\varphi}'\overline{\tilde{\theta}'} + \tilde{q}\tilde{\varphi}\overline{\tilde{\theta}}) = (s(a)\overline{s(b)})^{-1}\tilde{\varphi}'(t(b))$, which is therefore negative. Thus we have $s(a)\overline{s(b)} > 0$ so that $s(a)/s(b) > 0$.

We need to define t and s so that $s(x)\tilde{\varphi}(t(x))/s(a) = \varphi(x)$ and $s(x)\tilde{\theta}(t(x))/s(b) = \theta(x)$ for $x \in [a, b]$. The requirements are equivalent to the equations

$$s(a)\tilde{\theta}(t(x))/(\tilde{\varphi}(t(x))s(b)) = \theta(x)/\varphi(x)$$
$$s(x) = s(a)\varphi(x)/\tilde{\varphi}(t(x)).$$

Now, we saw above that $\mathcal{W}(\theta, \varphi) > 0$, and similarly $\mathcal{W}(\tilde{\theta}, \tilde{\varphi}) > 0$. Differentiating we obtain $(\theta/\varphi)' = \mathcal{W}(\theta, \varphi)/\varphi^2 > 0$, so θ/φ is strictly increasing with range $[0, \infty]$, and so is $\tilde{\theta}/\tilde{\varphi}$. Since also $s(a)/s(b) > 0$ the first equation defines t uniquely as a strictly increasing function mapping $[a, b]$ onto $[t(a), t(b)]$, so that s is uniquely defined by the second equation. □

We need one more lemma and can then finish the proof of Theorem 8.4.9.

Lemma 8.4.17. *The functions s and t satisfy the regularity claims of Theorem 8.4.9, and give the connections claimed between the coefficients of T and $\tilde{T}$.*

Also, the Jost solutions of T and $\tilde{T}$ are, for every $\lambda \in \mathbb{C}$, connected via $f_{\pm}(x, \lambda) = s(x)\tilde{f}_{\pm}(t(x), \lambda)$.

Proof. The regularity and asymptotic properties of s and t, and the formulas for the coefficients, follow from Theorem 7.2.1 and (7.2.3). Now, if $\hat{u} \in L^2_{\mathcal{J}}$ we have

$$\langle \hat{u}, F(x, \cdot)\rangle_{\mathcal{J}} = u(x) = s(x)\tilde{u}(t(x)) = s(x)\langle \hat{u}, \tilde{F}(t(x), \cdot)\rangle_{\mathcal{J}},$$

so it follows that $F(x, \cdot) = s(x)\tilde{F}(t(x), \cdot)$ as elements of $L^2_{\mathcal{J}}$. In particular, the two sides are continuous and therefore equal on (q_0, ∞). The difference is analytic in $\mathbb{C}_+$ with zero boundary values on (q_0, ∞), so the reflection principle shows that it extends to a function analytic on a neighborhood of (q_0, ∞). Since this function vanishes on the interval it is identically zero. Analytic continuation now proves the claim. $\square$

Proof (Hard direction of Theorem 8.4.9). We have already proved all claims except the asymptotic properties of s and t. However, by Lemma 8.4.17

$$e^{\pm ikx} \sim f_\pm(x, \lambda) = s(x)\tilde{f}_\pm(t(x), \lambda) \sim s(x)e^{\pm ikt(x)}$$

as $x \to \pm\infty$, so that $s(x)e^{\pm ik(t(x)-x)} \to 1$ as $x \to \pm\infty$. Since this is true for many k we find that $t(x) - x \to 0$ and $s(x) \to 1$ as $x \to \pm\infty$. This finishes the proof. $\square$

We finally turn to the corollaries.

Proof (Corollary 8.4.10). We have $t's^2 \equiv 1$ and $\tilde{w} \circ t = ws^2/t'$ so that $\sqrt{|\tilde{w} \circ t|}t' = \sqrt{|w|}$. Since $w - 1$ and $\tilde{w} - 1$ are integrable, since $|\sqrt{a} - 1| \le |a - 1|$ for any $a \ge 0$ and since $t(x) - x \to 0$ as $x \to -\infty$ we obtain

$$t(x) + \int_{-\infty}^{t(x)} (\sqrt{|\tilde{w}|} - 1) = x + \int_{-\infty}^{x} (\sqrt{|w|} - 1) .$$

Since $x \mapsto x + \int_{-\infty}^{x}(\sqrt{|w|} - 1)$ is strictly increasing in supp w and similarly with w replaced by $\tilde{w}$ this determines $x \mapsto t$ as a strictly increasing function in supp w with range supp $\tilde{w}$, using also that $t(x)$ is continuous. If $|w| = |\tilde{w}|$ it follows that $t(x) = x$ in supp w. $\square$

Proof (Corollary 8.4.11). For $\lambda = 0$, the functions $f_\pm(\cdot, 0)$ and $\tilde{f}_\pm(\cdot, 0)$ are real-valued, satisfy the same equation and have the same asymptotic behavior at $\pm\infty$. They are therefore equal. However, by Lemma 8.4.17, we also have $f_\pm(x, 0) = s(x)\tilde{f}_\pm(t(x), 0)$. Thus, setting $K = f_-(\cdot, 0)/f_+(\cdot, 0)$, we get $K(t(x)) = K(x)$. Now $K' = \mathcal{W}(f_-(\cdot, 0), f_+(\cdot, 0))/f_+(\cdot, 0)^2$, where the Wronskian is constant and, since $\lambda = 0$ is not an eigenvalue, non-zero. Thus K is strictly monotone. It follows that $t(x) = x$ and therefore $s = 1$ and $\tilde{w} = w$. $\square$

Exercises

Exercise 8.4.1 Prove that, with the assumptions of Theorem 8.4.3, the residues in $(-\infty, q_0)$ of the two terms defining $A_-(x, u, \lambda)$ in (8.4.1) cancel.

Hint: (8.3.3).

Exercise 8.4.2 Show that if $u \in \mathcal{H}(x_0, \infty)$, where $x_0 \in \mathrm{supp}\, w$ but is not the left endpoint of a gap in supp w, then u may be approximated arbitrarily well by functions which are in $\mathcal{H}(x, \infty)$ for some $x > x_0$.

Hint: Put $u_j = 0$ in $[x_0, x) + 1/j]$ and $u_j = u$ in $[x_0 + 2/j]$. In $[x_0 + 1/j, x_0 + 2/j]$ make u_j linear. Then project on $\mathcal{H}$.

Exercise 8.4.3 Prove that $\int_{-\infty}^{\infty} (\sqrt{|w|} - \sqrt{|\tilde{w}|}) = 0$ under the assumptions of Theorem 8.4.9.

Hint: The proof of Corollary 8.4.10.

Exercise 8.4.4 Prove Theorems 7.3.1 and 7.4.2 using the methods of this chapter.

Exercise 8.4.5 Deduce high energy asymptotics for Jost solutions similar to what was done for solutions with fixed initial data in Section 6.4. Then prove Theorem 8.4.9 using the methods of Section 7.4.

Exercise 8.4.6 State and prove a uniqueness theorem for inverse scattering for right-definite Sturm–Liouville equations analogous to Theorem 8.4.9.

8.5 The case of a compact resolvent

In the same way as above one may treat the inverse scattering problem when, for some constant $\kappa \neq 0$, $w - \kappa$ instead of $w - 1$ is integrable. The difference is only a scaling of w and thus the spectral parameter. However, when $\kappa = 0$ the situation is different. In this case the spectrum is discrete and one may perhaps not properly speak of scattering. Nevertheless, the corresponding inverse problem is of interest, as is explained in Appendix F. We will devote the present section to this case.

In the case of a discrete spectrum there is no scattering matrix so the spectral data are only the eigenvalues and their normalization constants. In order to use only intrinsic properties of the equation the normalization constants should refer to solutions determined by their asymptotic behavior at $\pm\infty$. In other words, we must look for an analogue of the Jost solutions. We shall consider the left-definite equation

$$-u'' + qu = \lambda wu \tag{8.5.1}$$

on $\mathbb{R}$ with locally integrable q and w. That w in some sense is negligible near $\pm\infty$ means that we should view $-u'' + qu = 0$ as a model equation for (8.5.1)

By Theorem 5.1.8 the equation is essentially self-adjoint, so there is a unique self-adjoint realization T in $\mathcal{H}_1$ and no boundary conditions are needed. By Theorem 5.1.9 compactly supported elements of $\mathcal{H}_1$ are dense in $\mathcal{H}_1$, and by Proposition 5.1.10 there are solutions $F_\pm$ of the model equation which are strictly positive, satisfy $\mathcal{W}(F_-, F_+) = 1$ and such that $|f'|^2 + q\,|f|^2$ is integrable over $(0, \infty)$ if $f = F_+$ and over $(-\infty, 0)$ if $f = F_-$. This means that the reproducing kernel is $g_0(x, y) = F_-(\min(x, y))F_+(\max(x, y))$.

Note that 0 is never an eigenvalue of T by Theorems 5.1.9 and 5.1.13, so if the spectrum is discrete, then $0 \in \rho(T)$. Since the minimal relation is a core of T it therefore follows that $0 \in \rho(T)$ precisely if the operator $\varphi \mapsto G_0(w\varphi)$, which is a core of the resolvent at 0 and defined on compactly supported elements of $\mathcal{H}_1$, is bounded. Now $\|G_0(w\varphi)\|^2 = \int_{-\infty}^{\infty} G(w\varphi)\overline{\varphi}w \le \int_{\mathrm{supp}\,\varphi} g_0(x, x)\,|w(x)|\,dx\,\|G_0(w\varphi)\|\|\varphi\|$. Thus the operator is bounded under the following assumption.

Assumption 8.5.1 *The function* $x \mapsto g_0(x, x)w(x)$ *is in* $L^1(\mathbb{R})$.

By Theorem 5.2.1 this assumption is even enough to guarantee a discrete spectrum. Under this assumption we can find *Jost solutions*, i.e., solutions $f_\pm$ of (8.5.1) such that $f_\pm(x, \lambda) \sim F_\pm(x)$ as $x \to \pm\infty$ for all $\lambda \in \mathbb{C}$. Thus we are viewing $-u'' + qu = 0$ as a model equation for (8.5.1).

Let us write $f_+(x, \lambda) = g(x, \lambda)F_+(x)$, so we are looking for g which tends to 1 at ∞. Setting $K = F_-/F_+$ we shall see that if there is a bounded solution to the integral equation

$$g(x, \lambda) = 1 - \lambda \int\limits_{x}^{\infty} (K(\cdot) - K(x))F_+^2 g(\cdot, \lambda)\, w, \qquad (8.5.2)$$

then it will have the desired properties. Now $K' = \mathcal{W}(F_-, F_+)F_+^{-2} = F_+^{-2} > 0$ so K is strictly increasing and $0 \le (K(t) - K(x))F_+^2(t) \le F_-(t)F_+(t) = g_0(t, t)$ for $x \le t$, so that

$$|g(x, \lambda)| \le 1 + |\lambda| \int\limits_{x}^{\infty} |g(t, \lambda)|\, g_0(t, t)\,|w(t)|\, dt . \qquad (8.5.3)$$

Therefore successive approximations in (8.5.2) starting with 0 will, like in the proof of Theorem 8.2.1, lead to a bounded solution. The convergence is uniform for $x \in (a, \infty)$ for any $a \in \mathbb{R}$, and locally so in λ, so our 'Jost solution' $f_+(x, \lambda) = g(x, \lambda)F_+(x)$ exists for all complex λ and is an entire function of λ, uniformly for $x \in (a, \infty)$ for any $a \in \mathbb{R}$, real-valued for real λ, and asymptotic to F_+ at infinity. Differentiating (8.5.2) we obtain

$$g'(x, \lambda) = \lambda F_+(x)^{-2} \int\limits_{x}^{\infty} F_+^2 g(\cdot, \lambda) w, \qquad (8.5.4)$$

so $f_+' = gF_+' + g'F_+ = \lambda F_+^{-1} \int_x^\infty F_+^2 g w + gF_+'$. Differentiating again shows that f_+ satisfies (8.5.1).

If $x \le t$, then $F_+^2(t) = g_0(t, t)/K(t) \le g_0(t, t)/K(x)$ so that $|F_+(x)g'(x, \lambda)| \le F_-(x)^{-1} \int_x^\infty g_0\,|w|$, which tends to 0 at infinity and is in $L^2(0, \infty)$. So is F_+' and g is bounded, so it follows that $f'_+ \in L^2(0, \infty)$. Now $q F_+^2 \in L^1(0, \infty)$, so also $q \, |f_+|^2$ and thus $|f'_+|^2 + q \, |f_+|^2$ are in $L^1(0, \infty)$.

Finally, F'_+ is increasing since $F''_+ = q F_+ \geq 0$ and in $L^2(0, \infty)$ so $F'_+ \to 0$ at infinity. Since g is bounded and we already showed that $g' F_+ \to 0$ it follows that $f'_+(x, \lambda) \to 0$ as $x \to \infty$.

Very similar arguments apply for solutions asymptotic to F_- at $-\infty$. We summarize as follows.

Lemma 8.5.2. *There are unique solutions $f_\pm$ of $-f'' + qf = \lambda w f$ for all $\lambda \in \mathbb{C}$ with the following properties:*

- $f_+(x, \lambda) \sim F_+(x)$ *as* $x \to \infty$ *and* $f_-(x, \lambda) \sim F_-(x)$ *as* $x \to -\infty$.

- $f'_+(x, \lambda) \to 0$ *as* $x \to \infty$ *and* $f'_-(x, \lambda) \to 0$ *as* $x \to -\infty$.

- *Any solution f of (8.5.1) for which $|f'|^2 + q|f|^2$ is integrable near ∞ is a multiple of f_+. Similarly, integrability near $-\infty$ implies that f is a multiple of f_-.*

- λ_k *is an eigenvalue precisely if $f_+(\cdot, \lambda_k)$ and $f_-(\cdot, \lambda_k)$ are linearly dependent, and all eigenfunctions with eigenvalue λ_k are multiples of $f_+(\cdot, \lambda_k)$.*

Thus eigenvalues are simple, and the eigenfunctions are multiples of $f_+(\cdot, \lambda)$. Although $f'_+(x, \lambda) \to 0$ as $x \to \infty$ one can in general not expect that $f'_+ \sim F'_+$. For $u \in \mathcal{H}_1$ and every eigenvalue λ_n we define the Fourier coefficients

$$u_\pm(\lambda_n) = \langle u, f_\pm(\cdot, \lambda_n) \rangle = \lambda_n \int_{\mathbb{R}} u f_\pm(\cdot, \lambda_n) w \,, \qquad (8.5.5)$$

where the second equality follows from integration by parts.

The estimates in the successive approximations (or an application of Gronwall's Lemma D.3 to (8.5.3)) gives

$$|g(x, \lambda)| \leq \exp\left(|\lambda| \int_x^\infty g_0 \, |w| \right),$$

$$|g'(x, \lambda)| \leq g_0(x, x) \left(\exp\left(|\lambda| \int_x^\infty g_0 \, |w| \right) - 1 \right),$$

where the second formula is easily obtained by inserting the first in (8.5.4). Thus $f_+(x, \cdot)$ and $f'_+(x, \cdot)$ are entire functions of exponential type $\int_x^\infty g_0 \, |w|$ at most. This is easily sharpened to yield the following lemma.

Lemma 8.5.3. *As functions of λ and uniformly for x in any interval bounded from below, $f_+(x, \lambda)$ is an entire function of zero exponential type. Similarly for $f_-(x, \lambda)$ in intervals bounded from above, while $f'_\pm(x, \lambda)$ are also entire functions of zero* *exponential type[5] locally uniformly in x. Also the Wronskian $\mathcal{W}(f_+, f_-)$ is an entire function of λ of zero exponential type.*

Proof. Consider first a solution f of (8.5.1) with λ-independent initial data at some point a. By Theorem D.6 such a solution and its derivative is, locally uniformly in x, an entire function of order at most $1/2$. If now the initial data of f are entire functions of λ of exponential type it follows that so are f and f', and at most of the same type as the initial data. It follows that locally uniformly in x the functions f_+ and f'_+ are entire of exponential type $\int_a^\infty g_0(\cdot, \cdot) |w|$ for any a. They are thus of zero type.

Similar arguments may be carried out for f_- and f'_-, which immediately implies the result for the Wronskian. □

We shall use the notation from Definition 8.4.6

Corollary 8.5.4. *For every $u \in \mathcal{H}(a, \infty)$ with $a \in \mathbb{R}$ the generalized Fourier transform $\hat{u}_+$ extends to an entire function of zero exponential type vanishing at 0 and defined by*

$$\hat{u}_+(\lambda) = \lambda \int_{\mathbb{R}} u f_+(\cdot, \lambda) w.$$

A similar statement is valid for $\hat{u}_-$ given any $u \in \mathcal{H}(-\infty, a)$.

We shall give a uniqueness theorem for the inverse spectral problem. In order to avoid the trivial non-uniqueness caused by the fact that translating the coefficients of the equation by an arbitrary amount does not change the spectral properties of the corresponding operator, we normalize $F_\pm$, and thus $f_\pm$, by requiring $F_+(0) = F_-(0)$. This means that $F_+(0) = F_-(0) = (g_0(0,0))^{1/2}$.

We will need the following lemma.

Lemma 8.5.5. *The Wronskian $\mathcal{W}(\lambda) = \mathcal{W}(f_-(\cdot, \lambda), f_+(\cdot, \lambda))$ is determined by the eigenvalues of T and if λ_k is an eigenvalue, then*

$$\lambda_k \mathcal{W}'(\lambda_k) = \langle f_-(\cdot, \lambda_k), f_+(\cdot, \lambda_k) \rangle. \tag{8.5.6}$$

Proof. The formula (8.5.6) is obtained exactly as in the proof of Lemma 8.3.5.

The zeros of the Wronskian are located precisely at the eigenvalues, and by (8.5.6) they are all simple. If we have two entire functions, both of exponential type zero and with the same zeros (counted with multiplicity), then their quotient is also entire of exponential type zero according to Corollary E.2.6 and has no zeros. It is therefore constant. Since the Wronskian equals 1 for $\lambda = 0$ it is determined by the eigenvalues. □

As before we introduce, for each eigenvalue λ_n, the corresponding *matching constant* α_n defined by $f_-(\cdot, \lambda_n) = \alpha_n f_+(\cdot, \lambda_n)$. Together with the eigenvalues the matching constants will be our data for the inverse spectral theory. Instead of the matching

---

[5] Uniformity here means that one can for every $\varepsilon > 0$ find a constant C_ε so that the function may be estimated by $e^{\varepsilon|\lambda|}$ for $|\lambda| \geq C_\varepsilon$, independently of x in the relevant interval.

constants one could use *normalization constants* $\|f_+(\cdot, \lambda_n)\|^2$ or $\|f_-(\cdot, \lambda_n)\|^2$. If λ_n is an eigenvalue, then by Lemma 8.5.5 the scalar product $\langle f_-(\cdot, \lambda_n), f_+(\cdot, \lambda_n)\rangle$ is determined by the Wronskian, in other words by the eigenvalues, and since

$$\langle f_-(\cdot, \lambda_n), f_+(\cdot, \lambda_n)\rangle = \alpha_n \|f_+(\cdot, \lambda_n)\|^2 = \alpha_n^{-1} \|f_-(\cdot, \lambda_n)\|^2$$

all three sets of data are equivalent if the eigenvalues are known. We therefore make the following definition.

Definition 8.5.6. By the *spectral data* of the operator T we mean the set of eigenvalues for T together with the corresponding matching constants and the two sets of normalization constants.

The spectral data of T are thus determined if the eigenvalues and for each eigenvalue either the matching constant or one of the normalization constants are known.

In our main result we will be concerned with two operators T and $\tilde{T}$ of the type we have discussed. Connected with $\tilde{T}$ there are then coefficients $\tilde{q}, \tilde{w}$ and solutions $\tilde{F}_\pm, \tilde{f}_\pm$, etc.

Theorem 8.5.7. *Suppose T and $\tilde{T}$ have the same spectral data. Then there are continuous functions s, t defined on $\mathbb{R}$ such that s is strictly positive with a locally integrable derivative, $t : \mathbb{R} \to \mathbb{R}$ is bijective and $t(x) = \int_0^x s^{-2}$. Moreover, $\tilde{q} \circ t = s^3(-s'' + qs)$ and $\tilde{w} \circ t = s^4 w$.*

Conversely, if the coefficients of T and $\tilde{T}$ are connected in this way, then T and $\tilde{T}$ have the same spectral data.

Given additional information one may even conclude that $T = \tilde{T}$.

Corollary 8.5.8. *Suppose in addition to the operators T and $\tilde{T}$ having the same spectral data that $|\tilde{w}| = |w|$. Then $\tilde{w} = w$ and $\tilde{q} = q$ on $\operatorname{supp} w$. In particular, if $\operatorname{supp} w = \mathbb{R}$ or if $\tilde{q} = q$ outside $\operatorname{supp} w$, then $T = \tilde{T}$.*

Corollary 8.5.9. *Suppose in addition to the operators T and $\tilde{T}$ having the same spectral data that $\tilde{q} = q$. Then $T = \tilde{T}$.*

These results are of interest for the spectral problem $-u'' + \frac{1}{4}u = \lambda w u$ on $\mathbb{R}$, which is associated with the Camassa–Holm equation (see Appendix F) in the so-called *zero dispersion case* $\kappa = 0$. As we saw in Section 8.4 the reproducing kernel in this case satisfies $g_0(x, x) = 1$ for all x, so that the spectrum is discrete if w is integrable.

We begin the proofs with the easy direction of Theorem 8.5.7.

Proof (Converse part of Theorem 8.5.7). Define $\varphi_\pm(\cdot, \lambda) = s\tilde{f}_\pm(t(\cdot), \lambda)$. Using that $s^2 t' = 1$ one easily checks that $\mathcal{W}(\varphi_-, \varphi_+) = \mathcal{W}(\tilde{f}_-, \tilde{f}_+)$. If we can prove that $\varphi_\pm = f_\pm$ it follows that eigenvalues and matching constants agree for the two equations.

Now $\varphi_\pm(x, \lambda)/\varphi_\pm(x, 0) = \tilde{f}_\pm(t(x), \lambda)/\tilde{F}_\pm(t(x)) \to 1$ as $x \to \pm\infty$ so we only need to prove that $\varphi_\pm$ solve the appropriate equation and that $\varphi_\pm(\cdot, 0) = F_\pm$. The first property follows as in Section 7.2, so it follows that $\varphi_\pm(\cdot, 0) = A_\pm F_+ + B_\pm F_-$ for constants $A_\pm$ and $B_\pm$. We have

$$\frac{A_- + B_- K}{A_+ + B_+ K} = \frac{\varphi_-(\cdot, 0)}{F_+(\cdot, 0)} = \tilde{K} \circ t,$$

so since $K(-\infty) = \tilde{K}(-\infty) = 0$, $K(0) = \tilde{K}(0) = 1$ and $K(\infty) = \tilde{K}(\infty) = \infty$ the Möbius transform $t \mapsto \frac{A_- + B_- t}{A_+ + B_+ t}$ has fixpoints 0, 1 and ∞, which means that $A_- = B_+ = 0$ and $B_- = A_+ \neq 0$. Thus $\varphi_\pm(\cdot, 0) = A F_\pm$ for some constant A which is > 0 since $\varphi_\pm(\cdot, 0)$ and $F_\pm$ are all positive. But $1 = \mathcal{W}(\tilde{F}_-, \tilde{F}_+) = \mathcal{W}(\varphi_-(\cdot, 0), F_+(\cdot, 0)) = A^2$ so $A = 1$ and the proof is finished. □

The proof of the hard direction of Theorem 8.5.7 is very similar to the proof of Theorem 8.4.9, so we will be as brief as possible. Lemmas 8.4.2, 8.4.4, and 8.4.5 remain valid under the present assumption, and the proof of the following lemma is identical to the proof of the main part of Lemma 8.4.7.

Lemma 8.5.10. *Suppose $a \in \operatorname{supp} w$ and $u \in \mathcal{H}(a, \infty)$. Then*

$$\hat{u}_+(\lambda)/\lambda f_+(a, \lambda) = o(|\lambda/\operatorname{Im} \lambda|) \text{ as } \lambda \to \infty \text{ outside } \mathbb{R}.$$

A similar estimate is valid for $\hat{u}_-(\lambda)/\lambda f_-(a, \lambda)$ if $u \in \mathcal{H}(-\infty, a)$.

We may expand every $u \in \mathcal{H}$ in a series $u(x) = \sum \hat{u}_+(\lambda_n) \frac{f_+(x, \lambda_n)}{\|f_+(\cdot, \lambda_n)\|^2}$, where the sequence $\{\hat{u}_\pm(\lambda_n)/\|f_\pm(\cdot, \lambda_n)\|\} \in \ell^2$. Conversely, any such series converges in $\mathcal{H}$ and thus locally uniformly. Similarly for $\tilde{u} \in \tilde{\mathcal{H}}$. If the eigenvalues and normalization constants for T and $\tilde{T}$ are the same we may therefore define a unitary map $\mathcal{U} : \mathcal{H} \to \tilde{\mathcal{H}}$ by setting

$$\mathcal{U}u(t) = \tilde{u}(t) = \sum \hat{u}_+(\lambda_n) \frac{\tilde{f}_+(t, \lambda_n)}{\|\tilde{f}_+(\cdot, \lambda_n)\|^2}.$$

Note that expanding with respect to the sequence $\{f_-(\cdot, \lambda_n)\}$ and similarly in $\tilde{\mathcal{H}}$, defining $\mathcal{U}$ by use of these expansions we obtain the same operator $\mathcal{U}$.

Assume now that the generalized Fourier transform $\hat{u}_\pm$ of $u \in \mathcal{H}$, which is defined on all eigenvalues λ_n, has an entire extension and define the auxiliary function

$$A_\pm(u, x, \lambda) = R_\lambda u(x) - \frac{\hat{u}_\pm(\lambda) f_\mp(x, \lambda)}{\lambda \mathcal{W}(\lambda)},$$

where R_λ is the resolvent at λ of T. Similar auxiliary functions $\tilde{A}_\pm$ may be defined related to $\tilde{T}$.

Lemma 8.4.12, the crucial Lemma 8.4.13 and Lemmas 8.4.14–8.4.16 remain valid with essentially the same proofs under our present assumptions. Together with Section 7.2 these are all the ingredients needed for a proof of Theorem 8.5.7. We leave the details to the reader (Exercise 8.5.1)

To prove Corollaries 8.5.8 and 8.5.9 we need the following proposition.

Proposition 8.5.11. *If x and y are in $\operatorname{supp} w$, then*

$$g_0(x, y) = s(x)s(y)\tilde{g}_0(t(x), t(y)).$$

Proof. Suppose $\tilde{u} \in \tilde{\mathcal{H}}$ and $u = \mathcal{U}^{-1}\tilde{u}$. Since $t(x) \in$ supp $\tilde{w}$ it follows that $\tilde{g}_0(t(x), \cdot) \in \tilde{\mathcal{H}}$ and, by Lemma 8.4.15, $u(x) = s(x)\tilde{u}(t(x))$ so that

$$\langle \tilde{u}, \mathcal{U}g_0(x, \cdot)\rangle = \langle u, g_0(x, \cdot)\rangle = u(x) = s(x)\tilde{u}(t(x)) = s(x)\langle \tilde{u}, \tilde{g}_0(t(x), \cdot)\rangle.$$

Thus $\mathcal{U}g_0(x, \cdot) = s(x)\tilde{g}_0(t(x), \cdot)$. Since $y \in$ supp w Lemma 8.4.15 also shows that $g_0(x, y) = s(y)\mathcal{U}g_0(x, \cdot)(t(y))$, and combining these formulas completes the proof. $\qquad \square$

The proposition has the following corollary.

Corollary 8.5.12. *If $x \in$ supp w, then*

$$F_\pm(x) = s(x)\tilde{F}_\pm(t(x)). \tag{8.5.7}$$

Proof. Suppose $x, y \in$ supp w and $y \leq x$. Then, by Proposition 8.5.11,

$$\frac{F_+(x)}{s(x)\tilde{F}_+(t(x))} = \frac{s(y)\tilde{F}_-(t(y))}{F_-(y)}.$$

This implies that both sides are independent of x and y and thus equal a constant C. The corollary is proved if we can prove that $C = 1$.

Let λ_n be an eigenvalue of $\tilde{T}$ so that $\tilde{f}_+(\cdot, \lambda_n)$ is an eigenfunction and according to Theorem 8.5.7 $f_+(x, \lambda_n) = s(x)\tilde{f}_+(t(x), \lambda_n)$ is the corresponding eigenfunction for T. We then have

$$C\frac{f_+(x, \lambda_n)}{F_+(x)} = \frac{\tilde{f}_+(t(x), \lambda_n)}{\tilde{F}_+(t(x))}$$

for all $x \in$ supp w. If supp w is bounded above, choose $x = \sup \text{supp } w$. Then we have $f_+(x, \lambda_n) = F_+(x)$ and $\tilde{f}_+(t(x), \lambda_n) = \tilde{F}_+(t(x))$ so that $C = 1$. If supp w is not bounded above we take a limit as $x \to \infty$ in supp w and arrive at the same conclusion. $\qquad \square$

Proof (Corollary 8.5.8). By Corollary 8.5.12 we have $F_+/F_- = K = \tilde{K} \circ t$, which determines $t(x)$ uniquely since K and $\tilde{K}$ are strictly increasing. By the normalization of $F_\pm$ and $\tilde{F}_\pm$ we have $K(0) = \tilde{K}(0) = 1$, so $1 = K(0) = \tilde{K}(t(0))$ shows that $t(0) = 0$.

Now $\tilde{w} \circ t = s^4 w$ so since $t' = 1/s^2 > 0$ we have $\sqrt{|\tilde{w} \circ t|}t' = \sqrt{|w|}$. We obtain

$$\int\limits_0^{t(x)} \sqrt{|\tilde{w}|} = \int\limits_0^x \sqrt{|w|}.$$

If $|\tilde{w}| = |w|$ it follows that $t(x) = x$ and therefore $s(x) = 1$ for $x \in$ supp w and we are done. $\qquad \square$

Proof (Corollary 8.5.9). If $q = \tilde{q}$, then $F_\pm = \tilde{F}_\pm$ so $F_-/F_+ = K = \tilde{K}$. Thus $K(x) = K(t(x))$, so since K is strictly monotone we have $t(x) = x$ and therefore $s(x) \equiv 1$. It follows that $\tilde{w} = w$. $\qquad \square$

Exercise

Exercise 8.5.1 Make a detailed proof of Theorem 8.5.7.

8.6 Notes and remarks

The first results on inverse scattering for the Schrödinger equation on $(0, \infty)$ seem to be those by Levinson [148] in 1949, but are of a rather different character from later results for the whole line. See also Marčenko [162] and Faddeev [71]. A related result for a left-definite equation is given by Bennewitz, Brown and Weikard [19] .

The problem of inverse scattering for the Schrödinger equation on the whole real line was first considered by Kay and Moses [123, 124, 125] in 1955. As mentioned in the introduction, the problem was essentially solved by Faddeev [69]. A new approach by Deift and Trubowitz [52] resolved some problems with Faddeev's theory at the expense of requiring a finite second moment of the potential for the characterization of scattering data (uniqueness only requires a finite first moment). In [166] Melin gave a third approach, using microlocal analysis, which only requires a finite first moment.

The uniqueness results for more general equations in this chapter are from Bennewitz, Brown and Weikard [20] and [21]. In [20] the proofs are carried out by use of a full Paley–Wiener type theorem and high energy asymptotics of the Jost solutions, as in Chapter 7. The method used here was introduced in [21] and has the advantage of not requiring high energy asymptotics, useful variants of which do not seem available unless the coefficients are locally integrable. In this way one may deal with coefficients that are just measures or even just in H_{loc}^{-1}. With the present methods these extensions are fairly routine.

In inverse scattering for the one-dimensional Schrödinger equation it is customary to use only the reflection coefficient $\mathfrak{R}_+$, the location of the eigenvalues (only finitely many in this case) and the norming constants as data for the inverse problem. However, in this case it is easy to see that $\mathfrak{T}(\lambda) \to 1$ as $\lambda \to \infty$. Since the eigenvalues are the poles of $\mathfrak{T}$ and $|\mathfrak{T}|^2 + |\mathfrak{R}_+|^2 = 1$ on $\mathbb{R}$ this is enough to determine $\mathfrak{T}$, and thus the full scattering matrix, uniquely. In our present situation the behavior of $\mathfrak{T}$ at infinity is more complicated, and we must assume knowledge of the full scattering matrix. For more information on this, see [20, Section 6].

The de Branges theory successfully used for inverse spectral theory does not seem applicable to genuine inverse scattering, since the Jost solutions are then not entire functions of the spectral parameter. However, the problem studied in Section 8.5 has been successfully dealt with by these methods in Eckhardt [58]. See also Eckhardt and Teschl [65] and Eckhardt and Kostenko [63].

If the spectral function of a half-line Schrödinger operator T exhibits a sharp increase in the vicinity of a point μ (a point of continuity), μ is called a point of spectral concentration of T. Points of spectral concentration were investigated by Brown, Eastham, and McCormack [37]. It turns out that, for fast decaying potentials, a point μ of spectral concentration indicates the presence of resonance near μ. Resonances can be defined in the case when the Jost solutions, as functions of $k = \sqrt{\lambda - q_0}$, have analytic continuations across the real axis to the lower half of the k-plane. While a zero k of the Jost function $f(0, \cdot)$ in the upper half-plane gives rise to an eigenvalue $\lambda = q_0 + k^2$, a zero in the lower half-plane (or its corresponding λ-value) is called a resonance. This notion gives rise to *inverse resonance problems*, i.e., to determine the potential from information about the resonances. Inverse resonance problems were investigated by Korotyaev [128], Zworski [229], and Brown, Knowles, and Weikard in [38] in the context of a half-line Schrödinger operator with a possibly complex potential. These results state that knowing the location of all eigenvalues and resonances determines the potential uniquely. Uniqueness theorems for inverse resonance problems were also proved in a variety of other contexts, viz., algebro-geometric potentials for Schrödinger equations (Brown and Weikard [42]), Jacobi and Hermite difference equations (Brown, Naboko, and Weikard [39, 40] and Weikard and Zinchenko [218]) and for left-definite equations (Bledsoe and Weikard [27]). Another line of results concerns the stability of inverse eigenvalue and inverse resonance problems: how much can be said if only a finite number of eigenvalues and resonances are known (perhaps within some error). Results along these lines were obtained by Bledsoe, Marletta, Naboko, Shterenberg, Weikard and Zinchenko in [159, 160, 158, 157, 26, 25, 203].

Appendix A
Functional analysis

In this appendix we shall give the proofs of some fundamental theorems as well as a few minor results from functional analysis needed in the text. They are all valid more generally than stated here, but we confine ourselves to what is needed in this book. Standard references for this material are Reed and Simon [185] and Rudin [196]. The proofs of the fundamental theorems are based upon the following important theorem. We have stated it for a Banach space, but the proof would be the same in any complete metric space.

Theorem A.1 (Baire). *Suppose $\mathcal{B}$ is a Banach space and $F_1, F_2, \dots$ a sequence of closed subsets of $\mathcal{B}$. If all F_n fail to have interior points, so does $\cup_{n=1}^{\infty} F_n$. In particular, the union is a proper subset of $\mathcal{B}$.*

Proof. Let $\overline{B}_0 = \{x \in \mathcal{B} : \|x - x_0\| \le R_0\}$ be an arbitrary closed ball. We must show that it cannot be contained in $\cup_{n=1}^{\infty} F_n$. We do this by first selecting a decreasing sequence of closed balls $\overline{B}_0 \supset \overline{B}_1 \supset \overline{B}_2 \supset \cdots$ such that the radii $R_n \to 0$ and $\overline{B}_n \cap F_n = \varnothing$ for each n. But if we have already chosen $\overline{B}_0, \dots, \overline{B}_n$ we can find a point $x_{n+1} \in B_n$ (in the interior of $\overline{B}_n$) which is not contained in F_{n+1}, since F_{n+1} has no interior points. Since F_{n+1} is closed we can choose a closed ball $\subset B_n$, centered at x_{n+1}, and which does not intersect F_{n+1}. If we also make sure that the radius R_{n+1} is at most half of the radius R_n of $\overline{B}_n$, it follows by induction that we may find a sequence of balls as required.

For $k > n$ we have $x_k \in \overline{B}_n$ so that $\|x_k - x_n\| \le R_n \to 0$. Thus $x_1, x_2, \dots$ is a Cauchy sequence, so it converges to a limit x. We have $x \in \overline{B}_n$ for every n since $x_k \in \overline{B}_n$ for $k > n$ and $\overline{B}_n$ is closed. Thus x is not contained in any F_n. $\overline{B}_0$ being arbitrary, it follows that no ball is contained in $\cup_{n=1}^{\infty} F_n$, which therefore has no interior points, and the proof is complete. $\qquad\square$

A set which is a subset of the union of a sequence of closed sets without interior points is said to be of the *first category*. More picturesquely such a set is said to be *meager*. Meager subsets of $\mathbb{R}^n$ have many properties in common with, or analogous to, sets of Lebesgue measure zero. There is no direct connection, however, since a meager set may have positive measure, and a set of measure zero does not have to

© Springer Nature Switzerland AG 2020
C. Bennewitz et al., *Spectral and Scattering Theory for Ordinary Differential Equations*,
Universitext, https://doi.org/10.1007/978-3-030-59088-8

be meager. A set which is not meager is said to be of the *second category*, or to be *non-meager* (how about *fat*?). The basic properties of meager sets are the following.

Proposition A.2. *A subset of a meager set is meager, a countable union of meager sets is meager, and no meager set has an interior point.*

Proof. The proofs of the first two claims are left as exercises for the reader; the third claim is Baire's theorem. □

The following theorem is one of the cornerstones of functional analysis.

Theorem A.3 (Banach). *Suppose $\mathcal{B}_1$ and $\mathcal{B}_2$ are Banach spaces and $T : \mathcal{B}_1 \to \mathcal{B}_2$ a bounded, injective (one-to-one) linear map. If the range of T is not meager it is all of $\mathcal{B}_2$ and T has a bounded inverse.*

Proof. We denote the norm in $\mathcal{B}_j$ by $\|\cdot\|_j$. Let

$$A_n = \{Tx : \|x\|_1 \leq n\}$$

be the image of the closed ball with radius n, centered at 0 in $\mathcal{B}_1$. The balls expand to all of $\mathcal{B}_1$ as $n \to \infty$, so the range of T is $\cup_{n=1}^\infty A_n \subset \cup_{n=1}^\infty \overline{A}_n$. The range not being meager, at least one $\overline{A}_n$ must have an interior point y_0. Thus we can find $r > 0$ so that $\{y_0 + y : \|y\|_2 < r\} \subset \overline{A}_n$. Since $\overline{A}_n$ is symmetric with respect to the origin, $-y_0 + y \in \overline{A}_n$ if $\|y\|_2 < r$. Furthermore, $\overline{A}_n$ is convex, as the closure of (the linear image of) a convex set. It follows that $y = \frac{1}{2}((y_0 + y) + (-y_0 + y)) \in \overline{A}_n$. Thus 0 is an interior point of $\overline{A}_n$.

Since all $\overline{A}_n$ are similar ($\overline{A}_n = n\overline{A}_1$), 0 is also an interior point of $\overline{A}_1$. This means that there is a number $C > 0$ such that any $y \in \mathcal{B}_2$ for which $\|y\|_2 \leq C$ is in $\overline{A}_1$. For such y we may therefore find $x \in \mathcal{B}_1$ with $\|x\|_1 \leq 1$ such that Tx is arbitrarily close to y. For example, we may find $x \in \mathcal{B}_1$ with $\|x\|_1 \leq 1$ such that $\|y - Tx\|_2 \leq \frac{1}{2}C$. For arbitrary non-zero $y \in \mathcal{B}_2$ we set $\tilde{y} = \frac{C}{\|y\|_2}y$, and then have $\|\tilde{y}\|_2 = C$, so we can find $\tilde{x}$ with $\|\tilde{x}\|_1 \leq 1$ and $\|\tilde{y} - T\tilde{x}\|_2 \leq \frac{1}{2}C$. Setting $x = \frac{\|y\|_2}{C}\tilde{x}$ we obtain

$$\|x\|_1 \leq \frac{1}{C}\|y\|_2 \quad \text{and} \quad \|y - Tx\|_2 \leq \frac{1}{2}\|y\|_2 . \tag{A.1}$$

Thus, to any $y \in \mathcal{B}_2$ we may find $x \in \mathcal{B}_1$ so that (A.1) holds (for $y = 0$, take $x = 0$).

We now construct two sequences $\{x_j\}_{j=0}^\infty$ and $\{y_j\}_{j=0}^\infty$, in $\mathcal{B}_1$ respectively $\mathcal{B}_2$, by first setting $y_0 = y$. If y_n is already defined, we define x_n and y_{n+1} so that $\|x_n\|_1 \leq \frac{1}{C}\|y_n\|_2$, $y_{n+1} = y_n - Tx_n$, and $\|y_{n+1}\|_2 \leq \frac{1}{2}\|y_n\|_2$. We obtain $\|y_n\|_2 \leq 2^{-n}\|y\|_2$ and $\|x_n\|_1 \leq \frac{1}{C}2^{-n}\|y\|_2$ from this. Furthermore, $Tx_n = y_n - y_{n+1}$, so adding we obtain $T(\sum_{j=0}^n x_j) = y - y_{n+1} \to y$ as $n \to \infty$. But the series $\sum_{j=0}^\infty \|x_j\|_1$ converges, since it is dominated by $\frac{1}{C}\|y\|_2 \sum_{j=0}^\infty 2^{-j} = \frac{2}{C}\|y\|_2$. Since $\mathcal{B}_1$ is complete, the series $\sum_{j=0}^\infty x_j$ therefore converges to some $x \in \mathcal{B}_1$ satisfying $\|x\|_1 \leq \frac{2}{C}\|y\|_2$, and since T is continuous we also obtain $Tx = y$. In other words, we can solve $Tx = y$ for any $y \in \mathcal{B}_2$, so the inverse of T is defined everywhere, and the inverse is bounded by $\frac{2}{C}$, so it is continuous. The proof is complete. □

In this book we do not use Banach's theorem, but often the following simple corollary (which is in fact equivalent to Banach's theorem). Recall that a linear map $T : \mathcal{B}_1 \to \mathcal{B}_2$ is called *closed* if the graph $\{(u, Tu) : u \in \mathcal{D}(T)\}$ is a closed subset of $\mathcal{B}_1 \oplus \mathcal{B}_2$. Equivalently, if $\mathcal{D}(T) \ni u_j \to u$ in $\mathcal{B}_1$ and $Tu_j \to v$ in $\mathcal{B}_2$ implies that $u \in \mathcal{D}(T)$ and $Tu = v$.

Corollary A.4 (Closed graph theorem). *Suppose T is a closed linear operator $T : \mathcal{B}_1 \to \mathcal{B}_2$, defined on all of $\mathcal{B}_1$. Then T is bounded.*

Proof. The graph $\{(u, Tu) : u \in \mathcal{B}_1\}$ is by assumption a Banach space with norm $\|(u, Tu)\| = \|u\|_1 + \|Tu\|_2$, where $\|\cdot\|_j$ denotes the norm of $\mathcal{B}_j$. The map $(u, Tu) \mapsto u$ is linear, defined in this Banach space, with range equal to $\mathcal{B}_1$, and it has norm ≤ 1. It is obviously injective, so by Banach's theorem the inverse is bounded, i.e., there is a constant so that $\|(u, Tu)\| \leq C \|u\|_1$. Hence also $\|Tu\|_2 \leq C \|u\|_1$, so that T is bounded. $\square$

In Section 2.1 we use the simplest version of the *Banach–Steinhaus theorem*. Since no extra effort is involved, we prove the following more general version.

Theorem A.5 (Banach–Steinhaus; uniform boundedness principle). *Let $\mathcal{B}$ be a Banach space, $\mathcal{V}$ a normed linear space, and M a subset of the set $\mathcal{L}(\mathcal{B}, \mathcal{V})$ of all bounded, linear maps from $\mathcal{B}$ into $\mathcal{V}$.*

Suppose M is pointwise bounded, i.e., for each $x \in \mathcal{B}$ there exists a constant C_x such that $\|Tx\|_\mathcal{V} \leq C_x$ for every $T \in M$. Then M is uniformly bounded, i.e., there is a constant C such that $\|Tx\|_\mathcal{V} \leq C\|x\|_\mathcal{B}$ for all $x \in \mathcal{B}$ and all $T \in M$.

Proof. Put $F_n = \{x \in \mathcal{B} : \|Tx\|_\mathcal{V} \leq n \text{ for all } T \in M\}$. Then F_n is closed, as the intersection of the closed sets which are inverse images of the closed interval $[0, n]$ under a continuous function $\mathcal{B} \ni x \mapsto \|Tx\|_\mathcal{V} \in \mathbb{R}$. The assumption means that $\cup_{n=1}^\infty F_n = \mathcal{B}$.

By Baire's theorem at least one F_n must have an interior point. Since F_n is convex[1] and symmetric with respect to the origin it follows, like in the proof of Banach's theorem, that 0 is an interior point in F_n. Thus, for some $r > 0$ we have $\|Tx\|_\mathcal{V} \leq n$ for all $T \in M$, if $\|x\|_\mathcal{B} \leq r$. By homogeneity it follows that $\|Tx\|_\mathcal{V} \leq \frac{n}{r}\|x\|_\mathcal{B}$ for all $T \in M$ and $x \in \mathcal{B}$. $\square$

In Section 3.2 we need the following two lemmas.

Lemma A.6. *Let A, B be subspaces of a Banach space $\mathcal{B}$, with $\dim B < \infty$. Then $\mathrm{Sp}(A, B)$ is closed if A is.*

Note that without the assumption that $\dim B$ is finite, or at least that the codimension of A in $\mathrm{Sp}(A, B)$ is finite, the claim of the lemma is false, even if B is assumed closed.

---

[1] That is, if $x, y \in F_n$ and $0 \leq t \leq 1$, then $\|tTx + (1-t)Ty\|_\mathcal{V} \leq t\|Tx\|_\mathcal{V} + (1-t)\|Ty\|_\mathcal{V} \leq n$.

Proof. We may assume $A \cap B = \{0\}$, for otherwise introducing a basis $e_1, \ldots, e_k$ in $A \cap B$ and completing it to a basis $e_1, \ldots, e_n$ in B we may replace B with the span of $e_{k+1}, \ldots, e_n$ without changing $\mathrm{Sp}(A, B)$. We are then dealing with a direct sum $V = A \dotplus B$, and it follows from Theorem 1.3.1 that B is closed.

We must prove that if $V \ni u_j \to u$, then $u \in V$. To do this we shall prove that the projection $P : V \to B$ along A is bounded. Then so is $I - P$, so setting $v_j = (I - P)u_j \in A$ and $f_j = Pu_j \in B$ it follows that $\{v_j\}_1^\infty$ and $\{f_j\}_1^\infty$ are Cauchy sequences in A and B respectively. Since A and B are closed these sequences have limits $v \in A$ and $f \in B$. It follows that $u = v + f \in A \dotplus B = V$, so V is closed.

Now, if P were not bounded we could find a bounded sequence $u_j \in V$ such that $u_j = v_j + f_j$ with $v_j \in A$, $f_j \in B$ and $\|f_j\| \to \infty$. Thus $u_j/\|f_j\| \to 0$ and $g_j = f_j/\|f_j\|$ are unit vectors in B. By the Bolzano–Weierstrass theorem there is a subsequence of $\{g_j\}_1^\infty$ converging to a unit vector $g \in B$. But then the corresponding subsequence of $-v_j/\|f_j\| \in A$ also converges to g. Since A is closed we obtain $0 \neq g \in A \cap B$, a contradiction. Thus P is necessarily bounded. $\qquad \square$

Lemma A.7. *Suppose $F : \mathcal{H}_1 \to \mathcal{H}_2$ is a bounded linear operator between Hilbert spaces. Assume the range $\mathcal{R}$ of F has finite codimension. Then $\mathcal{R}$ is closed.*

This is true also if $\mathcal{H}_1, \mathcal{H}_2$ are Banach spaces, but the space $\mathcal{H}_1 \ominus \mathcal{N}$ in the proof must then be replaced by $\mathcal{H}_1/\mathcal{N}$ appropriately normed. The lemma is also true if F is just closed, which is seen immediately on replacing $\mathcal{H}_1$ by the graph of F.

Proof. We may assume F injective, since we may otherwise replace F by its restriction to $\mathcal{H}_1 \ominus \mathcal{N}$, where $\mathcal{N}$ is the nullspace of F. This restriction is clearly injective with the same range as F.

Now suppose $v_1, \ldots, v_n \in \mathcal{H}_2$ have cosets $v_j + \mathcal{R}$ which are a basis in $\mathcal{H}_2/\mathcal{R}$. Then every $v \in \mathcal{H}_2$ may be written in a unique way as $v = \sum_1^n x_j v_j + Fu$ with $x(v) = (x_1, \ldots, x_n) \in \mathbb{C}^n$ and $u \in \mathcal{H}_1$.

Since both $\mathcal{H}_1$ and $\mathbb{C}^n$ are complete, so is $\mathcal{H}_1 \times \mathbb{C}^n$, if provided with the norm $\|(u, x)\| = \|u\| + \|x\|_{\mathbb{C}^n}$. Thus we have a linear map $G : \mathcal{H}_2 \ni v \mapsto (u, x) \in \mathcal{H}_1 \times \mathbb{C}^n$, which is obviously closed. The closed graph theorem shows that G is bounded. In particular, the map $\mathcal{H}_1 \ni v \mapsto x(v) \in \mathbb{C}^n$ is continuous, and since $\mathcal{R}$ is determined by $x(v) = 0$ it follows that $\mathcal{R}$ is closed. $\qquad \square$

Appendix B
Stieltjes integrals

B.1 Riemann–Stieltjes integrals

The *Riemann–Stieltjes integral* is a simple generalization of the (one-dimensional) Riemann integral. To define it, let f and g be two complex-valued functions defined on the compact interval $[a, b]$. Given a *partition* $\{x_j\}_{j=0}^n$ of $[a, b]$, i.e., $a = x_0 \le x_1 \le \cdots \le x_n = b$, we choose from each subinterval $[x_{k-1}, x_k]$ a point ξ_k. In this way we obtain a *tagged partition* Δ with *tags* ξ_k. We define the *mesh* of the partition Δ to be $|\Delta| = \max(x_k - x_{k-1})$. This is the length of the longest subinterval of $[a, b]$ in the partition. We next form the *Riemann–Stieltjes sum*

$$s(\Delta) = \sum_{k=1}^n f(\xi_k)(g(x_k) - g(x_{k-1})) \,.$$

Now suppose that $s(\Delta)$ tends to a limit as $|\Delta| \to 0$. The exact meaning of this is the following: There exists a number I with the property that for every $\varepsilon > 0$ there is a $\delta > 0$ such that $|s(\Delta) - I| < \varepsilon$ as soon as $|\Delta| < \delta$.

In this case we say that the *integrand* f is *Riemann–Stieltjes integrable* with respect to the *integrator* g and that the corresponding integral equals I. We denote this integral by $\int_a^b f(x)\, dg(x)$ or simply $\int_a^b f\, dg$. The choice $g(x) = x$ gives us, of course, the ordinary Riemann integral. Since only differences of values of g occur in the Riemann–Stieltjes sums, it is clear that adding a constant to g will change neither which f are integrable with respect to g, nor the value of the integral.

Proposition B.1.1. *A function f is integrable with respect to a function g if and only if for every $\varepsilon > 0$ there exists a $\delta > 0$ such that for any two tagged partitions Δ and Δ' we have $|s(\Delta) - s(\Delta')| < \varepsilon$ as soon as $|\Delta| < \delta$ and $|\Delta'| < \delta$.*

This is of course a version of the Cauchy convergence principle. We leave the proof as an exercise (Exercise B.1.1). The following calculation rules follow immediately from the definition (Exercise B.1.2).

© Springer Nature Switzerland AG 2020
C. Bennewitz et al., *Spectral and Scattering Theory for Ordinary Differential Equations*,
Universitext, https://doi.org/10.1007/978-3-030-59088-8

$$(1) \quad \int_a^b f_1 \, dg + \int_a^b f_2 \, dg = \int_a^b (f_1 + f_2) \, dg \, ,$$

$$(2) \quad C \int_a^b f \, dg = \int_a^b Cf \, dg \, ,$$

$$(3) \quad \int_a^b f \, dg_1 + \int_a^b f \, dg_2 = \int_a^b f \, d(g_1 + g_2) \, ,$$

$$(4) \quad C \int_a^b f \, dg = \int_a^b f \, d(Cg) \, ,$$

$$(5) \quad \int_a^b f \, dg = \int_a^d f \, dg + \int_d^b f \, dg \text{ for } a < d < b \, ,$$

where f, f_1, f_2, g, g_1 and g_2 are functions, C a constant and the formulas should be interpreted to mean that if the integrals to the left of the equality sign exist, then so do the integrals to the right, and equality holds.

Proposition B.1.2 (Change of variable). *Suppose the function h is continuous and increasing and f is integrable with respect to g over $[h(a), h(b)]$. Then the composite function $f \circ h$ is integrable with respect to $g \circ h$ over $[a, b]$ and*

$$\int_{h(a)}^{h(b)} f \, dg = \int_a^b f \circ h \, d(g \circ h) \, .$$

We also leave the proof of this proposition to the reader (Exercise B.1.3). The formula for integration by parts takes the following nicely symmetric form in the context of the Stieltjes integral.

Theorem B.1.3 (Integration by parts). *If f is integrable with respect to g, then g is also integrable with respect to f and*

$$\int_a^b g \, df = f(b)g(b) - f(a)g(a) - \int_a^b f \, dg \, .$$

Proof. Let $a = x_0 \le x_1 \le \cdots \le x_n = b$ be a tagged partition Δ of $[a, b]$ with tags $\{\xi_k\}_1^n$. Set $\xi_0 = a$, $\xi_{n+1} = b$. Then $a = \xi_0 \le \xi_1 \le \cdots \le \xi_{n+1} = b$ gives a tagged partition Δ' of $[a, b]$ with tags $\{x_k\}_0^n$ for which $|\Delta'| \le 2 |\Delta|$. Now define $s(\Delta)$ as

$$\sum_{k=1}^{n} g(\xi_k)(f(x_k) - f(x_{k-1})) = \sum_{k=1}^{n} g(\xi_k)f(x_k) - \sum_{k=0}^{n-1} g(\xi_{k+1})f(x_k)$$

$$= f(b)g(b) - f(a)g(a) - \sum_{k=0}^{n} f(x_k)(g(\xi_{k+1}) - g(\xi_k)) .$$

If $|\Delta| \to 0$ we have $|\Delta'| \to 0$, so the last sum converges to $\int_a^b f\, dg$. It follows that $s(\Delta)$ converges to $f(b)g(b) - f(a)g(a) - \int_a^b f\, dg$ and the theorem follows. $\square$

Note that Theorem B.1.3 is a statement about the Riemann–Stieltjes integral; for the more general (Lebesgue–Stieltjes) integrals we shall discuss later *it is not true* without further assumptions about f and g. The reason is that the Riemann–Stieltjes integrals cannot exist if f and g have discontinuities in common (Exercise B.1.4), whereas the Lebesgue–Stieltjes integrals exist as soon as f and g are, for example, both monotone. In such a case the integration by parts formula only holds under additional assumptions. This is discussed further on page 315 in Appendix C.

So far we don't know that any function is integrable with respect to any other.

Theorem B.1.4. *If g is non-decreasing on $[a, b]$, then every continuous function f is integrable with respect to g and we have*

$$\left| \int_a^b f\, dg \right| \le \max_{[a,b]} |f| \, (g(b) - g(a)) .$$

Proof. Let Δ' and Δ'' be tagged partitions of $[a, b]$ and consider the corresponding Riemann–Stieltjes sums

$$s(\Delta') = \sum_{k=1}^{m} f(\xi'_k)(g(x'_k) - g(x'_{k-1})) ,$$

$$s(\Delta'') = \sum_{k=1}^{n} f(\xi''_k)(g(x''_k) - g(x''_{k-1})) .$$

If we introduce the partition given by all x'_j and x''_j, supposing it to be $\{x_j\}_0^p$, we may write

$$s(\Delta') - s(\Delta'') = \sum_{j=1}^{p} (f(\xi'_{k_j}) - f(\xi''_{q_j}))(g(x_j) - g(x_{j-1})) ,$$

where $k_j = k$ for all j for which $[x_{j-1}, x_j] \subset [x'_{k-1}, x'_k]$ and $q_j = k$ for all j for which $[x_{j-1}, x_j] \subset [x''_{k-1}, x''_k]$ (check this carefully!). Thus, for all j, ξ'_{k_j} and x_j are in the same subinterval of the partition Δ', and ξ''_{q_j} and x_j in the same subinterval of the partition Δ''. It follows that

$$\left| \xi'_{k_j} - \xi''_{q_j} \right| \le \left| \xi'_{k_j} - x_j \right| + \left| \xi''_{q_j} - x_j \right| \le |\Delta'| + |\Delta''|$$

for all j. Since f is uniformly continuous on $[a, b]$, this means that given $\varepsilon > 0$, we have $\left| f(\xi'_{k_j}) - f(\xi''_{q_j}) \right| \leq \varepsilon$ if $|\Delta'|$ and $|\Delta''|$ are both small enough. It follows that $|s(\Delta') - s(\Delta'')| \leq \varepsilon \sum_{j=1}^{p} (g(x_j) - g(x_{j-1})) = \varepsilon(g(b) - g(a))$ for small enough $|\Delta'|$ and $|\Delta''|$. Thus f is integrable with respect to g according to Proposition B.1.1. We also have

$$|s(\Delta')| \leq \sum_{k=1}^{n} \left| f(\xi'_k) \right| (g(x'_k) - g(x'_{k-1})) \leq \max |f| (g(b) - g(a)) ,$$

and since $s(\Delta') \to \int_a^b f \, dg$ as $|\Delta'| \to 0$ this completes the proof.   □

Exercises

Exercise B.1.1 Prove Proposition B.1.1.

Hint: Consider first a sequence $\{\Delta_j\}_1^{\infty}$ of tagged partitions for which $\left| \Delta_j \right| \to 0$.

Exercise B.1.2 Prove the calculation rules (1)–(5) on page 273.

Hint: To prove (5) Proposition B.1.1 might be useful.

Exercise B.1.3 Prove Proposition B.1.2 and a similar formula for decreasing h.

Hint: The function h is uniformly continuous in $[a, b]$.

Exercise B.1.4 Show that if f and g has a common point of discontinuity in $[a, b]$, then f is not Riemann–Stieltjes integrable with respect to g over $[a, b]$.

B.2 Functions of bounded variation

As a generalization of Theorem B.1.4 we may of course take g to be any function which is the *difference* of two non-decreasing functions, or even a complex-valued function with real and imaginary parts which are such differences. We shall briefly discuss such functions; the main point is that they are characterized by having *bounded variation*.

Definition B.2.1. Let f be a complex-valued function defined on $[a, b]$. Then the *total variation* of f over $[a, b]$ is

$$V(f) = \sup_{\Delta} \sum_{k=1}^{n} |f(x_k) - f(x_{k-1})| , \qquad (B.2.1)$$

the supremum taken over all partitions $\Delta = \{x_0, x_1, \ldots, x_n\}$ of $[a, b]$. We have $0 \leq V(f) \leq +\infty$, and if $V(f)$ is finite, we say that f has *bounded variation* on $[a, b]$.

By $BV_{\mathrm{loc}}(a, b)$ we denote the set of functions which are of bounded variation on every compact subinterval of (a, b).

When the interval considered is not obvious from the context, one may write the total variation of f over $[a, b]$ as $V_a^b(f)$; another common notation is $\int_a^b |df|$. As mentioned above, a function of bounded variation can also be characterized as a function which is a linear combination of non-decreasing functions.

Theorem B.2.2.

(1) *The total variation $V_a^b(f)$ is an interval additive function, i.e., if $a < x < b$ we have $V_a^x(f) + V_x^b(f) = V_a^b(f)$.*

(2) *A function is of bounded variation if and only if its real and imaginary parts are and*

$$\frac{1}{\sqrt{2}}(V_a^b(\mathrm{Re}\, f) + V_a^b(\mathrm{Im}\, f)) \leq V_a^b(f) \leq V_a^b(\mathrm{Re}\, f) + V_a^b(\mathrm{Im}\, f) \ .$$

(3) *A real-valued function of bounded variation on an interval $[a, b]$ may be written as the difference of two non-decreasing functions. Conversely, any such difference is of bounded variation.*

(4) *If f is real-valued and of bounded variation on $[a, b]$, then there are non-decreasing functions P and N with $P(a) = N(a) = 0$, such that $f(x) = f(a) + P(x) - N(x)$, called the positive and negative variation functions of f on $[a, b]$, with the following properties: $P(x) + N(x) = V_a^x(f)$ and for any pair of non-decreasing functions u, v for which $f = u - v$ holds $u(x) \geq u(a) + P(x)$ and $v(x) \geq v(a) + N(x)$ for $a \leq x \leq b$.*

Proof. It is clear that if $a < x < b$ and Δ, Δ' are partitions of $[a, x]$ respectively $[x, b]$, then $\Delta \cup \Delta'$ is a partition of $[a, b]$; the corresponding sum is therefore $\leq V_a^b(f)$. Taking the supremum over Δ and then Δ' it follows that $V_a^x(f) + V_x^b(f) \leq V_a^b(f)$.

On the other hand, in calculating $V_a^b(f)$, we may restrict ourselves to partitions Δ containing x, since adding a point to Δ can only increase the sum (B.2.1). If $\Delta = \{x_0, \ldots, x_n\}$ and $x = x_p$ we have $\sum_{k=1}^p |f(x_k) - f(x_{k-1})| \leq V_a^x(f)$ respectively $\sum_{k=p+1}^m |f(x_k) - f(x_{k-1})| \leq V_x^b(f)$.

Taking the supremum over all Δ we obtain $V_a^b(f) \leq V_a^x(f) + V_x^b(f)$. The interval additivity of the total variation follows, and the second point is an immediate consequence of the fact that $\frac{1}{\sqrt{2}}(|\mathrm{Re}\, z| + |\mathrm{Im}\, z|) \leq |z| \leq |\mathrm{Re}\, z| + |\mathrm{Im}\, z|$ for any $z \in \mathbb{C}$.

Setting $T(x) = V_a^x(f)$ the function T is finite in $[a, b]$; it is called the *total variation function* of f over $[a, b]$. Since by interval additivity $T(y) - T(x) = V_x^y(f) \geq 0$ if $a \leq x \leq y \leq b$ it follows that T is non-decreasing. A particular partition of the interval $[x, y]$ is $\{x, y\}$ so we even have

$$T(y) - T(x) = V_x^y(f) \geq |f(y) - f(x)| \geq \pm(f(y) - f(x))$$

if f is real-valued. Thus $P = \frac{1}{2}(T + f - f(a))$ and $N = \frac{1}{2}(T - f + f(a))$ are non-decreasing and $P(a) = N(a) = 0$. A splitting of f into a difference of non-decreasing functions is then $f = (f(a) + P) - N$. Note also that $T = P + N$. Conversely, if u and v are non-decreasing functions on $[a, b]$ and $\{x_0, \ldots, x_n\}$ a partition of $[a, x]$, $a < x \le b$, then

$$\sum_{k=1}^{n} |(u(x_k) - v(x_k)) - (u(x_{k-1}) - v(x_{k-1}))|$$

$$\le \sum_{k=1}^{n} |u(x_k) - u(x_{k-1})| + \sum_{k=1}^{n} |v(x_k) - v(x_{k-1})|$$

$$= u(x) - u(a) + v(x) - v(a) ,$$

so that $V_a^x(u-v) \le u(x) + v(x) - (u(a) + v(a))$. In particular, for $x = b$ this shows that $u - v$ is of bounded variation on $[a, b]$. The inequality also shows that if $f = u - v$, then

$$P(x) = \tfrac{1}{2}(T(x) + f(x) - f(a))$$

$$\le \tfrac{1}{2}(u(x) - u(a) + v(x) - v(a) + f(x) - f(a)) = u(x) - u(a) .$$

Similarly one shows that $N(x) \le v(x) - v(a)$ so that the proof is complete. $\qquad\square$

Note that the integral of a function φ with respect to the total variation function of f is usually written as $\int \varphi \, |df|$. We also want to point out the following simple proposition.

Proposition B.2.3. *Functions of bounded variation have at most countably many discontinuities, all of them jump discontinuities. The sum of the absolute values of all jumps is less than or equal to the total variation.*

Proof. It is enough to prove this for an increasing function, which clearly can only have jump discontinuities. But if such a function has total variation V, the number of jumps of size at least $1/n$, $n \in \mathbb{N}$, is finite, in fact at most nV, so the total number of jumps is countable, and the sum of their sizes is clearly at most V. $\qquad\square$

It is also easy to see that the total variation function, and therefore the positive and negative variation functions, are continuous to the left or right wherever f is (Exercise B.2.2).

Corollary B.2.4. *If g is of bounded variation on $[a, b]$, then every continuous function f is integrable with respect to g and we have*

$$\left| \int_a^b f \, dg \right| \le \int_a^b |f| \, |dg| \le \max_{[a,b]} |f| \, V_a^b(g) .$$

Proof. The integrability and the last inequality follow immediately from Theorem B.1.4 and Proposition B.1.1 on writing the real and imaginary parts of g as the difference of non-decreasing functions. To obtain the first inequality, consider a Riemann–Stieltjes sum

$$s(\Delta) = \sum_{k=1}^{n} f(\xi_k)(g(x_k) - g(x_{k-1})) \, .$$

If T denotes the total variation function for g we obtain

$$|s(\Delta)| \leq \sum_{k=1}^{n} |f(\xi_k)| \, |g(x_k) - g(x_{k-1})|$$

$$\leq \sum_{k=1}^{n} |f(\xi_k)| \, (T(x_k) - T(x_{k-1})) \to \int_a^b |f| \, |dg|$$

as $|\Delta| \to 0$. The corollary follows. $\qquad\square$

The set of functions of bounded variation on $[a, b]$ is a linear space; in fact, since the product of functions of bounded variation is also of bounded variation we have an *algebra*, with the following properties:

(1) If f is of bounded variation and C constant, then Cf is of bounded variation and $V(Cf) = |C| V(f)$.

(2) If f, g are of bounded variation, then so is $f + g$ and $V(f + g) \leq V(f) + V(g)$.

(3) If f, g are of bounded variation, then so is fg and $V(fg) \leq V(f) \sup |g| + \sup |f| V(g)$.

These are all easy to verify. In particular, the total variation is a semi-norm, and setting $\|f\| = V(f) + |f(a)|$ we obtain a norm under which the linear space of functions of bounded variation on $[a, b]$ is a Banach space (Exercise B.2.3). We end the section with another useful formula.

Theorem B.2.5. *Suppose f, g are continuous and ρ is of bounded variation in $[a, b]$. Then $[a, b] \ni x \mapsto \int_a^x g \, d\rho$ is of bounded variation and*

$$\int_a^b f(x) d\left(\int_a^x g \, d\rho \right) = \int_a^b fg \, d\rho \, .$$

In particular, if φ is continuously differentiable, then $\int_a^b f \, d\varphi = \int_a^b f\varphi'$.

The proof of Theorem B.2.5 is left as an exercise (Exercise B.2.4).

Exercises

Exercise B.2.1 Show that if f is an integral function, i.e., $f(x) = C + \int_a^x g$ for some integrable (Riemann or Lebesgue) function g, then f is of bounded variation on $[a, b]$, and $V_a^b(f) = \int_a^b |g|$.

Hint: It is easy to see that $V_a^b(f) \le \int_a^b |g|$. To show the other direction, write the sum in (B.2.1) in the form $\int_a^b \varphi g$ for a step function φ.

Exercise B.2.2 Show that if a real function of bounded variation is continuous to the left (right) at a point, then so are its positive and negative variation functions, and that only if the function jumps up (down) will the positive (negative) variation function have a jump.

Hint: Proposition B.2.3.

Exercise B.2.3 Prove the properties (1)–(3) on page 279 and that the functions of bounded variation on $[a, b]$ is a Banach space with norm $\|f\| = V(f) + |f(a)|$.

Convergence in this norm is called *convergence in variation*. Also show that convergence in variation implies uniform convergence. Finally show that uniform convergence does not imply convergence in variation.

Hint: Bounded continuous functions are not necessarily of bounded variation.

Exercise B.2.4 Prove Theorem B.2.5.

Hint: Use carefully chosen tags in the Riemann–Stieltjes sums.

B.3 Radon measures

The *support* $\operatorname{supp} \varphi$ of a continuous function φ is the closure of the set of points where φ is different from zero. We denote by $C_0(\Omega)$ the set of compactly supported functions continuous in the open interval Ω, i.e., the supports are compact subsets of Ω. Now, if g is of locally bounded variation in Ω, i.e., of bounded variation on every compact subinterval of Ω, then according to Corollary B.2.4 we have

$$\left| \int_\Omega \varphi \, dg \right| \le C_K \sup |\varphi|$$

if φ vanishes outside the compact interval $K \subset \Omega$ and C_K is the total variation of g on K. The integral depends linearly on φ, so it is a *linear form* on $C_0(\Omega)$ and the fact that we have the estimate above means that the form is *bounded*.

Convergence in $C_0(\Omega)$ is defined by declaring that $\varphi_j \to \varphi$ if the supports of all φ_j are subsets of a fixed compact $K \subset \Omega$ and $\varphi_j \to \varphi$ uniformly. The bound on our form implies that it is a continuous function of φ. In fact, any continuous linear form is bounded. For, if μ is a linear form which is not bounded we can find a compact $K \subset \Omega$ such that for each j there exists a $\varphi_j \in C_0(\Omega)$ with supp $\varphi_j \subset K$ satisfying $|\mu(\varphi_j)| > j \sup |\varphi_j|$. Dividing φ_j by $\mu(\varphi_j)$ we may assume $\mu(\varphi_j) = 1$ and then obtain $\sup |\varphi_j| < 1/j$ so that $\varphi_j \to 0$ in $C_0(\Omega)$. Thus, if μ were continuous $\mu(\varphi_j)$ would converge to 0. Since it does not, any continuous form has to be bounded.

A bounded linear form on $C_0(\Omega)$ is also called a *Radon measure*,[1] or just a measure, on Ω. The ordinary (Riemann) integral on $C_0(\Omega)$ is thus the measure generated by $g(x) = x$. This is called *Lebesgue measure*. The linear space of continuous linear forms on a topological vector space is called the *dual* of the space, so the set of Radon measures on Ω is the dual of $C_0(\Omega)$. As we have seen, any Stieltjes integral with respect to a function of locally bounded variation is an example of a Radon measure, and we shall eventually show that all one-dimensional Radon measures are of this form.

Let us look at Radon measures in more detail. If μ is a Radon measure we define its conjugate by $\overline{\mu}(\varphi) = \overline{\mu(\overline{\varphi})}$, which is clearly a Radon measure. We have $\overline{\overline{\mu}} = \mu$ and a Radon measure is called *real* if $\overline{\mu} = \mu$. As is easily seen, this is equivalent to $\mu(\varphi)$ being real for real φ. We define Radon measures $\operatorname{Re} \mu = \frac{1}{2}(\mu + \overline{\mu})$ and $\operatorname{Im} \mu = \frac{1}{2i}(\mu - \overline{\mu})$, which are clearly real, and $\mu = \operatorname{Re} \mu + i \operatorname{Im} \mu$.

A Radon measure is called *positive* if $\mu(\varphi) \geq 0$ for all $\varphi \in C_0^+(\Omega)$, where $C_0^+(\Omega)$ denotes the set of non-negative functions in $C_0(\Omega)$. A positive Radon measure is real since any real $\varphi \in C_0(\Omega)$ may be written as a difference of elements of $C_0^+(\Omega)$, e.g. as the difference of $|\varphi|$ and $|\varphi| - \varphi$. If μ_1, μ_2 are real measures we write $\mu_1 \leq \mu_2$ if $\mu_2 - \mu_1$ is positive. Also note that if $\varphi_1 \leq \varphi_2$ are in $C_0(\Omega)$, then $\varphi_2 - \varphi_1 \in C_0^+(\Omega)$ so that $\mu(\varphi_1) \leq \mu(\varphi_2)$ if μ is positive.

Proposition B.3.1. *A positive measure μ on Ω satisfies the* triangle inequality $|\mu(\varphi)| \leq \mu(|\varphi|)$ *for any $\varphi \in C_0(\Omega)$*

Proof. We may find a real θ so that $|\mu(\varphi)| = e^{i\theta}\mu(\varphi) = \mu(e^{i\theta}\varphi)$. Since this is real we have $|\mu(\varphi)| = \mu(\operatorname{Re}(e^{i\theta}\varphi)) \leq \mu(|\varphi|)$. $\qquad\square$

We shall prove the following important theorem.

Theorem B.3.2. *Given any Radon measure μ on Ω there is a smallest positive Radon measure $|\mu|$ such that $|\mu(\varphi)| \leq |\mu|(|\varphi|)$ for $\varphi \in C_0(\Omega)$, called the* absolute value *or* total variation measure of μ.

Proof. For $\varphi \in C_0^+(\Omega)$ we define $|\mu|(\varphi) = \sup |\mu(\psi)|$, where the least upper bound is taken over all $\psi \in C_0(\Omega)$ with $|\psi| \leq \varphi$. Clearly $|\mu|(\varphi) \geq 0$ and $|\mu(\varphi)| \leq |\mu|(|\varphi|)$.

We also have $|\mu(\psi)| \leq C_K \sup |\psi| \leq C_K \sup \varphi$ if supp $\varphi \subset K$. Thus $0 \leq |\mu|(\varphi) \leq C_K \sup \varphi$. It is clear that $|\mu|(C\varphi) = C|\mu|(\varphi)$ for constants $C \geq 0$. We must now extend the definition of $|\mu|$ to all of $C_0(\Omega)$ so that it becomes linear.

If $|\psi_j| \leq \varphi_j$ we may choose real θ_j so that $|\mu(\psi_j)| = e^{i\theta_j}\mu(\psi_j)$. We have $|e^{i\theta_1}\psi_1 + e^{i\theta_2}\psi_2| \leq |\psi_1| + |\psi_2| \leq \varphi_1 + \varphi_2$ so that

---

[1] Radon measures may be defined in the same way if $\Omega \subset \mathbb{R}^n$, or is an n-dimensional manifold, with much the same proofs of the basic theorems.

$$|\mu(\psi_1)| + |\mu(\psi_2)| = \left|\mu(e^{i\theta_1}\psi_1 + e^{i\theta_2}\psi_2)\right| \leq |\mu|\,(\varphi_1 + \varphi_2)\,.$$

Varying first ψ_1 and then ψ_2 we obtain $|\mu|\,(\varphi_1) + |\mu|\,(\varphi_2) \leq |\mu|\,(\varphi_1 + \varphi_2)$.

On the other hand, if $|\psi| \leq \varphi_1 + \varphi_2$ and $\psi = e^{i\theta}\,|\psi|$ with a real-valued function θ, put $\psi_1 = e^{i\theta}\min(|\psi|, \varphi_1)$ and $\psi_2 = \psi - \psi_1$. Then

$$\psi_2 = e^{i\theta}\max(0, |\psi| - \varphi_1)\,, \quad |\psi_1| \leq \varphi_1, |\psi_2| \leq \varphi_2, \quad \text{and} \quad \psi_1 + \psi_2 = \psi\,,$$

so that $|\mu(\psi)| \leq |\mu(\psi_1)| + |\mu(\psi_2)| \leq |\mu|\,(\varphi_1) + |\mu|\,(\varphi_2)$. Thus $|\mu|\,(\varphi_1 + \varphi_2) \leq |\mu|\,(\varphi_1) + |\mu|\,(\varphi_2)$, so $|\mu|$ is additive on $C_0^+(\Omega)$.

Now, if $\varphi_j \in C_0^+$ and $\varphi_1 - \varphi_2 = \varphi_3 - \varphi_4$ then $\varphi_1 + \varphi_4 = \varphi_3 + \varphi_2$, so additivity shows that $|\mu|\,(\varphi_1) - |\mu|\,(\varphi_2) = |\mu|\,(\varphi_3) - |\mu|\,(\varphi_4)$. Given a real-valued $\varphi \in C_0(\Omega)$ and writing $\varphi = \varphi_1 - \varphi_2$ with $\varphi_j \in C_0^+(\Omega)$ we define $|\mu|\,(\varphi) = |\mu|\,(\varphi_1) - |\mu|\,(\varphi_2)$. The value is thus independent of how the splitting of φ into a difference of functions in $C_0^+(\Omega)$ is done.

Thus $|\mu|$ is real-valued and depends linearly on real-valued φ. Writing $\varphi = \max(0, \varphi) - \max(0, -\varphi)$ shows that $\|\,|\mu|(\varphi)\| \leq 2C_K \sup|\varphi|$. Defining $|\mu|\,(\varphi) = |\mu|\,(\operatorname{Re}\varphi) + i\,|\mu|\,(\operatorname{Im}\varphi)$ for complex-valued φ, clearly $|\mu|$ is linear and bounded and thus a positive measure.

Finally, if ν is a measure satisfying $|\mu(\varphi)| \leq \nu(|\varphi|)$ and $|\psi| \leq |\varphi|$ we have $|\mu(\psi)| \leq \nu(|\psi|) \leq \nu(|\varphi|)$ which implies $|\mu| \leq \nu$ and the proof is finished. $\square$

Corollary B.3.3. *Any real Radon measure μ may be written $\mu = \mu_+ - \mu_-$, where $\mu_\pm$ are positive Radon measures such that $|\mu| = \mu_+ + \mu_-$, and if $\mu = \mu_1 - \mu_2$ with positive Radon measures μ_j, then $\mu_+ \leq \mu_1$ and $\mu_- \leq \mu_2$.*

Proof. We define $\mu_\pm = \frac{1}{2}(|\mu| \pm \mu)$. Clearly $\mu_\pm$ are positive measures, $\mu = \mu_+ - \mu_-$ and $|\mu| = \mu_+ + \mu_-$ so $\mu_\pm \leq |\mu|$.

If $\mu = \mu_1 - \mu_2$ with positive measures μ_j, then $|\mu(\varphi)| \leq |\mu_1(\varphi)| + |\mu_2(\varphi)| \leq (\mu_1 + \mu_2)(|\varphi|)$ so that $|\mu| \leq \mu_1 + \mu_2$ from which $\mu_+ \leq \mu_1$ and $\mu_- \leq \mu_2$ immediately follows. $\square$

The last statement of the corollary means that $\mu_\pm$ are the *smallest* positive measures which allow a splitting $\mu = \mu_+ - \mu_-$. Note the analogy of these results with those for the total variation of a function of bounded variation and the splitting of a real function of bounded variation into a difference of non-decreasing functions. As we shall eventually see, this is more than an analogy.

Exercise

Exercise B.3.1 Show that a positive linear form on $C_0(\Omega)$ is automatically bounded.

Hint: Let $\psi \in C_0(\Omega)$ be positive and ≥ 1 on a compact $K \subset \Omega$. Then $|\varphi(x)| \leq \psi(x)\sup|\varphi|$ if $\varphi \in C_0(\Omega)$ with $\operatorname{supp}\varphi \subset K$. Now use the proof of Proposition B.3.1.

B.4 Integrable functions

We shall extend the domain of a Radon measure to a class of functions larger than $C_0(\Omega)$. Since every measure is a linear combination of positive measures it will be enough to do this extension for positive measures, so *all measures are assumed positive* in this section.

Suppose $\{\varphi_j\}_1^\infty$ is a (pointwise) non-decreasing sequence in $C_0^+(\Omega)$. This sequence tends pointwise to a function f, which is ≥ 0 and may have the value $+\infty$ at some (or all) points. The set of all such limit functions we denote by $I^+(\Omega)$. The following proposition lists some obvious properties of $I^+(\Omega)$.

Proposition B.4.1. *Suppose $C \geq 0$ is constant and $f, g \in I^+(\Omega)$. Then*

(1) $\max(f, g)$ *and* $\min(f, g)$ *are both in* $I^+(\Omega)$,

(2) Cf *is in* $I^+(\Omega)$ *if we define* Cf *identically zero for* $C = 0$,

(3) $f + g$ *is in* $I^+(\Omega)$,

(4) fg *is in* $I^+(\Omega)$ *if we define* fg *as zero wherever one of the factors is,*

(5) f *is* lower semi-continuous, *i.e., the set* $\{x \in \Omega : f(x) > C\}$ *is open for every* $C \in \mathbb{R}$.

Proof. The first four items are immediate consequences of the definition of I^+, as is the last item if we note that if $\varphi_j \in C_0^+(\Omega)$ increases to f, then the set where $f > C$ is the union of the sets where $\varphi_j > C$. These are all open since φ_j is continuous. $\square$

If μ is a positive measure the numerical sequence $\mu(\varphi_j)$ is also non-decreasing and has a limit ≥ 0 which may be $+\infty$. It is now tempting to define $\mu(f)$ to be this limit, but for this to make sense we must prove that if $\{\psi_j\}$ is another non-decreasing sequence in $C_0^+(\Omega)$ with the same pointwise limit f we have $\lim \mu(\varphi_j) = \lim \mu(\psi_j)$. In order to do this we need the following lemma.

Lemma B.4.2 (Dini). *If the sequence $\{\varphi_j\}_1^\infty$ in $C_0^+(\Omega)$ decreases pointwise to 0, then the convergence is uniform; in fact $\varphi_j \to 0$ as a sequence in $C_0(\Omega)$.*

Proof. We have $\operatorname{supp}\varphi_j \subset \operatorname{supp}\varphi_1$ for all j, and if $\varepsilon > 0$ and $x \in \Omega$ we may find an index $j(x)$ such that $0 \leq \varphi_{j(x)}(x) < \varepsilon$. Since $\varphi_{j(x)}$ is continuous there is a neighborhood $\omega(x)$ of x such that $0 \leq \varphi_{j(x)}(y) < \varepsilon$ if $y \in \omega(x)$, and since $\{\varphi_j\}_1^\infty$ decreases we have $0 \leq \varphi_j(y) < \varepsilon$ if $y \in \omega(x)$ and $j \geq j(x)$.

The neighborhoods $\omega(x)$ form an open cover of the compact set $\operatorname{supp}\varphi_1$. By the Heine–Borel lemma a finite subcover $\omega(x_1), \ldots, \omega(x_n)$ exists. Thus $0 \leq \varphi_j(x) < \varepsilon$ for all x if $j \geq \max(j(x_1), \ldots, j(x_n))$, so the convergence is uniform. $\square$

Lemma B.4.3. *Suppose μ is a positive Radon measure in Ω and $\{\varphi_j\}_1^\infty$, $\{\psi_j\}_1^\infty$ non-decreasing sequences in $C_0^+(\Omega)$ with limits f and g in $I^+(\Omega)$ respectively.*
Then if $f \leq g$ we also have $\lim \mu(\varphi_j) \leq \lim \mu(\psi_j)$.

Proof. Fix k and consider the sequence $f_j = \max(0, \varphi_k - \psi_j)$ in $C_0^+(\Omega)$. By assumption this sequence decreases pointwise to zero. According to Lemma B.4.2 $f_j \to 0$ in $C_0(\Omega)$, so $\mu(f_j) \to 0$ as $j \to \infty$.

Since $\varphi_k - \psi_j \le f_j$ we obtain $\mu(\varphi_k) - \mu(\psi_j) = \mu(\varphi_k - \psi_j) \le \mu(f_j) \to 0$ so that $\mu(\varphi_k) \le \lim \mu(\psi_j)$ for all k. The lemma follows on letting $k \to \infty$. $\qquad\square$

If $f = g$ Lemma B.4.3 shows that $\lim \mu(\varphi_j) = \lim \mu(\psi_j)$ so we may unambiguously define $\mu(f) = \lim \mu(\varphi_j)$, making μ well defined on $I^+(\Omega)$. A positive Radon measure is according to Lemma B.4.3 *monotone* on $I^+(\Omega)$, i.e., if $f \le g$ then $\mu(f) \le \mu(g)$. It also follows that μ is *positively homogeneous*, i.e. if $C \ge 0$ is constant then[2] $\mu(Cf) = C\mu(f)$, and *additive*, so that $\mu(f + g) = \mu(f) + \mu(g)$, on $I^+(\Omega)$. We also have the following fundamental fact about monotone convergence.

Lemma B.4.4. *Suppose μ is a positive Radon measure and $\{f_j\}_1^\infty$ a non-decreasing sequence in $I^+(\Omega)$ with pointwise limit f. Then $f \in I^+(\Omega)$ and $\mu(f_j) \to \mu(f)$ as $j \to \infty$.*

Proof. Suppose $C_0^+(\Omega) \ni \varphi_{jk}$ increases to f_j as $k \to \infty$ and put $\varphi_k = \max_{j \le k} \varphi_{jk}$. Then $\{\varphi_k\}_1^\infty$ is an increasing sequence in $C_0^+(\Omega)$, and $\varphi_k \le f_k \le f$. On the other hand, if $j \le k$, then $\varphi_{jk} \le \varphi_k$ so that $f_j \le \lim \varphi_k$ for all j.

Thus $f \le \lim \varphi_k$, but since $\varphi_k \le f$ in fact $\lim \varphi_k = f$. It follows that $f \in I^+(\Omega)$ and that $\mu(\varphi_k) \to \mu(f)$. But $\mu(\varphi_k) \le \mu(f_k) \le \mu(f)$, so it follows that $\mu(f_k) \to \mu(f)$. $\qquad\square$

Note that if $f_j \in I^+(\Omega)$, then $\sum_1^n f_j$, $n = 1, 2, \ldots$, is an increasing sequence in $I^+(\Omega)$. We therefore have the following corollary.

Corollary B.4.5. *If μ is a positive Radon measure and $f_j \in I^+(\Omega)$, then so is $\sum_1^\infty f_j$ and $\mu(\sum_1^\infty f_j) = \sum_1^\infty \mu(f_j)$.*

If we wish to extend μ to a class of functions which can take both signs we have to be careful since both functions $f \in I^+(\Omega)$ and $\mu(f)$ can sometimes have the value $+\infty$. We shall need the following definition.

Definition B.4.6. Let μ be a positive measure. We call $f \in I^+(\Omega)$ μ-*integrable* if $\mu(f) < \infty$.

A set $N \subset \Omega$ is called a μ-*nullset* if there is a μ-integrable $f \in I^+(\Omega)$ which equals $+\infty$ on N.

Any statement depending on a parameter $x \in \Omega$ is said to be true *almost everywhere with respect to* μ, abbreviated a.e.(μ), if for some μ-nullset N the statement is true for all $x \in \Omega \setminus N$. Note that the definition of nullset means that all μ-integrable functions in $I^+(\Omega)$ are finite a.e.(μ). By convention claims about integrability or nullsets without reference to a particular measure always refer to Lebesgue measure, i.e., integration with respect to $g(x) = x$.

Nullsets have the following general properties.

---

[2] Provided we define $C\mu(f) = 0$ if $C = 0$ even if $\mu(f) = +\infty$.

Proposition B.4.7.

(1) *If $M \subset N$ and N is a μ-nullset, then so is M.*

(2) *If N_j, $j = 1, 2, \ldots$ are μ-nullsets, then so is $\bigcup_{j=1}^{\infty} N_j$.*

Proof. The first point is obvious.

To prove the second point, note that if $f \in I^+(\Omega)$ is infinite in some set, so is Cf if $C > 0$. There are therefore functions $f_j \in I^+(\Omega)$ which are infinite on N_j and satisfy $\mu(f_j) \leq 2^{-j}$. Thus $f = \sum_1^{\infty} f_j$ is in $I^+(\Omega)$ and is μ-integrable by Corollary B.4.5 since $\mu(f) = \sum_1^{\infty} \mu(f_j) \leq 1$. But f is infinite on $\cup_1^{\infty} N_j$, so this is a μ-nullset. □

For another way of viewing nullsets, recall that the characteristic function χ_M of a set $M \subset \mathbb{R}$ is defined to be 1 in M and 0 outside.

Lemma B.4.8. *The characteristic function of an open set is in I^+.*

Proof. It is easily seen that any open subset of Ω is the union of a sequence of disjoint open intervals (Exercise B.4.2), so by Lemma B.4.4 it is enough to prove that the characteristic function of an open interval is in $I^+(\Omega)$.

Define $\varphi_j(x) = \max(0, \min(1, j(x - a) - 1, j(b - x) - 1))$, $j = 1, 2, \ldots$, if the interval (a, b) is bounded. If $a = -\infty$ replace a in the definition of φ_j by $-j$, and if $b = \infty$ replace b by j. Clearly $\varphi_j \in C_0^+(\Omega)$ and $\varphi_j \uparrow \chi_{(a,b)}$. Thus $\chi_{(a,b)} \in I^+(\Omega)$. □

See also Exercise B.4.3. We define the μ-measure of an open set ω as $\mu(\omega) := \mu(\chi_\omega)$ and then have the following characterization of nullsets.

Theorem B.4.9. *A set $N \subset \Omega$ is a μ-nullset if and only if for every $\varepsilon > 0$ there exists an open set $\omega \supset N$ with $\mu(\omega) < \varepsilon$.*

Proof. If $f \in I^+(\Omega)$ is infinite on N then $\omega_j = \{x \in \Omega : f(x) > j\}$ is an open set containing N, and $\mu(\omega_j) = \mu(\chi_{\omega_j}) \leq \frac{1}{j}\mu(f)$. If f is μ-integrable it follows that $\mu(\omega_j) \to 0$.

Conversely, if N may be enclosed in open sets of arbitrarily small measure, let ω_j be such a set with $\mu(\omega_j) < 2^{-j}$ and let $f = \sum_1^{\infty} \chi_{\omega_j}$, which is in $I^+(\Omega)$ and infinite on N. Furthermore $\mu(f) = \sum \mu(\omega_j) \leq \sum_1^{\infty} 2^{-j} = 1 < \infty$, so f is also μ-integrable. Thus N is a μ-nullset. □

Example B.4.10. Note that which sets are nullsets depends entirely on the measure considered. For example, if m is Lebesgue measure on $\mathbb{R}$, then any singleton $\{a\}$ is an m-nullset, since it is the center of an open interval of arbitrarily small length (= Lebesgue measure).

By Proposition B.4.7 (2) any countable set, for example the rational numbers $\mathbb{Q}$, is also an m-nullset. Note, however, that there are also uncountable Lebesgue nullsets (Exercise B.4.4). On the other hand, clearly no set with an interior point is a Lebesgue nullset.

For a radically different example, consider $\delta_a(\varphi) = \varphi(a)$, which is called the *Dirac measure* at a. This is a Radon measure since $|\varphi(a)| \leq \sup |\varphi|$. Clearly $\{a\}$

is not a δ_a-nullset, but all real sets which do not contain a are. For let ω be the complement of $\{a\}$. Then ω is open and $\delta_a(\varphi) = 0$ for every $\varphi \in C_0^+(\omega)$, so that also $\delta_a(\chi_\omega) = 0$.

Note that $\delta_a(\varphi) = \int \varphi \, dg$ where $g(x) = H(x - a)$ and H is the *Heaviside function* $H = \chi_{(0,\infty)}$ (Exercise B.4.5). $\qquad\qquad\qquad\qquad\qquad\qquad\qquad\qquad\qquad\qquad$ □

A measure, like Lebesgue measure, for which any singleton is a nullset is called *diffuse*. On the other hand, a measure, like the Dirac measure, for which there is a countable set the complement of which is a nullset, is called *atomic*.

One may show that if μ is atomic on Ω, and the complement of $\cup_1^\infty \{a_j\}$ is a μ-nullset, then $\mu = \sum_1^\infty c_j \delta_{a_j}$, where[3] the coefficients $\{c_j\}_1^\infty$ have the property that $\sum_{a_j \in K} |c_j| < \infty$ for every compact $K \subset \Omega$. Conversely, every such sum is an atomic Radon measure. One may also show that any measure μ on Ω may be written uniquely as $\mu = \mu_a + \mu_d$ where μ_a is atomic and μ_d is diffuse (Exercise B.4.6).

To extend the definition of a positive measure μ from $I^+(\Omega)$ to functions that can take both signs the simplest idea is to define $\mu(f)$ as $\mu(u) - \mu(v)$ if $f = u - v$ where $u, v \in I^+(\Omega)$. But note that $u - v$ is undefined at points where both u and v are infinite, and also that the difference of $\mu(u)$ and $\mu(v)$ is undefined unless at least one of them is finite.

We therefore require that u, v are μ-integrable, so that they are finite a.e.(μ) and only ask that $f = u - v$ a.e.(μ). Given f of this kind there are many ways of splitting f into such a difference, so we need the following lemma.

Lemma B.4.11. *If $u, v \in I^+(\Omega)$ and $u = v$ a.e.(μ), then $\mu(u) = \mu(v)$.*

Proof. Let $f \in I^+(\Omega)$ be μ-integrable and infinite on the set where $u \neq v$. Then $u + f = v + f$ so $\mu(u) + \mu(f) = \mu(v) + \mu(f)$. Since $\mu(f) < \infty$ we obtain $\mu(u) = \mu(v)$. $\qquad$ □

If $u_j \in I^+(\Omega)$ are μ-integrable and $u_1 - u_2 = u_3 - u_4$ a.e.(μ) we have $u_1 + u_4 = u_3 + u_2$ a.e.(μ) so the lemma shows that $\mu(u_1) - \mu(u_2) = \mu(u_3) - \mu(u_4)$.

If $f = u_1 - u_2$ a.e.(μ) we may therefore define $\mu(f) = \mu(u_1) - \mu(u_2)$, and the value of $\mu(f)$ is independent of the splitting a.e.(μ) of f into a difference of μ-integrable functions in $I^+(\Omega)$. This leads to the following definition.

Definition B.4.12. If μ is a positive Radon measure, then a μ-*integrable* function f is a function for which there are μ-integrable functions $u_j \in I^+(\Omega)$ such that $\text{Re } f = u_1 - u_2$ a.e.(μ) and $\text{Im } f = u_3 - u_4$ a.e.(μ). For a general Radon measure a μ-nullset means a $|\mu|$-nullset, and f is said to be μ-integrable if it is $|\mu|$-integrable. One defines the μ-integral of f as

$$\mu(f) = \mu(u_1) - \mu(u_2) + i(\mu(u_3) - \mu(u_4)) . \qquad\qquad\qquad □$$

Note that while $\mu(f)$ is defined for any $f \in I^+$ if μ is positive, for a general measure $\mu(f)$ is only defined if f is integrable($|\mu|$). Clearly sums and constant multiples of

---

[3] A sequence of measures $\{\mu_j\}_1^\infty$ on Ω is said to converge (weakly) to a measure μ on Ω if $\mu_j(\varphi) \to \mu(\varphi)$ for every $\varphi \in C_0(\Omega)$.

μ-integrable functions are μ-integrable, so the set of μ-integrable functions is a linear space. If μ is positive and $f \le g$ real-valued and μ-integrable we also have $\mu(f) \le \mu(g)$, so a *positive* measure gives a *monotone* integral.

It is also clear that every μ-integrable function has a finite value a.e.(μ). Note that although a difference $u_1 - u_2$ of μ-integrable elements of $I^+(\Omega)$ is not defined at points where u_1 and u_2 are both infinite, this can only happen in a μ-nullset, so assigning completely arbitrary values to the difference at such points defines a μ-integrable function. We give some calculation rules for integrable functions.

Theorem B.4.13. *Suppose f and g are μ-integrable and C constant. Then*

(1) *Cf and $f + g$ are integrable, so the set of integrable functions is a linear space.*

(2) *If f and g are real-valued, then $|f|$, $\max(f, g)$ and $\min(f, g)$ are integrable.*

Integrable complex-valued functions also have integrable absolute value, but the proof will have to be deferred to Theorem B.6.4.

Proof. (1) is an immediate consequence of the definition and similar rules for I^+. To prove (2), first note that if $f = u - v$ with non-negative u, v, then $|f| = \max(u, v) - \min(u, v)$, and if $u, v \in I^+$ so is $\max(u, v)$ and $\min(u, v)$. Furthermore, $0 \le \min(u, v) \le \max(u, v) \le u + v$ so $|f|$ is integrable if f is. Thus $\max(f, g) = \frac{1}{2}(f + g + |f - g|)$ and $\min(f, g) = \frac{1}{2}(f + g - |f - g|)$ are also integrable. $\qquad\square$

A measure μ on Ω has a *restriction* to an open interval $\omega \subset \Omega$, obtained by applying μ only to functions in $C_0(\omega)$. The *support* supp μ of the measure μ is the complement of the set of points which have a neighborhood such that the restriction of μ to the neighborhood equals zero. Thus supp μ is a closed set, and it is proved in Theorem C.8 that if $\varphi \in C_0(\Omega)$ and supp $\mu \cap$ supp $\varphi = \varnothing$, then $\mu(\varphi) = 0$.

A function f defined on Ω is said to be *locally μ-integrable* if its restriction to (a, b) is integrable with respect to the restriction of μ to (a, b) for any $[a, b] \subset \Omega$.

To show that every Radon measure is a Stieltjes integral we need the following lemma. Recall that the *characteristic function* χ_M of a set $M \subset \mathbb{R}$ is defined as 1 in M and 0 outside, and a *step function* in Ω is a finite linear combination of characteristic functions of intervals with compact closure in Ω.

Lemma B.4.14. *If μ is a Radon measure on Ω, then the characteristic function of an interval with compact closure contained in Ω is μ-integrable, and so is any step function in Ω.*

Proof. The second claim obviously follows from the first. Suppose $[a, b] \subset \Omega$. Then $\chi_{(a,b)} \in I^+(\Omega)$ by Lemma B.4.8. Now $\chi_{(a,b)}$ is also μ-integrable since for $k > 0$

$$\chi_{(a,b)} \le \max(0, \min(1, 1 + k(x - a), 1 + k(b - x))),$$

which is in $C_0^+(\Omega)$ for large k.

We have $\chi_{[a,b)} = \chi_{(a-\varepsilon,b)} - \chi_{(a-\varepsilon,a)}$, and $[a - \varepsilon, a] \subset \Omega$ if $\varepsilon > 0$ is small, so also $\chi_{[a,b)}$ and similarly $\chi_{(a,b]}$ and $\chi_{[a,b]}$ are μ-integrable. $\qquad\square$

If $c, x \in \Omega$ the lemma implies that the function

$$t \mapsto e_x(t) = \begin{cases} \chi_{[c,x)}(t) & \text{if } x > c, \\ 0 & \text{if } x = c, \\ -\chi_{[x,c)}(t) & \text{if } x < c, \end{cases}$$

is μ-integrable. Note that if $x < y$, then $e_y - e_x = \chi_{[x,y)} \geq 0$ so if μ is positive, then the function $g(x) = \mu(e_x)$, often called a *distribution function* for μ, is a non-decreasing function. For a general Radon measure μ the function g is therefore of locally bounded variation. We shall prove the following theorem, which is essentially the classical Riesz [187] representation theorem from 1909.

Theorem B.4.15. *Let μ be a one-dimensional Radon measure and put $g(x) = \mu(e_x)$. If $\varphi \in C_0(\Omega)$, then $\mu(\varphi) = \int \varphi \, dg$.*

The theorem states that every bounded linear form on $C_0(\Omega)$ is the Stieltjes integral with respect to any distribution function of μ. A function g such that $\mu(\varphi) = \int_\Omega \varphi \, dg$ for all $\varphi \in C_0(\Omega)$ is unique up to its definition at discontinuities and an additive constant, but we shall leave the proof of this to page 313 just after Theorem C.13.

Proof. We need only consider positive μ and real φ. If Δ is a tagged partition of a compact interval with supp φ in its interior the Riemann–Stieltjes sum corresponding to the integral is

$$s(\Delta) = \sum_{k=1}^{n} \varphi(\xi_k)(g(x_k) - g(x_{k-1})) = \mu(\varphi_\Delta),$$

where $\varphi_\Delta = \sum_1^n \varphi(\xi_k)(e_{x_k} - e_{x_{k-1}}) = \sum_1^n \varphi(\xi_k)\chi_{[x_{k-1},x_k)}$ is a step function equal to $\varphi(\xi_k)$ on $[x_{k-1}, x_k]$. Given $\varepsilon > 0$, and since φ is uniformly continuous, we therefore have $|\varphi - \varphi_\Delta| < \varepsilon$ if $|\Delta|$ is sufficiently small. Since φ and φ_Δ have supports in $[x_0, x_n)$ in fact

$$-\varepsilon\mu(\chi_{[x_0,x_n)}) < \mu(\varphi) - \mu(\varphi_\Delta) < \varepsilon\mu(\chi_{[x_0,x_n)})$$

if $|\Delta|$ is sufficiently small. Now $\mu(\varphi_\Delta) = s(\Delta) \to \int \varphi \, dg$ as $|\Delta| \to 0$ so the theorem follows.   □

Note that in many texts one writes the μ-integral of φ as $\mu(\varphi) = \int \varphi \, d\mu$. This is a very unfortunate notation since it directly clashes with the only reasonable notation for Schwarz distributions, where $\int \varphi \, dg$ may also be written as $g'(\varphi)$, as we shall see in Appendix C. If we write an integral $\int \varphi \, d\mu$ it is therefore more reasonable to denote the corresponding measure by $d\mu$ and let μ denote a distribution function for $d\mu$. Thus we should write $d\mu(\varphi) = \int \varphi \, d\mu$, which we shall do very rarely, but mostly not, to avoid confusion with standard notation.

Because of Theorem B.4.15 we also write $\int \varphi \, dg$ for $\mu(\varphi)$ when φ is any μ-integrable function. The μ-integral we have developed is usually called the *Lebesgue–Stieltjes integral* with respect to $g(x) = \mu(e_x)$.

Exercises

Exercise B.4.1 A function f is *lower semi-continuous* at a if $\underline{\lim}_{x \to a} f(x) \geq f(a)$. Define analogously *upper semi-continuity* and prove that f is continuous at a precisely if it is both upper and lower semi-continuous there.

Also show that f is lower semi-continuous at all points of Ω precisely if it is lower semi-continuous according to the definition given in Proposition B.4.1.

Exercise B.4.2 Show that any open subset ω of $\mathbb{R}$ is the disjoint union of a sequence of open subintervals of $\mathbb{R}$.

Exercise B.4.3 Show that $f_j(x) = \inf_z (f(z) + j\,|x - z|)$ is continuous and that $\{f_j\}_1^\infty$ increases pointwise to f for any non-negative, lower semi-continuous f defined in Ω and not identically $+\infty$. Use this to show that $I^+(\Omega)$ equals the set of non-negative lower semi-continuous function defined in Ω. The function f_j is called the *infimal convolution* of f and $j\,|\cdot|$.

Exercise B.4.4 Let N be all numbers in $[0, 1]$ which may be written as a decimal fraction with no decimals equal to 7. Show that N is a Lebesgue nullset. Also show that N is not countable.

Hint: Show N is not countable by proving that given any sequence of real numbers in $[0, 1]$ there is a real number in this interval where all decimals are either 1 or 2 and which is not in the sequence. For this, use Cantor's diagonal process, in the same way as one proves that $\mathbb{R}$ is not countable.

Exercise B.4.5 Show that the measure δ_a is given by $\int \varphi\, dg$ where $g(x) = H(x - a)$ and H is the Heaviside function (see Example B.4.10).

Exercise B.4.6 Show that an atomic measure on Ω is of the form $\sum_1^\infty c_j \delta_{a_j}$ where $a_j \in \Omega$, the series converges weakly as a measure and $\sum_{a_j \in K} |c_j| < \infty$ for any compact $K \subset \Omega$.

Also show that any measure μ on Ω may be split uniquely into $\mu = \mu_a + \mu_d$ where μ_a is an atomic and μ_d a diffuse measure on Ω.

Exercise B.4.7 Show that any Radon measure on Ω is the weak limit of a sequence of atomic measures.

Hint: Consider an increasing sequence of compacts K_j with union Ω and for each j a tagged partition Δ_j of K_j such that $|\Delta_j| \to 0$. Then show that any Riemann–Stieltjes sum of $\varphi \in C_0(\Omega)$ may be written as $\mu(\varphi)$ for some atomic measure μ.

Exercise B.4.8 Suppose μ is a Radon measure on Ω given by the distribution function g of locally bounded variation in Ω. Show that $|\mu|$ is given by the total variation function T of g, and that if μ is real, then μ_+ is given by P and μ_- by N, where P and N are the positive and negative variation functions of g. Thus Theorem B.2.2 (4) and Corollary B.3.3 state the same thing.

B.5 Convergence theorems

In this section we assume that μ is a positive Radon measure on Ω. We shall prove the basic theorems on taking limits under the integral sign, but first need a lemma.

Lemma B.5.1. *Suppose μ is a positive Radon measure, that $u \in I^+(\Omega)$ and that $\varphi \in C_0(\Omega)$ with $\varphi \le u$. Then $u - \varphi \in I^+(\Omega)$, and if u is μ-integrable we may choose φ so that $\mu(u - \varphi)$ is arbitrarily small.*

Proof. If $\varphi_j \in C_0^+(\Omega)$ increases to u and $u \ge \varphi \in C_0(\Omega)$, then $\max(0, \varphi_j - \varphi) \in C_0^+(\Omega)$ increases to $u - \varphi$, which is therefore in $I^+(\Omega)$. If u is μ-integrable, then $\mu(\varphi_j) \to \mu(u) < \infty$, so for large j we obtain $\mu(u - \varphi_j) = \mu(u) - \mu(\varphi_j)$ as small as we like.   □

Theorem B.5.2 (Beppo Levi). *Suppose $\mu \ge 0$ and that f_j are non-negative μ-integrable functions. Then $\sum_1^\infty f_j$ is μ-integrable if and only if $\sum_1^\infty \mu(f_j) < \infty$ and then $\mu(\sum_1^\infty f_j) = \sum_1^\infty \mu(f_j)$.*

Proof. If $f \ge 0$ is μ-integrable and $f = u - v$ with $u, v \in I^+(\Omega)$ we may assume $v \le u$, for if this is not true everywhere we may replace u, v by $\max(u, v)$ and $\min(u, v)$ respectively. This only changes $u - v$ on a μ-nullset.

Thus, by Lemma B.5.1 we may assume that $f_j = u_j - v_j$ with $u_j, v_j \in I^+(\Omega)$ and $\mu(v_j) \le 2^{-j}$, if necessary first subtracting the same $\varphi \in C_0(\Omega)$ from u_j and v_j. According to Corollary B.4.5 the functions $v = \sum v_j$ and $u = \sum u_j$ are in $I^+(\Omega)$, and $\mu(v) = \sum_1^\infty \mu(v_j) \le \sum_1^\infty 2^{-j} = 1 < \infty$. By the same corollary we therefore have

$$\mu(u) = \sum_1^\infty \mu(u_j) = \sum_1^\infty (\mu(f_j) + \mu(v_j)) = \mu(v) + \sum_1^\infty \mu(f_j) \,,$$

which is finite precisely if $\sum_1^\infty \mu(f_j)$ is. Since $\sum_1^\infty f_j = u - v$ a.e.(μ) this finishes the proof.   □

Corollary B.5.3 (Monotone convergence). *Suppose μ is a positive Radon measure. If g_j are real-valued, μ-integrable and increase pointwise a.e.(μ) to g, then g is μ-integrable if and only if $\lim \mu(g_j)$ is finite, and then $\lim \mu(g_j) = \mu(g)$. A similar statement is true for decreasing sequences.*

Proof. Setting $f_j = g_{j+1} - g_j$ for an increasing and $f_j = g_j - g_{j+1}$ for a decreasing sequence $\{g_j\}_1^\infty$ this is an obvious consequence of Beppo Levi's theorem. Actually, it is clear that the two theorems are equivalent.   □

By Theorem B.1.4 continuous functions are locally μ-integrable. This is also the case for functions of locally bounded variation.

Theorem B.5.4. *If f is of locally bounded variation on Ω it is locally μ-integrable for any Radon measure μ on Ω.*

Proof. We may assume μ is positive and since f is a linear combination of non-decreasing functions we may also assume that f is non-decreasing. If $[a, b] \subset \Omega$ and f is not continuous in $c_j \in (a, b)$ we set $A_j = f(c_j) - f(c_j-)$ and $B_j = f(c_j+) - f(c_j)$, where $f(c\pm)$ denotes one-sided limits of f at c. Thus $g_j = A_j \chi_{[c_j,b]} + B_j \chi_{(c_j,b]}$ is μ-integrable according to Lemma B.4.14 and $f - g_j$ is continuous at c_j.

Now suppose $\{c_j\}_1^\infty$ is an enumeration of all the discontinuities of f in $[a, b]$. Then $\sum_1^\infty (A_j + B_j) \leq f(b) - f(a) < \infty$ so the series $g = \sum_1^\infty g_j$ is uniformly convergent and thus continuous except at the points c_j, where it jumps exactly like g_j. It is also μ-integrable by Beppo Levi's theorem since $\mu(g_j) \leq (A_j + B_j)\mu(\chi_{[a,b]})$. It follows that $f - g$ is continuous and so μ-integrable on $[a, b]$. Thus $f = g + (f - g)$ is μ-integrable on $[a, b]$. $\qquad\square$

If one has a sequence of μ-integrable real-valued functions which is not monotone and thus may not have pointwise limits, one may consider $\underline{\lim}$ or $\overline{\lim}$ of the sequence[4] instead. This leads to the following basic theorem.

Theorem B.5.5 (Fatou's lemma). *If μ is a positive Radon measure and $0 \leq f_j$ are μ-integrable with $\underline{\lim}\,\mu(f_j) < \infty$, then $\underline{\lim} f_j$ is μ-integrable and $\mu(\underline{\lim} f_j) \leq \underline{\lim}\,\mu(f_j)$.*

Proof. Put $g_{jk} = \min(f_j, f_{j+1}, \ldots, f_{j+k})$. Fixing j the sequence $\{g_{jk}\}_{k=1}^\infty$ decreases and is non-negative, so by monotone convergence it has a μ-integrable pointwise a.e.(μ) limit g_j. The sequence $\{g_j\}_1^\infty$ is increasing and $g_j \to \underline{\lim} f_k$. By monotone convergence, and since $g_j \leq f_j$, we have $\mu(\underline{\lim} f_j) = \mu(\lim g_j) = \lim \mu(g_j) \leq \underline{\lim}\,\mu(f_j)$. $\qquad\square$

Example B.5.6. There can be strict inequality in the conclusion of Fatou's lemma. Consider for example Lebesgue measure on $(0, a)$, $0 < a \leq \infty$ and $f_j(x) = je^{-jx}$ or $f_j = (1 + (x - j)^2)^{-1}$. Then $\underline{\lim} f_j(x) = 0$ for all $x \in (0, a)$, but $\int_0^a f_j$ tends to 1 in the first case and, if $a = \infty$, π in the second as $j \to \infty$.

In addition to monotone convergence and Fatou's lemma the following theorem is of fundamental importance.

Theorem B.5.7 (Lebesgue's theorem; dominated convergence). *Suppose μ is a Radon measure, that f_j are μ-integrable, and $|f_j| \leq g$ a.e.(μ) for every j, with g μ-integrable. Assume also that $f_j \to f$ pointwise a.e.(μ).*

Then f is μ-integrable and $|\mu| (|f - f_j|) \to 0$ as $j \to \infty$. In particular, $\mu(f_j) \to \mu(f)$.

Proof. Since $0 \leq g + \mathrm{Re}\, f_j \leq 2g$ we may apply Fatou's lemma to conclude that $g + \mathrm{Re}\, f$ and therefore $\mathrm{Re}\, f$ is μ-integrable. Similarly $\mathrm{Im}\, f$ and hence also f is μ-integrable.

---

[4] Recall that for a real sequence $\{a_j\}$ we define $\underline{\lim} a_j = \lim_{n\to\infty} \inf_{j \geq n} a_j$ and $\overline{\lim} a_j = \lim_{n\to\infty} \sup_{j \geq n} a_j$. These are monotone limits and therefore always exist, finite or infinite, and $\lim a_j$ exists precisely if $\underline{\lim} a_j = \overline{\lim} a_j$ and then equals the common value.

Now apply Fatou's lemma to $2g - |f - f_j|$ to give

$$|\mu|\,(2g) \le \underline{\lim}\,|\mu|\,(2g - |f - f_j|) = |\mu|\,(2g) - \overline{\lim}\,|\mu|\,(|f - f_j|)\,.$$

Thus $|\mu|\,(|f - f_j|) \to 0$. Finally, $|\mu(f) - \mu(f_j)| \le |\mu|\,(|f - f_j|)$ so we also have
$\mu(f_j) \to \mu(f)$.   □

B.6 Measurability

It will be useful to introduce the concept of a *measurable function*.

Definition B.6.1. Suppose μ is a Radon measure on Ω. A function f defined a.e.(μ)
in Ω is called μ-*measurable* if it pointwise a.e.(μ) is the limit of a sequence in $C_0(\Omega)$.

The basic properties of measurable functions are as follows.

Theorem B.6.2.

(1) *If $f \in I^+(\Omega)$ or f is μ-integrable, then f is μ-measurable.*

(2) *If $f_1, \dots, f_k$ are μ-measurable and finite a.e.(μ), and $F : \mathbb{C}^k \to \mathbb{C}$ continuous,
   then $F(f_1, \dots, f_k)$ is μ-measurable.*

(3) *If C is constant and f, g are μ-measurable and finite a.e.(μ), then so are Cf,
   $f + g$, f/g if $g \ne 0$ a.e.(μ), $\max(f, g)$, $\min(f, g)$ and $|f|^p$ for any $p > 0$.*

(4) *If f is μ-measurable, g is μ-integrable and $|f| \le g$ a.e.(μ), then f is μ-integrable.*

(5) *If the f_j, $j = 1, 2, \dots$, are μ-measurable and $f_j \to f$ a.e.(μ), then f is μ-
   measurable.*

(6) *f is measurable if and only if it is the pointwise a.e.(μ) limit of a sequence of
   step functions.*

Proof. The first point follows from the definition of $I^+(\Omega)$ respective Definition B.4.12.

To prove the second point, suppose $\varphi_j \in C_0(\Omega)$ and $\varphi_j \to 1$ pointwise in[5] Ω
and $C_0(\Omega) \ni \varphi_{n,j} \to f_n$ a.e.(μ) as $j \to \infty$. Then $\varphi_j F(\varphi_{1,j}, \dots, \varphi_{k,j}) \in C_0(\Omega)$ and
tends a.e.(μ) to $F(f_1, \dots, f_k)$, which is therefore measurable.

Most claims of the third point are just special cases of the previous point. To deal
with f/g we note that if $\varphi_j, \psi_j \in C_0(\Omega)$ and $\varphi_j \to f$, $\psi_j \to g$, then $C_0(\Omega) \ni$
$\varphi_j \overline{\psi_j}/(|\psi_j|^2 + 1/j) \to f/g$ a.e.(μ).

For the fourth point, suppose first that f is real-valued and h is bounded, continuous, strictly positive, and μ-integrable; see Exercise B.6.1. Suppose $C_0(\Omega; \mathbb{R}) \ni$
$\varphi_j \to f$ a.e.(μ) and define for $j, k = 1, 2, \dots$ the functions

$$\psi_{kj} = \max(-kh, \min(kh, \varphi_j)) \in C_0(\Omega; \mathbb{R}).$$

---

[5] Such functions exist since $\chi_\Omega \in I^+(\Omega)$.

As $j \to \infty$ $\psi_{kj} \to f_k = \max(-kh, \min(kh, f))$. Now $|\psi_{kj}| \le kh$ so by dominated convergence f_k is μ-integrable.

Since $|f_k| \le g$ and $f_k \to f$ as $k \to \infty$ dominated convergence shows that f is μ-integrable. If f is not real-valued we may apply the above to $\operatorname{Re} f$ and $\operatorname{Im} f$ to draw the desired conclusion.

To prove the fifth point note that by (2) f is μ-measurable if and only if $T(f) = f/(|f| + 1)$ is, since T is continuous on $\mathbb{C}$ with a continuous inverse. Now $T(f_j) \to T(f)$ a.e.(μ) and $|T(f_j)| \le 1$ so by dominated convergence $hT(f)$, with h as above, is μ-integrable. Thus $T(f)$ and therefore also f is μ-measurable.

One direction of the last point follows from (5) since a step function is integrable and thus measurable. The other direction is a consequence of the fact that functions in $C_0(\Omega)$ are uniformly continuous and thus uniform limits of a sequence of step functions. If f is measurable, $f_j \in C_0(\Omega)$, and $f_j \to f$ pointwise a.e.(μ) we may choose step functions g_j with $\|f_j - g_j\|_\infty < 1/j$. It follows that $g_j \to f$ pointwise a.e.(μ). $\qquad\square$

It will also be convenient to introduce the idea of a measurable set.

Definition B.6.3. Suppose μ is a Radon measure on Ω. A set $M \subset \Omega$ is called μ-*measurable* if the characteristic function χ_M is μ-measurable. If χ_M is μ-integrable its μ-measure is defined to be $\mu(M) = \mu(\chi_M)$. Otherwise $\mu(M)$ is undefined unless $\mu \ge 0$, in which case we set $\mu(M) = +\infty$.

A number of rules for calculating with measures of sets follow from those for measurable functions, but we shall not dwell on them except to mention that M is a μ-nullset precisely if M is μ-measurable and $|\mu|(M) = 0$. This is an immediate consequence of Lemma B.7.1 which we shall prove in the next section.

It is clear from Theorem B.6.2 and its implications for measurable sets that it is not easy to construct a function or set which is *not* measurable, since doing arithmetic or taking pointwise limits will not lead out of the class of measurable functions. One might therefore ask whether perhaps *all* subsets of $\mathbb{R}$ are μ-measurable. Not surprisingly, the answer depends on the measure. But even if it is possible to construct a non-measurable set, as it is, for example, for Lebesgue measure, this requires the use of the so-called *axiom of choice*. On the other hand, every set is measurable with respect to an atomic measure. We shall not discuss this any further here.

Theorem B.6.4. *If f is μ-integrable, then so is $|f|$.*

This means that the theory of Lebesgue–Stieltjes integration is only about *absolutely convergent* integrals.

Proof. For real-valued f we proved this in Theorem B.4.13. If f is not real-valued at least $|\operatorname{Re} f|$ and $|\operatorname{Im} f|$ are integrable, and $|f|$ is μ-measurable by Theorem B.6.2 (3). Since $|f| \le |\operatorname{Re} f| + |\operatorname{Im} f|$, the claim follows from Theorem B.6.2 (4). $\qquad\square$

Borel functions. In addition to the concept of a measurable function the older notion of a *Borel function* is sometimes useful, since such a function is measurable with respect to *every* Radon measure on the relevant interval. Note that functions in $C_0(\Omega)$ are measurable with respect to any Radon measure on Ω.

Definition B.6.5. The set of Borel functions $B(\Omega)$ on an open interval Ω is the smallest class of complex-valued functions measurable with respect to every Radon measure on Ω which contains $C_0(\Omega)$ and for which pointwise limits of sequences in $B(\Omega)$ are also in $B(\Omega)$.

If $\sigma \subset \mathbb{R}$ is a closed set the Borel functions $B(\sigma)$ on σ are those functions in $B(\mathbb{R})$ which have support in σ.

A Borel set in Ω is a subset of Ω for which the characteristic function is in $B(\Omega)$.

To see that the definition makes sense, note that the set of all complex-valued functions on Ω measurable with respect to all Radon measures on Ω contains $C_0(\Omega)$, so it is non-empty, and is closed under pointwise limits by Theorem B.6.2 (5). Thus $B(\Omega)$ is the intersection of the non-empty class of all sets of functions with these properties. Note that it is easy to see that any function measurable with respect to a measure μ and a.e.(μ) finite is a.e.(μ) equal to a Borel function (Exercise B.6.2).

Proposition B.6.6. *If $f \in B(\Omega)$ the sets $\{x \in \Omega : f(x) > c\}$ and $\{x \in \Omega : f(x) = c\}$ are Borel sets for any $c \in \mathbb{R}$.*

For the proof, see Exercise B.6.3.

Exercises

Exercise B.6.1 Suppose μ is a positive Radon measure on Ω. Show that one may find a bounded, continuous and strictly positive function $f \in L^1(\mu)$.

Hint: Write Ω as the union of a sequence of open intervals with compact closure in Ω and construct for each interval a non-negative continuous function with support equal to the closure of the interval and bounded by 1 like in the proof of Lemma B.4.14. Then add up appropriate multiples of these functions.

Exercise B.6.2 Show that the function which equals 0 except on rational numbers, where it equals 1, is a Borel function.

Also show that a function measurable(μ) and a.e.(μ) finite is a.e.(μ) equal to a Borel function.

Hint: Prove the second claim first for bounded real-valued functions. In this case, the given function is the pointwise a.e.(μ) limit of a bounded sequence $\varphi_j \in C_0$. Consider $\overline{\lim}\, \varphi_j$.

Exercise B.6.3 If f and g are Borel functions, show that the sets where $f < g$, $f \leq g$ and $f = g$ are all Borel sets.

B.7 Some Banach spaces

Suppose μ is a positive measure. If f, and thus by Theorem B.6.4 $|f|$, is μ-integrable, we define $\|f\|_1 = \mu(|f|)$. It is clear that $\|f\|_1 \geq 0$, $\|Cf\|_1 = |C| \|f\|_1$ for constants C and that $\|f + g\|_1 \leq \|f\|_1 + \|g\|_1$ if g is also μ-integrable since $|f + g| \leq |f| + |g|$. Thus $\|\cdot\|_1$ is a *semi-norm* on the linear space of μ-integrable functions. It is not a norm since we may have $\|f\|_1 = 0$ even if f is not identically 0. We therefore need the following important lemma.

Lemma B.7.1. *If $\mu \geq 0$ and $f \geq 0$ is μ-integrable, then $\mu(f) = 0$ if and only if $f = 0$ a.e.(μ).*

Proof. Clearly $f = 0$ a.e.(μ) implies $\mu(f) = \mu(0) = 0$. To prove the converse let $f_j = jf$ so that $\mu(f_j) = j\mu(f) = 0$. Now $\{f_j\}_1^\infty$ is a non-decreasing sequence of μ-integrable functions with $\lim \mu(f_j) = 0 < \infty$, so by monotone convergence $f_j \to g$, where g is μ-integrable. Thus g is finite a.e.(μ). But $g(x) = \infty$ as soon as $f(x) > 0$, so the lemma follows.   □

The result of Lemma B.7.1 is a strong reason to identify two functions that are equal a.e.(μ).[6] More precisely, two μ-measurable functions f, g are said to be μ-*equivalent* if $f = g$ a.e.(μ). If one of them is μ-integrable so is the other and they have the same integral and the same semi-norm.

Definition B.7.2. The normed linear space $L^1(\mu)$ is the set of equivalence classes of μ-integrable functions, where the norm of an equivalence class is $\|f\|_1$, f being any element of the equivalence class.

In order to avoid unnecessarily cumbersome notation it is standard practice not to distinguish between a function and the equivalence class it belongs to, so the elements of $L^1(\mu)$ are usually thought of as functions, but one needs to keep in mind that they are not really. For example, to say that a function in $L^1(\mu)$ is continuous means that there is an element of the equivalence class which is continuous. Only if there is a non-empty open set which is a μ-nullset will any equivalence class contain more than one continuous function.

The following important fact is an immediate consequence of the definition of μ-integrability.

Theorem B.7.3. *Given $u \in L^1(\mu)$ there exists a sequence $\varphi_j \in C_0(\Omega)$, $j = 1, 2, \ldots$, such that $\varphi_j \to u$ a.e.(μ) and $\lim \|u - \varphi_j\|_1 = 0$. The second claim means that the space $C_0(\Omega)$ is dense in $L^1(\mu)$.*

Proof. If $u = u_1 - u_2 + i(u_3 - u_4)$ a.e.(μ) with integrable $u_j \in I^+(\Omega)$, there are increasing sequences $\{\varphi_{k,j}\}_{j=1}^\infty$, $k = 1, 2, 3, 4$, in $C_0^+(\Omega)$ such that $\varphi_{k,j} \to u_k$ pointwise and $0 \leq \mu(u_k) - \mu(\varphi_{k,j}) \to 0$ as $j \to \infty$. Setting $\varphi_j = \varphi_{1,j} - \varphi_{2,j} + i(\varphi_{3,j} - \varphi_{4,j})$ it follows that $\varphi_j \to u$ a.e.(μ) and $\|u - \varphi_j\|_1 \leq \sum_{k=1}^4 (\mu(u_k) - \mu(\varphi_{kj})) \to 0$.   □

---

[6] In other words, we consider a quotient space.

In the space $L^1(\mu)$ the Cauchy convergence principle applies, i.e., if $\{f_j\}_1^\infty$ is a Cauchy sequence in $L^1(\mu)$ then it converges in norm to an element of $L^1(\mu)$, i.e., $L^1(\mu)$ is *complete*. Recall that a sequence $\{f_j\}_1^\infty$ in a normed space is called a Cauchy sequence if the following is true: For every $\varepsilon > 0$ there is a number N such that $\|f_n - f_m\|_1 < \varepsilon$ if $n \geq N$ and $m \geq N$.

Recall that a complete normed linear space is called a *Banach space*.

Theorem B.7.4. $L^1(\mu)$ *is a Banach space.*

Proof. Let $\{f_j\}_1^\infty$ be a Cauchy sequence in $L^1(\mu)$. For each $k \in \mathbb{N}$ we can then find an index $j_k \geq k$ such that $\|f_n - f_m\|_1 < 2^{-k}$ if $n \geq j_k$ and $m \geq j_k$. In particular $\|f_{j_k} - f_{j_{k+1}}\|_1 < 2^{-k}$. Thus

$$\sum_{k=1}^\infty \mu(|f_{j_k} - f_{j_{k+1}}|) \leq \sum_{k=1}^\infty 2^{-k} = 1 < \infty,$$

so Beppo Levi's theorem implies that $\sum_1^\infty |f_{j_k} - f_{j_{k+1}}|$ is μ-integrable. Thus the series converges absolutely a.e.(μ), and the function $f_{j_n} = f_{j_1} - \sum_1^{n-1}(f_{j_k} - f_{j_{k+1}})$ converges to a measurable function f. Since $|f| \leq |f_{j_1}| + \sum_1^\infty |f_{j_k} - f_{j_{k+1}}|$, where the right-hand side is μ-integrable, we also have $f \in L^1(\mu)$. Now

$$\|f - f_{j_n}\|_1 \leq \|\sum_{k=n}^\infty |f_{j_k} - f_{j_{k+1}}|\|_1 = \sum_{k=n}^\infty \|f_{j_k} - f_{j_{k+1}}\|_1 \leq \sum_{k=n}^\infty 2^{-k} \to 0$$

as $n \to \infty$. We also have $\|f - f_n\|_1 \leq \|f - f_{j_n}\|_1 + \|f_{j_n} - f_n\|_1 \to 0$ as $n \to \infty$ and this completes the proof. $\square$

Note that we have also shown that if $f_j \to f$ in $L^1(\mu)$, then there is a subsequence of $\{f_j\}_1^\infty$ converging pointwise a.e.(μ) to f. It is important to note that for diffuse measures the original sequence does not have to converge pointwise at a single point. This is for example the case for Lebesgue measure. Finally note that one interpretation of the two last theorems is that $L^1(\mu)$ is the completion of $C_0(\Omega)$ in the norm $\|\cdot\|_1$.

Another important space is the Hilbert space $L^2(\mu)$, where again μ is a positive Radon measure on Ω. This is the set of (equivalence classes of) μ-measurable functions f for which $|f|^2 \in L^1(\mu)$. If $f \in L^2(\mu)$ clearly f is finite a.e.(μ), $\mu(|f|^2) = 0$ if and only if $f = 0$ a.e.(μ), and $C_0(\Omega) \subset L^2(\mu)$.

If also $g \in L^2(\mu)$, then fg is μ-measurable and we have $fg \in L^1(\mu)$ since $|fg| \leq \frac{1}{2}(|f|^2 + |g|^2)$. Thus $|f + g|^2 = |f|^2 + |g|^2 + 2\,\mathrm{Re}\,f\bar{g}$ is also in $L^1(\mu)$ so that $f + g \in L^2(\mu)$. It is also clear that $Cf \in L^2(\mu)$ for constants C, so $L^2(\mu)$ is a linear space in which $\langle f, g \rangle := \mu(f\bar{g})$ defines a scalar product. The corresponding norm is $\|f\|_2 = \sqrt{\mu(|f|^2)}$. We obtain the following theorem.

Theorem B.7.5. $L^2(\mu)$ *is a linear space with scalar product in which* $C_0(\Omega)$ *is dense.*

Proof. It remains only to prove the density of $C_0(\Omega)$. We may assume f is real-valued since we may otherwise treat real and imaginary parts separately.

Assume now that $h \in L^1(\Omega)$ is continuous and strictly positive (Exercise B.6.1) and put $f_k = \max(-k\sqrt{h}, \min(k\sqrt{h}, f))$. Then f_k is measurable and tends a.e.(μ) to f as $k \to \infty$ so that $|f_k - f|^2 \to 0$ a.e.(μ). Now $|f_k - f|^2 \le |f|^2$ so $f_k - f$ and thus $f_k \in L^2(\Omega)$, and $\mu(|f_k - f|^2) \to 0$ by dominated convergence. It is thus enough if for any $k > 0$ we can find a sequence in $C_0(\Omega)$ converging in $L^2(\mu)$ to f_k.

Since f is measurable there is a sequence $\varphi_j \in C_0(\Omega)$ converging pointwise a.e.(μ) to f. Setting $\psi_j = \max(-k\sqrt{h}, \min(k\sqrt{h}, \varphi_j))$ we have $\psi_j \to f_k$ a.e.(μ) and $|\psi_j - f_k|^2 \le 2|\psi_j|^2 + 2|f_k|^2 \le 4h$ so, again by dominated convergence, we obtain $\mu(|\psi_j - f_k|^2) \to 0$ as $j \to \infty$. The proof is finished. $\qquad\square$

Theorem B.7.6. *The space $L^2(\mu)$ is complete, and thus a Hilbert space. Any sequence in $L^2(\mu)$ converging to f has a subsequence converging a.e.(μ) to f.*

The proof follows that of Theorem B.7.4 closely.

Proof. Suppose $f_1, f_2, \ldots$ is a Cauchy sequence in $L^2(\mu)$. We may then find integers $j_k \ge k$ such that $\|f_{j_k} - f_{j_{k+1}}\|_2 \le 2^{-k}$. Thus

$$\left\| \sum_{k=1}^n |f_{j_k} - f_{j_{k+1}}| \right\|_2 \le \sum_{k=1}^n \|f_{j_k} - f_{j_{k+1}}\|_2 \le 1 .$$

Since $\left(\sum_1^\infty |f_{j_k} - f_{j_{k+1}}| \right)^2$ is the increasing limit of a sequence in $L^1(\Omega)$ with bounded norm it converges in $L^1(\Omega)$ by monotone convergence. In particular, the sum is finite a.e.(μ).

By absolute convergence the series $f_{j_n} = f_{j_1} - \sum_1^{n-1}(f_{j_k} - f_{j_{k+1}})$ converges a.e.(μ) to a measurable function f. Since $|f - f_{j_1}|^2$ is bounded by a function in $L^1(\mu)$ we have $f - f_{j_1}$ and thus f in $L^2(\mu)$. Now

$$\|f - f_{j_n}\|_2 \le \left\| \sum_{k=n}^\infty |f_{j_k} - f_{j_{k+1}}| \right\|_2 = \sum_{k=n}^\infty \|f_{j_k} - f_{j_{k+1}}\|_2 \le \sum_{k=n}^\infty 2^{-k} \to 0$$

as $n \to \infty$. Finally $\|f - f_n\|_2 \le \|f - f_{j_n}\|_2 + \|f_{j_n} - f_n\|_2 \to 0$ as $n \to \infty$, which completes the proof. $\qquad\square$

We shall finally briefly mention the Banach space $L^\infty(\mu)$. A function f is μ-*essentially bounded* if there is a constant C such that $|f(x)| \le C$ a.e.(μ). Let $\|f\|_\infty$ be the greatest lower bound of all such constants C.

Since we can then find a sequence C_j decreasing to $\|f\|_\infty$, and the set where $|f(x)| > \|f\|_\infty$ is the union of those for which $|f(x)| > C_j$, we have $|f(x)| \le \|f\|_\infty$ a.e.(μ). Thus the norm $\|f\|_\infty$ is the smallest C for which $|f(x)| \le C$ a.e.(μ).

Definition B.7.7. The set of all equivalence classes of μ-essentially bounded, μ-measurable functions is denoted by $L^\infty(\mu)$.

Proposition B.7.8. $L^\infty(\mu)$ *is a Banach space with norm* $\|\cdot\|_\infty$, *and if* $f \in L^\infty(\mu)$, $g \in L^1(\mu)$, *then* $fg \in L^1(\mu)$ *and* $\|fg\|_1 \leq \|f\|_\infty \|g\|_1$. *Similarly, if* $g \in L^2(\mu)$, *then* $fg \in L^2(\mu)$ *and* $\|fg\|_2 \leq \|f\|_\infty \|g\|_2$.

Proof. It is clear that $L^\infty(\mu)$ is a linear space with norm $\|\cdot\|_\infty$, and since convergence in this space implies convergence a.e.(μ) every Cauchy sequence in $L^\infty(\mu)$ converges pointwise a.e.(μ) to an essentially bounded, μ-measurable function which is thus in $L^\infty(\mu)$.

If $f \in L^\infty(\mu)$ and $g \in L^1(\mu)$, then fg is measurable and $|fg| \leq \|f\|_\infty |g|$, which is integrable, so $fg \in L^1(\mu)$ and $\|fg\|_1 \leq \|f\|_\infty \|g\|_1$. The case $g \in L^2(\mu)$ is treated similarly. $\square$

It should be noted that, in contrast to the situation for L^1 and L^2, C_0 is a closed subspace of L^∞, but dense only for very special measures (Exercise B.7.2).

Exercises

Exercise B.7.1 Show that if μ is a positive Radon measure on Ω then step functions on Ω are dense both in $L^1(\mu)$ an $L^2(\mu)$.

Exercise B.7.2 Show that $C_0(\Omega)$ is a closed subspace of $L^\infty(\mu)$, but dense if and only if μ is a finite linear combination of Dirac measures on Ω.

B.8 Relations between measures

If f is locally μ-integrable we may define $\nu(\varphi) = \mu(\varphi f)$ for $\varphi \in C_0(\Omega)$. By Proposition B.7.8 this is a Radon measure since $|\nu(\varphi)| \leq C_K \sup |\varphi|$, where C_K is the $|\mu|$-integral of $|f|$ over a compact $K \subset \Omega$ containing $\operatorname{supp} \varphi$. We say that the measure ν is *absolutely continuous* with respect to μ or that ν is a measure with *base* μ, and write $\nu = f\mu$.

By Theorem B.4.15 there is a function h of locally bounded variation such that $\nu(\varphi) = \int \varphi \, dh$. This is therefore also true if φ is replaced by any function integrable with respect to h, in particular the function e_x of page 288. If $\mu(\varphi) = \int \varphi \, dg$ this shows that for h we may use $\int e_x f \, dg$. We obtain the following corollary, generalizing Theorem B.2.5.

Corollary B.8.1. *If g is locally of bounded variation, f locally integrable with respect to g and $c \in \Omega$, then $h(x) = \int_c^x f \, dg$ is[7] locally of bounded variation. Also, u is integrable with respect to h if and only if uf is integrable with respect to g and*

---

[7] As we shall see in Theorem C.13, the measure dh of which h is the distribution function is the same whether the endpoints x and c are included in the interval of integration or not.

$$v(u) = \int uf \, dg = \int u(x) \, d\left(\int\limits_c^x f \, dg \right).$$

There is a classical notion of *absolutely continuous functions* which is related to the notion of a *measure* absolutely continuous with respect to Lebesgue measure but not precisely the same.

A function is called *locally absolutely continuous*, if it is the distribution function of a measure which is absolutely continuous with respect to Lebesgue measure. This means that it is of the form $F(x) = C + \int_a^x f$ for some constant C and a locally integrable function f. It is easy to see that F is of (locally) bounded variation as well as (locally) uniformly continuous in Ω (Exercise B.8.1). We shall see in Corollary B.8.10 that F is differentiable a.e. in Ω with derivative f and in Theorem C.13 that the distributional derivative of F is f.

To study the concept of absolute continuity for measures in some detail we make the following definition.

Definition B.8.2. Two Radon measures μ, v on Ω are said to be *foreign* or *mutually singular*, denoted $\mu \perp v$, if there is a μ-nullset E such that $\Omega \setminus E$ is a v-nullset.

Of course E is measurable with respect to both v and μ, because χ_E is 1 a.e.(v) and 0 a.e.(μ). In fact, E can be chosen measurable with respect to *all* Radon measures on Ω, namely as a Borel set (see Definition B.6.5).

One might say that $\mu \perp v$ means that μ and v live on disjoint sets. The simplest non-trivial example is when μ is diffuse and v atomic, but there are also non-trivial diffuse measures foreign to Lebesgue measure.[8] We can now prove the following fundamental theorem.

Theorem B.8.3 (Lebesgue decomposition). *Suppose $\mu \geq 0$ and v are Radon measures on Ω. Then one may find a measure σ and a locally μ-integrable function g such that $\mu \perp \sigma$ and $v = g\mu + \sigma$. The measures $g\mu$ and σ are uniquely determined and $g \geq 0$, $\sigma \geq 0$ if $v \geq 0$.*

Although $g\mu$ is uniquely determined as a measure the function g is of course only determined a.e.(μ). Note that the assumption that $\mu \geq 0$ is superfluous (except for the last claim), as we shall show later.

Proof. To prove uniqueness, suppose $v = g_j\mu + \sigma_j$ where $\mu \perp \sigma_j$, with associated sets E_j according to Definition B.8.2. Now $E_1 \cup E_2$ is a μ-nullset and $(\Omega \setminus E_1) \cap (\Omega \setminus E_2) = \Omega \setminus E_1 \cup E_2$ a nullset for σ_1 and σ_2 and hence for $\sigma_1 - \sigma_2$. Since $(g_2 - g_1)\mu = \sigma_1 - \sigma_2$ it follows that $\Omega = (E_1 \cup E_2) \cup (\Omega \setminus E_1 \cup E_2)$ is a $\sigma_1 - \sigma_2$-nullset. Thus $\sigma_1 = \sigma_2$ and $g_1\mu = g_2\mu$.

To show existence, assume first that v is positive and *finite*, i.e., $v(\Omega) < \infty$ or $1 \in L^1(v)$. Since $0 \leq v \leq \mu + v$ it follows by Cauchy–Schwarz that if $u \in L^2(\mu + v)$ then

---

[8] This follows from the fact that there are non-trivial, non-decreasing continuous functions with derivative 0 a.e.; such a function may even be strictly increasing. See [211, Section 5.2.2].

$$|\nu(u)|^2 \le \nu(1)\nu(|u|^2) \le \nu(1)(\mu + \nu)(|u|^2),$$

so that ν is a bounded linear form on $L^2(\mu + \nu)$. By Riesz' representation theorem there exists an $f \in L^2(\mu + \nu)$ such that $\nu(u) = (\mu + \nu)(uf)$, at least for $u \in C_0(\Omega)$. That is, we have $\nu = f(\mu + \nu)$ or

$$(1 - f)\nu = f\mu.$$

Clearly $0 \le f \le 1$ a.e.$(\mu + \nu)$ since the measures involved are positive, so $\sum_0^{n-1} f^j$ is locally $\mu + \nu$-integrable. Multiplying both sides by this we obtain $(1 - f^n)\nu = \sum_1^n f^j \mu$. By monotone convergence the left-hand side converges to $(1 - \chi_E)\nu$ where $E = \{x \in \Omega : f(x) = 1\}$. Again using monotone convergence it follows that $g = \sum_1^\infty f^j$ is locally μ-integrable and $\nu = g\mu + \chi_E \nu$. Since $g = \infty$ in E this is a μ-nullset, and the proof is complete for positive and finite ν.

If $\nu \ge 0$ is not finite we may find a bounded, continuous function $h \in L^1(\nu)$ which is strictly positive (Exercise B.6.1). Thus $h\nu$ is positive and finite and we have $h\nu = g_1\mu + \sigma_1$ with positive σ_1, $g_1 \in L^1_{\mathrm{loc}}(\mu)$ and $\mu \perp \sigma_1$. Thus $\nu = g\mu + \sigma$ where $\sigma = \sigma_1/h$ and $g = g_1/h$ is locally μ-integrable. Finally, if ν is not positive it is a finite linear combination of positive measures, and adding up the corresponding formulas gives the desired result. $\square$

The following fundamental theorem is an almost immediate consequence of this.

Theorem B.8.4 (Radon–Nikodym). *A Radon measure ν on Ω is absolutely continuous with respect to a positive Radon measure μ on Ω, i.e., $\nu = g\mu$ for some locally μ-integrable g, if and only if every μ-nullset is also a ν-nullset.*

As in the case of the Lebesgue decomposition, the assumption $\mu \ge 0$ will be shown to be superfluous later.

Proof. If $\nu = g\mu$ it is clear that any μ-nullset is a ν-nullset.

Conversely, by Theorem B.8.3 we may write $\nu = g\mu + \sigma$ or $\nu - g\mu = \sigma$ where $\mu \perp \sigma$. But any μ-nullset is a nullset for the left-hand side and thus for σ. Since there is such a nullset the complement of which is also a nullset for σ it follows that $\sigma = 0$, which finishes the proof. $\square$

Note that if $0 \le \nu \le \mu$, then any μ-nullset is also a ν-nullset, so $\nu = g\mu$ for some locally μ-integrable g satisfying $0 \le g \le 1$ a.e.(μ). We give two corollaries.

Corollary B.8.5. *If μ is a real Radon measure on Ω, then $\mu_\pm = g_\pm |\mu|$ where $g_\pm$ are characteristic functions of disjoint μ-measurable sets the union of which is Ω. In particular $\mu_+ \perp \mu_-$.*

Proof. $0 \le \mu_\pm \le |\mu|$ so $\mu_\pm$ are absolutely continuous with respect to $|\mu|$. Thus $\mu_\pm = g_\pm |\mu|$ with non-negative and μ-measurable $g_\pm$.

Now $\mu = \mu_+ - \mu_- = (g_+ - g_-)|\mu|$. Since $\mu_\pm$ are the smallest measures giving a splitting of μ into a difference of positive measures we must have $\min(g_+, g_-) = 0$ a.e.(μ), since otherwise we could subtract the positive function $\min(g_+, g_-)$ from $g_\pm$, leaving their difference unchanged but making $\mu_\pm$ smaller.

We also have $|\mu| = (g_+ + g_-)|\mu|$ so $g_+ + g_- = 1$ a.e.(μ). Changing $g_\pm$ on μ-nullsets leaves $\mu_\pm$ unchanged but makes $g_\pm$ characteristic functions of disjoint sets, and $\mu_+ \perp \mu_-$.   □

The following corollary shows that the assumption $\mu \geq 0$ in Theorems B.8.3 and B.8.4 is superfluous.

Corollary B.8.6. *If μ is a Radon measure on Ω, then there is a μ-measurable function g with $|g| = 1$ such that $|\mu| = g\mu$.*

Proof. By the Radon–Nikodym theorem we have $\mu = f|\mu|$ with a locally μ-integrable f. We shall show that $|f| = 1$ a.e.(μ) so that $g = \bar{f}$ will do.

If $\varphi \in C_0(\Omega)$ we have $|\mu(\varphi)| = |f| \big| |\mu|(\varphi) \big| \leq |f| |\mu|(|\varphi|)$, so since $|\mu|$ is the smallest measure with $|\mu(\varphi)| \leq |\mu|(|\varphi|)$ the measure $(|f| - 1)|\mu|$ is positive.

On the other hand, if $\varphi \geq 0$, then $|f| |\mu|(\varphi) = |\mu(\varphi)| \leq |\mu|(\varphi)$, so that $(1 - |f|)|\mu|$ is also positive. It follows that $|f| = 1$ a.e.(μ).   □

Example B.8.7. A typical application of the Radon–Nikodym theorem is the following. Suppose we are given a sequence of Radon measures μ_j on Ω. We wish to find a positive Radon measure μ on Ω with respect to which all the given measures are absolutely continuous.

We start by finding bounded, continuous and strictly positive functions f_j in $L^1(|\mu_j|)$ (Exercise B.6.1). Multiplying f_j by an appropriate strictly positive constant we may assume that $|\mu_j|(f_j) \leq 2^{-j}$. The measure $\mu = \sum_1^\infty f_j|\mu_j|$ is then finite and any μ-nullset is a nullset for all terms and therefore for every μ_j.

According to the Radon–Nikodym theorem we may therefore find locally μ-integrable functions g_j such that $\mu_j = g_j\mu$ for all j.

The function g in the Lebesgue decomposition $\nu = g\mu + \sigma$ is often called the *Radon–Nikodym derivative* of ν with respect to μ. For one-dimensional measures this actually has a pointwise meaning, in the following sense.

Theorem B.8.8 (Differentiation of measures). *Suppose μ, ν are one-dimensional Radon measures and that $\nu = g\mu + \sigma$ is the Lebesgue decomposition of ν with respect to μ. Then*

$$\lim_{\Delta \to \{x\}} \frac{|\nu - g(x)\mu|(\Delta)}{\mu(\Delta)} = 0 \qquad (B.8.1)$$

for a.e.(μ) $x \in \Omega$, where $\Delta \subset \Omega$ denotes non-trivial intervals containing x for which both endpoints tend to x. In particular, for all such x

$$\lim_{\Delta \to \{x\}} \frac{\nu(\Delta)}{\mu(\Delta)} = g(x) \ .$$

When μ is Lebesgue measure a point for which (B.8.1) holds is often called a *Lebesgue point* for ν, especially if ν is absolutely continuous with respect to μ. In the latter case (B.8.1) takes the form

$$\frac{1}{|\Delta|} \int_\Delta |g(t) - g(x)| \, dt \to 0 \text{ as } |\Delta| \to 0 \text{, while } x \in \Delta \,.$$

Here $|\Delta|$ denotes the length of Δ and x is also said to be a Lebesgue point for the function g. From this it follows that $\frac{d}{dx} \int_a^x g = g(x)$ a.e.

The proof uses a version of the *Hardy maximal theorem*. If ν and μ are *positive* measures in Ω we define the *maximal function* of ν with respect to μ as $\nu^\mu(x) = \sup_{\Delta \ni x} \frac{\nu(\Delta)}{\mu(\Delta)}$. Here Δ denotes non-trivial subintervals of Ω containing x.

Lemma B.8.9 (Hardy maximal theorem). *If μ and ν are positive measures in an interval Ω and $\omega_s = \{x : \nu^\mu(x) > s\}$ for $s > 0$, then $\mu(\omega_s) \leq 3\nu(\Omega)/s$.*

Proof. Assume $\nu(\Omega) < \infty$ since otherwise there is nothing to prove. It is clear that ω_s is the union of the family $\mathcal{F}$ of all open intervals $I \subset \Omega$ such that $\nu(I) > s\,\mu(I)$. Incidentally, this shows that ω_s is open so that ν^μ is lower semi-continuous. Suppose K is a compact subset of ω_s, so that $\mathcal{F}$ is an open cover of K. Thus there is a finite subcover $\mathcal{F}_K$.

We shall find disjoint intervals I_j in $\mathcal{F}_K$ and other open intervals $J_j \subset \Omega$ with $\mu(J_j) \leq 3\mu(I_j)$ such that $\mu(K \setminus \cup J_j) = 0$. From this the lemma follows since

$$\mu(K) \leq \sum \mu(J_j) \leq 3 \sum \mu(I_j) < 3s^{-1} \sum \nu(I_j) \leq 3s^{-1} \nu(\Omega) \,,$$

and we may choose K with $\mu(K)$ arbitrarily close to $\mu(\omega_s)$.

If $\mu(K) = 0$ there is nothing to prove, and if $\mu(K) > 0$ we choose I_1 to be an interval in $\mathcal{F}_K$ of maximal μ-measure.

If we have already found $I_1, \ldots, I_j$ and there are intervals in $\mathcal{F}_K$ disjoint from all of them with positive μ-measure, let I_{j+1} be such an interval of maximal μ-measure. This process will eventually stop since there are only finitely many intervals in $\mathcal{F}_K$, say we obtain the disjoint intervals $I_1, \ldots, I_k$ in $\mathcal{F}_K$.

Now, if $I_j = (a, b)$ put $J_k = (A, B)$ where $A = a$ if $\mu(\{a\}) > \mu(I_j)$ and otherwise $A = \inf\{t \in \Omega : \mu((t, a]) \leq \mu(I_j)\}$. Similarly, $B = b$ if $\mu(\{b\}) > \mu(I_j)$ and $B = \sup\{t \in \Omega : \mu[b, t) \leq \mu(I_j)\}$ if not. It follows that $\mu(J_j) \leq 3\mu(I_j)$.

The union of those $I \in \mathcal{F}_K$ for which $\mu(I) = 0$ is a μ-nullset. If x is outside this set but in K we have $x \in I$ for some $I \in \mathcal{F}_K$ with $\mu(I) > 0$. It is clear that I intersects some I_j since otherwise $\mu(I) = 0$ by construction. Let j be the smallest index such that $I \cap I_j \neq \emptyset$. Then $\mu(I) \leq \mu(I_j)$ so that $x \in I \subset J_j$. Thus $\mu(K \setminus \cup J_j) = 0$ and the proof is complete. $\qquad\square$

Proof (Theorem B.8.8). The second statement follows from the first since

$$\left| \frac{\nu(\Delta)}{\mu(\Delta)} - g(x) \right| = \left| \frac{\nu(\Delta) - g(x)\mu(\Delta)}{\mu(\Delta)} \right| \leq \frac{|\nu - g(x)\mu|\,(\Delta)}{|\mu(\Delta)|} \,.$$

If we can prove (B.8.1) for positive μ it follows in general, since $\mu = f\,|\mu|$ with $|f| = 1$ so that $|\mu(\Delta)/|\mu|\,(\Delta)| \to 1$ a.e. (μ) and $\nu = gf\,|\mu| + \sigma$. So, assume μ positive, and we may also assume $\nu(\Omega)$ finite, so that $g \in L^1(\mu)$, since the claim of the theorem is local.

First assume that $\sigma = 0$, i.e. $\nu = g\mu$. If we also assume that g is continuous at x the proof is elementary; in fact $\frac{|g\mu - g(x)\mu|(\Delta)}{\mu(\Delta)} \leq \sup_\Delta |g - g(x)|$ and therefore tends to 0 with the length of Δ.

If g is not continuous we may, for any $\varepsilon > 0$, find $\varphi \in C_0(\Omega)$ such that $\mu(|g - \varphi|) < \varepsilon$. Setting $h = g - \varphi$ we have just seen that

$$\overline{\lim_{\Delta \to \{x\}}} \frac{|g\mu - g(x)\mu|(\Delta)}{\mu(\Delta)} = \overline{\lim_{\Delta \to \{x\}}} \frac{|h\mu - h(x)\mu|(\Delta)}{\mu(\Delta)} \tag{B.8.2}$$

and the right-hand side can be estimated by $(|h|\mu)^\mu(x) + |h(x)|$. Clearly, we also have $\mu(\{t : |h(t)| > s\}) \leq \frac{1}{s}\mu(|h|)$. Lemma B.8.9 therefore shows that the set where the left-hand side of (B.8.2) is $> s$ has μ-measure less than $4\varepsilon/s$. Since $\varepsilon > 0$ is arbitrary this set is actually a μ-nullset. Taking the union of these sets as s runs through a sequence $\to 0$ proves the case when ν is absolutely continuous with respect to μ.

It remains to prove that $\lim_{\Delta \to \{x\}} \frac{\sigma(\Delta)}{\mu(\Delta)} = 0$ except for x in a μ-nullset if μ and σ are mutually singular. We can then find two disjoint sets M and N with $M \cup N = \Omega$ and such that $\mu(N) = |\sigma|(M) = 0$. Let $\mathcal{O} \subset \Omega$ be an open set containing M and define a positive measure $\sigma_{\mathcal{O}}$ by $\sigma_{\mathcal{O}} = \chi_{\mathcal{O}}|\sigma|$.

By Lemma B.8.9, $\mu(\{x \in \Omega : \sigma_{\mathcal{O}}^\mu(x) > s\}) \leq 3\sigma_{\mathcal{O}}(\Omega)/s$. If E_s denotes the set of those x for which $\overline{\lim}_{\Delta \to \{x\}} \frac{|\sigma(\Delta)|}{\mu(\Delta)} > s$, then $E_s \cap M$ is a subset of $\mathcal{O}$ and $\sigma_{\mathcal{O}}^\mu(x) > s$ for $x \in E_s$. Since N is a μ-nullset we have

$$\mu(E_s) = \mu(E_s \cap M) \leq 3\sigma_{\mathcal{O}}(\Omega)/s = 3|\sigma|(\mathcal{O})/s .$$

Now $\mathcal{O}$ may be chosen to make the right-hand side arbitrarily small, so it follows that E_s is a μ-nullset for every $s > 0$. The claim now follows by letting s run through a sequence tending to 0. $\qquad\square$

We have the following corollary.

Corollary B.8.10. *Any function of locally bounded variation is differentiable a.e. with a locally Lebesgue integrable derivative, and if the function is locally absolutely continuous it is an integral function of its derivative.*

Proof. If f is of locally bounded variation on Ω the corresponding Stieltjes integral generates a measure ν, and if μ is Lebesgue measure we may write $\nu = g\mu + \sigma$ for some locally integrable g and measure σ singular with respect to Lebesgue measure.

The differentiation theorem shows that for a.e. x the function f has derivative $g(x)$. If f is locally absolutely continuous we have $\sigma = 0$ and $f(x) = f(a) + \int_a^x df = f(a) + \int_a^x g(t)\,dt$ if $a \in \Omega$. $\qquad\square$

Exercises

Exercise B.8.1 Prove that a (locally) absolutely continuous function is of (locally) bounded variation and (locally) uniformly continuous.

Hint: Note that $f = |f| - (|f| - f)$. For the second claim, approximate by C_0 and show uniform convergence.

Exercise B.8.2 Suppose f is a locally absolutely continuous function with respect to Lebesgue measure and let φ be continuous.

Show that if a is a Lebesgue point for f, then it is also a Lebesgue point for φf.

Exercise B.8.3 Show that a strictly increasing, locally absolutely continuous function f has a locally absolutely continuous inverse if and only if $f' > 0$ a.e.

Hint: Use Theorem B.8.3.

B.9 Product measures

One may define Radon measures in an open set $\Omega \subset \mathbb{R}^n$, just as when Ω is a real interval, as bounded linear forms on $C_0(\Omega)$. Almost everything we have done then has a counterpart with essentially the same proofs, although an extension of Theorem B.8.8 is only true if μ is Lebesgue measure (or at least a measure behaving sufficiently 'uniformly' in different directions). The sets corresponding to the intervals Δ also need to have restrictive properties. We do not have much occasion to use higher-dimensional measures, but in a few cases need to consider so-called product measures made up from two one-dimensional measures.

Suppose $\Omega = \omega_1 \times \omega_2$ is a two-dimensional open interval and assume μ_1 and μ_2 are Radon measures on ω_1 respectively ω_2. If $\varphi \in C_0(\Omega)$, then $\omega_1 \ni x \mapsto \mu_2(\varphi(x, \cdot))$ is in $C_0(\omega_1)$. Continuity follows by dominated convergence as x converges to some point in ω_1, and that the support is compact in ω_1 is clear. Thus μ_1 is defined on this function and we define $\mu_1 \otimes \mu_2(\varphi) = \mu_1(\mu_2(\varphi(x, \cdot)))$, an iterated integral.

If $\operatorname{supp} \varphi \subset I_1 \times I_2$, where $I_j \subset \omega_j$ are compact intervals, then $|\mu_1 \otimes \mu_2(\varphi)| \leq C_1(I_1)C_2(I_2) \sup |\varphi|$, where C_j are the bounds associated with μ_j. Thus $\mu = \mu_1 \otimes \mu_2$ is a bounded linear form on $C_0(\Omega)$, in other words a Radon measure on Ω. It is called the *product measure* of μ_1 and μ_2.

The definition of a product measure is via iterated integration on $C_0(\Omega)$. We shall see that this formula generalizes to the case of μ-integrable functions, and also that the order of integration is irrelevant, in other words $\mu_1 \otimes \mu_2 = \mu_2 \otimes \mu_1$. We may then restrict ourselves to consider only positive measures, with no loss of generality since every measure is a finite linear combination of positive measures.

Lemma B.9.1. *Suppose μ_1 and μ_2 are positive and $f \in I^+(\Omega)$. Then $y \mapsto f(x, y)$ is for every fixed $x \in \omega_1$ in $I^+(\omega_2)$, $x \mapsto \mu_2(\varphi(x, \cdot))$ is in $I^+(\omega_1)$ and $\mu(f) = \mu_1(\mu_2(f(x, \cdot)))$.*

Furthermore, $f \in I^+(\Omega)$ is μ-integrable if and only if $y \mapsto f(x, y)$ is μ_2-integrable for a.e.(μ_1) fixed $x \in \omega_1$ and $x \mapsto \mu_2(\varphi(x, \cdot))$ is μ_1-integrable.

Proof. Suppose $\varphi_j \in C_0^+(\Omega)$ increases to f. Keeping $x \in \omega_1$ fixed $\varphi_j(x, \cdot) \in C_0^+(\omega_2)$ increases to $f(x, \cdot)$, which is therefore in $I^+(\omega_2)$, and $\mu_2(\varphi_j(x, \cdot))$ increases to $\mu_2(f(x, \cdot))$ for all $x \in \omega_1$ by definition. Since $\mu_2(\varphi_j(x, \cdot))$ clearly is in $C_0^+(\omega_1)$ this means that $\mu_2(f(x, \cdot)) \in I^+(\omega_1)$ and $\mu_1(\mu_2(f(x, \cdot)))$ is the limit of $\mu_1(\mu_2(\varphi_j(x, \cdot))) = \mu(\varphi_j)$. It follows that $\mu(f) = \mu_1(\mu_2(f(x, \cdot)))$.

From this it follows that if f is μ-integrable, then $x \mapsto \mu_2(\varphi(x, \cdot))$ is μ_1-integrable, and therefore finite a.e.(μ_1). For these values of x the function $y \mapsto f(x, y)$ is therefore μ_2-integrable. □

We have the following immediate consequence.

Corollary B.9.2. *Suppose N is a μ-nullset. Then the trace $N_x = \{y \in \omega_2 : (x, y) \in N\}$ is a μ_2-nullset for a.e.(μ_1) $x \in \omega_1$.*

Proof. A nullset is a set such that there is an integrable function in I^+ which is infinite on the set, so this follows from Lemma B.9.1. □

We can now prove *Fubini's theorem*.

Theorem B.9.3 (Fubini). *Suppose $f \in L^1(\mu)$. Then the function $y \mapsto f(x, y)$ is in $L^1(\mu_2)$ for a.e.(μ_1) x, the function $x \mapsto \mu_2(f(x, \cdot))$ is in $L^1(\mu_1)$ and $\mu(f) = \mu_1(\mu_2(f(x, \cdot)))$.*

Proof. If $f = u - v$ a.e.(μ), then for a.e.(μ_1) fixed $x \in \omega_1$ we have $f(x, \cdot) = u(x, \cdot) - v(x, \cdot)$ a.e.(μ_2), according to Corollary B.9.2. From Lemma B.9.1 we conclude: For a.e.(μ_1) fixed $x \in \omega_1$ the function $y \mapsto f(x, y)$ is a.e.(μ_2) the difference of two integrable functions in $I^+(\omega_2)$, so $y \mapsto f(x, y)$ is in $L^1(\omega_2)$. Furthermore, $x \mapsto \mu_2(f(x, \cdot))$ is in $L^1(\mu_1)$, since it is a.e.(μ_1) the difference $\mu_2(u(x, \cdot)) - \mu_2(v(x, \cdot))$ of two integrable functions in $I^+(\omega_1)$. The formula for iterated integration now follows from Lemma B.9.1. □

Given two measures μ_1, μ_2 we may define, in addition to $\mu_1 \otimes \mu_2$, the product measure $\mu_2 \otimes \mu_1$, the difference being that the order of the iterated integrations is interchanged. It is a very important fact that the order of integration is irrelevant.

Theorem B.9.4. $\mu_1 \otimes \mu_2 = \mu_2 \otimes \mu_1$.

Proof. Since $\varphi \in C_0(\Omega)$ is uniformly continuous it may be uniformly approximated by step functions, so it is enough to prove that $\mu_1 \otimes \mu_2(f) = \mu_2 \otimes \mu_1(f)$ for step functions f.

Since a step function is a finite linear combination of characteristic functions for bounded intervals it is in fact enough to prove this when f is such a characteristic function, i.e., of the form $f(x, y) = \chi_1(x)\chi_2(y)$, where χ_j is the characteristic function of a compact subinterval of ω_j. But in this case it is clear that both iterated integrals equal $\mu_1(\chi_1)\mu_2(\chi_2)$. □

In practice one often has an iterated integral where one would like to change the order of integration. Fubini's theorem tells us that this is allowed if we know that the integrand is integrable with respect to the product measure, but often all one knows is that the iterated integral exists, and this is not enough even to guarantee the existence of the iterated integral in the reverse order. The following theorem is very useful in such situations.

Theorem B.9.5 (Tonelli). *Given* $\mu = \mu_1 \otimes \mu_2$ *suppose that* f *is* μ-*measurable and the iterated integral* $\mu_1(\mu_2(f(x, \cdot)))$ *converges absolutely, i.e., for a.e.*(μ_1) *fixed* x *the function* $y \mapsto |f(x, y)| \in L^1(|\mu_2|)$ *and* $x \mapsto |\mu_2|(|f(x, \cdot)|) \in L^1(|\mu_1|)$. *Then* f *is integrable with respect to* $\mu_1 \otimes \mu_2$ *so that* $\mu_1(\mu_2(f(x, \cdot))) = \mu_2(\mu_1(f(\cdot, y)))$.

So, one may apply Fubini's theorem and change the order of integration if the integrand and measures are positive or the iterated integral is *absolutely* convergent and the integrand measurable with respect to the product measure.

Proof. We may assume μ_1 and μ_2 are positive, and by Theorem B.6.2 (4) we only need to verify that $|f|$ is μ-integrable.

Let χ_j, $j = 1, 2, \ldots$, be characteristic functions of a sequence of compact subintervals of Ω with union Ω and put $f_j = \min(|f|, j\chi_j)$ so that pointwise $0 \le f_j \le |f|$ and f_j increases to $|f|$. Furthermore, by Theorem B.6.2 (4), f_j is μ-integrable since $0 \le f_j \le j\chi_j$. Finally

$$\mu(f_j) = \mu_1(\mu_2(f_j(x, \cdot))) \le \mu_1(\mu_2(|f(x, \cdot)|)) < \infty,$$

so by monotone convergence $|f|$ is μ-integrable. $\qquad\qquad\qquad\qquad\qquad\square$

B.10 Notes and remarks

The theory of integration presented here is close to the (more general) *Daniell integral*. Our development is close to the theory of integration according to F. Riesz, which may be found for example in the classical functional analysis text by F. Riesz and B. Sz.-Nagy [191] or in an appendix to J. Weidmann [215]. Riesz used step functions instead of C_0 as the domain of a measure. Our choice of C_0 makes it quite obvious that *Schwartz distributions* are generalizations of Radon measures. This approach makes integrals of functions the central objects and not the measures of sets in a σ-algebra, although the approaches are of course more or less equivalent. But for most applications in analysis viewing a Radon measure as a functional on a set of test functions is more direct and to the point.

Appendix C
Schwartz distributions

The theory of Schwartz distributions is an essential tool for dealing with linear partial differential equations, but is usually not of much importance in the context of ordinary differential equations. However, in dealing with equations with low coefficient regularity it can be very helpful, and in this appendix we shall discuss those aspects of the theory of distributions of one variable that are useful to us, particularly in the later parts of Chapter 5. The fundamental reason for this usefulness is the fact that all normal spaces of functions defined on some open interval and their duals are dense subsets of the distributions on the interval.

Test functions

The support of a continuous function is the closure of the complement of the set where it vanishes. We denote by $C_0^k(\Omega)$ the linear space of k times continuously differentiable, complex-valued functions with compact support in an open interval Ω, where k is a non-negative integer or ∞. It is clear that if we extend a function in $C_0^k(\Omega)$ by 0 outside Ω we obtain a function in $C_0^k(\mathbb{R})$, so we may always think of $C_0^k(\Omega)$ as a subspace of $C_0^k(\mathbb{R})$. We define convergence in $C_0^\infty(\Omega)$ as follows.

Definition C.1. A sequence in $C_0^\infty(\Omega)$ tends to zero precisely if

(1) there is a fixed compact subset of Ω that contains the supports of all the elements in the sequence,

(2) the sequence and all its derivatives tend uniformly to 0.

Of course we say that $\varphi_j \to \varphi$ if $\varphi_j - \varphi \to 0$. Similarly we define convergence in $C_0^k(\Omega)$ for finite k, assuming (2) just for derivatives of order $\leq k$.

As a topological vector space $C_0^k(\Omega)$ is fairly complicated. It is *locally convex*, not a *Fréchet space* but the *inductive limit* of the Fréchet[1] spaces $C_0^k(K_j)$ where

---

[1] $C_0^k(K_j)$ (but not $C_0^k(\Omega)$) is a Banach space for finite k.

© Springer Nature Switzerland AG 2020
C. Bennewitz et al., *Spectral and Scattering Theory for Ordinary Differential Equations*,
Universitext, https://doi.org/10.1007/978-3-030-59088-8

$K_j \subset \Omega$ is compact and K_j increases to Ω. However, these facts are not really used in distribution theory. In most cases all one needs to know is the definition above of a convergent sequence.

We define a Cauchy sequence $\{\varphi_m\}_1^\infty$ in $C_0^k(\Omega)$, $0 \leq k \leq \infty$, as a sequence in $C_0^k(\Omega)$ for which the supports of all elements are subsets of a fixed compact subset of Ω and such that for every ν, $0 \leq \nu \leq k$, we have $\varphi_n^{(\nu)} - \varphi_m^{(\nu)} \to 0$ uniformly as $n, m \to \infty$. The exact meaning of the last statement is: Given any $\varepsilon > 0$ there exists an N such that $\left|\varphi_m^{(\nu)}(x) - \varphi_n^{(\nu)}(x)\right| < \varepsilon$ for every $x \in \Omega$ if $m > N$ and $n > N$.

Proposition C.2. *The space $C_0^k(\Omega)$, $0 \leq k \leq \infty$, is complete, i.e., every Cauchy sequence has a limit in the space.*

Proof. If $\{\varphi_m\}_1^\infty$ is a Cauchy sequence with supports in a compact $K \subset \Omega$ it follows that for any $x \in \Omega$ and any ν, $0 \leq \nu \leq k$, the numerical sequence $\{\varphi_m^{(\nu)}(x)\}_1^\infty$ is a Cauchy sequence and therefore has a limit, which we denote by $\varphi^{(\nu)}(x)$.

Letting $n \to \infty$ in the formula above we obtain $\left|\varphi_m^{(\nu)}(x) - \varphi^{(\nu)}(x)\right| \leq \varepsilon$ if $m > N$, so that $\varphi_m^{(\nu)}$ tends uniformly to $\varphi^{(\nu)}$ as $m \to \infty$. It follows that $\varphi^{(\nu)}$ is continuous with support in K and letting $m \to \infty$ in $\varphi_m^{(\nu-1)}(x) = \int_{-\infty}^x \varphi_m^{(\nu)}$ that $\varphi^{(\nu)}$ is the derivative of $\varphi^{(\nu-1)}$ if $\nu > 0$. It is now clear that setting $\varphi = \varphi^{(0)}$ we have $\varphi_m \to \varphi$ in $C_0^k(\Omega)$. $\qquad\square$

We shall show that even in $C_0^\infty(\Omega)$ there are many functions and then first construct a particular element of $C_0^\infty(\mathbb{R})$. To this end, let $g(x) = e^{1/x}$ for $x < 0$ and $g(x) = 0$ for $x \geq 0$, so that g is continuous. We need the following proposition.

Proposition C.3. *The function $x \mapsto g(x^2 - 1)$ is in $C_0^\infty(\mathbb{R})$, is even and > 0 in the interval $(-1, 1)$ and vanishes outside.*

Proof. It is clearly enough to show that $g^{(k)}(x) = 0$ for $x \geq 0$ and $g^{(k)}(x) = p_k(1/x)e^{1/x}$ if $x < 0$, where p_k is a polynomial. But this is clear if $k = 0$ and supposing it true for k we obtain $g^{(k+1)}(x) = 0$ for $x > 0$ and $g^{(k+1)}(x) = -(p_k(1/x) + p_k'(1/x))x^{-2}e^{1/x}$ for $x < 0$ so that $p_{k+1}(t) = -t^2(p_k(t) + p_k'(t))$ is a polynomial.

Also $(g^{(k)}(x) - g^{(k)}(0))/x = 0$ for $x > 0$ and equals $p_k(1/x)x^{-1}e^{1/x}$ for $x < 0$. The latter quantity tends to 0 as $x \to 0$ from the left, so it follows by induction that g is infinitely differentiable. $\qquad\square$

Setting $\rho(x) = g(x^2 - 1)/(\int_{-\infty}^\infty g(y^2 - 1)\,dy)$ and $\rho_\varepsilon(x) = \rho(x/\varepsilon)/\varepsilon$ for $\varepsilon > 0$ the function ρ_ε is even, strictly positive in the interval $(-\varepsilon, \varepsilon)$ and 0 outside, and is infinitely differentiable with $\int_{-\infty}^\infty \rho_\varepsilon = 1$. The function ρ_ε is a so-called *mollifier*, used to smooth irregular functions by *convolution* without changing them too much.

The convolution $f * g$ of two functions f and g is defined on $\mathbb{R}$ as

$$f * g(x) = \int_{-\infty}^\infty f(y)g(x - y)\,dy$$

whenever the integral makes sense. A change of variable $t = x - y$ shows that $f * g = g * f$ and we have the following lemma.

Lemma C.4. *Suppose $f \in C_0^k(\Omega)$. Then $f_\varepsilon = f * \rho_\varepsilon$ is ≥ 0 if $f \geq 0$ and in $C_0^\infty(\Omega)$ for small $\varepsilon > 0$ with support contained[2] in $[-\varepsilon, \varepsilon] + \text{supp } f$, and f_ε converges in $C_0^k(\Omega)$ to f as $\varepsilon \to 0$.*

Proof. Differentiating under the integral sign we have

$$f_\varepsilon^{(\nu)}(x) = \int\limits_{-\infty}^{\infty} f(y) \rho_\varepsilon^{(\nu)}(x - y)\, dy$$

so f_ε is infinitely differentiable. Now $f_\varepsilon(x)$ can only be different from 0 if there exists a $y \in \text{supp } f$ so that $x - y \in \text{supp } \rho_\varepsilon$, which proves the statement of support, and since $\int_{-\infty}^\infty \rho_\varepsilon = 1$ we have, if $\nu \leq k$,

$$\left| f_\varepsilon^{(\nu)}(x) - f^{(\nu)}(x) \right| = \left| \int\limits_{-\infty}^{\infty} \rho_\varepsilon(y)(f^{(\nu)}(x - y) - f^{(\nu)}(x))\, dy \right|$$

$$\leq \int\limits_{|y| < \varepsilon} \rho_\varepsilon(y) \left| f^{(\nu)}(x - y) - f^{(\nu)}(x) \right| dy \leq \sup_{|y| \leq \varepsilon} \left| f^{(\nu)}(x - y) - f^{(\nu)}(x) \right|.$$

If $f \in C_0^k(\Omega)$ then f and its first k derivatives are uniformly continuous, so the lemma follows. $\square$

The lemma implies that if $0 \leq j < k \leq \infty$, then $C_0^k(\Omega)$ is continuously and *densely* embedded[3] in $C_0^j(\Omega)$. Note that since $C_0(\Omega)$ is dense in $L^1(\mu)$ and in $L^2(\mu)$ according to Theorems B.7.3 and B.7.5 for any positive Radon measure μ on Ω it follows that $C_0^\infty(\Omega)$ is also dense in both $L^1(\mu)$ and $L^2(\mu)$. The functions in $C_0^\infty(\Omega)$, and sometimes also those in $C_0^k(\Omega)$ for finite k, are often called *test functions*.

Distributions

A *distribution* in $\mathcal{D}'(\Omega)$, where $\Omega \subset \mathbb{R}$ is an open interval, is a continuous linear form on $C_0^\infty(\Omega)$. Thus $\mathcal{D}'(\Omega)$ is a linear space and the (topological) *dual* of $C_0^\infty(\Omega)$. Schwartz denoted $C_0^\infty(\Omega)$ by $\mathcal{D}(\Omega)$, which is the reason for the notation $\mathcal{D}'(\Omega)$.

In $\mathcal{D}'(\Omega)$ we use the weak* topology on linear forms. This means 'pointwise convergence' or that $u_j \to u$ in $\mathcal{D}'(\Omega)$ means that $u_j(\varphi) \to u(\varphi)$ for every $\varphi \in C_0^\infty(\Omega)$. It follows from general principles that $\mathcal{D}'(\Omega)$ is complete, although we shall

---

[2] For sets $A, B \subset \mathbb{R}$ we define $A + B = \{x + y : x \in A, y \in B\}$.

[3] The embedding is even *compact*, i.e., a sequence bounded in $C_0^k(\Omega)$ has a subsequence converging in $C_0^j(\Omega)$, as follows from the Arzela–Ascoli theorem 1.4.4.

not use it. The topology on $C_0^\infty(\Omega)$ is very strong, so that the topology of $\mathcal{D}'(\Omega)$ is correspondingly weak.

Distributions in $\mathcal{D}'(\Omega)$ are precisely the bounded linear forms on $C_0^\infty(\Omega)$, in the following sense.

Proposition C.5. *A linear form u on $C_0^\infty(\Omega)$ is continuous if and only if it is bounded in the following sense:*

For every compact $K \subset \Omega$ there is a constant C_K and an index $j = j(K)$ such that

$$|u(\varphi)| \le C_K \sum_{\nu=0}^{j} \sup |\varphi^{(\nu)}| \tag{C.1}$$

for every $\varphi \in C_0^\infty(\Omega)$ with $\operatorname{supp} \varphi \subset K$.

Proof. Suppose u is a linear form on $C_0^\infty(\Omega)$ which is not bounded. This means that for some compact $K \subset \Omega$ and every j there is a test function φ_j with support in K such that

$$|u(\varphi_j)| > j \sum_{\nu=0}^{j} \sup |\varphi_j^{(\nu)}|.$$

Multiplying φ_j by a constant we may assume that $u(\varphi_j) = 1$ for all j. We then obtain $\sup |\varphi_j^{(\nu)}| < 1/j$ as soon as $j \ge \nu$. This means that $\varphi_j \to 0$ in $C_0^\infty(\Omega)$. Now, for all j we have $u(\varphi_j) = 1$, so u is not continuous. That a bounded form is continuous is obvious. $\square$

If $j(K)$ may be chosen to be independent of the compact $K \subset \Omega$ one says that u has *finite order*. The *order* of u is then the smallest j for which an estimate (C.1) remains valid for every compact $K \subset \Omega$. We denote the set of distributions of order $\le j$ on Ω by $\mathcal{D}_j'(\Omega)$. The notation $\mathcal{D}'^j(\Omega)$ is common in the literature but seems very unsightly. Note that (C.1) means that every distribution is *locally* of finite order.

Proposition C.6. *A distribution of order k extends uniquely to a bounded linear form on $C_0^k(\Omega)$.*

Proof. By Lemma C.4 any $\varphi \in C_0^k(\Omega)$ is the limit in $C_0^k(\Omega)$ of a sequence $\{\varphi_j\}_1^\infty$ in $C_0^\infty(\Omega)$. Now (C.1) holds for $\varphi_n - \varphi_m$ which implies that $\{u(\varphi_j)\}_1^\infty$ is a Cauchy sequence (of numbers). We define $u(\varphi)$ as the corresponding limit.

If $\{\psi_j\}_1^\infty$ is another sequence in $C_0^\infty(\Omega)$ converging in $C_0^k(\Omega)$ to φ, then $u(\varphi_j) - u(\psi_j) = u(\varphi_j - \psi_j) \to 0$ so the limit is independent of the particular sequence used, and taking limits of both sides it is clear that (C.1) continues to hold for $\varphi \in C_0^k(\Omega)$. $\square$

It follows that $\mathcal{D}_k'(\Omega)$ is the dual of $C_0^k(\Omega)$. Thus, a Radon measure is the same thing as a distribution of order zero, and we shall call a distribution of order 0 a measure from now on. In particular, if f is a locally integrable function with respect to Lebesgue measure in Ω, then $u(\varphi) = \int_\Omega \varphi f$ is a distribution of order 0. Clearly any other function which equals f a.e. yields the same distribution, which therefore depends only on the equivalence class of f with respect to Lebesgue measure. However, two different equivalence classes cannot give rise to the same distribution. This follows from the uniqueness, according to Theorem B.8.3, of the Lebesgue decomposition of the measure $f(x)\,dx$ with respect to Lebesgue measure.

If two locally integrable functions yield the same distribution their difference is therefore a.e. zero so they belong to the same equivalence class. We shall therefore not distinguish between a locally integrable function and the distribution it defines, and we shall use the same notation for the function and the corresponding distribution.

If $\omega \subset \Omega$ is open any distribution $u \in \mathcal{D}'(\Omega)$ has a restriction to $\mathcal{D}'(\omega)$ defined by applying u only to elements of $C_0^\infty(\omega)$. If the restriction of u to ω is zero u is said to vanish on ω. The *support* supp u of a distribution u is the complement of the union of all open sets on which it vanishes. Thus supp u is always a closed set (relative to Ω).

Theorem C.7. *Suppose* $\varphi \in C_0^\infty(\Omega)$ *and* $u \in \mathcal{D}'(\Omega)$ *such that* supp $\varphi \cap$ supp $u = \emptyset$. *Then* $u(\varphi) = 0$.

Proof. Every $x \in$ supp φ has a neighborhood $\omega(x)$ in which u vanishes. Choose $\psi_x \in C_0^\infty(\omega(x))$ non-negative such that $\psi_x(x) > 0$. Then the interiors of the supports of ψ_x give an open cover of supp φ. Since supp φ is compact there is a finite subcover, say given by the interiors of the supports of $\psi_{x_1}, \ldots, \psi_{x_n}$.

The function $\psi = \sum \psi_{x_j}$ is > 0 on supp φ, so setting $\varphi_j = \varphi \psi_{x_j}/\psi$ we have $\varphi_j \in C_0^\infty(\omega(x_j))$ so $u(\varphi_j) = 0$ and $\sum \varphi_j = \varphi$. Thus $u(\varphi) = \sum u(\varphi_j) = 0$. $\quad\square$

The set of compactly supported elements of $\mathcal{D}'(\Omega)$ is denoted $\mathcal{E}'(\Omega)$. Elements of $\mathcal{E}'(\Omega)$ are of finite order and extend uniquely to bounded linear forms on $C^\infty(\Omega)$; see Exercise C.2. A distribution u is called *positive* if $u(\varphi) \geq 0$ for all non-negative $\varphi \in C_0^\infty(\Omega)$.

Theorem C.8. *A positive distribution is a positive Radon measure.*

Proof. Suppose $K \subset \Omega$ is a compact interval and $\psi \in C_0^\infty(\Omega)$ is non-negative and ≥ 1 on K (Exercise C.1). If $\varphi \in C_0^\infty(\Omega)$ has support in K, then $\psi \sup |\varphi| + \mathrm{Re}(e^{i\theta}\varphi) \geq 0$ for every real θ. In particular, if $\theta = 0$ and φ real it follows that $u(\psi) \sup |\varphi| + u(\varphi) \geq 0$ so u is real.

In general we have $u(\psi) \sup |\varphi| + \mathrm{Re}(e^{i\theta}u(\varphi)) \geq 0$, and choosing θ so that $e^{i\theta}u(\varphi) = -|u(\varphi)|$ we obtain

$$|u(\varphi)| \leq u(\psi) \sup |\varphi| \ .$$

Thus u is a Radon measure. By Lemma C.4 it follows that extending the domain to $C_0(\Omega)$ u is still positive. $\quad\square$

Every distribution has a derivative.

Definition C.9. Given a distribution $u \in \mathcal{D}'(\Omega)$ its derivative $u' \in \mathcal{D}'(\Omega)$ is given by $u'(\varphi) = -u(\varphi')$.

If $u \in C^1(\Omega)$ and $\varphi \in C_0^\infty(\Omega)$ integration by parts shows that $\int u' \varphi = - \int u \varphi'$ so the definition is an extension of the usual derivative defined when u is a smooth function. Clearly u' is a distribution which may be of order 1 or 0 if u is of order 0. Otherwise differentiation increases the order by one. Note that differentiation is a continuous linear operation on $C_0^\infty(\Omega)$ and therefore also on $\mathcal{D}'(\Omega)$.

Every distribution also has a primitive.

Theorem C.10. *Given any distribution* $u \in \mathcal{D}'(\Omega)$ *there exists a* primitive $U \in \mathcal{D}'(\Omega)$ *of* u, *i.e.,* U *is a distribution such that* $U' = u$.

As we shall see in the next theorem the primitive is not unique but is determined up to an additive constant, as in the case of functions.

Proof. Fix $\psi \in C_0^\infty(\Omega)$ with $\int \psi = 1$ so that

$$\Phi(x) = \int_{-\infty}^x \varphi - \int_{-\infty}^\infty \varphi \int_{-\infty}^x \psi$$

is in $C_0^\infty(\Omega)$ whenever φ is.

If $u \in \mathcal{D}'(\Omega)$, define $U(\varphi) = -u(\Phi)$. It is easily checked that this is a distribution since $\sup |\Phi| \le |K| (1 + \int_{-\infty}^\infty |\psi|) \sup |\varphi|$, where $|K|$ is the length of a compact interval $K \subset \Omega$ containing $\operatorname{supp} \varphi$. Furthermore $\sup \left| \Phi^{(\nu)} \right| \le \sup \left| \varphi^{(\nu-1)} \right| + |K| \sup \left| \psi^{(\nu-1)} \right| \sup |\varphi|$ for $\nu > 0$. Since $\varphi(x) = \int_{-\infty}^x \varphi'$ for test functions φ we also have $U'(\varphi) = -U(\varphi') = u(\varphi)$ so that U is a primitive of u.   □

The proof shows that if the order of u is k, then U has order $\max(0, k-1)$, so that locally any distribution is the derivative to some order of a Radon measure. As we shall see in Theorem C.12 a Radon measure is the derivative of a function of locally bounded variation, which in turn is the derivative of a continuous function. Thus every distribution is locally the derivative to some order of a continuous function.

Theorem C.11 (du Bois-Reymond). *Suppose* $u \in \mathcal{D}'(\Omega)$ *and that* $u^{(k)} = 0$. *Then* u *is a polynomial of degree* $< k$.

Proof. Consider first the case $k = 1$. Given $\varphi \in C_0^\infty(\Omega)$ let Φ be defined as in the previous proof. Then if $u' = 0$

$$0 = u'(\Phi) = -u(\Phi') = u(\psi) \int \varphi - u(\varphi) .$$

It follows that $u(\varphi) = \int C\varphi$ where $C = u(\psi)$ so u is the constant C.

The proof is completed by induction, since if we know that $u^{(k)} = 0$ we have just proved that $u^{(k-1)}$ is some constant C. But then $v = u - Cx^{k-1}/(k-1)!$ satisfies $v^{(k-1)} = 0$.   □

In particular, if U_1 and U_2 are different primitives of the same distribution they can only differ by a constant since $U = U_1 - U_2$ has derivative 0.

Since a function of locally bounded variation is a linear combination of non-decreasing functions it is locally Riemann integrable (see also Theorem B.5.4) and may therefore be viewed as a distribution. We have the following important theorem.

Theorem C.12. *A function of locally bounded variation has a distributional derivative which is of order 0, i.e. it is a Radon measure, in fact the Stieltjes integral generated by the function.*

Conversely, any primitive of a distribution of order 0 is a function of locally bounded variation.

Note that by Corollary B.8.1 this implies that if f is locally integrable and F an integral function of f, then the distributional derivative $F' = f$. As we noted before, a locally integrable function determines a distribution which depends only on the Lebesgue equivalence class of the function. Thus two functions of locally bounded variation determine the same distribution precisely if they agree except at discontinuities. In view of du Bois-Reymond's lemma this proves the uniqueness claim on page 288 just after Theorem B.4.15.

Proof. If f is of locally bounded variation and $\varphi \in C_0^\infty$ we have

$$f'(\varphi) = -f(\varphi') = - \int \varphi' f = - \int f \, d\varphi = \int \varphi \, df \, ,$$

using Theorem B.2.5 and integrating by parts. This is a distribution of order 0.

Conversely, reading the calculation from the right, a distribution of order zero is the distributional derivative of any function f generating the corresponding Radon measure. By Theorem C.11 any other primitive differs as a distribution by a constant from f, but is pointwise only determined a.e. $\qquad\square$

Products of distributions

So far the calculus of distributions presents few difficulties. However, if one wants to multiply distributions, as in considering linear differential equations with distributional coefficients, there are difficulties. These are apparent already from the fact that the product of two locally integrable functions may not be locally integrable. However, note that if f is locally integrable and $g \in L^\infty_{loc}$, or if $f, g \in L^2_{loc}$, then f, g and fg are locally integrable, and thus distributions.

We may also multiply any distribution by a function of class C^∞.

Proposition C.13. *Suppose $u \in \mathcal{D}'(\Omega)$ and $f \in C^\infty(\Omega)$. Define fu by $fu(\varphi) = u(f\varphi)$. Then fu is a distribution, of at most the same order as u if this is finite, and the product rule $(fu)' = f'u + fu'$ applies for its derivative.*

Proof. Clearly $f\varphi \in C_0^\infty(\Omega)$, so $u(f\varphi)$ is well defined. Applying the Leibniz rule and using that all derivatives of f are bounded on $\operatorname{supp}\varphi$ it is clear that fu is a distribution, of at most the same order as u if this is finite. Finally

$$(fu)'(\varphi) = -fu(\varphi') = -u(f\varphi') = -u((f\varphi)') + u(f'\varphi) = fu'(\varphi) + f'u(\varphi) \,. \qquad \square$$

It is clear from the proof that we may use the same definition to multiply a distribution of order k by a function in $C^k(\Omega)$. However, if u is of order k and $f \in C^k(\Omega)$ the product rule for the derivative may fail, since the products $f'u$ and fu' may not be defined as distributions. One may of course use the product rule to differentiate the product of a distribution in $\mathcal{D}'_k(\Omega)$ by a function in $C^{k+1}(\Omega)$.

It follows easily from the above that linear ordinary differential equations with smooth coefficients have no more distributional than classical solutions, which is why distribution theory is not important for ordinary differential equations, except when the coefficients are sufficiently singular. This is in sharp contrast to the case of even constant coefficient, linear partial differential equations.

One cannot extend the multiplication of a distribution by a smooth function to an associative product of arbitrary distributions. Schwartz gave the example of the products $(uf)\delta_0$ and $u(f\delta_0)$ where u is the distributional derivative of $\log|x|$ (Exercise C.3) and $f(x) = x$. Here $f \in C^\infty$ as are $f\delta_0 = 0$ and $uf = 1$ so the first product is δ_0 and the second 0, as is easily verified.

Nevertheless, we shall have to discuss a few more cases where a product may be defined. The idea is always that lack of smoothness in one factor is compensated by an appropriate degree of smoothness in the other factor. For distributions in one dimension this is straightforward, but in higher dimensions this leads to the introduction of the *wave front set* of a distribution, which we shall not discuss here.

As in Corollary B.8.1 one may multiply a measure μ by a locally μ-integrable function f. The result is another measure, but a locally μ-integrable function is not necessarily locally Lebesgue integrable or even defined a.e. with respect to Lebesgue measure, so f may not be a distribution.

However, if f is of locally bounded variation then by Theorem B.5.4 it is locally μ-integrable for any μ, including Lebesgue measure. We obtain the following proposition.

Proposition C.14. *Suppose μ is a Radon measure and f locally of bounded variation on Ω. Then μ and f are well defined as distributions and the product $f\mu$ defined as above is a Radon measure on Ω.*

Note carefully, however, that knowing f only as a distribution does not necessarily define $f\mu$ uniquely, since changing the value of f at any point which is not a μ-nullset[4] will change the distribution $f\mu$. If $\mu = g'$, where g is of locally bounded variation, then f must be unambiguously defined at any point where g has different left- and right-hand limits to make $f\mu$ well defined. We must therefore introduce some convention for choosing a unique element of the equivalence class of f.

There are several possible such conventions, and we mention three. First of all, left- and right-hand limits of f exist everywhere. We may therefore use the *left-continuous* version f_- of f, defined as its left-hand limit everywhere, the *right-continuous* version f_+, using instead the right-hand limit, or the *balanced* version $\frac{1}{2}(f_+ + f_-)$. Often the balanced version of f is the most useful.

---

[4] More generally, changing f in a Lebesgue nullset which is not a μ-nullset.

Differentiating the product $f\mu$ the product rule does not in general apply, since the products $f'\mu$ and $f\mu'$ may not be defined as distributions. The most symmetric case, in that we require the same regularity of both factors, where the product rule *does* apply seems to be the following. It is important since it leads to more general forms of the integration by parts formula of Theorem B.1.3, although we shall not use them in this book.

Theorem C.15. *Suppose f, g are of locally bounded variation in Ω. Then the product fg is defined as a distribution and the product rules $(fg)' = f'g_+ + f_-g'$ and, if f, g are balanced, $(fg)' = f'g + fg'$ are valid.*

Proof. By (3) page 279 fg is of locally bounded variation so it is locally integrable and therefore a distribution. It will be enough to prove $(fg)'(\varphi) = f'g(\varphi) + fg'(\varphi)$ if $\varphi \in C_0^\infty(a, b)$ for any $[a, b] \subset \Omega$. We may then temporarily modify f and g to be constant outside (a, b) so we may assume they are of bounded variation on $\mathbb{R}$.

Let $\psi \in C_0^\infty(\mathbb{R})$ with $\int \psi = 1$ and for $\varepsilon > 0$ set $\psi_\varepsilon(x) = \psi(x/\varepsilon)/\varepsilon$. If $f_\varepsilon = f * \psi_\varepsilon$, then it follows as in the proof of Lemma C.4 that $f_\varepsilon \in C^\infty(\mathbb{R})$ so $(f_\varepsilon g)' = f_\varepsilon'g + f_\varepsilon g'$. Now $f_\varepsilon(x) = \int f(x - y)\psi_\varepsilon(y)\, dy = \int f(x - \varepsilon z)\psi(z)\, dz$, setting $z = y/\varepsilon$.

By dominated convergence we therefore get $f_\varepsilon \to f^*$ pointwise as $\varepsilon \downarrow 0$, where $f^* = tf_- + (1 - t)f_+$ and $t = \int_0^\infty \psi$. Note that $|f_\varepsilon| \le \sup|f| \int |\psi|$, so again by dominated convergence $(f_\varepsilon g)'(\varphi) = -\int g f_\varepsilon \varphi' \to -\int g f \varphi' = (fg)'(\varphi)$ and

$$f_\varepsilon g'(\varphi) = \int f_\varepsilon \varphi \, dg \to \int f^* \varphi \, dg = f^* g'(\varphi)$$

as $\varepsilon \downarrow 0$. Similarly

$$f_\varepsilon' g(\varphi) = g(f_\varepsilon' \varphi) = \int g(x)\varphi(x) \int f(y)\psi_\varepsilon'(x - y)\, dy\, dx$$

$$= -\int f(y) \frac{d}{dy} \int g(x)\varphi(x)\psi_\varepsilon(x - y)\, dx\, dy = f'((g\varphi) * \check{\psi}_\varepsilon) ,$$

where $\check{\psi}_\varepsilon(x) = \psi_\varepsilon(-x)$. Thus $(g\varphi) * \check{\psi}_\varepsilon \to g_* \varphi$ pointwise and boundedly, with support in $\varepsilon \operatorname{supp} \psi + \operatorname{supp} \varphi$, where $g_* = (1 - t)g_- + tg_+$. Thus $f_\varepsilon' g(\varphi) \to f'g_*(\varphi)$ as $\varepsilon \downarrow 0$ and we obtain $(fg)' = f'g_* + f^*g'$.

If $\operatorname{supp} \psi \subset \mathbb{R}_+$ we obtain $t = 1$ and the first formula follows. Similarly, if ψ is even we have $t = 1/2$ so f^* and g_* are balanced and the second formula follows. $\square$

These formulas lead immediately to integration by parts formulas for Lebesgue–Stieltjes integrals. For example, for $u = \chi_{[a,b]}$ the formula $(fg)'(u) = f'g(u) + fg'(u)$ for balanced f and g gives

$$\int_{(a,b)} f \, dg = f_-(b)g_-(b) - f_+(a)g_+(a) - \int_{(a,b)} g \, df . \qquad (C.2)$$

We shall give one more result on multiplication of distributions. If $n \in \mathbb{Z}_+$ we denote by $H_{\mathrm{loc}}^n(\Omega)$ distributions $u \in \mathcal{D}'(\Omega)$ for which $u^{(n)} \in L_{\mathrm{loc}}^2(\Omega)$. All lower derivatives are then continuous functions and therefore also belong to $L^2_{loc}(\Omega)$. Similarly we denote by $H^{-n}_{loc}(\Omega)$ the distributions in $\mathcal{D}'(\Omega)$ which are the nth derivative of a function in $L^2_{loc}(\Omega)$. It is clear that these are linear spaces, and that $H^k_{loc}(\Omega) \subset H^j_{loc}(\Omega)$ if $k > j$. We shall prove the following theorem (see also Exercise C.4).

Theorem C.16. *Suppose* $f \in H^n_{loc}(\Omega)$ *and* $u \in H^{-n}_{loc}(\Omega)$. *Then the product* fu *is defined as a distribution in* $H^{-n}_{loc}(\Omega)$. *If* $f \in H^{n+1}_{loc}$ *then the product formula* $(fu)' = f'u + fu'$ *is valid.*

For the proof we need the following lemma.

Lemma C.17. *Suppose* f, v *are distributions such that the products* fv *and* $f'v$ *are defined as distributions. Then also* fv' *is defined as a distribution and the product rule applies for the derivative of* fv.

Proof. Define $fv' = (fv)' - f'v$. By assumption this is a distribution. Clearly this definition is consistent with the product rule.   □

Proof (Theorem C.16). For $n = 1$ we have $u = v'$ where $v \in L^2_{loc}(\Omega)$ so since f is continuous and $f' \in L^2_{loc}(\Omega)$ we have $fv \in L^2_{loc}(\Omega)$ and $f'v$ locally integrable. Thus fv and $f'v$ are distributions. By Lemma C.17 $fu = fv' = (fv)' - f'v$ is well defined, and $(fv)'$ is in $H^{-1}_{loc}(\Omega)$, as is $f'v$ since it is locally integrable. Thus $fu \in H^{-1}_{loc}(\Omega)$.

We may now proceed by induction, assuming the claim for $n = k$. For $n = k + 1$ we then have f and f' in $H^k_{loc}(\Omega)$ and $u = v'$ where $v \in H^{-k}_{loc}(\Omega)$. Thus $f'v$ and fv are in $H^{-k}_{loc}(\Omega)$ so that $fu = fv' = (fv)' - f'v \in H^{-k-1}_{loc}(\Omega)$. This completes the induction.

The last claim is proved as in Theorem C.13, since clearly all products involved are well defined.   □

Exercises

Exercise C.1 Show that given a compact set $K \subset \Omega$ there exists a $\psi \in C^\infty_0(\Omega)$ with $\psi(x) = 1$ for $x \in K$.

Hint: Convolution with the characteristic function of a larger compact.

Exercise C.2 Prove that a compactly supported distribution u has finite order. If we provide $C^k(\Omega)$ with the topology of locally uniform convergence for all derivatives of order $\leq k$, show that a compactly supported distribution of order k has a unique extension to a bounded linear form on $C^k(\Omega)$. In particular all distributions in $\mathcal{E}'(\Omega)$ are bounded linear forms on $C^\infty(\Omega)$.

Hint: Find $\psi \in C^\infty_0$ which is 1 on a larger compact set than supp u and write $\varphi \in C^\infty_0$ as $\psi\varphi + (1 - \psi)\varphi$.

Exercise C.3 Show that the distributional derivative of $u(x) = \log|x|$ is given by the *principal value integral*

$$u'(\varphi) = \int\limits_{-a}^{a} \frac{\varphi(x) - \varphi(0)}{x} = \text{p.v.} \int \frac{\varphi(x)}{x}$$

if $\text{supp}\,\varphi \subset [-a, a]$ and $\text{p.v.} \int g = \lim_{\varepsilon \to 0} \int_{|x| > \varepsilon} g(x)\, dx$.

Exercise C.4 Show that the product rule for differentiation applies for uv if $u \in H_{\text{loc}}^n$ and $v \in H_{\text{loc}}^{1-n}$.

C.1 Notes and remarks

We have only presented a very limited view of the theory of distributions, suitable for our needs. In a more comprehensive treatment one must consider distributions of several variables and the Fourier transform as it applies to distributions. See for example the first chapter of Hörmander [109] or Friedlander [75]. A very comprehensive theory may be found in Hörmander [110] and a somewhat different viewpoint in Gel'fand and Shilov [81].

Appendix D
Ordinary differential equations

Here we shall prove some standard existence and uniqueness results for ordinary differential equations for functions with values in $\mathbb{C}^n$. We are primarily interested in linear equations, but particularly in Section 6.1 occasionally deal with non-linear equations.

We will also prove some standard results on analyticity of solutions of linear, homogeneous equations with analytic coefficients, including the most basic results on equations with *regular singular points*.

First order systems

Let (a, b) be an open, real interval and $f : (a, b) \times \mathbb{C}^n \to \mathbb{C}^n$. We shall consider a differential equation of the form $u' = f(\cdot, u)$, and shall assume:

(1) $(a, b) \ni t \mapsto f(t, u)$ is locally integrable for every $u \in \mathbb{C}^n$,

(2) f satisfies a *Lipschitz condition*

$$\|f(t, u) - f(t, v)\| \le M(t)\|u - v\|$$

for all $u, v \in \mathbb{C}^n$, where M is a locally integrable positive function.

This means that the map $\mathbb{C}^n \ni u \mapsto f(t, u)$ is a Lipschitz continuous map with values in $L^1_{\text{loc}}((a, b); \mathbb{C}^n)$.

Lemma D.1. *If $(a, b) \ni t \mapsto u(t) \in \mathbb{C}^n$ is measurable and locally bounded, then $t \mapsto f(t, u(t))$ is locally integrable.*

Proof. This is clear if u is a step function, since $f(t, u(t))$ is then a finite sum of terms which are characteristic functions of intervals with compact closure in (a, b) multiplied by $f(t, v)$ for some fixed $v \in \mathbb{C}^n$.

If u is not a step function we may find a sequence $\{u_j\}_1^\infty$ of step functions converging a.e. to u by Theorem B.6.2 (6). We then have

© Springer Nature Switzerland AG 2020

C. Bennewitz et al., *Spectral and Scattering Theory for Ordinary Differential Equations*, Universitext, https://doi.org/10.1007/978-3-030-59088-8

$$\|f(t, u_j(t)) - f(t, u(t))\| \leq M(t)\|u_j(t) - u(t)\| \, ,$$

so that $f(t, u_j(t)) \to f(t, u(t))$ a.e., which therefore is measurable. If u is locally bounded we may assume a common bound for all u_j, and then $f(t, u(t))$ is locally integrable by dominated convergence.   □

We shall consider the initial value problem

$$\begin{cases} u' = f(\cdot, u) \, , \\ u(c) = u_0 \, , \end{cases} \tag{D.1}$$

where $c \in (a, b)$ and $u_0 \in \mathbb{C}^n$. The sense in which the $\mathbb{C}^n$-valued function u should satisfy the initial value problem is traditionally taken to be the following: The entries of u should be locally absolutely continuous functions such that $u(c) = u_0$ and the differential equation is satisfied a.e. One could also require the equation to be satisfied in distributional sense. For $f(\cdot, u)$ to be defined as a $\mathbb{C}^n$-valued distribution one should then at least require the elements of u to be in L^{∞}_{loc} so that $f(\cdot, u)$ is locally integrable by Lemma D.1. Thus the equation implies that u' must be locally integrable. It follows that a distributional solution u would be locally absolutely continuous, so the two interpretations would give exactly the same solutions.

To prove the existence of a unique solution it turns out to be convenient to consider an integral equation equivalent to (D.1). The integral equation is

$$u(t) = u_0 + \int_c^t f(\cdot, u) \, . \tag{D.2}$$

Integration of $\mathbb{C}^n$-valued functions are carried out componentwise. We are looking for a solution u with components that are at least in L^{∞}_{loc} so $f(\cdot, u)$ is locally integrable. Thus a solution is locally absolutely continuous and satisfies the initial condition. Differentiating we find than $u' = f(\cdot, u)$ a.e. Conversely, if u is locally absolutely continuous and satisfies (D.1) we may integrate from c to x, which gives (D.2), so the two equations are equivalent.

We shall prove the existence of a solution of (D.2) by a method called *successive approximations*. Define inductively for $j = 0, 1, \ldots,$

$$u_{j+1}(x) = u_0 + \int_c^x f(\cdot, u_j) \, .$$

Now $f(\cdot, u_0)$ is locally integrable on (a, b) so u_1 is locally absolutely continuous on (a, b). It follows by Lemma D.1 that $f(\cdot, u_1)$ is locally integrable in (a, b). By induction, all u_j are well defined and have the same properties. We put $h(t) = \int_c^t M$ and shall prove that for $t \geq c$

$$\|u_{j+1}(t) - u_j(t)\| \le \frac{h^j(t)}{j!} \int\limits_c^t \|f(\cdot, u_0)\|, \ j = 0, 1, 2, \dots \qquad (D.3)$$

To do this we need the triangle inequality $\|\int_a^b f\| \le \int_a^b \|f\|$, valid for vector-valued integrable functions f; see Exercise D.1. The inequality (D.3) is then clear for $j = 0$, and assuming it true for $j = k - 1$ we obtain

$$\|u_{k+1}(t) - u_k(t)\| \le \int\limits_c^t \|f(\cdot, u_k) - f(\cdot, u_{k-1})\| \le \int\limits_c^t M\|u_k - u_{k-1}\|$$

$$\le \frac{1}{(k-1)!} \int\limits_c^t \Big(\int\limits_c^s \|f(\cdot, v)\| \Big) h^{k-1}(s) M(s) \, ds \le \frac{h^k(t)}{k!} \int\limits_c^t \|f(\cdot, u_0)\| \ .$$

It follows that each component of $u_n = u_0 + \sum_{j=0}^{n-1}(u_{j+1} - u_j)$ converges locally uniformly, so if the limit is u it follows that u is continuous. Similar estimates prove this for $a < t \le c$. Now

$$\|f(t, u_j(t)) - f(t, u(t))\| \le M(t)\|u_j(t) - u(t)\| \ ,$$

where $\|u_j(t) - u(t)\|$ tends locally uniformly to zero, so $f(\cdot, u_j)$ tends pointwise to $f(\cdot, u)$. Let j be so large that $\|u_j - u\| \le 1$ in I, where I is a compact subinterval of (a, b) containing c. Then $\|f(t, u_j)\| \le \|f(t, u)\| + M(t)$ in I, so by dominated convergence it follows that u satisfies the equation (D.2) in (a, b).

We have shown the existence part of the following theorem.

Theorem D.2. *For every $u_0 \in \mathbb{C}^n$ the equation (D.2) has a unique solution defined in (a, b) which is locally absolutely continuous and depends continuously on the initial value u_0, locally uniformly in t.*

To prove uniqueness we shall use the following important lemma.

Lemma D.3 (Gronwall). *Suppose f and g are locally integrable and non-negative scalar functions in (a, b) and C a non-negative constant. Then, if $c \in (a, b)$ and $f(x) \le C + \left|\int_c^x gf\right|$ in (a, b) it follows that for $a < x < b$*

$$f(x) \le C \exp\Big(\Big|\int\limits_c^x g\Big|\Big) .$$

Proof. We shall carry out the proof for $x \ge c$ and leave the case $x \le c$ to the reader. Put $h(x) = C + \int_c^x gf$ so that $h' = gf \le gh$ in $[c, b)$. Multiplying by the non-negative function $\exp(-\int_c^x g)$ we obtain $(h(x) \exp(-\int_c^x g))' \le 0$. Integrating over (c, x) and dividing by the exponential we obtain $f(x) \le h(x) \le C \exp(\int_c^x g)$. $\qquad\square$

Proof (Theorem D.2). It remains to prove uniqueness and continuous dependence on the initial value. Suppose $u(\cdot, v)$ is a solution with initial value v and put $U(t) = \|u(t, v) - u(t, w)\|$. We then obtain, for $c \leq t < b$,

$$U(t) \leq \|v - w\| + \int_c^t \|f(\cdot, u(\cdot, v)) - f(\cdot, u(\cdot, w))\| \leq \|v - w\| + \int_c^t UM$$

and similarly for $a < t \leq c$. Thus Gronwall's inequality (Lemma D.3) shows that

$$\|u(t, v) - u(t, w)\| \leq \|v - w\| \exp\left|\int_c^t M\right|,$$

so the solution satisfies a locally uniform Lipschitz condition with respect to its initial value. In particular, u is uniquely determined by its initial value and depends continuously on it, locally uniformly in t. $\square$

Similar arguments may be used to prove continuous dependence of the solution jointly on the initial point, initial value and the function f, but we will not need this.

Linear first order systems

In discussing existence and uniqueness for solutions of the initial value problem for a Sturm–Liouville equation $-(pu')' + qu = wf$ it is convenient to write the equation as an equivalent first order 2×2 system $U' = AU + F$, where A is a 2×2 matrix with entries which are locally integrable functions, and F a similar matrix of type 2×1. This is done by setting $A = \begin{pmatrix} 0 & 1/p \\ q & 0 \end{pmatrix}$ and $F = \begin{pmatrix} 0 \\ -wf \end{pmatrix}$. The two equations are then easily seen to be equivalent by setting $U = \begin{pmatrix} u \\ pu' \end{pmatrix}$. Local integrability of the elements in A is clearly equivalent to the standard assumptions that $1/p$ and q are locally integrable, local integrability of F the same as local integrability of wf, and the existence of a solution U which is locally absolutely continuous equivalent to the existence of u which is locally absolutely continuous for which pu' is also locally absolutely continuous. Giving initial values for u and pu' at some point c is the same as giving an initial value at c for U. We shall therefore first discuss linear first order systems.

Let (a, b) be an open, real interval and suppose A is an $n \times n$ matrix-valued function with entries that are locally integrable functions in (a, b). Similarly, let F be an $n \times 1$ matrix-valued function with such entries, and consider the initial value problem

$$\begin{cases} u' = Au + F, \\ u(c) = u_0, \end{cases} \tag{D.4}$$

where $c \in (a, b)$ and u_0 is a fixed $n \times 1$ matrix. This system is called *homogeneous* if $F \equiv 0$ and is a special case of (D.1), where $f(\cdot, u) = Au + F$ so that

$$\|f(t, u) - f(t, v)\| = \|A(t)(u - v)\| \le \|A(t)\| \|u - v\| .$$

The assumptions of Theorem D.2 are therefore satisfied and there is a unique solution of (D.4).

Before moving on, let us note that it is also possible to consider the case when F is an $n \times n$ matrix-valued function with locally integrable entries. In this case we are looking for a solution u which is an $n \times n$ matrix-valued function and u_0 is a fixed $n \times n$ matrix. However, u is clearly a solution in this sense precisely if each column of u satisfies the equation, with F and u_c replaced by the corresponding columns. The two kinds of initial value problems are therefore essentially equivalent. We will stay with the case of $n \times 1$ solutions.

Corollary D.4. *The set of solutions of* (D.4) *for* $F = 0$ *is an n-dimensional linear space. Any basis for the solutions is pointwise a basis for* $n \times 1$*-matrices.*

Proof. Clearly the solutions of the homogeneous equation form a linear space. If $e_1, \ldots, e_n$ is a basis for the $n \times 1$-matrices and u_j the solution of (D.4) with initial value $u_0 = e_j$, then every solution is a unique linear combination of $u_1, \ldots, u_n$ since a solution is uniquely determined by its (arbitrary) initial value. Thus we have a basis for the solutions. Because $u_1, \ldots, u_n$ span all solutions and the initial value problem is solvable for any initial value and any initial point, $u_1(x), \ldots, u_n(x)$ span the $n \times 1$ matrices, and is therefore a basis, for any $x \in (a, b)$. $\qquad \square$

We often have to deal with linear equations containing a parameter r, usually a scalar varying in an interval Ω, but Ω could be an arbitrary metric space. The equation is then $u' = A(\cdot, r)u + F(\cdot, r)$, where the components of $A(\cdot, r)$ and $F(\cdot, r)$ depend continuously on r as locally integrable functions. This means that $\int_I \|A(\cdot, r) - A(\cdot, r_0)\| \to 0$ as $r \to r_0$ for every $r_0 \in \Omega$ and every compact subinterval I of (a, b), and similarly for F.

A solution with r-independent initial value then depends continuously on r, locally uniformly with respect to t. This follows from the proof of Theorem D.2, since it is clear by a trivial induction that the approximate solutions u_j depend continuously on r, locally uniformly in t. Since $u_j \to u$ locally uniformly in (t, r) the continuous dependence of r follows. We have proved the first part of the following theorem.

Theorem D.5. *Suppose Ω is a metric space and $r \mapsto A(\cdot, r) \in L^1_{\text{loc}}((a, b), \mathbb{C}^{n \times n})$ and $r \mapsto F(\cdot, r) \in L^1_{\text{loc}}((a, b), \mathbb{C}^n)$ are continuous functions defined on Ω. Then a solution of $u' = Au + F$ with r-independent initial value depends continuously on $r \in \Omega$, locally uniformly for $t \in (a, b)$.*

Similarly, if $\Omega \ni \lambda \mapsto A(\cdot, \lambda)$ and $\Omega \ni \lambda \mapsto F(\cdot, \lambda)$ are analytic in an open domain $\Omega \subset \mathbb{C}$ with values in $L^1_{\text{loc}}((a, b), \mathbb{C}^{n \times n})$ and $L^1_{\text{loc}}((a, b), \mathbb{C}^n)$ respectively, [1]

---

[1] This means that the difference quotients converge in the corresponding spaces, or equivalently that there are power series expansions around any $\lambda \in \Omega$ with t-dependent coefficients and converging in the appropriate spaces.

then a solution of $u' = Au + F$ with λ-independent initial value depends analytically on $\lambda \in \Omega$, locally uniformly for $t \in (a, b)$.

Proof. It remains to prove the second part. However, it is clear by induction that in this case all the approximate solutions u_j depend analytically on λ, and since they converge locally uniformly with respect to (t, λ) to the solution u, the solution depends analytically on λ, locally uniformly in t.   □

Of particular interest is the analytical dependence on λ of the solutions of a system $u' = (A + \lambda B)u + F$, where A and B are $n \times n$ matrix-valued and F $n \times 1$ matrix-valued, and A, B, and F all independent of λ. The theorem then states that a solution with initial value independent of λ is, locally uniformly with respect to t, an entire function of λ. This function is even of exponential type, locally uniformly in t. For the concepts of exponential type and order of an entire function, see page 341. The proof of exponential type is left to Exercise D.2.

For systems derived from scalar equations of order > 1 the order of the solution as a function of λ is strictly less than one. In particular, for Sturm–Liouville equations we have the following theorem.

Theorem D.6. *Consider the solution of $-(pu')' + qu = \lambda wu$ with $u(c)$ and $pu'(c)$ independent of λ, $c \in (a, b)$, where $1/p$, q, and w are in $L^1_{\mathrm{loc}}(a, b)$. Then, as a function of λ, $u(x, \lambda)$ and $pu'(x, \lambda)$ are entire of order $\leq 1/2$, locally uniformly with respect to $x \in (a, b)$.*

Unless $w = 0$ a.e. between c and x the order is in fact precisely $1/2$, but we will not need this.

Proof. For right or left-definite equations this follows from Corollary 6.4.2, but a simpler proof without assuming this is as follows.

Put $h = |pu'|^2 + |\lambda|\,|u|^2$. Differentiating and using the arithmetic-geometric inequality we obtain

$$
\begin{aligned}
h' = 2\,\mathrm{Re}((pu')'\overline{pu'}) + 2\,|\lambda|\,\mathrm{Re}((u\overline{u'})) &= 2\,\mathrm{Re}((q - \lambda w + |\lambda|\,/p)u\overline{pu'}) \\
&\leq (|q|\,|\lambda|^{-1/2} + (|w| + 1/|p|)\,|\lambda|^{1/2})2\,|\lambda|^{1/2}\,|u|\,|pu'| \\
&\leq (|q|\,|\lambda|^{-1/2} + (|w| + 1/|p|)\,|\lambda|^{1/2})h \,.
\end{aligned}
$$

By the Gronwall inequality (Theorem D.3) we obtain

$$
|h(x)| \leq |h(c)|\exp\left(\left|\int_c^x |q|\,|\lambda|^{-1/2} + (|w| + 1/|p|)\,|\lambda|^{1/2}\right|\right),
$$

which is the requisite estimate.   □

Analytic coefficients

Here we will only need to consider the linear, homogeneous equation $u' = Au$ where A is $n \times n$ matrix-valued with analytic components in some open domain $\Omega \subset \mathbb{C}$. The set of such functions is denoted $\mathcal{O}(\Omega, \mathbb{C}^{n \times n})$, and similar for matrices of other types. As before we are looking for an $n \times 1$ matrix-valued solution u. We first consider the case when Ω is simply connected.

A function f defined in a domain $\Omega \subset \mathbb{C}$ with values in $\mathbb{C}^{n \times k}$ with analytic components has a power series expansion $f(z) = \sum_{n=0}^{\infty} (z - z_0)^n f_n$ with all $f_n \in \mathbb{C}^{n \times k}$ for every $z_0 \in \Omega$. This is proved just as for complex-valued functions. The series is norm-convergent in the largest circle $|z - z_0| < R$ contained in Ω and $f_n = f^{(n)}(z_0)/n!$. If f has no analytic extension outside Ω the series diverges if $|z - z_0| > R$.

Remark D.7. A power series $\sum_{k=0}^{\infty} (z - z_0)^k f_k$ with coefficients $f_k \in \mathbb{C}^{n \times k}$ has a radius of convergence $0 \le R \le \infty$, such that the series is *norm-convergent*, i.e., $\sum_{k=0}^{\infty} |z - z_0|^k \|f_k\| < \infty$ if $|z - z_0| < R$ but does not converge if $|z - z_0| > R$. To see this, let

$$R = \sup \left\{ r \ge 0 : \{r^j \|f_j\|\}_{j=0}^{\infty} \text{ is bounded} \right\}.$$

If $0 < \rho < r < R$ and $r^j \|f_j\| \le C$ for all j we have $\rho^j \|f_j\| \le C(\rho/r)^j$, which is the jth term in a convergent geometric series.

It follows that the sequence of partial sums of $\sum_{k=0}^{\infty} (z - z_0)^k f_k$ is a Cauchy sequence in $\mathbb{C}^{n \times k}$ if $|z - z_0| \le \rho < R$ and therefore converges uniformly there. On the other hand, if $|z - z_0| > R$ the terms of this series do not tend to zero since they are not even bounded, and the series diverges.

Now consider the initial value problem

$$\begin{cases} u' = Au \text{ in } \Omega, \\ u(z_0) = u_0, \end{cases} \tag{D.5}$$

where Ω is simply connected, $z_0 \in \Omega$, $u_0 \in \mathbb{C}^{n \times 1}$ and $A \in \mathcal{O}(\Omega, \mathbb{C}^{n \times n})$.

Theorem D.8. *For simply connected Ω there is a unique solution $u \in \mathcal{O}(\Omega, \mathbb{C}^{n \times 1})$ satisfying* (D.5).

Proof. Suppose the open disk B of radius r and center z_0 is in Ω. For $u \in \mathcal{O}(B, \mathbb{C}^{n \times 1})$ the function $B \ni z \mapsto \int_{z_0}^{z} Au$ is in $\mathcal{O}(B, \mathbb{C}^{n \times 1})$ and independent of the path in Ω from z_0 to z. Setting

$$u_{j+1}(z) = u_0 + \int_{z_0}^{z} Au_j, \quad j = 1, 2, \ldots,$$

it follows by a trivial induction that all $u_j \in \mathcal{O}(B, \mathbb{C}^{n \times 1})$. Using a radial path from z_0 to z it follows as in the proof of Theorem D.2 that u_j converges locally uniformly in B, so that $u = \lim_{j\to\infty} u_j$ is also in $\mathcal{O}(B, \mathbb{C}^{n\times 1})$. It follows immediately that u is a solution of (D.5). Successive differentiation of $u' = Au$ shows that all derivatives of u at z_0 are determined by A and u_0. This determines u uniquely as a power series near z_0 and in the rest of Ω by analytic continuation. $\square$

Note that near z_0 we have $A(z) = \sum_{j=0}^{\infty} (z-z_0)^j A_j$ for some coefficients $A_j \in \mathbb{C}^{n\times n}$ so assuming $u(z) = \sum_{k=0}^{\infty} (z - z_0)^k u_k$ the equation becomes

$$\sum_{k=0}^{\infty} (k + 1)(z - z_0)^k u_{k+1} = \sum_{k=0}^{\infty} (z - z_0)^k \sum_{j=0}^{k} A_j u_{k-j} \,,$$

so that $u_{k+1} = \frac{1}{k+1} \sum_{j=0}^{k} A_j u_{k-j}$, $k = 0, 1, 2, \dots$ Since u_0 is given, this is a recursion formula from which all u_k may be successively determined.

If we consider a Sturm–Liouville equation $-(pu')' + qu = \lambda w u$, where $1/p$, q, and w are in $\mathcal{O}(\Omega, \mathbb{C})$ we may write the equation as a first order system in the usual way. This shows that given initial values $u(z_0)$, $pu'(z_0)$ there is a unique solution $u \in \mathcal{O}(\Omega, \mathbb{C})$ such that also $pu' \in \mathcal{O}(\Omega, \mathbb{C})$.

Regular singular points. If Ω is not simply connected the solution u may not be single-valued. We shall consider the case when A has an isolated singularity which we place at 0. To see what may be expected, consider the case of a scalar equation $u'(z) = z^{-k} A u(z)$ where k is a positive integer and $A \in \mathbb{C}$ constant. One immediately finds that all solutions are multiples of $\exp(Az^{1-k}/(k - 1))$ except if $k = 1$ when we obtain multiples of z^A. We therefore have an essential singularity at 0 for $k > 1$ but nothing worse than a pole or branch point if $k = 1$. If we consider non-scalar equations, in general we can hardly expect anything better. We shall therefore only consider the case when the coefficient of the equation has a simple pole at 0. We may then write the equation

$$zu'(z) = A(z)u(z) \,,$$

where $A \in \mathcal{O}(\Omega, \mathbb{C}^{n\times n})$ and $A(0) \neq 0$. Such an equation is said to have a (Fuchsian) *regular singular point* at 0. A less self-contradictory term is *weakly singular*, but we will stay with the traditional terminology. We assume $A(z) = \sum_{j=0}^{\infty} z^j A_j$ with $A_0 \neq 0$, and, guided by our example, attempt to find a solution of the form $u(z) = z^\alpha \sum_{k=0}^{\infty} z^k u_k$ with $u_0 \neq 0$ for some number α. Inserting this into the equation we obtain

$$\sum_{k=0}^{\infty} (\alpha + k)z^{\alpha+k} u_k = \sum_{k=0}^{\infty} z^{\alpha+k} \sum_{j=0}^{k} A_j u_{k-j} \,,$$

so that $(\alpha + k)u_k = \sum_{j=0}^{k} A_j u_{k-j} = A_0 u_k + \sum_{j=1}^{k} A_j u_{k-j}$. Thus

$$(\alpha - A_0)u_0 = 0 \quad \text{and} \quad (k - (A_0 - \alpha))u_k = \sum_{j=1}^{k} A_j u_{k-j}, \ k = 1, 2, \dots \,.$$

We must therefore choose α as an eigenvalue and u_0 a corresponding eigenvector of A_0. If we do this the second formula determines all other u_k by recursion provided the matrix $k - (A_0 - \alpha)$ is invertible for $k = 1, 2, \ldots$. This will certainly be the case unless A_0 has eigenvalues that differ by an integer.

For the moment we assume $k - (A_0 - \alpha)$ is invertible for all integers $k > 0$. It then remains to prove convergence of the series for u. We shall need a lemma.

Lemma D.9 (Neumann series). *Suppose $\mathcal{B}$ is a normed space, $B \in \mathcal{L}(\mathcal{B})$ and let $I \in \mathcal{L}(\mathcal{B})$ be the identity. Then, if the Neumann series $\sum_{j=0}^{\infty} B^j$ is norm-convergent,[2] i.e., $\sum_{j=0}^{\infty} \|B^j\| < \infty$, the sum of the Neumann series is the inverse of $I - B$. The series is norm-convergent if $\|B\| < 1$.*

We remark that the series is norm-convergent also if $\|B^n\| < 1$ for some positive integer n, in particular if B is *nilpotent*, i.e., $B^n = 0$ for some $n \in \mathbb{Z}_+$.

Proof. If the series is norm-convergent we have $\|B^j\| \to 0$ as $j \to \infty$. A simple calculation shows that $(I - B) \sum_{j=0}^{k} B^j = \sum_{j=0}^{k} B^j (I - B) = I - B^{k+1} \to I$ as $k \to \infty$ if the series is norm-convergent, which it is if $\|B\| < 1$ since $\|B^j\| \le \|B\|^j$. □

Now note that $k - (A_0 - \alpha) = k(1 - (A_0 - \alpha)/k)$ and if $\|A_0 - \alpha\|/k < 1$ the Neumann series $\sum_{j=0}^{\infty}((A_0 - \alpha)/k)^j$ converges in norm to the inverse of the operator $1 - (A_0 - \alpha)/k$. For large k we therefore have

$$\|(k - (A_0 - \alpha))^{-1}\| \le \frac{1}{k} \sum_{j=0}^{\infty} (\|A_0 - \alpha\|/k)^j = 1/(k - \|A_0 - \alpha\|) .$$

Now assume $0 < r < R$ where R is the radius of convergence for $\sum_0^{\infty} z^j A_j$. Thus $M(r) = \sum_{j=0}^{\infty} r^j \|A_j\|$ is finite, and we have $M(r)\|(k - (A_0 - \alpha))^{-1}\| \le 1$ for $k \ge n$ if n is sufficiently large. Multiplying $u_k = (k - (A_0 - \alpha))^{-1} \sum_{j=1}^{k} A_j u_{k-j}$ by r^k for $k \ge n$ we obtain

$$r^k \|u_k\| \le \|(k - (A_0 - \alpha))^{-1}\| \sum_{j=1}^{k} r^j \|A_j\| r^{k-j} \|u_{k-j}\|$$

$$\le \max_{0 \le j < k} (r^j \|u_j\|) \frac{M(r)}{k - \|A_0 - \alpha\|} \le \max_{0 \le j < k} (r^j \|u_j\|) .$$

Setting $C = \max_{0 \le j < n}(r^j \|u_j\|)$ this estimate shows by induction that $r^j \|u_j\| \le C$ for all j so that the radius of convergence of the series for u is at least r for any $r < R$, thus in fact at least R.

Picking α as an eigenvalue such that $\beta - \alpha$ is not a positive integer for any other eigenvalue β there is an at least one-dimensional set of solutions of (D.5) of the form $u(z) = \sum_{k=0}^{\infty} z^{\alpha+k} u_k$, and if A_0 has a basis of eigenvectors to eigenvalues none of which differ by an integer, then all solutions are linear combinations of solutions of this kind. The eigenvalues are found by solving the *indicial equation* $\det(A_0 - x) = 0$.

---

[2] $B^0 = I$ by definition.

This approach to finding solutions near a regular singular point is often named the *method of Frobenius*.

If for some eigenvalue the algebraic multiplicity (the multiplicity as a root of the indicial equation) is larger than the geometric multiplicity (the dimension of the eigenspace) not all solutions can be linear combinations of solutions of this form. We also need to consider the case when there are eigenvalues that differ by an integer. It turns out that if one cannot find a basis of solutions of the form considered so far, then there are solutions containing logarithms.

An up-to-date and detailed treatment of the general case may be found in Hille [105, Chapters 5 and 9] and slightly more down to earth treatments may be found in Coddington and Levinson [46, Chapter 4] and Ince [112, Chapter XVI]. Instead of treating the intricacies of the general case here we will limit our discussion to the case of a 2×2 system, since this is all we need in Section 6.6.

If our equation is a 2×2 system the indicial equation has either two distinct roots α and β or a double root α, so we need to find a second solution if α is a double root but the eigenspace is one-dimensional, or if $n = \alpha - \beta$ is a positive integer. Consider first the latter case.

Attempting to find a solution $v(z) = \sum_{k=0}^{\infty} z^{\beta+k} v_k$ where $v_0 \neq 0$ is an eigenvector with eigenvalue β to A_0 the difficulty is to determine v_n from $(\alpha - A_0)v_n = \sum_{j=1}^{n} A_j v_{n-j}$ since $A_0 - \alpha$ is not invertible. If there is a solution it will only be determined up to addition of multiples of u_0, but this should not concern us since it only means that the second solution will only be determined up to addition of multiples of u. But the range of $\alpha - A_0$ is one-dimensional and spanned by v_0 so unless we are lucky we cannot find a second solution of this form. This is also not possible if $\alpha = \beta$ but the corresponding eigenspace has dimension one. In these cases we will find a second solution $w(z) = u(z) \log(z) + v(z)$ where $v(z) = \sum_{k=0}^{\infty} z^{\beta+k} v_k$.

Inserting w into the equation we see that it will be a solution precisely if $zv'(z) = A(z)v(z) - u(z)$, which gives $(\beta - A_0)v_0 = 0$ and the recursion formula

$$(\alpha - n + k - A_0)v_k = \sum_{j=1}^{k} A_j v_{k-j} - u_{k-n} , \quad k = 1, 2, \ldots, \tag{D.6}$$

for the coefficients of v, where $u_j = 0$ for $j < 0$. Thus we must choose v_0 as an eigenvector of A_0 to the eigenvalue $\beta = \alpha - n$. Clearly the recursion formula now determines all v_k except possibly v_n since $\alpha - A_0$ is not invertible. Now u_0, v_0 is a basis for $\mathbb{C}^2$ and the range of $\alpha - A_0$ is spanned by v_0 since $(\alpha - A_0)(xu_0 + yv_0) = yn b_0$. The case when $\sum_{j=1}^{n} A_j v_{n-j}$ is a multiple of v_0 was taken care of earlier, and in other cases replacing v_0 by an appropriate multiple ensures that the right-hand side of (D.6) for $k = n$ is a multiple of v_0 so that all v_n may be determined.

Now move on to the case when α is a double root of the indicial equation but with a one-dimensional eigenspace, so that α is the only eigenvalue of A_0. The recursion formula is now

$$(\alpha + k - A_0)v_k = \sum_{j=1}^{k} A_j v_{k-j} - u_k, \quad k = 1, 2, \ldots,$$

and the equation for v_0 is $(A_0 - \alpha)v_0 = u_0$. If we can solve this equation all other v_k are uniquely determined by the recursion formula. Now, if f, u_0 is a basis for $\mathbb{C}^2$ we have $(A_0 - \alpha)f = xu_0 + yf \neq 0$ for some numbers x, y. Here $y = 0$ since otherwise $xu_0 + yf$ would be an eigenfunction with eigenvalue y of $A_0 - \alpha$. Thus $v_0 = \frac{1}{x}f$ satisfies $(A_0 - \alpha)v_0 = u_0$.

It remains to prove that the series expansion obtained for v converges, but the proof of this is much the same as for the expansion of u and will be left to the reader in Exercise D.3.

One may also speak of a regular singular point at ∞. This means, by definition, that the equation obtained by the change of variable $\zeta = 1/z$ has a regular singular point at 0. The change of variable gives the equation $\zeta \frac{du}{d\zeta} = -A(1/\zeta)u$, so this means that $A(1/\zeta) = \sum_{j=0}^{\infty} \zeta^j A_j$ with $A_0 \neq 0$ so that $A(z) = \sum_{j=0}^{\infty} z^{-j} A_j$. Thus we have a regular point at ∞ if $A(z) \to 0$ at ∞, and a regular singular point there if the limit is finite but non-zero.

We will be particularly interested in the case of a second order scalar equation. Such an equation is said to have a regular singular point at 0 if it is of the form $z^2 w''(z) + a(z)zw'(z) + b(z)w(z) = 0$ with a, b analytic near 0. It may be written as a first order system with a regular singular point at 0 by setting $u(z) = \begin{pmatrix} w(z) \\ zw'(z) \end{pmatrix}$. We then have $A(z) = \begin{pmatrix} 0 & 1 \\ -b(z) & 1-a(z) \end{pmatrix}$ and the indicial equation is $\begin{vmatrix} x & -1 \\ b(0) & x+a(0)-1 \end{vmatrix} = 0$. We obtain the following theorem.

Theorem D.10. *A scalar equation $z^2 u''(z) + a(z)zu'(z) + b(z)u(z) = 0$ with a and b analytic near 0 has indicial equation $x^2 + (a(0) - 1)x + b(0) = 0$.*

If the indicial roots are α and β with $\mathrm{Re}(\alpha - \beta) \geq 0$, then there is a solution $u(z) = \sum_{k=0}^{\infty} a_k z^{k+\alpha}$ with $a_0 \neq 0$ and a similar solution with α replaced by β except if $\alpha = \beta$ is a double root and possibly if $\alpha - \beta$ is a positive integer. In this case a second solution may be found of the form $u(z) \log z + v(z)$, where $v(z) = \sum_{k=0}^{\infty} b_k z^{k+\beta}$.

Exercises

Exercise D.1 Prove the triangle inequality $\|\int_a^b f\| \leq \int_a^b \|f\|$, valid for functions f with values in $\mathbb{C}^n$ and components integrable in (a, b).

Hint: If $v \in \mathbb{C}^n$ is constant, $\langle \int_a^b f, v \rangle = \int_a^b \langle f, v \rangle$. Use the Cauchy–Schwarz inequality and a judicious choice of v.

Exercise D.2 Prove that solutions of $u' = (A + \lambda B)u + F$ with λ-independent initial data, which are entire functions of λ by Theorem D.5, are of exponential type, locally uniformly.

Hint: The Gronwall inequality.

Exercise D.3 Prove that the radius of convergence for the function v in (D.6) is at least the same as that for A.

Hint: Copy the convergence proof for u.

Appendix E
Analytic functions

In this appendix we will discuss some aspects of the theory of analytic functions which go beyond the most basic, but are nevertheless needed in various parts of the book.

E.1 Nevanlinna functions

Our proofs of the spectral theorem and the various expansion theorems given in Chapters 4, 5, and 8 are based on the following fundamental representation theorem.

Theorem E.1.1. *F is analytic in $\mathbb{C} \setminus \mathbb{R}$, $F(\overline{\lambda}) = \overline{F(\lambda)}$ and maps the upper half-plane into itself if and only if there exists a unique positive measure $d\eta$ with $\int_{-\infty}^{\infty} \frac{d\eta(t)}{1+t^2} < \infty$, and unique real constants A and $B \geq 0$, such that*

$$F(\lambda) = A + B\lambda + \int_{-\infty}^{\infty} \frac{1+t\lambda}{t-\lambda} \frac{d\eta(t)}{t^2+1}$$

$$= A + B\lambda + \int_{-\infty}^{\infty} \left(\frac{1}{t-\lambda} - \frac{t}{1+t^2} \right) d\eta(t) . \quad \text{(E.1.1)}$$

The integral is absolutely convergent.

For the meaning of such an integral, see Appendix B. Often the measure $d\eta$ is called the *spectral measure* for F, because of the applications of this formula to spectral theory. Functions with the properties of F are usually called Nevanlinna, Herglotz or Pick functions. We are not sure who first proved the theorem, but results of this type play an important role in the classical book [174] by Rolf Nevanlinna (1930). We shall tackle the proof through a sequence of lemmas.

Lemma E.1.2 (H. A. Schwarz). *Let G be analytic in the unit disk, and put $u(R, \theta) =$ Re $G(Re^{i\theta})$. For $|z| < R < 1$ we then have:*

© Springer Nature Switzerland AG 2020
C. Bennewitz et al., *Spectral and Scattering Theory for Ordinary Differential Equations*,
Universitext, https://doi.org/10.1007/978-3-030-59088-8

$$G(z) = i \operatorname{Im} G(0) + \frac{1}{2\pi} \int_{-\pi}^{\pi} \frac{Re^{i\theta} + z}{Re^{i\theta} - z} u(R, \theta)\, d\theta \;. \tag{E.1.2}$$

Proof. According to Poisson's integral formula (see almost any textbook on complex analysis, *e.g.* Ahlfors [1, Chapter 4.6]), we have

$$\operatorname{Re} G(z) = \frac{1}{2\pi} \int_{-\pi}^{\pi} \frac{R^2 - |z|^2}{|Re^{i\theta} - z|^2} u(R, \theta)\, d\theta \;. \tag{E.1.3}$$

The integral here is easily seen to be the real part of the integral in (E.1.2). The latter is obviously analytic in z for $|z| < R < 1$, so the two sides of (E.1.2) can only differ by an imaginary constant. However, for $z = 0$ the integral is real, so (E.1.2) follows.   □

The formula (E.1.2) is not applicable for $R = 1$, since we do not know whether $\operatorname{Re} G$ has reasonable boundary values on the unit circle. However, if one assumes that $\operatorname{Re} G \geq 0$ the boundary values exist at least in the sense of measure, and one has the following theorem.

Theorem E.1.3 (Riesz–Herglotz). *Let G be analytic in the (open) unit disk with non-negative real part. Then there exists an increasing function ξ on $[-\pi, \pi]$ such that*

$$G(z) = i \operatorname{Im} G(0) + \frac{1}{2\pi} \int_{[-\pi,\pi]} \frac{e^{i\theta} + z}{e^{i\theta} - z}\, d\xi(\theta) \;.$$

With a suitable normalization the function ξ will also be unique, but we shall not use this. To prove Theorem E.1.3 we need a compactness result, so that we can obtain the theorem as a limiting case of Lemma E.1.2. What is needed is weak* compactness in the dual of the continuous functions on a compact interval, provided with the maximum norm. This is the classical Helly theorem. Since we assume minimal knowledge of functional analysis we shall give the classical proof.

Lemma E.1.4 (Helly).

(1) *Suppose $\{\xi_j\}_1^\infty$ is a uniformly bounded[1] sequence of increasing functions on an interval I. Then there is a subsequence converging pointwise to an increasing function.*

(2) *Suppose $\{\xi_j\}_1^\infty$ is a uniformly bounded sequence of increasing functions on a compact interval I, converging pointwise to ξ. Then*

$$\int_I f\, d\xi_j \to \int_I f\, d\xi \text{ as } j \to \infty \;, \tag{E.1.4}$$

for any function f continuous on I.

---

[1] That is, all functions in the sequence are bounded by the same constant.

Proof. Let $r_1, r_2, \ldots$ be a dense sequence in I, for example an enumeration of the rational numbers in I. By the Bolzano–Weierstrass theorem we may choose a subsequence $\{\xi_{1j}\}_1^\infty$ of $\{\xi_j\}_1^\infty$ such that $\xi_{1j}(r_1)$ converges. Similarly, we may choose a subsequence $\{\xi_{2j}\}_1^\infty$ of $\{\xi_{1j}\}_1^\infty$ such that $\xi_{2j}(r_2)$ converges; as a subsequence of $\xi_{1j}(r_1)$ the sequence $\xi_{2j}(r_1)$ also converges. Continuing in this fashion, we obtain a sequence of sequences $\{\xi_{kj}\}_{j=1}^\infty$, $k = 1, 2, \ldots$, such that each sequence is a subsequence of those preceding it, and such that $\xi(r_n) = \lim_{j\to\infty} \xi_{kj}(r_n)$ exists for $n \le k$. Thus $\xi_{jj}(r_n) \to \xi(r_n)$ as $j \to \infty$ for every n, since $\xi_{jj}(r_n)$ is a subsequence of $\xi_{nj}(r_n)$ from $j = n$ on. Clearly ξ is increasing, so if $x \in I$ but $\ne r_n$ for all n, we may choose an increasing subsequence $\{r_{j_k}\}_{k=1}^\infty$ of $\{r_j\}_1^\infty$ converging to x, and define $\xi(x) = \lim_{k\to\infty} \xi(r_{j_k})$.

Suppose x is a point of continuity of ξ. If $r_k < x < r_n$ we get $\xi_{jj}(r_k) - \xi(r_n) \le \xi_{jj}(x) - \xi(x) \le \xi_{jj}(r_n) - \xi(r_k)$. Given $\varepsilon > 0$ we may choose k and n such that $\xi(r_n) - \xi(r_k) < \varepsilon$. We then obtain

$$-\varepsilon \le \varliminf_{j\to\infty} (\xi_{jj}(x) - \xi(x)) \le \varlimsup_{j\to\infty} (\xi_{jj}(x) - \xi(x)) \le \varepsilon .$$

Hence $\{\xi_{jj}\}_1^\infty$ converges pointwise to ξ, except possibly at points of discontinuity of ξ. But there are at most countably many such discontinuities, ξ being increasing. Hence repeating the trick of extracting subsequences, and then using the 'diagonal' sequence, we get a subsequence of the original sequence which converges everywhere in I. We now obtain (1).

If f is the characteristic function of a compact interval whose endpoints are points of continuity for ξ and all ξ_j it is obvious that (E.1.4) holds. It follows that (E.1.4) holds if f is a step function with all discontinuities at points where ξ and all ξ_j are continuous. If f is continuous and $\varepsilon > 0$ we may, by uniform continuity, choose such a step function g such that $\sup_I |f - g| < \varepsilon$. If C is a common bound for all ξ_j we then obtain $\left| \int_I (f - g)\, d\xi \right| < 2C\varepsilon$ and similarly with ξ replaced by ξ_j. It follows that $\varlimsup_{j\to\infty} \left| \int_I f\, d\xi_j - \int_I f\, d\xi \right| \le 4C\varepsilon$ and since ε is arbitrary positive (2) follows. $\square$

Proof (Theorem E.1.3). According to Lemma E.1.2, if $0 < R < 1$ and $|z| < 1$,

$$G(Rz) = \mathrm{i}\,\mathrm{Im}\, G(0) + \frac{1}{2\pi} \int_{-\pi}^{\pi} \frac{e^{i\theta} + z}{e^{i\theta} - z}\, d\xi_R(\theta) ,$$

where $\xi_R(\theta) = \int_{-\pi}^{\theta} \mathrm{Re}\, G(Re^{i\varphi})\, d\varphi$. Hence ξ_R is increasing, ≥ 0 and bounded from above by $\xi_R(\pi)$. Now $\mathrm{Re}\, G$ is a harmonic function so it has the mean value property, which means that $\xi_R(\pi) = 2\pi\, \mathrm{Re}\, G(0)$. This is independent of R, so by Helly's theorem we may choose a sequence $R_j \uparrow 1$ such that ξ_{R_j} converges to an increasing function ξ. Use of the second part of Helly's theorem completes the proof. $\square$

To prove the uniqueness of the function η of Theorem E.1.1 we need the following simple, but important, lemma.

Lemma E.1.5 (Stieltjes' inversion formula). *Let η be complex-valued of locally bounded variation, and such that $\int_{-\infty}^{\infty} |d\eta|\,(t)/(t^2+1) < \infty$. Suppose $F(\lambda)$ is given by* (E.1.1). *Then if $y < x$ are points of continuity of η we have*

$$\eta(x) - \eta(y) = \lim_{\varepsilon \downarrow 0} \frac{1}{2\pi i} \int_{y}^{x} (F(s+i\varepsilon) - F(s-i\varepsilon))\,ds$$

$$= \lim_{\varepsilon \downarrow 0} \frac{1}{\pi} \int_{y}^{x} \int_{-\infty}^{\infty} \frac{\varepsilon\,d\eta(t)}{(t-s)^2+\varepsilon^2}\,ds\ .$$

Proof. The second equality is an elementary calculation. By absolute convergence and Tonelli's Theorem B.9.5 we may change the order of integration in the last integral. The inner integral is then easily calculated to be

$$\frac{1}{\pi}(\arctan((x-t)/\varepsilon) - \arctan((y-t)/\varepsilon))\ .$$

This is bounded by 1, and also by a constant multiple of $1/t^2$ if ε is bounded (Exercise E.1.1.). Furthermore it converges pointwise to 0 outside $[y, x]$, and to 1 in (y, x) (and to $\frac{1}{2}$ for $t = x$ and $t = y$). The theorem follows by dominated convergence.
□

Proof (Theorem E.1.1). The uniqueness of η follows immediately on applying the Stieltjes inversion formula to (E.1.1).

We obtain (E.1.1) from the Riesz–Herglotz theorem by a change of variable. The mapping $z = \frac{1+i\lambda}{1-i\lambda}$ maps the upper half-plane bijectively to the unit disk, so $G(z) = -iF(\lambda)$ is defined for z in the unit disk and has positive real part. Applying Theorem E.1.3 we obtain, after simplification,

$$F(\lambda) = \operatorname{Re} F(i) + \frac{1}{2\pi} \int_{[-\pi,\pi]} \frac{1 + \lambda \tan(\theta/2)}{\tan(\theta/2) - \lambda}\,d\xi(\theta)\ .$$

Setting $t = \tan(\theta/2)$ maps the open interval $(-\pi, \pi)$ onto the real axis. For $\theta = \pm\pi$ the integrand equals λ, so any mass of $d\xi$ at $\pm\pi$ gives rise to a term $B\lambda$ with $B \geq 0$. After the change of variable we get

$$F(\lambda) = A + B\lambda + \int_{-\infty}^{\infty} \frac{1+t\lambda}{t-\lambda}\,d\tau(t)\ ,$$

where we have set $A = \operatorname{Re} F(i)$ and $\tau(t) = \xi(\theta)/(2\pi)$. Since

$$\frac{1+t\lambda}{t-\lambda} = \left(\frac{1}{t-\lambda} - \frac{t}{1+t^2}\right)(1+t^2)$$

we now obtain (E.1.1) by setting $d\eta(t) = (1 + t^2)\, d\tau(t)$.

It remains to show the uniqueness of A and B. However, setting $\lambda = i$, it is clear that $A = \operatorname{Re} F(i)$, and since we already know that $d\eta$ is unique, so is B.   □

One can also calculate B directly from F since by dominated convergence $\operatorname{Im} F(i\nu)/\nu \to B$ as $\nu \to \infty$. It is usual to refer to B as the 'mass at infinity', an expression explained by our proof. Note, however, that it is the mass of τ at infinity and not that of η.

If η is the left-continuous distribution function of the measure $d\eta$ in (E.1.1) with $\eta(0) = 0$, then η is often called the *spectral function* for F. We next show that the growth and decay of Nevanlinna functions is strictly limited.

Theorem E.1.6. *Let $d\eta$ be a Radon measure with $\int_{-\infty}^{\infty} |d\eta(t)| / (t^2 + 1) < \infty$. Then any function F represented as in (E.1.1) is bounded by $K(1 + |\lambda|^2)/|\operatorname{Im}\lambda|$ for some non-negative constant K.*

Furthermore, $F(\lambda)/\lambda = o(|\lambda/\operatorname{Im}\lambda|)$ as $\lambda \to \infty$, and given any double sector $|\operatorname{Re}\lambda| \le C\,|\operatorname{Im}\lambda|$ we have $F(\lambda)/\lambda = o(1)$ as $\lambda \to \infty$ in the double sector.

If F is a Nevanlinna function $\not\equiv 0$ similar estimates are valid for $1/(\lambda F(\lambda))$, and in this case $\log|F(\lambda)| = o(\log|\lambda|)$ as $\lambda \to \infty$ in the double sector.

Proof. Since $1 + t\lambda = \lambda(t - \lambda) + 1 + \lambda^2$ we may write (E.1.1) as

$$
F(\lambda) = A + B\lambda + (1 + \lambda^2) \int_{-\infty}^{\infty} \frac{1}{t - \lambda} \frac{d\eta(t)}{1 + t^2} \,,
$$

with a B different from that in (E.1.1), from which the first estimate follows. Now $1/(t - \lambda)$ tends pointwise (in t) to 0 as $\lambda \to \infty$ and is bounded by $1/|\operatorname{Im}\lambda|$, while $|\lambda| \le (1 + C^2)^{1/2}\,|\operatorname{Im}\lambda|$ in the double sector, so by dominated convergence the estimates of $F(\lambda)/\lambda$ follow.

If $F \not\equiv 0$ is a Nevanlinna function so is $-1/F$ so $1/(\lambda F(\lambda))$ may be estimated in the same way, which implies the last estimate.   □

The Stieltjes inversion formula tells us how to retrieve $d\eta$ when F is given, but sometimes one needs to consider pointwise limits as λ approaches the real axis. A simple result of this type is the following theorem.

Theorem E.1.7. *Suppose η is of locally bounded variation such that the integral $\int_{-\infty}^{\infty} |d\eta(t)| / (t^2 + 1)$ converges, and let F be given by the formula (E.1.1).*

Writing $\mu = \operatorname{Re}\lambda$, $\nu = \operatorname{Im}\lambda$ we then have

$$
\lim_{\nu \to 0} \nu F(\mu + i\nu) = i(\eta(\mu+) - \eta(\mu-)) \,.
$$

Proof. We have $\nu F(\mu + i\nu) - \int_{\mu-1}^{\mu+1} \frac{\nu\, d\eta(t)}{t - \mu - i\nu} \to 0$ as $\nu \to 0$, while the integrand $\nu/(t - \mu - i\nu)$ is bounded by 1 and tends pointwise to i times the characteristic function of the singleton $\{\mu\}$ as $\nu \to 0$. The theorem follows by dominated convergence.   □

Exercise

Exercise E.1.1 Show that $t \mapsto \frac{1}{\pi}(\arctan((x-t)/\varepsilon) - \arctan((y-t)/\varepsilon))$ is bounded by 1 and also by a multiple of $1/t^2$ if $0 < \varepsilon < 1$.

E.2 Entire functions

In this section we will examine some properties of entire functions needed from Chapter 6 onwards. We will only prove what we need and refer to the literature for a more comprehensive treatment.

Analytic functions, particularly entire functions, used to be thought of as 'polynomials of infinite degree' because they could be expanded in power series, and had other properties reminiscent of polynomials. One important property of polynomials is given by the fundamental theorem of algebra and the factor theorem, which imply that a polynomial $p(z) = a_n z^n + a_{n-1} z^{n-1} + \cdots + a_0$ may be factored as $p(z) = a_n(z - z_1)(z - z_2) \cdots (z - z_n)$, where $z_1, \ldots, z_n$ are the zeros of p, counted according to multiplicity. This formula does not generalize to entire functions for the simple reason that an entire function does not have a highest order coefficient unless it is a polynomial.

However, one may also write $p(z) = a_k z^k \prod_{j=1}^{n-k}(1 - z/z_j)$ if $a_0 = \cdots = a_{k-1} = 0$, i.e., 0 is a zero with multiplicity k, $0 \le k \le n$. The question is then: If f is an entire function, is there a constant C such that $f(z) = C z^k \prod_{j \ge 1}(1 - z/z_j)$ where 0 is a zero of multiplicity $k \ge 0$ and $z_1, z_2, \ldots$ are the other zeros of f, repeated according to multiplicity?

It is immediately clear that such a formula cannot be true in general, whatever the meaning of the product if there are infinitely many zeros, because there are non-constant entire functions without a single zero, for example $f(z) = e^z$. Now, if f is entire with no zeros one may define a branch of $\log f(z)$ which is also entire, since f'/f is entire and therefore has an entire primitive g. Thus we may write $f(z) = e^{g(z)}$ for some entire function g. It is clear that in general we will have to allow such a function as a factor in any product expansion of an entire function. But we first need to discuss infinite products.

Infinite products

What meaning should we attach to an infinite product $\prod_{j=1}^{\infty} a_j$? The obvious answer would be to say that $\prod_{j=1}^{\infty} a_j$ converges to P if $\prod_{j=1}^{n} a_j \to P$ as $n \to \infty$. In general this would be an unsuitable definition, because P would be zero if just one a_j is zero, regardless of the values of all other factors. We shall therefore assume that all $a_j \ne 0$. Of course one can allow a finite number of zero factors if the product without these factors converges. Now, if the partial products converge to $P \neq 0$ we have $a_n = \prod_{j=1}^{n} a_j / \prod_{j=1}^{n-1} a_j \to P/P = 1$, but if $P = 0$ this may not be so; for example, if all $a_j = a$ with $|a| < 1$ we obtain $P = 0$ while $a_j \to a \neq 1$. To avoid this complication we make the following definition.

Definition E.2.1. If all $a_j \neq 0$ the infinite product $\prod_{j=1}^{\infty} a_j$ is said to converge to P if the partial products converge to P and $P \neq 0$.

If a finite number of the factors a_j equal zero we say that the product converges to 0 if the product without the zero factors converges. In all other cases the product is said to diverge.

Because the factors in a convergent product tend to 1 it is convenient to write $a_j = 1 + u_j$, and then $u_j \to 0$ if the product $\prod_{j=1}^{\infty}(1 + u_j)$ converges. It is also easy to see that if log denotes the branch of the logarithm which is real on the positive half-axis, then $\sum_{j=N}^{\infty} \log(1 + u_j)$ converges for large N precisely if the product converges; see Exercise E.2.1. To examine convergence of the infinite product we may therefore instead consider convergence of the series $\sum_{j=N}^{\infty} \log(1 + u_j)$. We also make the following definition.

Definition E.2.2. The product $\prod_{j=1}^{\infty}(1 + u_j)$, where all but finitely many $u_j \neq 1$, is said to converge *absolutely* if the series $\sum_{j=N}^{\infty} \log(1 + u_j)$ does for large N. If u_j is a function of a complex variable the product is said to converge (absolutely) uniformly in some set $\Omega \subset \mathbb{C}$ if the corresponding series does.

If there are (a finite number of) zero factors in the product a convergent product converges to zero, respectively uniformly to zero.

The derivative of $\log(1 + z)$ at $z = 0$ is 1, so $\log(1 + u_j)/u_j \to 1$ if $u_j \to 0$. Thus, the comparison theorem for positive series shows that $\sum_{j=N}^{\infty} \left| \log(1 + u_j) \right|$ converges if and only if $\sum_{j=1}^{\infty} |u_j|$ does. We obtain:

Proposition E.2.3. *The product* $\prod_{j=1}^{\infty}(1 + u_j)$ *converges absolutely if and only if* $\sum_{j=1}^{\infty} u_j$ *does. If* u_j *are functions, the product converges absolutely uniformly if and only if the series does.*

Note that the proposition implies that $\prod_{j=1}^{\infty}(1 + u_j)$ converges absolutely precisely if $\prod_{j=1}^{\infty}(1 + |u_j|)$ converges.

We are now ready to return to the product expansion of an entire function. Proposition E.2.3 shows that $\prod_{j=1}^{\infty}(1 - z/z_j)$ converges absolutely for every $z \in \mathbb{C}$ precisely if $\sum_{j=1}^{\infty} 1/|z_j|$ converges, and that the convergence is then locally uniform (uniform on any compact set).

To see for which functions $\sum 1/|z_j|$ converges we must discuss the *growth order* of an entire function.

Functions of finite order

If f is an entire function we define $M(r; f) = \sup_{|z|=r} |f(z)|$. The maximum principle then shows that $r \mapsto M(r; f)$ is increasing, strictly unless f is constant, so that

$|f(0)| = M(0; f) \le M(r; f)$ for all $r \ge 0$. The *order* $\rho(f)$ of a non-constant entire function f (which is unbounded by Liouville's Theorem E.2.12) is defined by

$$\rho(f) = \varlimsup_{r \to \infty} \frac{\log \log M(r; f)}{\log r},$$

so $0 \le \rho(f) \le \infty$. By convention[2] a constant function has order 0. If $0 < \rho(f) < \infty$ one also considers the *type* $\gamma(f)$ of f, defined by

$$\gamma(f) = \varlimsup_{r \to \infty} \frac{\log M(r; f)}{r^\rho}.$$

We have $0 \le \gamma(f) \le \infty$, and a function of order 1 and type $0 \le \gamma < \infty$ is often said to be of *exponential type* γ. A function of order < 1 is also considered to be of exponential type 0.

A non-constant entire function f has order ρ if and only if $\log M(r; f) = O(r^\sigma)$ as $r \to \infty$ as soon as $\sigma > \rho$ but not for any $\sigma < \rho$; see Exercise E.2.2. If $0 < \rho(f) < \infty$, then f has type $0 < \gamma < \infty$ if and only if $\log M(r; f) < \delta r^\rho$ for large r if $\delta > \gamma$ but not for any $\delta < \gamma$.

Example E.2.4. Any polynomial is of order zero and e^{az} is of exponential type $|a|$. The functions $\cos(a\sqrt{z})$ and $\sin(a\sqrt{z})/\sqrt{z}$ are of order $1/2$ and type $|a|$ and the function $\exp(e^z)$ has infinite order. There are entire functions of any order and type which is most easily seen by characterizing order and type in terms of the coefficients in a Taylor expansion. For this, see Boas [30, Theorems 2.2.2. and 2.2.10] or Levin [145, Section 1.3].

It is nearly obvious that if f and g are entire, then $\rho(fg) \le \max(\rho(f), \rho(g))$. It is a little harder to prove the following theorem, which is Theorem 12 and its corollary in Chapter 1 of Levin [144].

Theorem E.2.5. *If f and g are entire so is fg and $\rho(fg) \le \max(\rho(f), \rho(g))$, with equality if $\rho(f) \ne \rho(g)$, and then $\gamma(fg)$ equals the type of the factor with highest order.*

If $\rho(fg) = \rho(f) = \rho(g)$ then $\gamma(fg)$ is infinite if $\max(\gamma(f), \gamma(g)) = \infty$ and $\min(\gamma(f), \gamma(g)) < \infty$ and $\gamma(fg) = \max(\gamma(f), \gamma(g))$ if $\min(\gamma(f), \gamma(g)) = 0$.

If f/g is entire the estimates of $\rho(fg)$ are also valid for $\rho(f/g)$ and those of $\gamma(fg)$ are also valid for $\gamma(f/g)$.

In particular, if f and g are of zero exponential type and f/g entire, then also f/g is of zero exponential type. Derivatives and primitives of entire functions have the same order, in fact even the same type, as the original function. We shall be content to prove the following since this is all we need.

Theorem E.2.6. *If f is entire, then $\rho(f') \le \rho(f)$.*

---

[2] If log is replaced by $\log_+ = \max(\log, 0)$ the formula also gives the correct order for constants.

Proof. We have

$$f'(z) = \frac{1}{2\pi i} \oint_{|\zeta-z|=1} \frac{f(\zeta)\,d\zeta}{(\zeta-z)^2}$$

so we obtain $M(r, f') \leq M(r+1, f)$. Since $\log(r+1)/\log r \to 1$ as $r \to \infty$ the theorem follows. $\qquad\square$

Product expansions

We shall need the following theorem in Chapter 6.

Theorem E.2.7. *If f is entire of order $\rho(f) < 1$, then*

$$f(z) = Cz^k \prod_{j\geq 1}(1 - z/z_j),$$

where C is constant, $k \geq 0$ the multiplicity of 0 as a zero of f and $z_1, z_2, \ldots$ are the zeros $\neq 0$ of f, repeated according to multiplicity. The product is absolutely and locally uniformly convergent.

Note that an important implication of this theorem is that a function of order < 1 which is not a polynomial will always have infinitely many zeros.

Since polynomials have order 0, and by Corollary E.2.6, we may, and will, in proving Theorem E.2.7 assume that $f(0) \neq 0$. For the proof we need several simple lemmas in which we keep the notation of the theorem. We order the zeros z_j of f by increasing absolute values so that $|z_1| \leq |z_2| \leq \ldots$ and define the counting function $N(r)$ as the number of zeros in $|z| \leq r$. We then have:

Lemma E.2.8. *If $\sigma > \rho(f)$, then $N(r) = O(r^\sigma)$ as $r \to \infty$.*

Proof. Suppose $N(r) = n$ and put

$$h(z) = f(z)/((1 - z/z_1) \cdots (1 - z/z_n)).$$

Then h is entire with $h(0) = f(0)$. Thus $M(3r; h) \geq |h(0)| = |f(0)|$ and if $|z| = 3r$, then $|1 - z/z_j| \geq |z/z_j| - 1 \geq 2$. It follows that $M(3r; f)/|f(0)| \geq 2^n$ so taking logarithms $N(r) \leq ((3r)^\sigma - \log|f(0)|)/\log 2 = O(r^\sigma)$. $\qquad\square$

Lemma E.2.9. *If $\sigma > \rho(f)$, then $\sum |z_j|^{-\sigma}$ converges.*

Proof. If $|z_n| = r$, then $N(r) \geq n$ so if $\rho(f) < \tilde{\sigma} < \sigma$, then by Lemma E.2.8 $n = O(|z_n|^{\tilde{\sigma}})$ or $|z_n|^{-\tilde{\sigma}} = O(1/n)$. Thus, $|z_n|^{-\sigma} = O(n^{-\sigma/\tilde{\sigma}})$ so since $\sigma/\tilde{\sigma} > 1$ the claim follows. $\qquad\square$

Lemma E.2.10. *If $\rho(f) < 1$, then the product $\prod_{j\geq 1}(1 - z/z_j)$ converges locally uniformly and is entire of order $\leq \rho(f)$.*

Proof. The product converges absolutely and locally uniformly by Proposition E.2.3 and Lemma E.2.9. Suppose $\rho(f) < \sigma < 1$. Then $\log(1 + |z|)/|z|^{\sigma}$ is bounded, say by C, since it tends to 0 at 0 and ∞.

Now $\log|\prod(1 - z/z_j)| = \sum \log|1 - z/z_j| \leq \sum \log(1 + |z/z_j|)$, which is less than $C \sum |z/z_j|^{\sigma}$. This is $O(|z|^{\sigma})$ since $\sum |z_j|^{-\sigma} < \infty$ by the previous lemma, so the order of the product is at most $\rho(f)$. $\qquad\square$

Lemma E.2.11. *If g is entire and $\operatorname{Re} g(z) = O(|z|^{\sigma})$ as $z \to \infty$, then $M(r; g) = O(r^{\sigma})$ as $r \to \infty$.*

Proof. From (E.1.2) we obtain, with $R = 2r$ and $|z| = r$,

$$M(r; g) \leq |g(0)| + 3 \sup_{|\zeta| = 2r} |\operatorname{Re} g(\zeta)| = O(r^{\sigma}). \qquad\square$$

We shall also need *Liouville's theorem*, which is standard but sometimes stated in a less general form than we need, so we give a proof.

Theorem E.2.12 (Liouville). *Suppose f is an entire function satisfying $f(z) = O(|z|^{\sigma})$ as $z \to \infty$. Then f is a polynomial of degree at most σ. In particular, if f is bounded it is constant.*

Proof. Suppose $f(z) = \sum_{k=0}^{\infty} a_k z^k$. Then $a_k = f^{(k)}(0)/k! = \frac{1}{2\pi i} \oint_{|z|=r} \frac{f(z)\,dz}{z^{k+1}}$ so $|a_k| = O(r^{\sigma-k}) \to 0$ as $r \to \infty$ if $k > \sigma$. Thus all a_k for $k > \sigma$ vanish and f is a polynomial of degree $\leq \sigma$. $\qquad\square$

Proof (Theorem E.2.7). $F(z) = f(z)/\prod_{j \geq 1}(1 - z/z_j)$ is entire without zeros if $f(0) \neq 0$ and by Lemma E.2.10 and Corollary E.2.6 at most of order $\rho(f)$. We may therefore define a branch of $\log F$ which is entire, so that $F(z) = e^{g(z)}$ and $1/F(z) = e^{-g(z)}$ for some entire function g, where $\rho(F)$ and $\rho(1/F) \leq \rho(f)$.

If $\rho(f) < \sigma < 1$ we therefore have $\operatorname{Re} g(z) = O(|z|^{\sigma})$ as $z \to \infty$, so we obtain $|g(z)| = O(|z|^{\sigma})$ as $z \to \infty$ by Lemma E.2.11. Since $\sigma < 1$ it follows from Theorem E.2.12 that g is constant. $\qquad\square$

Remark E.2.13. Theorem E.2.7 is a special case of the Hadamard theorem, which gives a similar, more complicated representation valid for entire functions of finite order; see [1, Chapter 5.3.2] and Exercise E.2.3. This is in turn a refinement of the Weierstrass theorem, a simpler result valid for any entire function; see [1, Chapter 5.1.1].

Exercises

Exercise E.2.1 Show that $\prod_{j=1}^{\infty}(1 + u_j)$ converges if and only if $\sum_{j=N}^{\infty} \log(1 + u_j)$ does for large N. Here log denotes the logarithm with argument in $(-\pi, \pi]$.

Hint: Clearly the product converges if the series does. If the product converges then $u_j \to 0$ so the terms in the series tend to zero.

Consider $p_n = \prod_{j=N}^{n}(1 + u_j)$ and $s_n = \sum_{j=N}^{n} \log(1 + u_j)$ and show that $s_n - \log p_n = 2i k_n \pi$ for some integer k_n. Conclude that $k_n = k_{n+1}$ for large n.

Exercise E.2.2 Show that if f is entire then $\log M(r; f) = O(r^\sigma)$ if $\sigma > \rho(f)$ but not if $\sigma < \rho(f)$. Use this to prove Proposition E.2.5.

Exercise E.2.3 Let $E_n(z) = (1 - z)\exp(\sum_{k=1}^{n} z^k/k)$. Show that E_n is entire with just the simple zero 1 and that $\log|E_n(z)| = O(|z|^{n+1})$ as $z \to 0$.

Also show that if $\rho(f) < \infty$ and the zeros $\neq 0$ of f are $z_1, z_2, \ldots$, then the product $\prod_{j \geq 1} E_n(z/z_j)$ converges locally uniformly if $n + 1 > \rho(f)$, so if n is the smallest integer for which the product is locally uniformly convergent, then $n \leq \rho(f)$. The product is then called the *canonical product* for f. Show that the order of the canonical product is $\leq \min(n + 1, \rho(f))$.

Finally, show that $f(z) = e^{p(z)} z^k \prod_{j \geq 1} E_n(z/z_j)$ where $k \geq 0$ is the multiplicity of 0 as a zero of f and p is a polynomial of degree $\leq \rho(f)$.

This is the Hadamard theorem of Remark E.2.13.

Exercise E.2.4 Use the result of the previous exercise to show that an entire function of non-integer finite order has infinitely many zeros. Use this to prove that such a function takes every complex value at infinitely many points.

E.3 Bounded type

Functions of *bounded type* in a simply connected domain Ω, or equivalently of the Nevanlinna class in Ω, are those that are the quotient of two functions bounded and analytic in Ω (the denominator is assumed not to be identically zero). Using this definition functions of bounded type in Ω are thus meromorphic in Ω. In the literature it is usually assumed that they are analytic in Ω, but to begin with it seems natural to allow poles, which gives a class of functions with nicer properties.

It follows from Liouville's Theorem E.2.12 that only constants are of the Nevanlinna class in $\mathbb{C}$. Another immediate consequence of the definition is that linear combinations, products and quotients (with non-trivial denominator) of functions of bounded type are of bounded type. Since constants are obviously of bounded type this amounts to saying that the set of functions of bounded type on Ω is a *field* under the usual operations of sum and product. Not allowing poles gives us only an algebra.

Now, if g is analytic in Ω and bounded by 1, then $\log|g(z)| \leq 0$. If f is of bounded type in Ω we have $f = g_1/g_2$ with analytic functions g_j, which we may assume bounded by 1. If this is not so, we simply replace g_j by $g_j/\max(\sup|g_1|, \sup|g_2|)$. We then have

$$\log|f| = \log|g_1| - \log|g_2| \leq -\log|g_2| \ .$$

If g_2 has no zeros the right-hand side is positive and harmonic (otherwise just positive and superharmonic), so we have proved one direction of the following theorem, where $\log_+ = \max(0, \log)$.

Theorem E.3.1. *Suppose f is analytic in a simply connected domain Ω. Then* $\log_+ |f|$ *has a harmonic majorant if and only if f may be written as $f = g_1/g_2$ where g_j are analytic and bounded in Ω, and g_2 has no zeros.*

Proof. It remains to prove the converse. But if $u \geq 0$ is a harmonic majorant it has a conjugate function v in the simply connected domain Ω. Setting $g_2 = e^{-(u+iv)}$ we have g_2 analytic and bounded by 1. Then $g_1 = fg_2$ is analytic and $\log |g_1| = \log |f| + \log |g_2| = \log |f| - u \leq 0$, so that g_1 is bounded by 1. Thus $f = g_1/g_2$ is of bounded type. □

This theorem is the reason why textbooks do not allow poles for functions of bounded type. It is also the reason why the Nevanlinna class for domains which are not simply connected is defined by requiring $\log_+ |f|$ to have a harmonic majorant.

It is clear that if Ω has an exterior point p, then $z \mapsto 1/(z - p)$ is bounded and analytic in Ω and thus of bounded type. Consequently also the reciprocal $z \mapsto z - p$ and thus $z \mapsto z$ is of bounded type. It follows that polynomials and thus all rational functions are of bounded type in such a domain.

We shall only deal with functions of bounded type in the upper half-plane $\mathbb{C}_+$, since this is the case of interest to us. We shall therefore often not mention the domain in the sequel. Also note that f is of bounded type in the lower half-plane precisely if $\lambda \mapsto \overline{f(\overline{\lambda})}$ is of bounded type in the upper half-plane.

$\mathbb{C}_+$ is simply connected and since it has exterior points it is clear that any rational function is of bounded type. Functions of bounded type may also have infinitely many poles and zeros, a consequence of the following fundamental theorem.

Theorem E.3.2. *If f is a bounded analytic function in $\mathbb{C}_+$ and $\{z_n\}_1^\infty$ its zeros, repeated according to multiplicity, then*

$$\sum_{n=1}^{\infty} \frac{\operatorname{Im} z_n}{1 + |z_n|^2} < \infty .$$

If a sequence $\{z_n\}_1^\infty$ in $\mathbb{C}_+$ satisfies this condition, then the product

$$B(z) = \prod_{n=1}^{\infty} \frac{|z_n^2 + 1|}{z_n^2 + 1} \frac{z - z_n}{z - \overline{z_n}} ,$$

where the convergence factor $\frac{|z_n^2+1|}{z_n^2+1}$ is to be interpreted as 1 if $z_n = i$, is absolutely and locally uniformly convergent in $\mathbb{C}_+$ outside the closure of $\{z_n\}_1^\infty$ and $|B(z)| \leq 1$.

The zeros of B are precisely all z_n, counted with multiplicity, and the absolute value $|B(z)|$ has non-tangential[3] boundary value 1 at almost all real points.

A product such as $B(z)$ is called a *Blaschke product* in $\mathbb{C}_+$ and is thus in particular of bounded type. If the sequence $\{z_n\}_1^\infty$ is bounded the convergence factors[4]

---

[3] The limit when approaching a point in $\mathbb{R}$ along a curve in $\mathbb{C}_+$ not tangent to $\mathbb{R}$.

[4] Required to guarantee convergence of the arguments of the partial products.

$\left|z_n^2 + 1\right|/(z_n^2 + 1)$ may all be replaced by 1, and if $\{1/z_n\}_1^\infty$ is bounded they may be replaced by $\overline{z_n}/z_n$, so that the product becomes $\prod(1 - z/z_n)/(1 - z/\overline{z_n})$. Note that the various forms of the product differ only by a constant factor of absolute value 1 in the cases when more than one form converges.

We shall not prove the theorem here. A proof may be found for example in Rosenblum and Rovnyak [193, Theorem 5.13]. For us the most important consequence of the theorem is that it allows us to factor out zeros and poles from a function of bounded type, leaving us with a function of bounded type without zeros or poles. That such a function need not be constant follows from the following proposition.

Proposition E.3.3. *If f is of bounded type and m is a non-trivial Nevanlinna function, then $f \circ m$ is of bounded type. In particular, any Nevanlinna function is of bounded type.*

Proof. The first statement is an obvious consequence of the definition of bounded type and the fact that a non-trivial Nevanlinna function maps $\mathbb{C}_+$ into itself. Since $z \mapsto z$ and real constants are of bounded type also the second statement follows. $\square$

Another, equally simple proof that a Nevanlinna function m is of bounded type is to write m as the quotient of $m/(m + i)$ and $1/(m + i)$, both of which are obviously analytic in $\mathbb{C}_+$ and bounded by 1.

Of course, similar arguments show that any analytic function the range of which has an exterior point is of bounded type. Thus the range of any non-constant entire function is dense in $\mathbb{C}$ (the Casorati–Weierstrass theorem). We shall also need the following theorem, similar to Theorem E.1.6.

Theorem E.3.4. *Suppose $d\mu$ is a Radon measure and that $h(\lambda) = \int_{-\infty}^\infty \frac{d\mu(t)}{t-\lambda}$ is absolutely convergent for $\operatorname{Im}\lambda \neq 0$. Then h is of bounded type in the upper and lower half-planes, and as $\lambda \to \infty$ we have $h(\lambda) = o(|\lambda|/|\operatorname{Im}\lambda|)$. In particular, for any $C > 0$ we have $h(\lambda) = o(1)$ as $\lambda \to \infty$ in the double sector $|\operatorname{Re}\lambda| \leq C|\operatorname{Im}\lambda|$.*

Proof. Writing $d\mu$ as a linear combination of the positive and negative parts of its real and imaginary parts it is clear that h is a linear combination of Nevanlinna functions and thus of bounded type. We have

$$|h(\lambda)| \leq \int_{\mathbb{R}} \left|\frac{t-i}{t-\lambda}\right| \frac{|d\mu|(t)}{|t-i|}.$$

Here the first factor tends pointwise to 0 as $\lambda \to \infty$ and may be easily estimated by $(2|\lambda| + 1)/|\operatorname{Im}\lambda|$ so the first estimate follows by dominated convergence. Since $|\lambda/\operatorname{Im}\lambda|$ is bounded in the double sector the second estimate follows. $\square$

The representation formula

To further analyze the properties of a function of bounded type we may, in view of the previous section, first factor out the zeros and poles in the form of Blaschke products. We shall need the following lemma.

Lemma E.3.5. *Suppose f is a bounded analytic function in $\mathbb{C}_+$ and let B be the Blaschke product formed by it zeros. Then $f = Bg$ where g is analytic, bounded and without zeros in $\mathbb{C}_+$.*

Proof. Suppose z_1 is a zero of f and consider the set of z such that $|z - \overline{z_1}| / |z - z_1| \leq 1 + \varepsilon$. If $\varepsilon > 0$ this is everything outside a circle containing z_1 and contained in $\mathbb{C}_+$. Outside this circle the analytic function $g_1(z) = f(z)(z - \overline{z_1})/(z - z_1)$ is bounded by $(1 + \varepsilon) \sup |f|$.

By the maximum principle the same bound is valid inside the circle, and since $\varepsilon > 0$ is arbitrary we have $\sup |g_1| \leq \sup |f|$. Letting[5] B_n be the nth partial product of B and $g_n = f/B_n$ we obtain $\sup |g_n| \leq \sup |f|$ by induction, so letting $n \to \infty$ we obtain $\sup |g| \leq \sup |f|$.

Actually $\sup |g| = \sup |f|$ since $\sup |B| \leq 1$ (which also shows that $\sup |B| = 1$). $\qquad\square$

This leads to the following fundamental theorem.

Theorem E.3.6. *Suppose f is of bounded type. Then we may write*

$$f(z) = \frac{B_1(z)}{B_2(z)} e^{iG(z)},$$

where B_1 and B_2 are Blaschke products without common zeros and G is a difference of Nevanlinna functions, determined up to addition by an integer multiple of 2π.

Conversely, every function of this form is of bounded type.

Proof. Suppose first that f is analytic in $\mathbb{C}_+$ without zeros and bounded by 1. We may then define a branch of $\log f$ in $\mathbb{C}_+$, and it follows that its real part $\log |f|$ is non-positive. The analytic function $m = -i \log f$ is therefore a Nevanlinna function, and $f(z) = e^{im(z)}$.

We write a general f as a quotient of bounded analytic functions and factor out the zeros of these by Blaschke products. What remains is then by Lemma E.3.5 a quotient of bounded analytic functions without zeros which we may as before assume bounded by 1. This gives the desired representation. Uniqueness follows since the Blaschke products are determined by their zeros, and then G will be determined up to addition by an integer multiple of 2π.

The converse follows immediately from the fact that Blaschke products are bounded analytic functions and e^{im} is bounded and analytic for any Nevanlinna function m. $\qquad\square$

---

[5] We have just proved a version of the Schwarz lemma.

This is essentially the Nevanlinna factorization. In textbooks one normally goes a few steps further. The representation formula (E.1.1) for a Nevanlinna function shows that for a difference G of Nevanlinna functions we have

$$G(z) = A + Bz + \int_{-\infty}^{\infty} \frac{1+tz}{t-z} \frac{d\eta(t)}{1+t^2},$$

where the real constants A and B and the measure $d\eta$ are still uniquely determined, although B may now be negative and $d\eta$ is a signed (but real) measure such that $\int_{-\infty}^{\infty} |d\eta(t)|/(t^2+1)$ is convergent. The uniqueness of $d\eta$ follows from the Stieltjes inversion formula, and then the linear part is also determined. Since G is in any case only determined up to addition by an integer multiple of 2π one normally writes

$$e^{iG(z)} = Ce^{-i\tau z} \exp\left(i \int_{-\infty}^{\infty} \frac{1+tz}{t-z} \frac{d\eta(t)}{1+t^2}\right),$$

where now $C = e^{iA}$ is a uniquely determined constant with $|C| = 1$ and $\tau = -B$ is real, uniquely determined and called the *mean type* of f. By splitting the uniquely determined measure $d\eta$ into its positive and negative parts the integral is uniquely split into the difference of two Nevanlinna functions without the linear part and with mutually singular spectral measures. In the textbooks one goes one step further and splits these measures into their absolutely continuous and singular parts, but this will be of no use to us.

The mean type will be important for us, since if $\neq 0$ it determines the growth of e^{iG} in double sectors $|\operatorname{Re}\lambda| \leq C|\operatorname{Im}\lambda|$, the imaginary part of the integral being always $o(|z|)$ as $z \to \infty$ in such sectors, as follows from Theorem E.1.6.

We have the following useful criterion for bounded type.

Theorem E.3.7. *Suppose f is continuous in $\overline{\mathbb{C}_+}$ and analytic in $\mathbb{C}_+$, that $|f(z)| \leq M \exp(C|z|)$ and that $\int_{-\infty}^{\infty} \frac{\log_+|f(t)|\,dt}{t^2+1} < \infty$. Then f is of bounded type in $\mathbb{C}_+$.*

Given a function f we define $f^*(z) = \overline{f(\bar{z})}$. Clearly a function is of bounded type in the lower half-plane $\mathbb{C}_-$ precisely if f^* is of bounded type in $\mathbb{C}_+$. We shall say that f has mean type τ in $\mathbb{C}_-$ if f^* has mean type τ in $\mathbb{C}_+$. In particular, if $f^* = f$ in $\mathbb{C}_+$, with mean type < 0, then f tends exponentially to 0 as $\lambda \to \infty$ in any double sector $|\operatorname{Re}\lambda| \leq C|\operatorname{Im}\lambda|$.

Theorem E.3.8. *Suppose f is entire and that f and f^* are of bounded type in $\mathbb{C}_+$. Then f is of exponential type equal to the maximum of the mean types of f and f^*.*

Combining Theorems E.3.7 and E.3.8 we obtain a theorem of M. G. Kreĭn [132]. See also Levin [145, Lecture 16, Theorem 3] or Rosenblum and Rovnyak [193, Theorems 6.17 and 6.18].

E.4 A lemma of de Branges

The main result in this section is the following theorem, which is a very slight extension of a lemma by de Branges. We shall give a full proof, however, since there is an oversight in the proof by de Branges which will be corrected below. We are not aware of the oversight being noted in the literature, but a correct proof may also be found in the master's thesis of Koliander [127]. See also Heins [100].

Theorem E.4.1. *Suppose F_j are entire functions of zero exponential type, and assume that for some $\alpha \geq 0$ we have*

$$\min(|F_1(\lambda)|, |F_2(\lambda)|) = o(|\lambda/\operatorname{Im}\lambda|^{\alpha})$$

uniformly in $\operatorname{Re}\lambda$ as $|\operatorname{Im}\lambda| \to \infty$. Then F_1 or F_2 vanishes identically.

This is a simple consequence of the following lemma, which is essentially de Branges' [34, Lemma 8, p. 108].

Lemma E.4.2. *Let F_j be entire functions of zero exponential type, and assume that $\min(|F_1(z)|, |F_2(z)|) = o(1)$ uniformly in $\operatorname{Re}z$ as $|\operatorname{Im}z| \to \infty$. Then F_1 or F_2 is identically zero.*

For the proof of Theorem E.4.1 we also need the following theorem, which is one of the simplest versions of an elaboration of the maximum principle known as the *Phragmén–Lindelöf principle*. Proofs are given in many textbooks on complex analysis, but for the reader's convenience we also give the standard proof here.

Theorem E.4.3 (Phragmén–Lindelöf). *Suppose f is analytic in an open sector bounded by two rays from the origin making an angle $< \pi/\rho$, $\rho > 0$, that it is continuous in the closed sector, that it is bounded by C on the rays, and that $|f(z)| \leq Ae^{B|z|^{\rho}}$ in the sector, for some constants A and B. Then f is bounded by C in the sector.*

Proof. Replacing f by $z \mapsto f(e^{i\theta}z)$ for an appropriate $\theta \in \mathbb{R}$ we may without loss of generality assume that the rays make the angles $\pm\beta$, $0 < \beta < \pi/(2\rho)$, with the positive real axis. Let $\varepsilon > 0$ and $F(z) = e^{-\varepsilon z^{\gamma}} f(z)$, where $\rho < \gamma < \pi/(2\beta)$ and the branch of z^{γ} is chosen to be positive real for positive real z.

Now, for $z = re^{\pm i\beta}$ we have $|F(z)| = e^{-\varepsilon r^{\gamma}\cos(\beta\gamma)} |f(z)|$, where $\cos(\beta\gamma) > 0$. We then have $|F(z)| \leq C$ on the rays, and for $z = Re^{i\delta}$ with $|\delta| \leq \beta$ we have

$$|F(z)| \leq A\exp(BR^{\rho} - \varepsilon R^{\gamma}\cos(\beta\gamma)),$$

which tends to 0 as $R \to \infty$. Thus, on all circular sectors bounded by the rays we have $|F(z)| \leq C$ on the boundary if the radius R is sufficiently large. By the maximum principle this also holds in the interior of the circular sector. Since R can be chosen arbitrarily large, the bound is valid in the entire sector bounded by the rays. It follows that if z is in this domain, then $|f(z)| \leq Ce^{\varepsilon|z|^{\gamma}}$ for every $\varepsilon > 0$, and letting $\varepsilon \to 0$ we obtain the desired result. □

Proof (Theorem E.4.1). Suppose first that F_1 is a polynomial not identically zero. Then, by assumption, $F_2(\lambda) = o(1)$ as $\lambda \to \infty$ in any double sector $|\operatorname{Re}\lambda| \leq C |\operatorname{Im}\lambda|$, so by the Phragmén–Lindelöf Theorem E.4.3 and Liouville's Theorem E.2.12 it follows that F_2 vanishes identically. Similarly, $F_1 \equiv 0$ if F_2 is a polynomial.

In all other cases F_1, F_2 both have infinitely many zeros, so if $n \geq \alpha$ and $z_1, \ldots, z_n$ are zeros of F_1 we put $G_1(\lambda) = F_1(\lambda)/\prod_1^n (\lambda - z_j)$. Defining G_2 similarly we now have $\min(|G_1(\lambda)|, |G_2(\lambda)|) = o(1)$ uniformly in $\operatorname{Re}\lambda$ as $\operatorname{Im}\lambda \to \pm\infty$, while G_1, G_2 are still entire of zero exponential type. By Lemma E.4.2 it follows that G_1 or G_2 is identically zero, and the lemma follows. $\qquad\square$

To prove Lemma E.4.2 we need some additional lemmas. If u is continuous on $|z| = R$ the *Poisson integral*[6]

$$P(u; R)(z) = \frac{1}{2\pi} \int_{-\pi}^{\pi} \frac{R - |z|^2}{|Re^{i\theta} - z|^2} u(Re^{i\theta}) \, d\theta$$

is harmonic in $|z| < R$, continuous in $|z| \leq R$, and equals u on $|z| = R$. We shall need the following lemma.

Lemma E.4.4. *Suppose F is analytic in $|z| < R$ and continuous in $|z| \leq R$ and let $u = \log_+ |F|$. Then $u(z) \leq P(u, R)(z)$ in $|z| \leq R$.*

Proof. Since $u(z) \geq 0$ on $|z| = R$ the maximum principle for harmonic functions shows that $P(u; R) \geq 0$ throughout $|z| \leq R$.

In the open set $M = \{z : |z| < R \text{ and } u(z) > 0\}$, u is harmonic, being the real part of (any branch of) $\log F$, so $P(u; R) - u$ is harmonic in M and $P(u; R) - u \geq 0$ on ∂M. Using the maximum principle again we find that $P(u; R) - u \geq 0$ in M and therefore throughout $|z| \leq R$. $\qquad\square$

Remark E.4.5. A function which is below all its Poisson integrals is called *subharmonic*, so Lemma E.4.4 states that $\log_+ |F|$ is subharmonic if F is analytic. There are various, more or less equivalent, definitions of subharmonicity, the most general and concise requiring that $\Delta u \geq 0$ in the distributional sense, i.e., Δu is a positive measure, where Δ is the Laplacian.

Lemma E.4.6. *Suppose F is entire of exponential type. If there is a constant C and a sequence $r_j \to \infty$ such that $|F(z)| = \mathcal{O}(1)$ for $|\operatorname{Im} z| \geq C$ and $|z| = r_j$ as $j \to \infty$, then F is constant.*

Proof. Setting $u = \log_+ |F|$ we have $u(z) = \mathcal{O}(1)$ for $|\operatorname{Im} z| \geq C$ and $|z| = r_j$ as $j \to \infty$. If $z = r_j e^{i\theta}$ the condition $|\operatorname{Im} z| \leq C$ means $|\sin\theta| \leq C/r_j$, and the measure of the set of $\theta \in [0, 2\pi]$ satisfying this is $\mathcal{O}(1/r_j)$ as $j \to \infty$, whereas $|F(z)| \leq e^{\mathcal{O}(|z|)}$ so that $u(r_j e^{i\theta}) = \mathcal{O}(r_j)$. Thus $\int_{-\pi}^{\pi} u(r_j e^{i\theta}) \, d\theta = \mathcal{O}(1)$ as $j \to \infty$.

Now, since u is subharmonic and non-negative we have, by the Poisson integral formula (E.1.3),

---

[6] For claims about harmonic functions and Poisson integrals we refer to Ahlfors [1, Chapter 4.6].

$$0 \le u(z) \le \frac{1}{2\pi} \int\limits_{-\pi}^{\pi} \frac{r_j^2 - |z|^2}{\left| r_j e^{i\theta} - z \right|^2} u(r_j e^{i\theta}) \, d\theta \le \frac{3}{2\pi} \int\limits_{-\pi}^{\pi} u(r_j e^{i\theta}) \, d\theta$$

if $|z| \le r_j/2$, since then

$$0 \le \frac{r_j^2 - |z|^2}{\left| r_j e^{i\theta} - z \right|^2} \le \frac{r_j^2 - |z|^2}{(r_j - |z|)^2} = \frac{r_j + |z|}{r_j - |z|} \le 3 \, .$$

It follows that $u(z) = \mathcal{O}(1)$ if $|z| \le r_j/2$, so $|F(z)| = \mathcal{O}(1)$ for large $|z|$. Thus F is bounded, and thus constant by Liouville's Theorem E.2.12. □

Next we prove a version of de Branges' [34, Lemma 7, p. 108], with the added assumption that $0 < p < 1$, with p defined as below. Without the extra assumption the lemma is not true. If F is an entire function, put $f(x, \theta) = \log_+ \left| F(e^{x+i\theta}) \right|$ and

$$v(x) = \int\limits_{-\pi}^{\pi} (f(x, \theta))^2 \, d\theta.$$

Since $F(e^z)$ is analytic the function f is continuous, subharmonic and non-negative, with period 2π in θ. Let $M = \{(x, \theta) : f(x, \theta) > 0\}$. The set M is open and 2π-periodic in θ and f is harmonic in M. Define $p(x)$ so that $2\pi p(x)$ is the measure of the trace

$$M(x) = \{\theta \in [-\pi, \pi) : (x, \theta) \in M\}.$$

The function p is lower semi-continuous[7] and $p(x) \le 1$. Now assume one may choose a so that $p(x) > 0$ for $x \ge a$. Then p is locally in $[a, \infty)$ bounded away from 0, so that $1/p$ is upper semi-continuous, positive and locally bounded. We may therefore define the strictly increasing function

$$s(x) = \int\limits_{a}^{x} \exp\left(\int\limits_{a}^{t} 1/p \right) dt.$$

The function s is continuously differentiable with derivative > 0, so it has an inverse of class C^1. Furthermore, s' is locally absolutely continuous, so the inverse also has a locally absolutely continuous derivative.

Lemma E.4.7. *Suppose $0 < p(x) < 1$ for all $x \ge a$. Then the quantity v is a convex function of $s > 0$.*

Proof. We may think of f as defined on a cylindrical manifold $\mathcal{C}$ with coordinates $(x, \theta) \in \mathbb{R} \times [-\pi, \pi)$ of which M is an open subset, and the boundary ∂M is a level set of $|F|$. The boundary is therefore of class C^1 except where the gradient of $|F|$ vanishes. However, the length of the gradient equals $|F'|$, as is easily seen, and the

---

[7] By Lemma B.9.1 $p \in I^+(\mathbb{R})$ since M is open.

exceptional points are therefore locally finite in number. At such points ∂M has a corner. We may therefore use integration by parts (the divergence theorem or the general Stokes theorem; see Rudin [195, Theorem 10.33]) for the set M.

Denote by n_x the x-component of the outer normal and by $\partial/\partial n$ the normal derivative on ∂M. Assuming $\varphi \in C_0^\infty(\mathcal{C})$ and integrating by parts we obtain

$$\int_M f^2 \varphi_x = \int_{\partial M} f^2 \varphi\, n_x - \int_M (f^2)_x \varphi = -\int_M 2 f f_x \varphi\,,$$

$$\int_M f^2 \Delta\varphi = \int_{\partial M} \left(f^2 \frac{\partial \varphi}{\partial n} - 2\varphi f \frac{\partial f}{\partial n}\right) + \int_M \varphi \Delta f^2 = 2\int_M \varphi\, |\mathrm{grad}\, f|^2\,,$$

since f vanishes on ∂M and is harmonic in M. Now suppose φ is independent of θ. Then we may write the above as

$$\int_R \varphi' v = -\int_R \varphi(x)\Big(2\int_{M(x)} f f_x\Big)\, dx\,,$$

$$\int_R \varphi'' v = \int_R \varphi(x)\Big(2\int_{M(x)} (f_x^2 + f_\theta^2)\Big)\, dx\,,$$

so in the sense of distributions $v'(x) = 2\int_{M(x)} f f_x$ and $v''(x) = 2\int_{M(x)} (f_x^2 + f_\theta^2)$.

The function s has a C^1 inverse, so we define $g = v \circ s^{-1}$ and must then show that g is convex. Differentiating we obtain $v' = s'g' \circ s$ and $v'' = (s')^2 g'' \circ s + s''g' \circ s$. Thus $(s')^2 g'' \circ s = v'' - v's''/s' = v'' - v'/p$. We need to prove that this is non-negative. Now

$$v''(x) - v'(x)/p(x) = 2\int_{M(x)} (f_x^2 + f_\theta^2 - f f_x/p)$$

$$= 2\int_{M(x)} ((f_x - f/2p)^2 + f_\theta^2 - f^2/4p^2)\, d\theta \geq 2\Big(\int_{M(x)} f_\theta^2 - \frac{1}{4p^2}\int_{M(x)} f^2\Big).$$

We are therefore done if for all x we have the inequality

$$\int_{M(x)} f_\theta^2 \geq \frac{1}{4p^2(x)}\int_{M(x)} f^2. \tag{E.4.1}$$

Since $p(x) < 1$ the function $\theta \mapsto f(x,\theta)$ has a zero, so that $f(x,\cdot)$ vanishes at the endpoints of all components of the open set $M(x)$. If I is such a component we therefore have $\int_I (f_\theta)^2 \geq (\pi/|I|)^2 \int_I f^2$ where $|I|$ is the length of I.

By the theory of Section 1.4 this just expresses the fact that the smallest eigenvalue of $-f'' = \lambda f$ with Dirichlet boundary conditions on I is $(\pi/|I|)^2$. We have $(\pi/|I|)^2 \geq$

$(2p)^{-2}$ since $|I| \leq 2\pi p$, so adding the inequalities for the various components of $M(x)$ we obtain (E.4.1), and the proof is finished.   □

Proof (Lemma E.4.2). Suppose first that F_1 is bounded and therefore constant. If this constant is not zero the assumption implies that $F_2(\lambda) \to 0$ as $\lambda \to \infty$ in any double sector $|\mathrm{Re}\,\lambda| \leq C\,|\mathrm{Im}\,\lambda|$. Since F_2 is of exponential type the Phragmén–Lindelöf principle (Theorem E.4.3) shows that F_2 is bounded and thus constant by Liouville's Theorem E.2.12. The limit is zero in the double sector and therefore $F_2 \equiv 0$. Similarly, $F_1 \equiv 0$ if F_2 is bounded. We may thus assume that F_1 and F_2 are both unbounded.

If there is a sequence $r_j \to \infty$ such that $F_k(z)$ satisfies the assumptions of Lemma E.4.6, then F_k is constant according to Lemma E.4.6.

We may thus also assume that for $k = 1, 2$ and every large r the inequality $|F_k(z)| \leq 1$ is violated for some z with $|z| = r$ and $|\mathrm{Im}\,z| > C$. Since F_k is analytic and thus continuous, the opposite inequalities must hold for every large r on some open θ-sets if $z = re^{i\theta}$.

But for large $|z|$ and $|\mathrm{Im}\,z| > C$ we must have $|F_2(z)| \leq 1$ if $|F_1(z)| > 1$ and *vice versa*. It follows that there exists an a such that $0 < p_k(x) < 1$, $k = 1, 2$, for $x > a$. Now associate with F_k functions f_k, v_k, p_k and s_k as before.

By Cauchy–Schwarz $\frac{1}{2\pi}\int_{-\pi}^{\pi} f_k(x,\theta)\,d\theta \leq \left(\frac{1}{2\pi}\int_{-\pi}^{\pi} f_k^2(x,\theta)\,d\theta\right)^{1/2} = v_k^{1/2}(x)$, so it follows that if v_k is bounded, then so is F_k, using the Poisson integral formula in much the same way as in the proof of Lemma E.4.6. Thus v_k must be unbounded, and since it is non-negative and convex as a function of s_k there is for $k = 1, 2$ a constant $c > 0$ such that $v_k(x) \geq cs_k(x)$ for large x. We shall show that this contradicts the assumption of order for F_k, $k = 1, 2$.

Using the convexity of the exponential function we obtain for large $x > a$

$$(v_1(x) + v_2(x))/2 \geq c \int_a^x \exp\left(\int_a^t (1/p_1 + 1/p_2)/2\right) dt. \qquad \text{(E.4.2)}$$

Now, by assumption $\min(f_1(x,\theta), f_2(x,\theta)) = 0$ for large x and $C \leq e^x\,|\sin\theta|$ so that then f_1 or f_2 equal zero. The measure of the θ-set not satisfying $e^x\,|\sin\theta| \geq C$ for a given x is less than $2\pi Ce^{-x}$. It follows that $p_1 + p_2 \leq 1 + 2Ce^{-x}$. Since

$$\frac{1}{2}\left(\frac{1}{p_1} + \frac{1}{p_2}\right) = \frac{p_1 + p_2}{2p_1 p_2} \geq \frac{2}{p_1 + p_2} \geq \frac{2e^x}{e^x + 2C}$$

the integral in (E.4.2) is at least $\frac{1}{2}(e^{2x} - e^{2a})/(e^a + 2C)^2$. Thus $v_1(x) + v_2(x) \geq De^{2x}$ for some constant $D > 0$ and large x. The assumption of order for F_k means, however, that $f_k(x,\theta) = o(e^x)$ for large x so that $v_k(x) = o(e^{2x})$ and thus $v_1(x) + v_2(x) = o(e^{2x})$. This contradiction proves the lemma.   □

E.5 Notes and remarks

Most of the contents of this appendix is classical by now and may be found in one or another of the texts already referred to, for example Ahlfors [1] and for material related to functions of bounded type in Rosenblum and Rovnyak [193].

The material in Section E.4 is more specialized. The crucial parts may be found in de Branges [34, Lemmas 7 and 8, p. 107] (note the oversight pointed out in the text) or Bennewitz, Brown and Weikard [21]. The idea used in the fundamental Lemma E.4.2 is due to Carleman [45]. Carleman's idea was elaborated by Heins [100], which led de Branges to the results quoted.

Appendix F
The Camassa–Holm equation

This appendix is of a different character to the rest of the book. Many arguments given are not completely rigorous, and the aim is simply to give a motivation for our study of scattering and inverse scattering for left-definite Sturm–Liouville equations in Chapter 8.

We shall give an informal explanation of the classical concept of an integrable system and then explain why the term is nowadays applied to certain non-linear partial differential equations and its connection with spectral theory. We will then focus on the Camassa–Holm equation. For further details we refer the reader to the book [84] by Gesztesy and Holden.

Integrable systems

The term *integrable system* (or *completely integrable system*) was coined in the 19th century in connection with Hamilton's formulation of classical mechanics. To explain it[1] we first consider a system of particles moving freely in space under the influence of a potential force. If there are k particles their positions are given by $3k$ coordinates $x = (x_1, \ldots, x_{3k})$, giving the system $3k$ *degrees of freedom*, and their *momenta* (mass times velocity) by another $3k$ coordinates $p = (p_1, \ldots, p_{3k})$. Thus the velocity $\dot{x} = pM$, where M is a diagonal matrix with the reciprocals of the various particle masses on the diagonal. The coordinates x vary in the *configuration space* and the variables (x, p) in the *phase space* with dimensions $3k$ and $6k$ respectively.

A potential force is given as $-\operatorname{grad} U$ where the *potential* U is a function of x, so Newton's second law Force=Mass×Acceleration, where the acceleration is the second derivative $\ddot{x}$, takes the form $\dot{p} = -\operatorname{grad} U$. Introducing the *kinetic energy* $\frac{1}{2} pMp^*$ (p^* denoting the transpose of p) and adding it to the potential energy U gives the *total energy* $H(x, p) = \frac{1}{2} pMp^* + U(x)$, and we obtain *Hamilton's equations*

---

[1] More detailed and rigorous explanations are to be found in Arnold [6] and the excellent survey Palais [177].

© Springer Nature Switzerland AG 2020

C. Bennewitz et al., *Spectral and Scattering Theory for Ordinary Differential Equations*,
Universitext, https://doi.org/10.1007/978-3-030-59088-8

$$\begin{cases} \dot{x} = H_p \\ \dot{p} = -H_x\,, \end{cases} \tag{F.1}$$

where H_x and H_p are the gradients of H with respect to the x and p variables. This is a system of $6k$ ordinary differential equations with $6k$ unknown, so under reasonable conditions on H the corresponding initial value problem has a unique solution (see Appendix D). Thus we can specify initial positions and momenta and obtain a unique solution of the system (F.1).

Now, one also considers particles moving under so-called holonomic constraints. The simplest example would be a (plane) pendulum, where a single particle is constrained to move on a circle centered at the fulcrum of the pendulum. The position of the pendulum is then determined by just one variable, for example the central angle with a fixed radius in the circle, so the pendulum would have just one degree of freedom.

Hamilton showed that even if our system of particles (in the case of a rigid body they can even be infinitely many) is constrained to a lower-dimensional manifold of dimension n, then the system can be described by 'generalized' positions $x = (x_1, \ldots, x_n)$ and momenta $p = (p_1, \ldots, p_n)$ still satisfying Hamilton's equations for some function H of (x, p) called the *Hamiltonian* of the system. Variables (x, p) for which this is true are called *canonical coordinates*, and transformations between two sets of canonical coordinates are called canonical transformations.

The Hamiltonian is a constant of the motion, i.e., if (x, p) solve Hamilton's equations, then $t \mapsto H(x(t), p(t))$ is constant and thus determined by the initial value of (x, p). This is clear, since using the chain rule and Hamilton's equations we obtain $\frac{d}{dt} H(x, p) = H_x \cdot \dot{x} + H_p \cdot \dot{p} = H_x \cdot H_p + H_p \cdot (-H_x) = 0$. Interpreting H as representing the energy of the system we therefore have conservation of energy as the system evolves in time, and this means that any orbit of (F.1) is contained in a level set of H determined by the initial data.

In many cases one may find other conserved quantities, often caused by configuration symmetries, and for each such constant of the motion the orbits are constrained to a level set of the conserved quantity. If there are n independent conserved quantities the system of particles is called *completely integrable*, and according to a fundamental theorem due to Liouville and Arnold's modern formulation [6, p. 272] any completely integrable system has a set of canonical coordinates, called *action angle variables*, with p constant on the orbit.

This means that the Hamiltonian does not depend on x and that H_p is constant on the orbit, so that x evolves linearly in time. In turn, this means that we can solve Hamilton's equations explicitly as soon as we have found action angle coordinates. In practice completely integrable systems are rare, and the cases where one can determine action angle coordinates explicitly even rarer. Nevertheless, this circle of ideas is central to classical mechanics.

If one considers infinite systems of particles, or rather a body in continuum mechanics, mathematical models tend to yield partial differential equations instead of a system of ordinary differential equations. There are then obvious difficulties in generalizing the idea of an integrable system; it is clearly not enough to have infinitely many conserved quantities to obtain an analogue of an integrable system since even if we do, we may still be missing crucial conserved quantities.

It would perhaps seem that the existence of some general form of action angle coordinates would be a reasonable definition of an integrable system in infinite dimensions, but there is no generally accepted definition in this case. However, in the late 1960s a discovery was made which turned this circle of ideas into an area of very active research.

The Korteweg–de Vries equation

The remarkable story of the Korteweg–de Vries (KdV) equation starts in 1834 when a British engineer, J. Scott Russell, noticed a curious phenomenon which he called a *solitary wave*. He had watched a barge pulled by horses along a canal suddenly stopping. This caused a smooth hump of water to move for miles with constant speed along the canal without changing shape. Since the experts believed such a wave to be unstable, no satisfactory explanation could be given until in 1895 the Dutch mathematicians Diederik Korteweg and Gustav de Vries [129] managed to find an approximation to the Euler equations for (one-dimensional) incompressible flow which had stable solutions of the form

$$u(x,t) = \frac{c}{2\cosh^2(\sqrt{c}(x - ct + a)/2)}$$

for any constants a and $c > 0$. These solutions correspond well to the wave observed by Russell, even to the point that the amplitude of the wave is proportional to the wave speed c.

The function u given is a solution of the non-linear partial differential equation $u_t + 6uu_x + u_{xxx} = 0$, which is one of the standard forms of the KdV equation. However, any equation $u_t + Auu_x + Bu_{xxx} = 0$ for non-zero constants A and B is also called a KdV equation, since the difference is just a scaling of dependent and independent variables. No other solutions than the one given which decay as $x \to \pm\infty$ were known until 1965, when Zabusky and Kruskal [226] carried out some remarkable numerical experiments inspired by earlier experiments by Fermi, Pasta and Ulam (the article may be found in [72]), contradicting a general belief that in a nonlinear, conservative system energy would eventually distribute evenly over the system.

They found that with initial data consisting of a sum of spatially well separated solitary waves with different wave speeds the solution, after the expected nonlinear interaction, eventually separates again into (approximatively) such a sum, with the same wave speeds but changed phases. This is behavior normally associated with (quantum) particles, and for this reason the authors coined the term *soliton* for such solutions.

In a famous 1967 paper Gardner, Green, Kruskal and Miura [80] made another totally unexpected discovery, this time connected with spectral theory. They discovered that using a solution u of $u_t - 6uu_x + u_{xxx} = 0$ as a potential in the one-dimensional Schrödinger[2] equation $-f_{xx} + uf = \lambda f$ in $L^2(\mathbb{R})$ (in this equation t enters just as a parameter on which u depends), any eigenvalue of this equation for a certain value of t is also an eigenvalue for all other values of t. Such an eigenvalue is therefore a conserved quantity for the time evolution given by the KdV equation. The fact that eigenvalues are conserved under the KdV flow indicates that they could be part of a set of action angle variables. Exploring this in detail, Zakharov and Faddeev [227] argued that the KdV equation should be viewed as an integrable system of infinite dimension.

The spectral problem has two singular endpoints, but the theory of Section 4.4 is unsuitable for this application, because of the artificial choice of an anchor point. A more intrinsic normalization is obtained by use of an expansion theorem with respect to solutions of the equation with prescribed asymptotic behavior at $\pm\infty$. We are thus led to attempt to interpret scattering data, as discussed in Chapter 8, as action angle variables. Doing this one finds that eigenvalues and the transmission coefficient are independent of the time evolution, that reflection coefficients evolve like $\mathfrak{R}_\pm(\lambda; t) = \mathfrak{R}_\pm(\lambda; 0) \exp(\mp 8ik\lambda t)$ where $k = \sqrt{\lambda}$, and the norming constants like $\|f_\pm(\cdot, \lambda_n; t)\|^2 = \|f_\pm(\cdot, \lambda_n; 0)\|^2 \exp(\mp 8ik_n\lambda_n t)$ if λ_n is an eigenvalue (so that $ik_n = \sqrt{-\lambda_n}$ is real). We shall carry out similar calculations below, for the spectral problem associated with the Camassa–Holm equation. Returning to the KdV equation, the known direct and inverse scattering for the Schrödinger equation on $\mathbb{R}$ is then used as follows:

Given initial data u_0 for the KdV equation first determine the scattering data for $-f'' + u_0 f = \lambda f$ and then evolve them as described above. Inverse scattering theory then determines u for future values of t, allowing one, for example, to conclude that the solution to KdV exists for all future times. Thus it clearly makes sense to consider the scattering data associated with a solution of KdV as action angle variables for the time evolution (the logarithms evolve linearly in time).

Lax pairs and compatibility conditions

A basic mechanism explaining the isospectral property of the KdV was found by Peter Lax [139] in 1968. The starting point of Lax was the well-known fact that if two operators are *similar* they have the same spectra. The isospectral property of one-dimensional Schrödinger operators when the potential is a solution of KdV might therefore be a consequence of the operators being similar.

Two operators L and K are called similar if there is an invertible operator S such that $LS = SK$. If $\lambda \in \mathbb{C}$ we also have $(L - \lambda)S = S(K - \lambda)$. It follows that $\sigma(L) = \sigma(K)$; in particular f is an eigenvector of K if and only if Sf is an eigenvector of L with the same eigenvalue. In fact, the multiplicity of an eigenvalue (geometric and algebraic) is the same for L and K.

---

[2] Using the equation $u_t + 6uu_x + u_{xxx} = 0$ corresponds to the eigenvalue equation $-f_{xx} - uf = \lambda u$.

Now suppose we have a differentiable family $t \mapsto S(t)$ of invertible operators, and define $L(t)$ by $L(t)S(t) = S(t)K$ where K is some given (t-independent) operator. Differentiating[3] we obtain $L_t S + LS_t = S_t K$ so $L_t + LS_t S^{-1} = S_t K S^{-1} = S_t S^{-1} L$. Setting $P = S_t S^{-1}$ we obtain the *Lax equation* $L_t = [P, L]$, where $[P, L] = PL - LP$ is the *commutator* of P and L.

Conversely, suppose $L(t)$ is differentiable and satisfies $L_t = [P, L]$ for some operator P. Let S satisfy the linear differential equation $S_t = PS$ with invertible initial value for $t = 0$. The solution will then be invertible for all t and we find, setting $K = S^{-1}LS$ and differentiating $SK = LS$, that

$$S_t K + SK_t = L_t S + LS_t = [P, L]S + LS_t = S_t S^{-1} LS - LS_t + LS_t = S_t K \,.$$

Thus $SK_t = 0$, so K is independent of t and all $L(t)$ are similar to K and thus to each other.

Remark F.1. All of the above are formal calculations which are easily justified if all operators involved are bounded. Since we want to apply all this when L is a differential operator the arguments above must be considered just heuristics. However, if all operators $L(t)$ have a common point in their resolvent set, *e.g.* if all $L(t)$ are self-adjoint, we may apply all the above to their resolvents and so allow unbounded L. Furthermore, in this case it is natural to assume all $S(t)$ to be unitary, which implies that P is skew-self-adjoint (iP is self-adjoint; see Example 3.1.7) and also simplifies the rigorous justification of our calculation.

We now introduce the abbreviation D for the operator of differentiation with respect to x and consider the case $L = -D^2 + u$ where the potential u depends on a parameter t. Thus L_t is multiplication by u_t. To find P such that $[P, L]$ is a multiplication operator we may then try different skew-symmetric differential operators. The simplest such operator is D. In this case the Lax equation becomes $u_t = u_x$, a linear (uni-directional) wave equation. Next we may try a third order operator, and simple calculations show that we should take $P = 4D^3 - 6uD - 3u_x$. The Lax equation then becomes the KdV equation $u_t - 6uu_x + u_{xxx} = 0$. For each odd order one may similarly determine an operator P which gives a non-linear evolution equation. This sequence of equations is called the *KdV hierarchy*.

The study of the KdV equation led to the discovery of a large number of other non-linear evolution equations with more or less similar properties which we will not discuss here.

The Lax equation can also be formulated as a *compatibility condition*. Suppose f is a common solution of the equations $Lf = \lambda f$ and $f_t = Pf$, where λ does not depend on t. Then $PLf = \lambda Pf = \lambda f_t = (\lambda f)_t = (Lf)_t = L_t f + LPf$. Since $L_t f = u_t f$ this is $[P, L]f = u_t f$, which gives the KdV equation.

Since $D^3 = D(u - L)$ we may rewrite P as $P = 4D(u - L) - 6uD - 3u_x$, so if $Lf = \lambda f$ we have $Pf = u_x f - (4\lambda + 2u)f_x$. The KdV equation is therefore also the compatibility condition for $Lf = \lambda f$ and $f_t = u_x f - (4\lambda + 2u)f_x$.

---

[3] A lower index t represents a time differentiation.

One may carry this idea a step further by writing both equations as first order systems $F_x = XF$, $F_t = TF$, in terms of $F = \begin{pmatrix} f \\ f_x \end{pmatrix}$. We obtain

$$[X, T]F = XF_t - TF_x = (XF)_t - X_tF - (TF)_x + T_xF = F_{xt} - F_{tx} - (X_t - T_x)F ,$$

so the compatibility condition becomes $X_t - T_x + [X, T] = 0$. Here it is of course expected that both X and T depend on the eigenvalue parameter λ, and the condition is assumed to be satisfied for all λ. This form of the compatibility condition is often called the *zero curvature condition* because of its similarity to a condition found in differential geometry.

The Camassa–Holm equation

As we have seen the KdV equation had some success as a model for water waves, but a very obvious feature of real water waves is missing. These waves often *break*, i.e., while staying bounded the front of the wave becomes vertical, perhaps then toppling over. This behavior cannot be modelled by the KdV equation, since it may be proved that given smooth initial data (periodic or decaying sufficiently fast at $\pm\infty$) there is a corresponding unique solution which stays smooth for all time.

Applied mathematicians searched for a long time for a reasonably simple equation that could model breaking waves, and in 1993 Camassa and Holm [43] suggested an equation which was able to do this and still was an integrable system. The equation was first found by Fokas and Fuchssteiner [77] as one of many examples of integrable systems, but it was later re-derived by Camassa and Holm as a model for shallow water waves.

The Camassa–Holm equation (CH) is

$$u_t - u_{xxt} + 2\kappa u_x + 3uu_x = 2u_x u_{xx} + uu_{xxx} , \tag{F.2}$$

where κ is a real constant related to dispersion. This looks quite unstructured and not like an evolution equation. However, if we introduce $w = u_{xx} - u - \kappa$ the equation becomes

$$w_t + 2u_x w + uw_x = 0 , \tag{F.3}$$

which is an equation of evolution for w (sometimes called the *momentum*). But to express u and u_x in terms of w we have to invoke the operator K, which is the inverse of $D^2 - 1$. This inverse exists in $L^2(-1, 1)$ for periodic boundary conditions and also in $L^2(\mathbb{R})$, and $u = K(w + \kappa)$, $u_x = Kw_x$. Thus the equation for w is *non-local* (not a differential equation). Nevertheless, it is an integrable system in the sense that it is the compatibility condition between a spectral problem and a linear evolution equation. The spectral problem is

$$-f_{xx} + \tfrac{1}{4}f = \lambda wf , \tag{F.4}$$

and CH is the compatibility condition between this and the equation

$$f_t = (\tfrac{1}{2\lambda} - u)f_x + \tfrac{1}{2}u_x f ,\qquad\qquad (F.5)$$

as is easily verified. The spectral problem (F.4) is a Sturm–Liouville problem, but it is sufficiently different from the one-dimensional Schrödinger equation so that no complete inverse scattering theory is known for this equation. One may sometimes use a Liouville transform to convert it into the Schrödinger equation, as in Constantin [47], but this requires a certain amount of smoothness for w and, more importantly, that w does not change sign. However, it is known that no breaking waves can occur if w is of one sign. Clearly the equation is left-definite, so this points to a need for inverse scattering theory for left-definite equations. Although no complete inverse scattering theory is so far available for (F.4), in Chapter 8 we report on some available uniqueness results.

When analyzing the Camassa–Holm equation on $\mathbb{R}$ one may restrict oneself to the cases $\kappa = 1$ and $\kappa = 0$ since u solves the equation for some $\kappa \neq 0$ if and only if $v(x, t) = u(x, t/\kappa)/\kappa$ solves the equation for $\kappa = 1$. Note that $u(x)$ decays as $x \to \pm\infty$ if and only if v does.

We shall end this discussion with a somewhat heuristic derivation of the time evolution of the scattering data for (F.4), and start with the case when $w = u_{xx} - u - 1$ and u solves (F.2) for $\kappa = 1$. Setting

$$F(f) = f_t + (u - 1/(2\lambda))f_x - \tfrac{1}{2}u_x f$$

and assuming that f solves (F.4) for $\lambda \neq 0$ we find that

$$-(F(f))'' + (\tfrac{1}{4} - \lambda w)F(f) = \lambda(w_t + 2u_x w + u w_x)f ,$$

so that $F(f)$ also solves (F.4) if and only if w solves (F.3). Taking for f the Jost solution f_+ of Section 8.2 for $q = q_0 = 1/4$ we have $f_+(x, \lambda; t) \sim e^{ikx}$, $f'_+(x, \lambda; t) \sim ike^{ikx}$ as $x \to \infty$. Since the asymptotics of f_+ is independent of t it seems reasonable to expect $\tfrac{d}{dt}f_+$ to be much smaller at ∞ than f_+ and f'_+. We are also at present interested in the case when u and u_x decay at $\pm\infty$. Thus we expect that $F(f_+) \sim -\tfrac{ik}{2\lambda} e^{ikx}$ at ∞. Since $F(f_+)$ also solves (F.4) this means that $F(f_+) = -\tfrac{ik}{2\lambda} f_+$. It follows that $\psi_+(x, \lambda; t) = e^{-ikt/(2\lambda)} f_+(x, \lambda; t)$ satisfies $F(\psi_+) = 0$. Similarly $\psi_-(x, \lambda; t) = e^{ikt/(2\lambda)} f_-(x, \lambda; t)$ satisfies $F(\psi_-) = 0$. If f_j satisfies both (F.4) and (F.5) it is easily verified that

$$\mathcal{W}(\tfrac{d}{dt}f_1, f_2) = (\tfrac{1}{2\lambda} - u)(f'_1 f'_2 + (\lambda w - 1/4)f_1 f_2) + \tfrac{1}{2}u_x(f'_1 f_2 + f_1 f'_2) - \tfrac{1}{2}u_{xx} f_1 f_2.$$

This expression is symmetric in f_1, f_2 so that

$$\tfrac{d}{dt}\mathcal{W}(f_1, f_2) = \mathcal{W}(\tfrac{d}{dt}f_1, f_2) - \mathcal{W}(\tfrac{d}{dt}f_2, f_1) = 0 .$$

The Wronskian is therefore independent not only of x but also of t. We obtain the following theorem, using the notation of Sections 8.2 and 8.3:

Theorem F.2. *The evolution of scattering data for* (F.3) *when* $w = u_{xx} - u - 1$ *where* u *satisfies* (F.2) *for* $\kappa = 1$ *is as follows:*

(1) $\mathfrak{T}(\lambda; t) = \mathfrak{T}(\lambda; 0)$.

(2) $\mathfrak{R}_+(\lambda; t) = e^{-ikt/\lambda}\mathfrak{R}_+(\lambda; 0)$.

(3) $\mathfrak{R}_-(\lambda; t) = e^{ikt/\lambda}\mathfrak{R}_-(\lambda; 0)$.

(4) *Eigenvalues do not depend on* t.

(5) *If* λ_n *is an eigenvalue and* α_n *the corresponding matching constant, then*
$\|f_+(\cdot, \lambda_n; t)\| = e^{-ikt/\lambda_n}\|f_+(\cdot, \lambda_n; 0)\|$, $\|f_-(\cdot, \lambda_n; t)\| = e^{ikt/\lambda_n}\|f_-(\cdot, \lambda_n; 0)\|$,
and $\alpha_n(t) = e^{-ikt/\lambda_n}\alpha_n(0)$.

Proof. Since $\mathcal{W}(\psi_+, \psi_-)$ is independent of t and $\mathcal{W}(\psi_+(\cdot, \lambda; t), \psi_-(\cdot, \lambda; t)) = \mathcal{W}(f_+(\cdot, \lambda; t), f_-(\cdot, \lambda; t))$ we obtain (1) from (8.2.5), and since the eigenvalues are the poles of $\mathfrak{T}$, also (4). If $\lambda > q_0 = 1/4$ then $\mathcal{W}(\psi_-(\cdot, \lambda; t), \overline{\psi_+(\cdot, \lambda; t)})$ is also independent of t and equal to $e^{ikt/\lambda}\mathcal{W}(f_-(\cdot, \lambda; t), \overline{f_+(\cdot, \lambda; t)})$. Using (8.2.5) again we obtain (2) and (3).

Finally, if $f_+(\cdot, \lambda_n; t) = \alpha_n(t)f_-(\cdot, \lambda_n; t)$, then

$$0 = F(\alpha_n f_- - f_+) = \tfrac{d}{dt}\alpha_n f_- + \alpha_n F(f_-) - F(f_+)$$
$$= (\tfrac{d}{dt}\alpha_n - \tfrac{ik}{2\lambda_n})f_- - \tfrac{-ik}{2\lambda_n}f_+ = (\tfrac{d}{dt}\alpha_n + \tfrac{ik}{\lambda_n})f_-,$$

from which follows $\frac{d}{dt}\alpha_n + \frac{ik}{\lambda_n}\alpha_n = 0$, which gives the final equation in (5). The other equations in (5) follow from this and the arguments just before Definition 8.3.6. $\square$

Similar results are valid for the case of (F.4) when $w = u_{xx} - u$ and u solves (F.2) for $\kappa = 0$, and we obtain the following theorem.

Theorem F.3. *The evolution of spectral data for* (F.3) *when* $w = u_{xx} - u$ *where* u *satisfies* (F.2) *for* $\kappa = 0$ *is as follows:*

(1) *Eigenvalues do not depend on* t.

(2) *If* λ_n *is an eigenvalue and* $\kappa = 0$, *then* $\|f_+(\cdot, \lambda_n; t)\| = e^{t/(2\lambda_n)}\|f_+(\cdot, \lambda_n; 0)\|$,
$\|f_-(\cdot, \lambda_n; t)\| = e^{-t/(2\lambda_n)}\|f_-(\cdot, \lambda_n; 0)\|$, *and* $\alpha_n(t) = e^{t/(2\lambda_n)}\alpha_n(0)$.

Proof. In the present case the spectrum is discrete so there is no scattering matrix, and the asymptotics of the Jost solutions is $f_\pm(x, \lambda) \sim e^{\mp x/2}$. This leads to defining $\psi_\pm(\cdot, \lambda; t) = e^{\mp t/(2\lambda)}f_\pm(\cdot, \lambda; t)$. Apart from this the proof is much the same as for the relevant part of Theorem F.2. $\square$

Notes and remarks

For most of the topics covered here we refer to the already mentioned survey by Palais [177] and the references given there. The evolution of the scattering data for the Camassa–Holm equation is discussed in Constantin [47].

References

1. Ahlfors, L.V.: Complex analysis, third edn. McGraw-Hill Book Co., New York (1978). An introduction to the theory of analytic functions of one complex variable, International Series in Pure and Applied Mathematics
2. Akhiezer, N.I., Glazman, I.M.: Theory of linear operators in Hilbert space. Vol. II, *Monographs and Studies in Mathematics*, vol. 10. Pitman (Advanced Publishing Program), Boston, Mass.-London (1981). Translated from the third Russian edition by E. R. Dawson, Translation edited by W. N. Everitt
3. Akhiezer, N.I., Glazman, I.M.: Theory of linear operators in Hilbert space. Dover Publications, Inc., New York (1993). Translated from the Russian and with a preface by Merlynd Nestell, Reprint of the 1961 and 1963 translations, Two volumes bound as one
4. Ambarzumian, V.: Über eine Frage der Eigenwerttheorie. Zeitschrift für Physik **53**, 690–695 (1929)
5. Arens, R.: Operational calculus of linear relations. Pacific J. Math. **11**, 9–23 (1961)
6. Arnol'd, V.I.: Mathematical methods of classical mechanics. Springer-Verlag, New York-Heidelberg (1978). Translated from the Russian by K. Vogtmann and A. Weinstein, Graduate Texts in Mathematics, 60
7. Atkinson, F.V.: On second-order linear oscillators. Univ. Nac. Tucumán. Revista A. **8**, 71–87 (1951)
8. Atkinson, F.V.: A class of limit-point criteria. In: Spectral theory of differential operators (Birmingham, Ala., 1981), *North-Holland Math. Stud.*, vol. 55, pp. 13–35. North-Holland, Amsterdam-New York (1981)
9. Atkinson, F.V.: On bounds for Titchmarsh–Weyl m-coefficients and for spectral functions for second-order differential operators. Proc. Roy. Soc. Edinburgh Sect. A **97**, 1–7 (1984).
10. Atkinson, F.V., Evans, W.D.: On solutions of a differential equation which are not of integrable square. Math. Z. **127**, 323–332 (1972).
11. Atkinson, F.V., Mingarelli, A.B.: Asymptotics of the number of zeros and of the eigenvalues of general weighted Sturm–Liouville problems. J. Reine Angew. Math. **375/376**, 380–393 (1987)
12. Beals, R., Coifman, R.R.: Scattering and inverse scattering for first order systems. Comm. Pure Appl. Math. **37**(1), 39–90 (1984).
13. Beals, R., Wong, R.: Special functions and orthogonal polynomials, *Cambridge Studies in Advanced Mathematics*, vol. 153. Cambridge University Press, Cambridge (2016).
14. Bennewitz, C.: Symmetric relations on a Hilbert space. In: Conference on the Theory of Ordinary and Partial Differential Equations (Univ. Dundee, Dundee, 1972), pp. 212–218. Lecture Notes in Math., Vol. 280. Springer, Berlin (1972)
15. Bennewitz, C.: Spectral theory for pairs of differential operators. Ark. Mat. **15**(1), 33–61 (1977)

© Springer Nature Switzerland AG 2020

C. Bennewitz et al., *Spectral and Scattering Theory for Ordinary Differential Equations*,
Universitext, https://doi.org/10.1007/978-3-030-59088-8

16. Bennewitz, C.: Spectral asymptotics for Sturm–Liouville equations. Proc. London Math. Soc. (3) **59**(2), 294–338 (1989)

17. Bennewitz, C.: A proof of the local Borg–Marchenko theorem. Comm. Math. Phys. **218**(1), 131–132 (2001).

18. Bennewitz, C.: On the spectral problem associated with the Camassa–Holm equation. J. Nonlinear Math. Phys. **11**(4), 422–434 (2004).

19. Bennewitz, C., Brown, B.M., Weikard, R.: Inverse spectral and scattering theory for the half-line left-definite Sturm–Liouville problem. SIAM J. Math. Anal. **40**(5), 2105–2131 (2008/09).

20. Bennewitz, C., Brown, B.M., Weikard, R.: Scattering and inverse scattering for a left-definite Sturm–Liouville problem. J. Differential Equations **253**(8), 2380–2419 (2012).

21. Bennewitz, C., Brown, B.M., Weikard, R.: The spectral problem for the dispersionless Camassa–Holm equation. In: Operator theory, function spaces, and applications, *Oper. Theory Adv. Appl.*, vol. 255, pp. 67–90. Birkhäuser/Springer, Cham (2016)

22. Bennewitz, C., Everitt, W.N.: The Titchmarsh–Weyl eigenfunction expansion theorem for Sturm–Liouville differential equations. In: Sturm–Liouville theory, pp. 137–171. Birkhäuser, Basel (2005).

23. Bers, L., John, F., Schechter, M.: Partial differential equations. American Mathematical Society, Providence, R.I. (1979). With supplements by Lars Gårding and A. N. Milgram, With a preface by A. S. Householder, Reprint of the 1964 original, Lectures in Applied Mathematics, 3A

24. Bingham, N.H., Goldie, C.M., Teugels, J.L.: Regular variation, *Encyclopedia of Mathematics and its Applications*, vol. 27. Cambridge University Press, Cambridge (1989)

25. Bledsoe, M.: Stability of the inverse resonance problem for Jacobi operators. Integral Equations Operator Theory **74**(4), 481–496 (2012).

26. Bledsoe, M.: Stability of the inverse resonance problem on the line. Inverse Problems **28**(10), 105003, 20 (2012).

27. Bledsoe, M., Weikard, R.: The inverse resonance problem for left-definite Sturm–Liouville operators. J. Math. Anal. Appl. **423**(2), 1753–1773 (2015).

28. Bliss, G.A.: Lectures on the Calculus of Variations. University of Chicago Press, Chicago, Ill. (1946)

29. Bloh, A.v.: On the determination of a differential equation by its special matrix function. Doklady Akad. Nauk SSSR (N.S.) **92**, 209–212 (1953)

30. Boas Jr., R.P.: Entire functions. Academic Press Inc., New York (1954)

31. Bois-Reymond, P.d.: Erläuterungen zu der Anfangsgründen der Variationsrechnung. Math. Ann. **15**, 283—314 (1879)

32. Borg, G.: Eine Umkehrung der Sturm–Liouvilleschen Eigenwertaufgabe. Bestimmung der Differentialgleichung durch die Eigenwerte. Acta Math. **78**, 1–96 (1946).

33. Borg, G.: Uniqueness theorems in the spectral theory of $y'' + (\lambda - q(x))y = 0$. In: Den 11te Skandinaviske Matematikerkongress, Trondheim, 1949, pp. 276–287. Johan Grundt Tanums Forlag, Oslo (1952)

34. de Branges, L.: Hilbert spaces of entire functions. Prentice-Hall, Inc., Englewood Cliffs, N.J. (1968)

35. Brillouin, L.: Remarques sur la mécanique ondulatoire. J. Phys. Radium **7**(12), 353–368 (1926).

36. Brinck, I.: Self-adjointness and spectra of Sturm–Liouville operators. Math. Scand. **7**, 219–239 (1959).

37. Brown, B.M., Eastham, M.S.P., McCormack, D.K.R.: Spectral concentration and rapidly decaying potentials. J. Comput. Appl. Math. **81**(2), 333–348 (1997).

38. Brown, B.M., Knowles, I., Weikard, R.: On the inverse resonance problem. J. London Math. Soc. (2) **68**(2), 383–401 (2003)

39. Brown, B.M., Naboko, S., Weikard, R.: The inverse resonance problem for Jacobi operators. Bull. London Math. Soc. **37**(5), 727–737 (2005)

40. Brown, B.M., Naboko, S., Weikard, R.: The inverse resonance problem for Hermite operators. Constr. Approx. **30**(2), 155–174 (2009).

41. Brown, B.M., Peacock, R.A., Weikard, R.: A local Borg–Marchenko theorem for complex potentials. J. Comput. Appl. Math. **148**(1), 115–131 (2002). On the occasion of the 65th birthday of Professor Michael Eastham

42. Brown, B.M., Weikard, R.: The inverse resonance problem for perturbations of algebro-geometric potentials. Inverse Problems **20**(2), 481–494 (2004)

43. Camassa, R., Holm, D.D.: An integrable shallow water equation with peaked solitons. Phys. Rev. Lett. **71**(11), 1661–1664 (1993).

44. Carleman, T.: Sur les équations intégrales singulières à noyau réel et symétrique. Uppsala Universitets Årsskrift 1923. Matematik och Naturvetenskap. 3; 228 S. 8° (1923).

45. Carleman, T.: Sur une inégalité différentielle dans la théorie des functions analytiques. C. R. Acad. des Sci. de Paris **196**, 995–997 (1933)

46. Coddington, E.A., Levinson, N.: Theory of ordinary differential equations. McGraw-Hill Book Company, Inc., New York-Toronto-London (1955)

47. Constantin, A.: On the scattering problem for the Camassa–Holm equation. R. Soc. Lond. Proc. Ser. A Math. Phys. Eng. Sci. **457**(2008), 953–970 (2001).

48. Courant, R., Hilbert, D.: Methoden der Mathematischen Physik. Vols. I, II. Interscience Publishers, Inc., N.Y. (1943)

49. Daho, K., Langer, H.: Some remarks on a paper: "Some remarks on a differential expression with an indefinite weight function" (*spectral theory and asymptotics of differential equations* (Proc. Conf., Scheveningen, 1973), pp. 13–28, North-Holland Math. Studies, Vol. 13, North-Holland, Amsterdam, 1974) by W. N. Everitt. Proc. Roy. Soc. Edinburgh Sect. A **78**(1-2), 71–79 (1977/78).

50. Davies, E.B.: A hierarchical method for obtaining eigenvalue enclosures. Math. Comp. **69**(232), 1435–1455 (2000).

51. Davies, E.B.: Singular Schrödinger operators in one dimension. Mathematika **59**(1), 141–159 (2013).

52. Deift, P., Trubowitz, E.: Inverse scattering on the line. Comm. Pure Appl. Math. **32**(2), 121–251 (1979)

53. Delsarte, J.: Sur certaines transformations fonctionnelles rélatives aux équations linéaires aux dérivées partielles du second ordre. C. R. Acad. Sci., Paris **206**, 1780–1782 (1938)

54. Delsarte, J.: Sur une extension de la formule de Taylor. J. Math. Pures Appl. (9) **17**, 213–231 (1938)

55. *NIST Digital Library of Mathematical Functions.* http://dlmf.nist.gov/, Release 1.0.21 of 2018-12-15. URL `http://dlmf.nist.gov/`. F. W. J. Olver, A. B. Olde Daalhuis, D. W. Lozier, B. I. Schneider, R. F. Boisvert, C. W. Clark, B. R. Miller and B. V. Saunders, eds.

56. Doob, J.L., Koopman, B.O.: On analytic functions with positive imaginary parts. Bull. Amer. Math. Soc. **40**(8), 601–605 (1934).

57. Dym, H., McKean, H.P.: Gaussian processes, function theory, and the inverse spectral problem. Academic Press [Harcourt Brace Jovanovich, Publishers], New York-London (1976). Probability and Mathematical Statistics, Vol. 31

58. Eckhardt, J.: Direct and inverse spectral theory of singular left-definite Sturm–Liouville operators. J. Differential Equations **253**(2), 604–634 (2012).

59. Eckhardt, J.: Inverse uniqueness results for Schrödinger operators using de Branges theory. Complex Anal. Oper. Theory **8**(1), 37–50 (2014).

60. Eckhardt, J.: Two inverse spectral problems for a class of singular Krein strings. Int. Math. Res. Not. IMRN **2014**(13), 3692–3713 (2014).

61. Eckhardt, J.: An inverse spectral problem for a star graph of Krein strings. J. Reine Angew. Math. **715**, 189–206 (2016).

62. Eckhardt, J., Kostenko, A.: The inverse spectral problem for indefinite strings. Invent. Math. **204**(3), 939–977 (2016).

63. Eckhardt, J., Kostenko, A.: Quadratic operator pencils associated with the conservative Camassa–Holm flow. Bull. Soc. Math. France **145**(1), 47–95 (2017).

64. Eckhardt, J., Kostenko, A.: The classical moment problem and generalized indefinite strings. Integral Equations Operator Theory **90**(2), Paper No. 23, 30 (2018).

65. Eckhardt, J., Teschl, G.: On the isospectral problem of the dispersionless Camassa–Holm equation. Adv. Math. **235**, 469–495 (2013).

66. Edmunds, D.E., Evans, W.D.: Spectral theory and differential operators. Oxford Mathematical Monographs. Oxford University Press, Oxford (2018). Second edition of [MR0929030]

67. Engel, K.J., Nagel, R.: One-parameter semigroups for linear evolution equations, *Graduate Texts in Mathematics*, vol. 194. Springer-Verlag, New York (2000). With contributions by S. Brendle, M. Campiti, T. Hahn, G. Metafune, G. Nickel, D. Pallara, C. Perazzoli, A. Rhandi, S. Romanelli and R. Schnaubelt

68. Everitt, W.N.: On a property of the m-coefficient of a second-order linear differential equation. J. London Math. Soc. (2) **4**, 443–457 (1971/72).

69. Faddeev, L.D.: Properties of the S-matrix of the one-dimensional Schrödinger equation. Trudy Mat. Inst. Steklov. **73**, 314–336 (1964)

70. Faddeev, L.D.: The inverse problem in the quantum theory of scattering. II. In: Current problems in mathematics, Vol. 3 (Russian), pp. 93–180, 259. (loose errata). Akad. Nauk SSSR Vsesojuz. Inst. Naučn. i Tehn. Informacii, Moscow (1974)

71. Faddeyev, L.D.: The inverse problem in the quantum theory of scattering. J. Mathematical Phys. **4**, 72–104 (1963).

72. Fermi, E.: Collected papers (Note e memorie). Vol. II: United States, 1939–1954. The University of Chicago Press, Chicago, Ill.; Accademia Nazionale dei Lincei, Rome (1965)

73. Folland, G.B.: Introduction to partial differential equations, second edn. Princeton University Press, Princeton, NJ (1995)

74. Fourier, J.: Théorie analytique de la chaleur. Firmin Didot, père et fils (1822)

75. Friedlander, F.G.: Introduction to the theory of distributions, second edn. Cambridge University Press, Cambridge (1998). With additional material by M. Joshi

76. Friedrichs, K.: Spektraltheorie halbbeschränkter Operatoren und Anwendung auf die Spektralzerlegung von Differentialoperatoren. Math. Ann. **109**(1), 465–487 (1934).

77. Fuchssteiner, B., Fokas, A.S.: Symplectic structures, their Bäcklund transformations and hereditary symmetries. Phys. D **4**(1), 47–66 (1981/82).

78. Fulton, C., Langer, H., Luger, A.: Mark Krein's method of directing functionals and singular potentials. Math. Nachr. **285**(14-15), 1791–1798 (2012).

79. Gårding, L.: Applications of the theory of direct integrals of Hilbert spaces to some integral and differential operators. The Institute for Fluid Dynamics and Applied Mathematics, Lecture series no. 11. University of Maryland, College Park, Md. (1954)

80. Gardner, C.S., Greene, J.M., Kruskal, M.D., Miura, R.M.: Method for Solving the Korteweg-de Vries Equation. Phys. Rev. Lett. **19**, 1095–1097 (1967).

81. Gel' fand, I.M., Shilov, G.E.: Generalized functions. Vol. 1. AMS Chelsea Publishing, Providence, RI (2016). Properties and operations, Translated from the 1958 Russian original [MR0097715] by Eugene Saletan, Reprint of the 1964 English translation [MR0166596]

82. Gel'fand, I.M., Levitan, B.M.: On the determination of a differential equation from its spectral function (Russian). Izvestiya Akad. Nauk SSSR. Ser. Mat. **15**, 309–360 (1951)

83. Gel'fand, I.M., Levitan, B.M.: On the determination of a differential equation from its spectral function. Amer. Math. Soc. Transl. (2) **1**, 253–304 (1955)

84. Gesztesy, F., Holden, H.: Soliton equations and their algebro-geometric solutions. Vol. I, *Cambridge Studies in Advanced Mathematics*, vol. 79. Cambridge University Press, Cambridge (2003). $(1 + 1)$-dimensional continuous models

85. Gesztesy, F., Simon, B.: Inverse spectral analysis with partial information on the potential. II. The case of discrete spectrum. Trans. Amer. Math. Soc. **352**(6), 2765–2787 (2000).

86. Gesztesy, F., Simon, B.: A new approach to inverse spectral theory. II. General real potentials and the connection to the spectral measure. Ann. of Math. (2) **152**(2), 593–643 (2000).

87. Gesztesy, F., Weikard, R.: Picard potentials and Hill's equation on a torus. Acta Math. **176**(1), 73–107 (1996)

88. Gesztesy, F., Weikard, R.: Elliptic algebro-geometric solutions of the KdV and AKNS hierarchies—an analytic approach. Bull. Amer. Math. Soc. (N.S.) **35**(4), 271–317 (1998)

89. Gesztesy, F., Weikard, R.: Some remarks on the spectral problem underlying the Camassa–Holm hierarchy. In: Operator theory in harmonic and non-commutative analysis, *Oper. Theory Adv. Appl.*, vol. 240, pp. 137–188. Birkhäuser/Springer, Cham (2014).

90. Gesztesy, F., Weikard, R., Zinchenko, M.: Initial value problems and Weyl–Titchmarsh theory for Schrödinger operators with operator-valued potentials. Oper. Matrices **7**(2), 241–283 (2013).

91. Gesztesy, F., Weikard, R., Zinchenko, M.: On spectral theory for Schrödinger operators with operator-valued potentials. J. Differential Equations **255**(7), 1784–1827 (2013).

92. Gesztesy, F., Zinchenko, M.: On spectral theory for Schrödinger operators with strongly singular potentials. Math. Nachr. **279**(9-10), 1041–1082 (2006)

93. Ghatasheh, A., Weikard, R.: On Leighton's comparison theorem. J. Differential Equations **262**(12), 5978–5989 (2017).

94. Gilbert, D.: Subordinacy and spectral analysis of schrödinger operators. Ph.D. thesis, University of Hull (1984)

95. Gilbert, D.J., Pearson, D.B.: On subordinacy and analysis of the spectrum of one-dimensional Schrödinger operators. J. Math. Anal. Appl. **128**(1), 30–56 (1987).

96. Glazman, I.M.: Direct methods of qualitative spectral analysis of singular differential operators. Translated from the Russian by the IPST staff. Israel Program for Scientific Translations, Jerusalem, 1965; Daniel Davey & Co., Inc., New York (1966)

97. Green, G.: On the motion of waves in a variable canal of small depth and width. Trans. Cambridge Philos. Soc. **6**, 457–462 (1837)

98. Halmos, P.R.: Finite-dimensional vector spaces, second edn. Springer-Verlag, New York-Heidelberg (1974/87). Undergraduate Texts in Mathematics

99. Hartman, P.: On the eigenvalues of differential equations. Amer. J. Math. **73**, 657–662 (1951).

100. Heins, M.: On a notion of convexity connected with a method of Carleman. J. Analyse Math. **7**, 53–77 (1959).

101. Hilbert, D.: Grundzüge einer allgeminen Theorie der linaren Integralrechnungen. (Erste Mitteilung). Göttinger Nachrichten pp. 49–91 (1904)

102. Hilbert, D.: Grundzüge einer allgeminen Theorie der linaren Integralrechnungen. (Fünfte Mitteilung). Göttinger Nachrichten pp. 439–476 (1906)

103. Hilbert, D.: Grundzüge einer allgeminen Theorie der linaren Integralrechnungen. (Vierte Mitteilung). Göttinger Nachrichten pp. 157–228 (1906)

104. Hille, E.: Lectures on ordinary differential equations. Addison-Wesley Publ. Co., Reading, Mass.-London-Don Mills, Ont. (1969)

105. Hille, E.: Ordinary differential equations in the complex domain. Dover Publications, Inc., Mineola, NY (1997). Reprint of the 1976 original

106. Hille, E., Phillips, R.S.: Functional analysis and semi-groups. American Mathematical Society, Providence, R. I. (1974). Third printing of the revised edition of 1957, American Mathematical Society Colloquium Publications, Vol. XXXI

107. Hochstadt, H.: On the determination of a Hill's equation from its spectrum. Arch. Rational Mech. Anal. **19**, 353–362 (1965)

108. Hochstadt, H., Lieberman, B.: An inverse Sturm–Liouville problem with mixed given data. SIAM J. Appl. Math. **34**(4), 676–680 (1978).

109. Hörmander, L.: Linear partial differential operators. Die Grundlehren der mathematischen Wissenschaften, Bd. 116. Academic Press, Inc., Publishers, New York; Springer-Verlag, Berlin-Göttingen-Heidelberg (1963)

110. Hörmander, L.: The analysis of linear partial differential operators. I, *Grundlehren der Mathematischen Wissenschaften [Fundamental Principles of Mathematical Sciences]*, vol. 256. Springer-Verlag, Berlin (1983). Distribution theory and Fourier analysis

111. Hörmander, L.: The analysis of linear partial differential operators. I. Classics in Mathematics. Springer-Verlag, Berlin (2003). Distribution theory and Fourier analysis, Reprint of the second (1990) edition [Springer, Berlin; MR1065993 (91m:35001a)]

112. Ince, E.L.: Ordinary Differential Equations. Dover Publications, New York (1944)

113. Jeffreys, H.: On certain approximate solutions of linear differential equations of the second order. Proc. Lond. Math. Soc. (2) **23**, 428–436 (1924)

114. John, F.: Partial differential equations, *Applied Mathematical Sciences*, vol. 1, fourth edn. Springer-Verlag, New York (1991)

115. Jordan, P., von Neumann, J.: On inner products in linear, metric spaces. Ann. of Math. (2) **36**(3), 719–723 (1935).

116. Kac, I.S.: Power asymptotic behavior of the spectral functions of generalized second order boundary value problems. Dokl. Akad. Nauk SSSR **203**, 752–755 (1972)

117. Kac, I.S.: A generalization of the asymptotic formula of V. A. Marčenko for the spectral functions of a second order boundary value problem. Izv. Akad. Nauk SSSR Ser. Mat. **37**, 422–436 (1973)

118. Kamke, E.: Zum Entwicklungssatz bei polaren Eigenwertaufgaben. Math. Z. **45**, 706–718 (1939).

119. Karamata, J.: Neuer Beweis und Verallgemeinerung der Tauberschen Sätze, welche die Laplacesche und Stieltjessche Transformation betreffen. J. Reine Angew. Math. **164**, 27–39 (1931).

120. Karlsson, A.: Sturm–Liouville egenvärdesproblem. Master's thesis, Lund University (2009)

121. Kasahara, Y.: Spectral theory of generalized second order differential operators and its applications to Markov processes. Japan. J. Math. (N.S.) **1**(1), 67–84 (1975/76).

122. Kato, T.: Perturbation theory for linear operators. Classics in Mathematics. Springer-Verlag, Berlin (1995). Reprint of the 1980 edition

123. Kay, I., Moses, H.E.: The determination of the scattering potential from the spectral measure function. I. Continuous spectrum. Nuovo Cimento (10) **2**, 917–961 (1955)

124. Kay, I., Moses, H.E.: The determination of the scattering potential from the spectral measure function. II. Point eigenvalues and proper eigenfunctions. Nuovo Cimento (10) **3**, 66–84 (1956)

125. Kay, I., Moses, H.E.: The determination of the scattering potential from the spectral measure funtion. III. Calculation of the scattering potential from the scattering operator for the one-dimensional Schrödinger equation. Nuovo Cimento (10) **3**, 276–304 (1956)

126. Kneser, A.: Untersuchungen über die Darstellung willkürlicher Funktionen in der mathematischen Physik. Math. Ann. **58**(1-2), 81–147 (1903).

127. Koliander, G.: Hilbert Spaces of Entire Functions in the Hardy Space Setting. Master's thesis, Technische Universität Wien (2011)

128. Korotyaev, E.: Inverse resonance scattering on the half line. Asymptot. Anal. **37**(3-4), 215–226 (2004).

129. Korteweg, D.J., de Vries, G.: On the change of form of long waves advancing in a rectangular canal, and on a new type of long stationary waves. Philos. Mag. (5) **39**(240), 422–443 (1895).

130. Kostenko, A., Sakhnovich, A., Teschl, G.: Weyl–Titchmarsh theory for Schrödinger operators with strongly singular potentials. Int. Math. Res. Not. IMRN **2012**(8), 1699–1747 (2012).

131. Kramers, H.A.: Wellenmechanik und halbzahlige Quantisierung. Z. Phys. **39**, 828–840 (1926)

132. Kreĭn, M.: A contribution to the theory of entire functions of exponential type. Bull. Acad. Sci. URSS. Sér. Math. [Izvestia Akad. Nauk SSSR] **11**, 309–326 (1947)

133. Kreĭn, M.G.: The theory of self-adjoint extensions of semi-bounded Hermitian transformations and its applications. II. Mat. Sbornik N.S. **21(63)**, 365–404 (1947)

134. Kreĭn, M.G.: Determination of the density of a nonhomogeneous symmetric cord by its frequency spectrum. Doklady Akad. Nauk SSSR (N.S.) **76**, 345–348 (1951)

135. Kreĭn, M.G.: Solution of the inverse Sturm–Liouville problem. Doklady Akad. Nauk SSSR (N.S.) **76**, 21–24 (1951)

136. Kreĭn, M.G.: On inverse problems for a nonhomogeneous cord. Doklady Akad. Nauk SSSR (N.S.) **82**, 669–672 (1952)

137. Kreĭn, M.G.: On some cases of effective determination of the density of an inhomogeneous cord from its spectral function. Doklady Akad. Nauk SSSR (N.S.) **93**, 617–620 (1953)

138. Kreĭn, M.G.: On the transfer function of a one-dimensional boundary problem of the second order. Doklady Akad. Nauk SSSR (N.S.) **88**, 405–408 (1953)

139. Lax, P.D.: Integrals of nonlinear equations of evolution and solitary waves. Comm. Pure Appl. Math. **21**, 467–490 (1968).

140. Lax, P.D.: Functional analysis. Pure and Applied Mathematics (New York). Wiley-Interscience [John Wiley & Sons], New York (2002)

141. Lax, P.D.: Linear algebra and its applications, second edn. Pure and Applied Mathematics (Hoboken). Wiley-Interscience [John Wiley & Sons], Hoboken, NJ (2007)

142. Leighton, W.: Comparison theorems for linear differential equations of second order. Proc. Amer. Math. Soc. **13**, 603–610 (1962)

143. Lejeune-Dirichlet, P.G.: Sur la convergence des séries trigonométriques qui servent à représenter une fonction arbitraire entre des limites données. J. Reine Angew. Math. **4**, 157–169 (1829).

144. Levin, B.J.: Distribution of zeros of entire functions, *Translations of Mathematical Monographs*, vol. 5, revised edn. American Mathematical Society, Providence, R.I. (1980). Translated from the Russian by R. P. Boas, J. M. Danskin, F. M. Goodspeed, J. Korevaar, A. L. Shields and H. P. Thielman

145. Levin, B.Y.: Lectures on entire functions, *Translations of Mathematical Monographs*, vol. 150. American Mathematical Society, Providence, RI (1996). In collaboration with and with a preface by Yu. Lyubarskii, M. Sodin and V. Tkachenko, Translated from the Russian manuscript by Tkachenko

146. Levinson, N.: Criteria for the limit-point case for second order linear differential operators. Časopis Pěst. Mat. Fys. **74**, 17–20 (1949)

147. Levinson, N.: The inverse Sturm–Liouville problem. Mat. Tidsskr. B. **1949**, 25–30 (1949)

148. Levinson, N.: On the uniqueness of the potential in a Schrödinger equation for a given asymptotic phase. Danske Vid. Selsk. Mat.-Fys. Medd. **25**(9), 29 (1949)

149. Levitan, B.M.: Obratnye zadachi Shturma–Liuvillya. "Nauka", Moscow (1984)

150. Levitan, B.M.: Inverse Sturm–Liouville problems. VSP, Zeist (1987). Translated from the Russian by O. Efimov

151. Liouville, J.: Mémoire sur l'usage que l'on peut faire de la formule de Fourier, dans le calcul des différentielles à indices quelconques. J. Reine Angew. Math. **13**, 219–232 (1835)

152. Liouville, J.: Mémoire sur le d'eveloppement des fonctions ou parties de fonctions en séries dont les divers termes sont assujettis à satisfaire á une même équation différentielle du second ordre, contenant un paramètre variable. J. Math. Pures Appl. **1**, 253–265 (1836)

153. Liouville, J.: Second mémoire sur le d'eveloppement des fonctions ... J. Math. Pure Appl. **2**, 16–35 (1837)

154. Liouville, J.: Troisième mémoire sur le d'eveloppement des fonctions ... J. Math. Pure Appl. **2**, 418–437 (1837)

155. Lützen, J.: Sturm and Liouville's work on ordinary linear differential equations. The emergence of Sturm–Liouville theory. Arch. Hist. Exact Sci. **29**(4), 309–376 (1984)

156. Marchenko, V.A.: Sturm–Liouville operators and applications, revised edn. AMS Chelsea Publishing, Providence, RI (2011).

157. Marletta, M., Naboko, S., Shterenberg, R., Weikard, R.: On the inverse resonance problem for Jacobi operators—uniqueness and stability. J. Anal. Math. **117**, 221–247 (2012).

158. Marletta, M., Shterenberg, R., Weikard, R.: On the inverse resonance problem for Schrödinger operators. Comm. Math. Phys. **295**(2), 465–484 (2010).

159. Marletta, M., Weikard, R.: Weak stability for an inverse Sturm–Liouville problem with finite spectral data and complex potential. Inverse Problems **21**(4), 1275–1290 (2005)

160. Marletta, M., Weikard, R.: Stability for the inverse resonance problem for a Jacobi operator with complex potential. Inverse Problems **23**(4), 1677–1688 (2007)

161. Marčenko, V.A.: Concerning the theory of a differential operator of the second order. Doklady Akad. Nauk SSSR. (N.S.) **72**, 457–460 (1950)

162. Marčenko, V.A.: On reconstruction of the potential energy from phases of the scattered waves. Dokl. Akad. Nauk SSSR (N.S.) **104**, 695–698 (1955)

163. Mason, M.: The expansion of a function in terms of normal functions. Trans. Amer. Math. Soc. **8**(4), 427–432 (1907).

164. McKean, H.P., van Moerbeke, P.: The spectrum of Hill's equation. Invent. Math. **30**(3), 217–274 (1975).

165. McKean, H.P., Trubowitz, E.: Hill's operator and hyperelliptic function theory in the presence of infinitely many branch points. Comm. Pure Appl. Math. **29**(2), 143–226 (1976).

166. Melin, A.: Operator methods for inverse scattering on the real line. Comm. Partial Differential Equations **10**(7), 677–766 (1985).

167. Milne, W.E.: On the degree of convergence of expansions in an infinite interval. Trans. Amer. Math. Soc. **31**(4), 907–918 (1929).

168. Miura, R.M.: Korteweg-de Vries equation and generalizations. I. A remarkable explicit non-linear transformation. J. Mathematical Phys. **9**, 1202–1204 (1968).

169. Molčanov, A.M.: On conditions for discreteness of the spectrum of self-adjoint differential equations of the second order (Russian). Trudy Moskov. Mat. Obšč. **2**, 169–199 (1953)

170. Naĭmark, M.A.: Linear differential operators. Part II: Linear differential operators in Hilbert space. With additional material by the author, and a supplement by V. È. Ljance. Translated from the Russian by E. R. Dawson. English translation edited by W. N. Everitt. Frederick Ungar Publishing Co., New York (1968)

171. Neumann, J.v.: Eigenwertproblem symmetrischer Funktionaloperatoren. Jahresber. Dtsch. Math.-Ver. **37**, 11–14 (1928)

172. Neumann, J.v.: Allgemeine Eigenwerttheorie Hermitescher Funktionaloperatoren. Math. Ann. **102**(1), 49–131 (1930)

173. Neumann, J.v.: On rings of operators. Reduction theory. Ann. of Math. (2) **50**, 401–485 (1949).

174. Nevanlinna, R.: Eindeutige analytische Funktionen. Die Grundlehren der mathematischen Wissenschaften in Einzeldarstellungen mit besonderer Berücksichtigung der Anwendungs-gebiete, Bd XLVI. Springer-Verlag, Berlin-Göttingen-Heidelberg (1953). 2te Aufl

175. Olver, F.W.J.: Asymptotics and special functions. AKP Classics. A K Peters, Ltd., Wellesley, MA (1997). Reprint of the 1974 original [Academic Press, New York; MR0435697 (55 #8655)]

176. Orcutt, B.C.: Canonical differential equations. ProQuest LLC, Ann Arbor, MI (1969). Thesis (Ph.D.)–University of Virginia

177. Palais, R.S.: The symmetries of solitons. Bull. Amer. Math. Soc. (N.S.) **34**(4), 339–403 (1997).

178. Paley, R.E.A.C., Wiener, N.: Fourier transforms in the complex domain, *American Mathematical Society Colloquium Publications*, vol. 19. American Mathematical Society, Providence, RI (1987). Reprint of the 1934 original

179. Pearson, D.B.: Quantum scattering and spectral theory, *Techniques of Physics*, vol. 9. Academic Press, Inc. [Harcourt Brace Jovanovich, Publishers], London (1988)

180. Picone, M.: Sui valori eccezionali di un parametro da cui dipende un'equazione differenziale lineare del secondo ordine. Ann. Scuola Norm. Sup. Pisa **11**, 1—141 (1910)

181. Pleijel, A.: Some remarks about the limit point and limit circle theory. Ark. Mat. **7**, 543–550 (1969).

182. Pleijel, Å.: Spectral theory for pairs of ordinary formally self-adjoint differential operators. J. Indian Math. Soc. (N.S.) **34**(3-4), 259–268 (1971) (1970)

183. Povzner, A.: On differential equations of Sturm–Liouville type on a half-axis. Mat. Sbornik N.S. **23(65)**, 3–52 (1948)

184. Prüfer, H.: Neue Herleitung der Sturm–Liouvilleschen Reihenentwicklung stetiger Funktio-nen. Math. Ann. **95**(1), 499–518 (1926).

185. Reed, M., Simon, B.: Methods of modern mathematical physics. I, second edn. Academic Press, Inc. [Harcourt Brace Jovanovich, Publishers], New York (1980). Functional analysis

186. Remling, C.: Schrödinger operators and de Branges spaces. J. Funct. Anal. **196**(2), 323–394 (2002).

187. Riesz, F.: Sur les opératines functionelles linéaires. Comptes Rendus **149**, 974–977 (1909)

188. Riesz, F.: Über lineare Funktionalgleichungen. Acta math. **41**, 71–98 (1918)

189. Riesz, F.: Zur Theorie des Hilbertschen Raumes. Acta Sci. Math. **7**, 34–38 (1934–35)

190. Riesz, F., Lorch, E.R.: The integral representation of unbounded self-adjoint transformations in Hilbert space. Trans. Amer. Math. Soc. **39**(2), 331–340 (1936).

191. Riesz, F., Sz.-Nagy, B.: Functional analysis. Dover Books on Advanced Mathematics. Dover Publications, Inc., New York (1990). Translated from the second French edition by Leo F. Boron, Reprint of the 1955 original

192. Rofe-Beketov, F.S.: The spectral matrix and the inverse Strum–Liouville problem on the axis $(-\infty, \infty)$. Teor. Funkciĭ Funkcional. Anal. i Priložen. Vyp. **4**, 189–197 (1967)

193. Rosenblum, M., Rovnyak, J.: Topics in Hardy classes and univalent functions. Birkhäuser Advanced Texts: Basler Lehrbücher. [Birkhäuser Advanced Texts: Basel Textbooks]. Birkhäuser Verlag, Basel (1994).

194. Rosenfeld, N.S.: The eigenvalues of a class of singular differential operators. Comm. Pure Appl. Math. **13**, 395–405 (1960).

195. Rudin, W.: Principles of mathematical analysis, third edn. McGraw-Hill Book Co., New York-Auckland-Düsseldorf (1976). International Series in Pure and Applied Mathematics

196. Rudin, W.: Functional analysis, second edn. International Series in Pure and Applied Mathematics. McGraw-Hill, Inc., New York (1991)

197. Schäfke, F.W., Schneider, A.: S-hermitesche Rand-Eigenwertprobleme. I. Math. Ann. **162**, 9–26 (1965/66).

198. Schäfke, F.W., Schneider, A.: S-hermitesche Rand-Eigenwertprobleme. II. Math. Ann. **165**, 236–260 (1966).

199. Schäfke, F.W., Schneider, A.: S-hermitesche Rand-Eigenwertprobleme. III. Math. Ann. **177**, 67–94 (1968).

200. Schneider, A., Niessen, H.D.: Linksdefinite singuläre kanonische Eigenwertprobleme. I. J. Reine Angew. Math. **281**, 13–52 (1976)

201. Schneider, A., Niessen, H.D.: Linksdefinite singuläre kanonische Eigenwertprobleme. II. J. Reine Angew. Math. **289**, 62–84 (1977).

202. Schrödinger, E.: Quantisierung als Eigenwertproblem. Annalen der Physik **384**, 361–376 (1926).

203. Shterenberg, R., Weikard, R., Zinchenko, M.: Stability for the inverse resonance problem for the CMV operator. In: Spectral analysis, differential equations and mathematical physics: a festschrift in honor of Fritz Gesztesy's 60th birthday, *Proc. Sympos. Pure Math.*, vol. 87, pp. 315–326. Amer. Math. Soc., Providence, RI (2013)

204. Simon, B.: A new approach to inverse spectral theory. I. Fundamental formalism. Ann. of Math. (2) **150**(3), 1029–1057 (1999).

205. Stekloff, W.: Sur le problème de refroidissement d'une barre hétérogène. Comp. Rend. Acad. Sci. Paris. **126**, 215–218 (1898)

206. Stone, M.H.: On one-parameter unitary groups in Hilbert space. Ann. of Math. (2) **33**(3), 643–648 (1932).

207. Stone, M.H.: Linear transformations in Hilbert space, *American Mathematical Society Colloquium Publications*, vol. 15. American Mathematical Society, Providence, RI (1990). Reprint of the 1932 original

208. Sturm, C.F.: Mémoire sur les équations différentielles linéaires du second ordre. J. Math. Pures Appl. **1**, 106–186 (1836)

209. Sturm, C.F.: Mémoire sur une classe d'équations à différences partielles. J. Math. Pures Appl. **1**, 373–444 (1836)

210. Swanson, C.: Comparison and oscillation theory of linear differential equations. Academic Press (1968)

211. Sz.-Nagy, B.: Introduction to real functions and orthogonal expansions. Oxford University Press, New York (1965)

212. Titchmarsh, E.C.: Eigenfunction expansions associated with second-order differential equations. Part I. Second Edition. Clarendon Press, Oxford (1962)

213. Trefethen, L.N., Bau III, D.: Numerical linear algebra. Society for Industrial and Applied Mathematics (SIAM), Philadelphia, PA (1997).

214. Walter, J.: Regular eigenvalue problems with eigenvalue parameter in the boundary condition. Math. Z. **133**, 301–312 (1973).

215. Weidmann, J.: Linear operators in Hilbert spaces, *Graduate Texts in Mathematics*, vol. 68. Springer-Verlag, New York-Berlin (1980). Translated from the German by Joseph Szücs

216. Weidmann, J.: Lineare Operatoren in Hilberträumen. Teil 1. Mathematische Leitfäden. [Mathematical Textbooks]. B. G. Teubner, Stuttgart (2000). Grundlagen. [Foundations]

217. Weikard, R.: A local Borg–Marchenko theorem for difference equations with complex coefficients. In: Partial differential equations and inverse problems, *Contemp. Math.*, vol. 362, pp. 403–410. Amer. Math. Soc., Providence, RI (2004)

218. Weikard, R., Zinchenko, M.: The inverse resonance problem for CMV operators. Inverse Problems **26**(5), 055012, 10 (2010)

219. Wentzel, G.: Eine Verallgemeinerung der Quantenbedingungen für die Zwecke der Wellenmechanik. Z. Phys. **38**, 518–529 (1926)

220. Wet, J.S.d., Mandl, F.: On the asymptotic distribution of eigenvalues. Proc. Roy. Soc. London. Ser. A. **200**, 572–580 (1950).

221. Weyl, H.: Über beschränkte quadratische Formen, deren Differenz vollstetig ist. Rend. d. Circ. Mat. di Palermo **27**(1), 373–392 (1909)

222. Weyl, H.: über gewöhnliche Differentialgleichungen mit Singularitäten und die zugehörigen Entwicklungen willkürlicher Funktionen. Math. Ann. **68**(2), 220–269 (1910)

223. Weyl, H.: über gewöhnliche lineare differentialgleichungen mit singulären stellen und ihre eigenfunktionen (2. note). Gött. Nachr. **29**, 442–467 (1910)

224. Weyl, H.: Das asymptotische Verteilungsgesetz der Eigenwerte linearer partieller Differentialgleichungen (mit einer Anwendung auf die Theorie der Hohlraumstrahlung). Math. Ann. **71**(4), 441–479 (1912).

225. Weyl, H.: Ramifications, old and new, of the eigenvalue problem. Bull. Amer. Math. Soc. **56**, 115–139 (1950).

226. Zabusky, N.J., Kruskal, M.D.: Interaction of "Solitons" in a Collisionless Plasma and the Recurrence of Initial States. Phys. Rev. Lett. **15**, 240–243 (1965).

227. Zaharov, V.E., Faddeev, L.D.: The Korteweg-de Vries equation is a fully integrable Hamiltonian system. Funkcional. Anal. i Priložen. **5**(4), 18–27 (1971)

228. Zettl, A.: Sturm–Liouville theory, *Mathematical Surveys and Monographs*, vol. 121. American Mathematical Society, Providence, RI (2005)

229. Zworski, M.: A remark on isopolar potentials. SIAM J. Math. Anal. **32**(6), 1324–1326 (2001).

Symbol Index

$A^{\perp}$, 34
D_λ, 48
E_Δ, 60
E_t, 60
L^2, 32
R_λ, 48
S_λ, 48
Sp, 6
χ, 287
ℓ^2, 31
$\mathbb{C}$, 4

$\mathbb{R}$, 4
$\mathbf{D}_{\pm i}$, 52
$\mathcal{L}$, 38
$\mathcal{W}$, 87
$\rho(T)$, 48
a.e., 284
$\sigma(T)$, 48
$n_\pm$, 53
$\langle \cdot, \cdot \rangle$, 10
$\|\cdot\|$, 8, 38

© Springer Nature Switzerland AG 2020
C. Bennewitz et al., *Spectral and Scattering Theory for Ordinary Differential Equations*,
Universitext, https://doi.org/10.1007/978-3-030-59088-8

Subject Index

B^*-algebra, 39
C^*-algebra, 39
μ-equivalent, 295
μ-essentially bounded, 297
μ-integrable, 284, 286
μ-measurable, 292, 293
μ-nullset, 284
m-function, *see* Weyl–Titchmarsh
    function

abelian group, 4
action angle variables, 354
additivity
    of total variation, 277
adjoint, 38, 41
algebra, 279
algebraic dual, 34
almost everywhere, 284
almost periodic function, 15
anchor point, 113
anti-linear, 10
Arzela–Ascoli theorem, 19, 20, 309
asymptotic estimates, 178
axiom of choice, 293

Baire's theorem, 269
balanced, 314
Banach algebra, 38, 39
Banach space, 14, 34, 36, 279, 296
Banach's theorem, 271

Banach–Steinhaus theorem, 271
base of a measure, 298
basis, 5
Beppo Levi's theorem, 290
Bessel's inequality, 12, 36
bijective, 31
bilinear, 10
Blaschke product, 342
Bolzano–Weierstrass theorem, 9, 14,
    20, 36, 74, 272
Borel function, 61, 293
Borel set, 64
Borg–Marčenko theorem, 217
boundary conditions, 18, 46, 90, 129
    coupled, 26
    Dirichlet, 25, 82, 92, 134, 142, 158
    Neumann, 25, 92
    periodic, 25, 91, 92, 130
    semi-periodic, 25, 91, 92, 130
    separated, 1, 26, 91
boundary form, 55
boundary operator, 42
bounded from below, 43
bounded type, 251, 341
bounded variation, 276
boundedness
    uniform, 36

calculus of variations, 17
Camassa–Holm equation, 353

© Springer Nature Switzerland AG 2020
C. Bennewitz et al., *Spectral and Scattering Theory for Ordinary Differential Equations*,
Universitext, https://doi.org/10.1007/978-3-030-59088-8

Printed in the United States
By Bookmasters